LANTHANIDES SERIES DETERMINATION BY VARIOUS ANALYTICAL METHODS

LANTHANIDES SERIES DETERMINATION BY VARIOUS ANALYTICAL METHODS

MOHAMMAD REZA GANJALI
Center of Excellence in Electrochemistry
University of Tehran, Tehran, Iran

VINOD KUMAR GUPTA
Department of Chemistry
Indian Institute of Technology Roorkee, Roorkee, India;
Department of Applied Chemistry
University of Johannesburg, Johannesburg, South Africa

FARNOUSH FARIDBOD
Center of Excellence in Electrochemistry
University of Tehran, Tehran, Iran

PARVIZ NOROUZI
Center of Excellence in Electrochemistry
University of Tehran, Tehran, Iran

ELSEVIER

AMSTERDAM • BOSTON • HEIDELBERG • LONDON • NEW YORK • OXFORD
PARIS • SAN DIEGO • SAN FRANCISCO • SINGAPORE • SYDNEY • TOKYO

Elsevier
Radarweg 29, PO Box 211, 1000 AE Amsterdam, Netherlands
The Boulevard, Langford Lane, Kidlington, Oxford OX5 1GB, UK
50 Hampshire Street, 5th Floor, Cambridge, MA 02139, USA

British Library Cataloguing-in-Publication Data
A catalogue record for this book is available from the British Library

Library of Congress Cataloging-in-Publication Data
A catalog record for this book is available from the Library of Congress

ISBN: 978-0-12-804704-0

Typeset by Thomson Digital

Contents

Preface

Lanthanides, or as suggested by IUPAC, lanthanoids, are f-block elements that are also referred to as rare earths, due to their scarcity in the earth's crust. Lanthanides, the discovery of which mainly dates back to the 18th century, were found comparatively later than many other chemical elements mostly due to the similarity of the characteristics of their naturally occurring compounds that are mainly oxides, the semblance of their physico-chemical features, unavailability of expedient purification, separation, and detection means, and above all, lack of awareness of their number and nature.

The expansion of human knowledge and the expansion of information on the chemistry of these elements have gradually increased the appearance of these elements in various aspects of human life, ranging from chemical and automotive industries to glass and ceramics, optics, electronics, and agriculture. Although lanthanides are customarily considered as being rather similar in nature, their applications, very much, dependent on certain peculiar properties of each element and its compounds, which arise from differences in the arrangements of their f-block electrons. The similarities in the properties of these elements, with those of main elements present in the biological systems like calcium, have increased the role and significance of lanthanides in pharmaceutical industry and medicine, and these elements and their compounds are currently the focus of various biological research.

Consequently, the increased applications of these elements have intensified the environmental concerns about them, leading to amplified interest in research on the biology of rare earths. These developments have intensified the need for techniques and procedures that are applicable to the analysis of lanthanides, and hence, a range of complex as well as simple and fast methods have been developed to this end.

Although there exists a number of books on the nature and applications of these elements, the need for a comprehensive text, further covering details on their analysis and quantification, has been felt for a long time. In this accord and in the light of our long years of experience on the analysis of lanthanides, we decided to strike a blow in filling in this gap through compiling a rather comprehensive book covering the history, chemistry, and varied techniques applied to the analysis of lanthanides in geological, industrial, agricultural, environmental, biological, and food samples.

The result of this endeavor, *Lanthanides Series Determination by Various Analytical Methods*, tends to cover the diverse spectroscopic and electrochemical procedures developed and used for the analysis of rare earths. The text begins with a synopsis on the nature of the elements, and then provides an extensive review on the varied properties and applications of lanthanides in the context of chemical, agricultural, environmental, clinical, and pharmaceutical industries and biological research. And then, the text provides a brief description on the methods of determination used in the analysis of real samples, the principles of their instrumentation, and their application, accompanied by a wealth of experimental examples on the analysis of various samples.

Based on its content, the present text can be useful for researchers and academics, as well as those active in different areas of industry. Graduate and undergraduate students interested in the analysis of lanthanides can also find this book useful. The authors hope that this work can meet the endless need among the researchers in various areas of science and technology.

In the end, I would like to appreciate all the sincere efforts invested into this project by Professor Vinod K. Gupta from the Indian Institute of Technology Roorkee (India), Professor Parviz Norouzi and Dr Farnoush Faridbod from the University of Tehran (Iran), and Dr Morteza Rezapour from the Research Institute of Petroleum Industry (Iran), without whose constructive work and heartwarming comments this work could not have been composed.

Mohammad Reza Ganjali

Professor in Analytical Chemistry,
University of Tehran, Tehran, Iran
2016

CHAPTER

1

Introduction to Lanthanide Series (From Lanthanum to Lutetium)

OUTLINE

Lanthanides Series Determination by Various Analytical Methods. http://dx.doi.org/10.1016/B978-0-12-804704-0.00001-3

A quick look at the structure of the periodic table of elements (Fig. 1.1) usually draws one's attention toward the two rows of boxes containing elements with rather unfamiliar names, conventionally placed under the main body of the table. The elements in the first row (i.e., elements 57 to 71), which we shall focus on here, are named "lanthanides," after the first element in the row, that is, lanthanum, which is the 57th element in the table.

These elements possess unique photogenic, magnetic, mechanical, and nuclear properties and have found widespread applications in glass and ceramic manufacturing, metallurgy, electronics, and agricultural industries.

The name lanthanum originates from the Greek word λανθανω [lanthanō], which literally means to lie hidden. The International Union of Pure and Applied Chemistry (IUPAC) has accepted the collective names lanthanides or lanthanoids for these elements, and in the light of the fact that the term lanthanoids means similar to lanthanum (while the element lanthanum is a member of the group, the first name, i.e., lanthanides, has become more common). A point to be considered is that the suffix "-ide" is used for making the names of several anions (i.e., sulfide, bromide, etc.) in chemistry, which might lead to misconceptions about the nature of these elements. However, the application of the term has become commonplace in the chemical literature [1].

The history of lanthanides initiated with Lieutenant Carl Axel Arrhenius in 1787, who although an artillery officer at the time, remained an amateur geologist. As he was studying a mine located near the town Ytterby in Switzerland, he came across a blackish, very dense mineral that he decided to entitle Ytterbite. Initially, he suspected that Ytterbite contained tungsten, which was a newly discovered element at the time. But the later

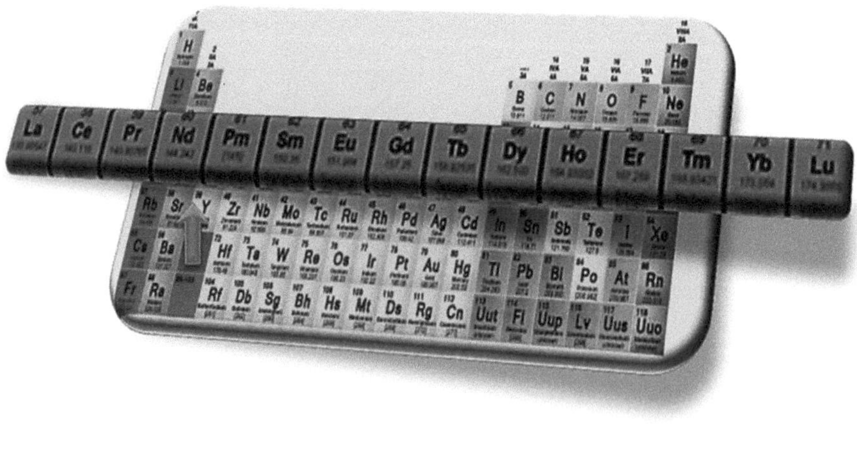

FIGURE 1.1 The lanthanide series in the periodic table.

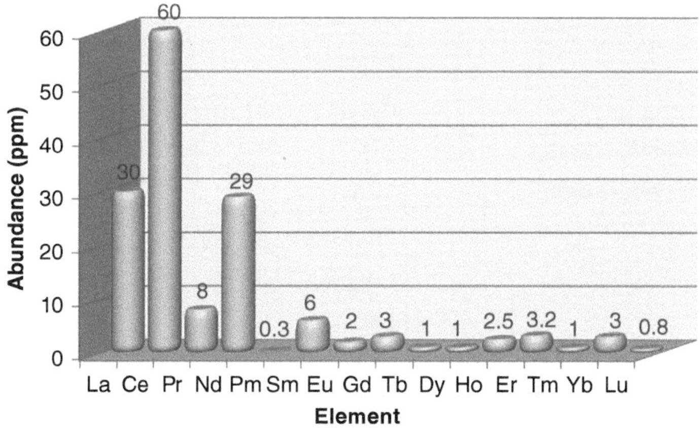

FIGURE 1.2 Abundance of lanthanides in the earth's crust.

studies by the Finnish chemist Johan Gadolin in 1794, using conventional wet chemistry techniques, showed that the mineral was composed of the oxides of beryllium, iron, silicon, and something unidentified. This unidentified "element" was named "yttria" by Gadolin. Further scrutiny in the 19th century showed that this new "element" was actually a mixture of the oxides of six other elements. During this work, pure dysprosium (Dy) and thulium (Tm) were obtained after 58 and 11,000 recrystallizations steps, respectively. The last of the naturally occurring elements was lutetium (Lu), which was discovered in 1907, but the last member of the lanthanide series (i.e., element number 61, named promethium), which is a synthetic element, was characterized in 1947 [2].

A classification of the lanthanides (Ln) is based on dividing the elements into the so-called light (i.e., La, Ce, Pr, Nd, Pm, and Sm) and heavy Lns (i.e., Eu, Gd, Tb, Dy, Ho, Er, Tm, Yb, and Lu), the former group being more abundant (Fig. 1.2).

Lanthanides are sometimes considered to behave as group 3 elements because of the formation of trivalent cations. An interesting fact is that the elements are not as rare as their names might imply. Lanthanides are naturally found in some minerals, including but not limited to xenotime, monazite, and bastnaesite.

Xenotime and monazite are orthophosphate ($LnPO_4$) salts of the elements. The mineral monazite further contains yttrium and the radioactive thorium, making it difficult and dangerous to deal with. Bastnaesite is a mixed fluoride carbonate mineral ($LnCO_3F$) (Table 1.1). As a rule of thumb, one can say that lanthanides with even atomic numbers are more common than those with odd ones, and the abundance of the most common ones decreases from Ce to La, Nd, and Pr.

TABLE 1.1 Common Minerals Containing Lanthanides [3]

Mineral	Formula
Aeschynite	$(Ln,Ca,Fe,Th)(Ti,Nb)_2(O,OH)_6$
Allanite (orthite)	$(Ca,Ln)_2(Al,Fe)_3(SiO_4)_3(OH)$
Anatase	TiO_2
Ancylite	$SrLn(CO_3)_2(OH)\bullet H_2O$
Apatite	$Ca_5(PO_4)_3(F,Cl,OH)$
Bastnaesite	$LnCO_3F$
Brannerite	$(U,Ca,Ln)(Ti,Fe)_2O_6$
Britholite	$(Ln,Ca)_5(SiO_4,PO_4)_3(OH,F)$
Cerianite	$(Ce,Th)O_2$
Cheralite	$(Ln,Ca,Th)(P,Si)O_4$
Churchite	$YPO_4\bullet 2H_2O$
Eudialyte	$Na_{15}Ca_6(Fe,Mn)_3Zr_3(Si,Nb)Si_{25}O_{73}(OH,Cl,H_2O)_5$
Euxenite	$(Ln,Ca,U,Th)(Nb,Ta,Ti)_2O_6$
Fergusonite	$Ln(Nb,Ti)O_4$
Florencite	$LnAl_3(PO_4)_2(OH)_6$
Gadolinite	$LnFeBe_2Si_2O_{10}$
Huanghoite	$BaLn(CO_3)_2F$
Hydroxylbastnaesite	$LnCO_3(OH,F)$
Kainosite	$Ca_2(Y,Ln)_2Si_4O_{12}CO_3\bullet H_2O$
Loparite	$(Ln,Na,Ca)(Ti,Nb)O_3$
Monazite	$(Ln,Th)PO_4$
Mosandrite	$(Ca,Na,Ln)_{12}(Ti,Zr)2Si_7O_{31}H_6F_4$
Parisite	$CaLn_2(CO_3)_3F_2$
Samarskite	$(Ln,U,Fe)_3(Nb,Ta,Ti)_5O_{16}$
Synchisite	$CaLn(CO_3)_2F$
Thalenite	$Y_3Si_3O_{10}(OH)$
Xenotime	YPO_4
Yttrotantalite	$(Y,U,Fe)(Ta,Nb)O_4$

It is noteworthy that although the isolation of pure lanthanides is rather cumbersome, fortunately it is not required for many applications. The conventional separation routines are based on the extraction of the elements from the ores using sulfuric and hydrochloric acids and/or sodium hydroxide, whereas the modern routines are conducted through selective complexation, solvent extractions, and ion exchange techniques. The main source of the ores is China, with almost 43% of the global deposit, whereas India, Kazakhstan, Kyrgyzstan, Malaysia, Russia, Thailand, and the United States also have deposits of these elements [3].

The terms rare earths (REs), rare earth elements (REEs), or rare earth metals (REMs) are also used to refer to lanthanides in addition to scandium (Sc) and yttrium (Y), which have a range of strategic applications including medical and energy technologies, lasers, batteries, magnets, magnetic resonance imaging (MRI) contrast agents, catalysts, and alloys. The estimates of worldwide rare earth reserves for different countries are illustrated in Fig. 1.3.

The US Geological Survey estimated the total rare earth reserves worldwide to be about 130 million metric tons (mmt), most of which is deposited in China, which is also the biggest commercial supplier of rare earths. Brazil also owns almost 22 mmt of these reserves, and other countries with considerable reserves of rare earths include Australia, India, USA, and Malaysia. The lanthanides, cerium, lanthanum, and neodymium, are the

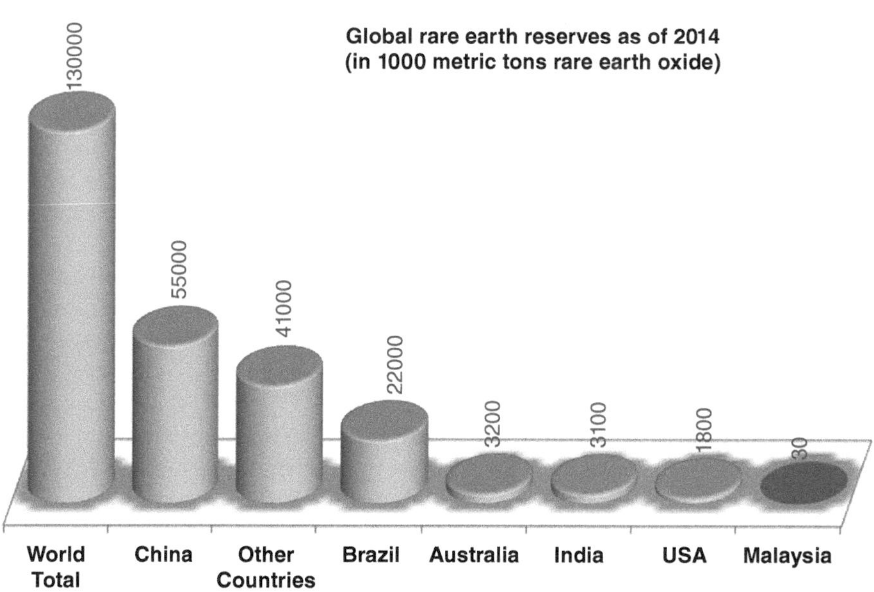

FIGURE 1.3 The global rare earth reserves as estimated as of 2014, by country.

TABLE 1.2 Price of Some Metallic Lanthanides [5]

Metal	Purity	$/kg
La	Min. 99%	9.6
Ce	Min. 99%	10
Pr	Min. 99%	150
Nd	Min. 99%	83
Sm	Min. 99%	25
Eu	Min. 99%	1000
Gd	Min. 99%	132.5
Tb	Min. 99%	825
Dy	Min. 99%	477

most highly produced rare earth elements [4]. Information on the prices of metallic lanthanides are also provided in Table 1.2.

CHEMICAL PROPERTIES OF LANTHANIDES

In fact, Lns are silvery and shiny when cut, but they tarnish quickly in air. These elements react slowly with cold water, and heating or decreasing the pH increases the reaction rate considerably. These f-block elements share many characteristics, including similar physical properties, with 3+ as the most common oxidation state, in addition to the less common states of 2+ and 4 +. Formation of crystalline structures with coordination numbers greater than 6 (usually 8–9), which tends to decrease across the series; minor crystal-field effects; a preference for reacting with the most electronegative elements such as oxygen (the reaction with oxygen is slow at the ambient temperature, but the elements ignite if heated up to 150–200°C), or halogens (upon heating) and S, H, C, and N upon heating. This similarity is the reason why lutetium, which is a d-block element, is considered a lanthanide [6,7].

The electron configurations of lanthanides (Table 1.3) were experimentally determined through the study of the lines in their emission spectra, which basically indicate the energy change for electron transfer between energy levels. A look at the electron configuration of these elements simply shows the reason behind the similarity of the properties of these elements to be due to the fact that in the majority of cases, the fifth and sixth energy levels of the elements contain identical numbers of electrons.

TABLE 1.3 Electron Configuration and Basic Properties of Lanthanides

Symbol	Atomic number	Atomic mass	No. of Stable isotopes	Idealized	Observed	Unpaired electron in ion (3+)
La	57	138.9	139	$5d^1 6s^2$	$5d^1 6s^2$	0
Ce	58	140.1	136,138,142	$4f^1 5d^1 6s^2$	$4f^1 5d^1 6s^2$	1
Pr	59	140.9	141	$4f^2 5d^1 6s^2$	$4f^3 6s^2$	2
Nd	60	144.2	142,143,145	$4f^3 5d^1 6s^2$	$4f^4 6s^2$	3
Pm	61	145.0	145	$4f^4 5d^1 6s^2$	$4f^5 6s^2$	4
Sm	62	150.4	144,149,150,154	$4f^5 5d^1 6s^2$	$4f^6 6s^2$	5
Eu	63	151.9	151,153	$4f^6 5d^1 6s^2$	$4f^7 6s^2$	6
Gd	64	157.3	154-158,160	$4f^7 5d^1 6s^2$	$4f^7 5d^1 6s^2$	7
Tb	65	158.9	159	$4f^8 5d^1 6s^2$	$4f^9 6s^2$ or $4f^8 5d^1 6s^2$	6
Dy	66	162.5	156, 160-164	$4f^9 5d^1 6s^2$	$4f^{10} 6s^2$	5
Ho	67	164.9	165	$4f^{10} 5d^1 6s^2$	$4f^{11} 6s^2$	4
Er	68	167.3	162-168,170	$4f^{11} 5d^1 6s^2$	$4f^{12} 6s^2$	3
Tm	69	168.9	169	$4f^{12} 5d^1 6s^2$	$4f^{13} 6s^2$	2
Yb	70	173.0	168, 170-174	$4f^{13} 5d^1 6s^2$	$4f^{14} 6s^2$	1
Lu	71	174.9	175	$4f^{14} 5d^1 6s^2$	$4f^{14} 5d^1 6s^2$	0

The atomic and ionic radii of the elements, which are determined considering the nuclear charge as opposed to the number of electrons, decreases from La to Lu. This is referred to as Lanthanide contraction, a phenomenon according to which 5s and 5p orbitals penetrate 4f. This way, the 4f orbital is not shielded from the nuclear change. This effect increases from La to Lu, leading to the rather unanticipated decrease in the atomic and ionic radii throughout the series.

As already mentioned, lanthanides prefer a trivalent oxidation state, because of the adoption of the stable $[Xe]4f^n$ configuration (n = 0–14). However, because equivalent configurations can be reached for di- and tetra-valent cations (naturally with a different n value), 2+ and 4+ cations are also observed as the less preferred oxidation states.

Lns are very electropositive based on their electronegativity and ionization energies, which apart from some irregularities, grow along the series. Examples of such irregularities, which arise from the stability of half-filled and filled orbitals, include the anomaly in the third ionization energies of gadolinium and lutetium to higher than expected values, as well as those of europium and ytterbium [8,9].

Another indicator is the considerable hydration enthalpies of the lanthanide cations, which increase along the group, indicating that the dissolution energy of the ions is high. This means the lanthanide cations can only form bonds with strongly coordinating ligands, such as hard bases like carboxylates. Further to the formation of strong bonds with such species in the complexes is due to entropy changes. An additional property of these compounds is the considerable stability of their hydroxy species. This seriously limits the pH ranges, in which lanthanide cations can remain dissolved.

The coordination of the lanthanide ions in nonaqueous solvents is a further prevalent phenomenon, particularly in large donating solvents like dimethylsulfoxide (DMSO) and dimethylformamide (DMF). In aqueous media, formation of complexes with anions is not a considerable phenomenon and does not hence interfere with complexation reactions with ligands. However, such phenomena are so predominant with nonprotic solvents that even noncoordinating anions such as perchlorate can form strong complexes with lanthanides. For instance in rather dilute solutions in acetonitrile, the major species include $[Ln(NO_3)_3(MeCN)_x]$, $[Ln(SO_3CF_3)_2(MeCN)_x]^+$, and $[Ln(ClO_4)(MeCN)_x]^{2+}$ [9].

The rather large radii of lanthanides ions (the radii of the ions of the middle elements of this family is close to that of Ca^{2+}) gives them the capability to host rather large number of ligands and hence have large coordination numbers. As already mentioned about the lanthanide contraction, the gradual decrease from right to left in the row leads to increase in the charge density of the Ln cations from right to left. The coordination chemistry of the family has been reviewed recently [9].

A property of these elements, which influences the way they react with other elements, is referred to as basicity, which is a measure of their tendency to lose electrons or anions. The basicity of lanthanides changes in the order of $La^{3+} > Ce^{3+} > Pr^{3+} > Nd^{3+} > Pm^{3+} > Sm^{3+} > Eu^{3+} > Gd^{3+} > Tb^{3+} > Dy^{3+} > Ho^{3+} > Er^{3+} > Tm^{3+} > Yb^{3+} > Lu^{3+}$. This means the basicity decreases with increasing atomic number along the row. The differences in the property are illustrated by the solubility of the salts and the formation of the complexes.

The magnetic properties of Lns ranges from being paramagnetic (i.e., having some unpaired electrons, e.g., La^{3+}, Lu^{3+}, Yb^{2+}, and Ce^{4+}) to being diamagnetic (i.e., having no unpaired electrons, like the rest of the family) [10-12]. The following lines shall cover the history and features of each Lns element in some detail.

LANTHANUM

Lanthanum, which is the first element in the series, was first discovered in mixture with a cerium compound in the late 1830s, when the Swedish chemist Carl Gustav Mosander thermally decomposed samples of cerium nitrate. The pure form of the element was, however, not isolated until 1923 [13]. Regarding the isotopes of the element, almost 99.91% of the naturally occurring lanthanum is ^{139}La, whereas ^{38}La radioisotopes have been characterized including the most stable ^{138}La (half-life = 1.05×10^{11} years) and ^{137}La (half-life = 60,000 years). The rest of the isotopes have half-lives of less than hours and even minutes, and are synthesized through the bombardment of other atoms. The element adopts three allotropic forms of α, β, and γ based on the temperature and pressure. The α-phase has a double close-packed hexagonal structure, with a = 3.7740 A° and c = 12.171 A°, and is prevalent at room temperature, whereas the β-phase, which forms at 325°C, has a face-centered cubic structure with a = 5.303 A°. The γ-phase, which is formed at 887°C, is body-centered cubic with a = 4.26 A°.

The silvery white element has an atomic number of 57, is soft and can be cut with a knife and is very ductile. La is a rather reactive rare-earth and oxidizes upon exposure to air. Like the other members of the series, La reacts slowly with cold and rather fast with hot water [14], adopts are 3+ and 2+ oxidation states, the latter being less stable, as discussed earlier [11-13]. The element is rather electropositive and readily dissolves in dilute sulfuric acid.

The metallic form of this element directly reacts with carbon, boron, silicon, phosphorus, nitrogen, sulfur, selenium, and halogens, and is the major component of the alloy known as Misch metal. Misch metal is composed of Ce and La as well as small amounts of Nd and Pr, and is used in lighter flints. La has the highest basicity as opposed to the rest of the family, which was the reason Mosander could isolate and purify its salt.

The earth's crust contains 32 ppm of the element, which shows that the element is as abundant as copper or zinc (contrary to common belief). The element is, however, very much distributed and is hence not found in one place. This makes the mining and preparation of La very difficult and increases its price [13]. The most common ores containing lanthanum are monazite and bastnaesite, which are actually the phosphate and mixed carbonates-fluoride of Ce, La, Th, Nd, Pr, and Y, respectively. A third mineral used for preparing pure lanthanum is loparite. United States and China have the world's largest deposits of bastnaesite, but the major monazite deposits are found in Australia, Brazil, China, India, Malaysia, South Africa, Sri Lanka, Thailand, and the United States. Loparite sources have also been discovered in Russia [15].

The extraction of the metal starts with crushing the minerals. One advantage of monazite, in this regard, is its magnetic properties, which makes its separation much more facile. After separating the crushed phase from the residues, they are treated with hot concentrated sulfuric acid so as to dissolve their rare earth content. The resulting solution is then filtered and its pH is partially neutralized to around 3-4 using sodium hydroxide or other basic species. At this stage, the thorium content of the minerals forms precipitates, which are easily removed. In the next step, an ammonium oxalate solution is used to treat the solution to form rare earth oxalates, which are next converted to their oxides through annealing. These oxides are then dissolved in nitric acid. Cerium is separated at this stage, as a result of the o = insolubility of its oxide under the operating conditions. At this stage, a mixed salt of La and ammonium nitrate is separated through crystallization, based on the fact that it is relatively less soluble than the equivalent species of the other REs [13].

A more efficient procedure for the separation of La is based on the ion exchange or liquid–liquid extraction mechanisms. In the former technique, the target ions are adsorbed onto a proper ion exchanger, which leads to their separation. Next they are desorbed through complexing with ligands such as ammonium citrate or nitrilotriacetate. The latter procedure, on the other hand, involves the application of organic solvents like tributyl phosphate, bis-2-ethylhexyl phosphate, or 2-ethylhexyl ester of 2-ethylhexylphosphonic acid.

Metallic La is formed through heating its oxide up to 300–400°C in the presence of an ammonium halide (F or Cl) and hydrofluoric acid

$$La_2O_3 + 6NH_4Cl \rightarrow 2LaCl_3 + 6NH_3 + 3H_2O$$

and further reducing the resulting chloride salt with an alkali or alkaline earth metal in vacuum or inert atmosphere according to the reaction below or through the electrolysis of the molten chloride salt.

$$LaCl_3 + 3Li \rightarrow La + 3LiCl$$

CERIUM

The element was simultaneously discovered by two independent research teams in Bastnäs in Sweden and in Germany 1803 [16]. The Swiss team named it Ceres, after a planet that had been discovered earlier in 1801 [17]. To qualitatively evaluate the presence of Ce in REs mixtures, NH_3 and H_2O_2 are added to the sample. The formation of a dark brown color indicates the presence of the element. Cerium metal is highly pyrophoric and can ignite if ground or scratched [18].

Because the metal itself is too electropositive to freely exist in nature, the originally isolated form of the element was its oxide, which was called ceria, and the later development of electrochemical techniques helped prepare the metallic form of Ce as well as the other lanthanides. Having an atomic number of 58, the element is a silvery metal with the same color and luster as iron, but like the other members of its family, it is soft, malleable, and ductile [19]. The element has four naturally occurring isotopes (i.e., ^{140}Ce [88.5%], ^{142}Ce [11%], ^{138}Ce [0.3%], and ^{136}Ce [0.2%]), among, which only ^{140}Ce is stable and the rest have half-lives of 5×10^{16} years, 1.5×10^{15} years, and 70 trillion years respectively [17].

Apart from these four naturally occurring isotopes, cerium has 35 synthetic isotopes with mass numbers in the range of 119 to 157 [17], all of which have very short half-lives of some thousandths to 1 seconds. In terms of the nuclear reactions, it can be said that the heavier Ce isotopes undergo β-decays, whereas the lighter ones (lighter than ^{140}Ce) prefer electron capture or β+ decay reactions [20]. Furthermore, Ce has four allotropes of α to δ under standard pressures. The α-cerium allotrope exists below −150°C and has a density of 8.16 g/cm^3; the β form that has a double hexagonal close-packed structure exists from −150°C to room temperature; γ-cerium with a face-centered cubic structure exist from 25°C to 726°C; and finally the δ with a body-centered cubic structure exists above 726°C [21-24].

The electronic structure of cerium is rather variable, since the energy level of the 4f level is almost the same as the valence levels. This means the energy barrier required for the electrons of these levels to transfer between the levels is very low, and therefore so-called "dual-valency states" are viable. As a result, even the physical properties of cerium changes on changing the environmental conditions. The volume of the metallic cerium, for instance, decreases about 10% as a result of high pressures or low temperatures, or even its valency changes from about 3 to 4.

The salts of the weakly basic Ce^{4+} or ceric cation tend to be orange whereas those of the strongly basic Ce^{3+} or cerous salts are white or colorless. Yet both oxidation states have high absorption profiles in the UV area. Because cerous salts have no color, some cerous salts are used in UV-absorbing glasses [17,25]. In addition to the two oxidation states of 3+ and

4 + , the element also undergo a rather rare 2+ state in compounds like CeH_2, CeI_2, and CeS [24].

The most abundant cerium compound is CeO_2, which is used as "Jeweller's rouge" or in some self-cleaning ovens. The other common compound of the 4+ cation are its sulfate ($Ce(SO_4)_2$); ammonium Ce^{4+} sulfate ($(NH_4)_2Ce(SO_4)_3$); and ammonium Ce^{4+} nitrate ($(NH_4)_2Ce(NO_3)_6$). Ce^{3+} chloride ($CeCl_3$) has been reported to catalyze carbonyl group reactions. The other 3+ compounds include the carbonate ($Ce_2(CO_3)_3$); fluoride (CeF_3); oxide (Ce_2O_3); and triflate ($Ce(OSO_2CF_3)_3$).

Cerium has a rather different basic property as compared to the other members of its family, which makes it possible to separate it very easily. The procedure for the separation is based on the different basicities of the two oxidation states. According to this procedure, a dilute nitric acid solution is used to leach the mixed hydroxides. Next the Ce^{4+} cations are precipitated from a nitrate or chloride solution of lanthanides using potassium permanganate and sodium carbonate in a 1:4 molar ratio. Finally the solution of REs nitrate is boiled with potassium bromate and marble chips.

Older methods worked based on addition of a Ce^{4+} solution in nitric acid, to a dilute sulfuric acid solution. This led to the precipitation of Ce^{4+}, whereas the rest of Ln, which were in a 3+ state, did not react. The method, whoever, suffered the drawback of high corrosion of the reaction media, which made the separation of the precipitates a formidable task.

There are of course isolation routines that do not totally rely on the basicity difference, like the crystallization of ceric ammonium nitrate [ammonium hexanitratocerate(IV)] crystallization from nitric acid, or the extraction of Ce^{4+} nitrate (hexanitratoceric acid), which is readily extractable into certain organic solvents (e.g., tri-*n*-butyl phosphate); However, the phenomenon does not occur for the trivalent Lns. One difference between the tri- and tetravalent ions is that the latter oxidation state is a strong oxidizing agent in acidic media, whereas the former does the opposite and is a strong reducing agent, which is readily oxidized by atmospheric oxygen (O_2).

Because of its electropositivity, cerium reacts slowly with cold water as below:

$$2\,Ce(s) + 6\,H_2O(l) \rightarrow 2\,Ce(OH)_3\,(aq) + 3\,H_2\,(g)$$

The rate of this reaction increases on increasing the temperature. The metallic form also reacts with all the halogens according to the overall reaction below:

$$2\,Ce(s) + 3\,X_2\,(g) \rightarrow 2\,CeX_3\,(s)$$

The reaction products are white, except for the iodide salt, which has a yellowish color.

The metallic form of Ce is readily oxidized to Ce^{3+} and dissolved in dilute sulfuric acid, where it forms $[Ce(OH_2)_9]^{3+}$ complexes [26]. The solubility of cerium in methanesulfonic acid is, however, not as high [27].

Cerium is the most abundant element among the lanthanides and is present in the earth's crust and sea water at about 0.0046 wt% and 1.5 parts per trillion (ppt), respectively [17]. The minerals containing Ce are allanite (orthite), monazite, bastnaesite, hydroxylbastnaesite, rhabdophane, zircon, and synchysite. The first two ores are also the most important sources of cerium as well as thorium [23]. The major deposits of monazite are found in Travancore in India, and Brazil, and those of bastnaesite and allanite are abundant is the United States.

To isolate cerium, the minerals are crushed, ground before being subjected to hot concentrated sulfuric acid, filtered, and partially neutralized to pH values of 3–4 using NaOH. At this stage, thorium hydroxide salts out and can be separated, and next the thorium-free solution is treated with ammonium oxalates to yield the oxalate salts of the rare earths present therein, which are next annealed to the oxides. The oxides of these elements are then dissolved in HNO_3 to isolate the insoluble cerium salts that are further treated through the thermal reduction routines by reducing cerium fluoride or chloride with calcium. As mentioned in the case of lanthanum, the molten salts can also be electrolyzed to produce high-purity cerium [24,28].

PRASEODYMIUM

The name of the element comes from two Greek terms "prasios" meaning "green" and "didymos" meaning "twin." After the discovery of an element that was named didymia from lanthana by Mosander in 1879, and the separation of a new element called samaria from the didymia samples obtained from samarskite ores in 1879 by Lecoq de Boisbaudran, in 1885, von Welsbach could distinguish and isolate two differently colored elements from didymium. He called the green one praseodymia, and the other one neodymia. As with other rare earths, compounds of these elements in solution phase have distinctive sharp spectral absorption bands or lines, some of which are only a few Angstroms wide.

Praseodymium (Pr) has the atomic number 59, and like the Ce and La, it is soft, silvery, malleable and ductile, and enjoys good magnetic, electrical, chemical, and optical properties [29]. The element has isotopes with mass numbers between 121 and 159 [23], with the most abundant one being ^{141}Pr. Lighter isotopes undergo $\beta+$ decay, which does not change the nature of Pr, whereas the product of the $\beta-$ decay that is common among the heavier isotope are isotopes of neodymium (60 protons) [23].

The element cannot be found in elemental form in nature, because of its high reactivity and even after the artificial preparation of its pure form, gradually a green oxide layer covers its surface, so like its other fellow elements, its pure form should be stored in a light mineral oil. The metallic Pr gradually tarnishes in an air atmosphere and starts burning at 150°C according to the reaction below:

$$12\,Pr + 11\,O_2 \rightarrow 2\,Pr_6O_{11}$$

The element also reacts with cold water to form a hydroxide, and has the same behavior as La and Ce in hot water and in contact with halogens (except for the green product in the case of all halogens):

$$2\,Pr(s) + 6\,H_2O(l) \rightarrow 2\,Pr(OH)_3\,(aq) + 3\,H_2\,(g)$$

$$2\,Pr(s) + 3\,X_2\,(g) \rightarrow 2\,PrX_3\,(s)[green]$$

Its reaction with dilute sulfuric acid, which leads to the formation of green Pr^{3+} in the form of $[Pr(OH_2)_9]^{3+}$ complexes, is the same as cerium [30].

Like the two previously mentioned Lns, the oxidation states of Pr are 2+ , 3+ , or 4+ . Pr^{4+} is a very strong oxidizing agent, which can oxidize water and hydrochloric acid to form molecular O_2 and Cl_2. As a result, the tetravalent Pr cation is not stable in aqueous media and changes to the stable Pr^{3+} with yellow-green color. The oxide of praseodymium oxide, which is formed through the ignition of its oxalate or carbonate salts, and contains a mixture of Pr^{3+} and Pr^{4+} and is illustrated as Pr_6O_{11}, is black. Other common salts of Pr include PrF_2, PrF_3, PrF_4; $PrCl_3$; $PrBr_3$, Pr_2Br_5; PrI_2, PrI_3, Pr_2I_5; PrO_2, Pr_2O_3, Pr_6O_{11}; PrS, Pr_2S_3; $Pr_2(SO_4)_3$; PrSe; PrTe, Pr_2Te_3; and PrN.

Naturally occurring praseodymium is composed of one stable isotope, ^{141}Pr [31], which is of use in nuclear magnetic resonance (NMR) and electron paramagnetic resonance (EPR) spectroscopies [32]. Overall, 38 radioisotopes have been characterized, with the most stable being ^{143}Pr with a half-life ($t_{1/2}$) of 13.57 days, and ^{142}Pr with a $t_{1/2}$ of 19.12 hours [23]. All of the remaining radioactive isotopes have half-lives less than 6 hours, and the majority of these have half-lives that are less than 10 minutes [23]. Although not as abundant as the two previous elements (9.5 ppm in earth's crust), Pr is a rather abundant rare earth, and is more abundant in monazite and bastnaesite, which are common in China, United States, Brazil, India, Sri Lanka, and Australia [33].

NEODYMIUM

As already mentioned, neodymium was discovered at the same time as praseodymium in 1885 by von Welsbach who could separate the two elements from the so-called didymium. The separation process included multiple fractionation steps using ammonium didymium nitrate, and the name of the element was derived from the Greek "neos" in addition to "didumous" which means the new twin.

Having an atomic number of 60, its properties are very similar to those of praseodymium, which is why they could not be separated for a long time. Nd is a silvery-white metal of medium reactivity. On contact with air, the surface of the metal is oxidized, leaving a yellowish layer on the surface that peels off. The metallic form of the element has a hexagonal structure with a specific weight of 7.004 g/cm^3 [14]. The other allotrope of Nd has a body-centered cubic structure that forms around 863°C [23].

Natural neodymium samples are composed of ^{142}Nd (27%), ^{143}Nd, ^{145}Nd, ^{146}Nd, ^{148}Nd, and two radioisotopes, ^{144}Nd and ^{150}Nd. All in all, the element has 24 other synthetic radioisotopes, the most stable of which is ^{144}Nd (with α decay and a half-life of 2.29×10^{15} years) and ^{150}Nd (double β-decay, and $t_{1/2}$ of about 7×10^{18} years). The remaining radioactive isotopes have $t_{1/2}$ ranging from some days to less than 70 seconds.

Metallic neodymium reacts with atmospheric oxygen at a slow rate but starts burning at about 150°C:

$$4\,Nd + 3\,O_2 \rightarrow 2\,Nd_2O_3$$

Similar to the other Lns discussed so far, neodymium is an electropositive element, which reacts with water at different rates based on the temperature and forms neodymium(III) hydroxide according to the reaction below:

$$2\,Nd\,(s) + 6\,H_2O\,(l) \rightarrow 2\,Nd(OH)_3\,(aq) + 3\,H_2\,(g)$$

It has identical reactions with halogens, leaving reaction products of different colors (NdF_3 and $NdBr_3$ are violet, $NdCl_3$ is mauve, and NdI_3 is green) which are formed according to the general reaction below:

$$2\,Nd\,(s) + 3\,X_2\,(g) \rightarrow 2\,NdX_3\,(s)$$

Neodymium dissolves in dilute sulfuric acid forming Nd^{3+} ions in the form of $[Nd(OH_2)_9]^{3+}$ complexes [34]:

$$2\,Nd\,(s) + 3\,H_2SO_4\,(aq) \rightarrow 2\,Nd^{3+}\,(aq) + 3\,SO_4^{2-}(aq) + 3\,H_2\,(g)$$

Other common neodymium compounds include are NdS, Nd_2S_3, NdN, $Nd(OH)_3$, NdP, NdC_2, $Nd(NO_3)_3$, and $Nd_2(SO_4)_3$.

In terms of abundance, neodymium is second to cerium, with an abundance of almost 28 ppm in the earth's crust. The minerals containing Nd include monazite and bastnaesite, which are mostly located in Brazil, China, USA, India, Sri Lanka, Thailand, and Australia [35]. Another mineral containing Nd is loparite, which is abundant in Russia [15,35,36].

The conventional process for the preparation of neodymium metal is through heating the halide salts of Nd^{3+} in the presence of excess amounts of calcium, during which process CaF_2 is formed and Nd is produced in the form of a pure metal (calciothermic reduction). The most commonly used halide salt is NdF_3, which is prepared by steaming a mixture of dry HF acid and 60% Ar over Nd^{2+} oxide at 700°C for 16. The rather impure fluoride salt, which still contains about 300 ppm of oxygen, is purified in a second stage through heating to high temperatures [37].

PROMETHIUM

The discovery of promethium, named after Prometheus, the Greek mythological titan who stole fire from the gods and brought it to the Earth, was the result of the works of an American team at Oak Ridge National Laboratory in 1945 [38]. Lawrence E. Glendenin, Jacob A. Marinsky, and Charles D. Coryell separated the element from the fission products of uranium, but did not announce the discovery for 2 years because of the secrecy considerations of their project [39].

Metallic promethium was first prepared in 1963, through the reduction of promethium(III) fluoride with lithium vapor [40]. However, the metal is currently prepared through the calciothermic reduction of the trihalides of the metal (anhydrous $PmCl_3$ is the commonest) [41]. Unlike the other Lns, Pm does not have stable isotopes, and even the element itself is not very stable. That is why the element has been entitled the least stable among the first 84 elements of the periodic table [42]. Through its decay, the element converts to its neighboring elements that is, neodymium and samarium (more specifically, its positron decay and electron capture decays convert it to neodymium, whereas its beta decay changes it into samarium) [42].

The most stable isotope of this element is ^{145}Pr, which has a half-life of 17.7 years via electron capture [23]. Because it has 84 neutrons (2 more than the magic 82 that is synonymous with a stable neutron configuration), it may emit an alpha particle (i.e., 2 neutrons) to form ^{141}Pr with 82 neutrons. Thus, it is the only promethium isotope with proven alpha decay. Its half-life for alpha decay is about 6.3×10^9 years, and the relative probability for a ^{145}Pm nucleus to decay through this route is 2.8×10^{-7}%. Several other Pm isotopes (^{144}Pm, ^{146}Pm, ^{147}Pm, etc.) also have a positive energy release for alpha decay. These alpha decays, though anticipated to occur, have not been experimentally observed.

With an atomic number of 61, all promethium isotopes are radioactive (similar to technetium). A distinction between Pm and the previously described Lns is that it does not form tetravalent cations and it only forms compounds by losing two 6s and 14f electrons. In terms of the atomic radius, Pm is the third largest lanthanide; however, it is a bit larger than the neighboring elements, which makes it the only exception to the general lanthanide contraction trend [43,44].

The properties of promethium, like its melting point, first three ionization, and hydration energies are an average of those of the elements before or after it in the table (i.e., neodymium and samarium) [43], and the same pattern holds about its chemical properties [45]. It should be noted that because the element is rather unstable, it has not been thoroughly studied in terms of chemical properties [17].

By dissolving the element in HCl, or HNO_3, the soluble yellow $PmCl_3$ or pink $Pm(NO_3)_3$ are produced [45]. Increasing the pH of Pm^{3+} solutions with NH_3, on the other hand, leads to the formation of the light-brown water-insoluble $Pm(OH)_3$. As in the case of $Ce_2(SO_4)_3$, $Pm_2(SO_4)_3$ is slightly water-soluble, and the oxalate of Pm has the lowest solubility as compared to the oxalates of the rest of lanthanides [45]. Once heated, the oxalate salt is converted to the white or lavender amorphous oxide salt, with properties that are closer to those of the oxide of samarium, and rather different from those of neodymium. This oxide salt holds a cubic lattice up to 600°C, but its further annealing to 800°C followed by heating up to 1750°C leads to the irreversible formation of a structure that is a combination of monoclinic and hexagonal phases [46]. Because of the low concentration of the element in the earth's crust, its mining is not economical. The mineral containing Pm is pitchblende, which is actually a uranium that contains very small amounts of the element (4 parts per quintillion) as a result of the fission of its uranium-238 [45,47].

SAMARIUM

The element was isolated from the mineral samarskite by Paul Émile Lecoq de Boisbaudran in 1879. The mineral itself carried the name of Vasili Samarsky-Bykhovets, who was a Russian mining official. Samarium has an atomic number of 62 and is a relatively harder metal as compared to the lanthanides discussed so far, but it oxidizes in contact with air. The most stable oxidation state adopted by the element is 3+ , but compounds where it forms a 2+ oxidation state (like SmO, SmS, SmSe and SmTe, and SmI_2) are also known. The hardness and density of samarium are close to those of zinc.

Samarium is radioactive (128 Bq/g), and its natural samples are composed of 4 stable isotopes ^{144}Sm, ^{150}Sm, ^{152}Sm and ^{154}Sm, and radioisotopes with very long half-lives, plus ^{147}Sm ($t_{1/2} = 1.06 \times 10^{11}$ years), ^{148}Sm ($t_{1/2} = 7 \times 10^{15}$ years) and ^{149}Sm ($t_{1/2} > 2 \times 10^{15}$ years). ^{152}Sm is the most abundant isotope (26.75 wt%) [42]. Some references consider ^{149}Sm as stable [42] or radioactive [48]. The half-lives of the rest of the isotopes are as follows: ^{151}Sm, 90 years; ^{145}Sm, 340 days; and those of the remaining isotopes fall in the order of minutes to seconds. ^{146}Sm, ^{147}Sm, and ^{148}Sm, with rather long $t_{1/2}$ values, undergo α decays and form isotopes of Nd. The lighter isotopes, however, undergo electron capture decays and form isotopes of Pm, whereas the heavier have beta decays and form isotopes of Eu [42]. The alpha-decay of ^{147}Sm to ^{143}Nd, which has a $t_{1/2}$ of 1.06×10^{11} years, is used as the basis for samarium-neodymium dating technique [42].

The element and its 3+ oxide (Sm_2O_3, also called sesquioxide) are paramagnetic under ambient conditions, but if cooled down to 14.8 K the metal shows antiferromagnetic properties [49,50]. There have been reports on the isolation of individual samarium atoms through encapsulation in fullerene molecules [51] or doping C60 with them to create superconducting behaviors below 8 K [52]. Iron doped with Sm has also been reported to show superconducting properties at temperatures as high as 56 K [53].

Similar to the other Lns when fresh, samarium has a silvery luster, but tarnishes in the air and even ignites above 150°C [23,54]. The metallic form is so active that it tarnishes even if stored under mineral oil, developing a grayish-yellow oxide–hydroxide mixture at the surface. The behavior of the metallic form is identical with the previously discussed lanthanides both in water, halogens, and dilute sulfuric acid solutions except for the color of the complex formed in sulfuric acid, which is pale green [43,55]. In the 2+ state, the cation turns the aqueous solutions to blood-red colors [43].

Sesquioxide (Sm_2O_3) is the most stable samarium oxide with several crystalline phases, including the trigonal structure that is formed upon cooling the molten salt. Another Sm_2O_3 crystalline structure is the monoclinic structure, where Sm_2O_3 is formed through the flame fusion method (Verneuil process), which yields transparent cylindrical crystals of up to several centimeters length. Another approach includes thermal treatment of the trigonal structure to 1900°C [56]. There have also been reports on the cubic Sm_2O_3 [57]. Sm(III) also forms sulfide, selenide, and telluride salts.

Samarium is among the exceptional lanthanides that form monoxides. SmO, which has a cubic structure, is a shiny, golden-yellow compound that can be prepared through the reduction of Sm_2O_3 in the presence of metallic samarium at 1000°C above 50 kbar [58]. Other divalent Sm compounds include SmS, SmSe, and SmTe, all of which have a cubic rock-salt crystal structure [59]. The trihalides of samarium (SmX_3) can be further reduced at about 700–900°C, with elemental samarium, lithium, or sodium to yield dihalides (SmX_2). In the case of the diiodide salt, the product can be achieved by reacting metallic Sm with 1,2-diiodoethane in anhydrous tetrahydrofuran under ambient conditions [43]:

$$\mathrm{Sm(s)} + \mathrm{ICH_2\text{-}CH_2I} \rightarrow \mathrm{SmI_2} + \mathrm{CH_2{=}CH_2}.$$

The reduction of the SmX_3 can also give rise to the formation of several nonstoichiometric crystalline halides like Sm_3F_7, $Sm_{14}F_{33}$, and $Sm_{27}F_{64}$ [60] or $Sm_{11}Br_{24}$, Sm_5Br_{11}, and Sm_6Br_{13} [61]. A rather uncommon behavior of the halides of Sm is their tendency to change their crystal structures upon changing the halide. This has been elaborated by discovering that many halides take two major crystal phases with different stabilities. The less stable one of these can convert to the more stable one through pressure or thermal treatment. An example is the conversion of the monoclinic SmI_2 and SmI_3 to $PbCl_2$-type orthorhombic structures with densities of 5.90 g/cm^3 under pressure and 5.97 g/cm^3 respectively. Sintering powders of samarium oxide and boron yield several samarium boride phases in the form of powders in vacuum [62]. Larger samarium bromide crystals with different stoichiometries can be prepared from the powder using arc or zone melting techniques under different temperature conditions [63]. The methods, which require rather harsh temperature and pressure conditions, cannot be used for the preparation of $SmBr_2$ [64]. At temperatures higher than 50 K, samarium hexaboride is a so-called Kondo insulator. This means it shows metallic electrical conductivity with strong electron scattering above this temperature, whereas at lower temperatures, the compound acts as a nonmagnetic insulator [65,66]. New research suggests the compound be used as topological insulator [67-69].

By melting the metallic form of the element in the presence of graphite under an inert atmosphere, one can prepare unstable samarium carbides, which be worked with under inert atmospheres because of their instability [70]. Samarium monophosphide (SmP) is another compound of the element showing semiconductive behavior with a band gap of 1.10 eV, which is identical to that of silicon. The compound can be prepared through the annealing mixed powders of phosphorus and samarium in a quartz ampule at 1100°C [71]. The same procedure can be applied to the preparation of SmAs at 1800°C [72]. A large number of crystalline binary compounds are known for samarium, and one of the group 4, 5, or 6 elements (X), where X is Si, Ge, Sn, Pb, Sb or Te, and metallic alloys of samarium form another large group. They are all prepared by annealing mixed powders of the corresponding elements. Many of the resulting compounds are non-stoichiometric and have nominal compositions Sm_aX_b, where the b/a ratio varies between 0.5 and 3 [73-75].

Samarium cyclopentadienide $Sm(C_5H_5)_3$ and chloro cyclopentadienide, for example, $Sm(C_5H_5)_2Cl$ and $Sm(C_5H_5)Cl_2$, can be prepared through reacting $SmCl_3$ with NaC_5H_5 in tetrahydrofuran [43,76]. Samarium alkyls and aryls can be prepared through the reactions illustrated below, in tetrahydrofuran or ether as solvents:

$$SmCl_3 + 3\,LiR \rightarrow SmR_3 + 3\,LiCl$$

$$Sm(OR)_3 + 3\,LiCH(SiMe_3)_2 \rightarrow Sm\{CH(SiMe_3)_2\}_3 + 3\,LiOR$$

where R is a hydrocarbon and Me stands for methyl.

Samarium, being more abundant that tin, is the 40th most abundant element in the earth's crust and can be found at concentrations up to 2.8%

in cerite, gadolinite, samarskite, monazite, and bastnaesite. Monazite and bastnaesite are the most common commercially used ores and are reportedly abundant in China, the United States, Brazil, India, Sri Lanka, and Australia.

The commercial purification of the element was not possible before the advent of ion-exchange technology in the 1950s, and only a mixture of samarium and gadolinium, which was a by-product of the fractional crystallization of neodymium, was available under the trade name of Lindsay Mix, which is a similar mixture titled samarium–europium–gadolinium (SEG) [77].

The element is currently prepared through solvent extraction methods on mixed lanthanides isolated from bastnaesite (or monazite). This is possible based on the fact that because the heavier Lns in the extraction mixture have greater affinity for the solvent used and is hence easily extracted from the bulk using relatively small proportions of solvent. Not all rare-earth producers using bastnaesite as theory feed do so. Further use of this process is not economical for the producers, and hence they do not further purify the already marketable SEG.

EUROPIUM

The existence of europium was assumed by P. É. Lecoq de Boisbaudran in 1890, who was of the opinion that samples of samarium-gadolinium mixtures contained an unknown contaminant based on the spectral lines that he could not attribute to samarium or gadolinium. It was not until 1896 when Eugène-Anatole Demarçay, who suspected if the samples of a recently discovered elemental samarium sample was contaminated with an unknown element, began experimenting. He was finally able to isolate the new element in 1901 and named it europium. Like samarium, this element is a rather hard, silvery, and ductile metal that is oxidized on contact with air and water. Eu has a body-centered cubic crystal lattice [78]. Many of the properties of Eu are due to its half-filled electron shell [78]. When cooled down to 1.8 K and compressed to above 80 GPa, europium shows superconducting behaviors. This is due to the fact that the divalent state of the element in the metallic form [79] converts into its trivalent state under the mentioned conditions. This is further indicated by the strong local magnetic moment ($J = 7/2$) of the element in the divalent state, which has a suppressing effect on the superconductivity and is eliminated in the trivalent state ($J = 0$ in Eu^{3+}) [80].

Like the other Lns, Eu adopts 3+ and 2+ oxidation states, the compounds corresponding to the latter being mildly reducing. The element does not have any biological effects and can be considered as nontoxic as opposed to the other heavy metals. Eu is the most reactive rare earth (it is actually as reactive as calcium) and hence reacts with air to the extent that a centimeter cube of the element oxidizes completely in a matter of some days [14]. The behavior of the element in the presence of water, dilute sulfuric acid, and air is identical to those of the other Lns except for the fact that it ignites at a rather higher temperature range of 150 to 180°C [81].

As mentioned above, apart from the trivalent oxidation state, europium also forms divalent compounds, which is rather uncommon to most lanthanides. The divalent state is similar to barium ion regarding the size and coordination number [82]. Eu^{2+} ions are rather unstable in the air and easily oxidize to Eu^{3+}, but in the absence of oxygen, it is rather stable and is hence exchanged with the minerals of calcium and the other alkaline earths. This phenomenon is referred to as the "negative europium anomaly," or in other terms, lower than expected europium content in several lanthanide minerals. Geochemistry and petrology studies often use the data of the analysis of trace amounts of Eu for understanding the processes, which created igneous rocks based on the nature of the europium anomaly [83].

Naturally occurring europium samples contain ^{151}Eu and ^{153}Eu, the latter being more stable and the most abundant (52.2 wt%). ^{151}Eu has been found to be unstable and undergo α decay ($t_{1/2} = 5 \times 10^{18}$ years) [84]. The element has 35 more artificial radioisotopes, with the most stable ones being ^{150}Eu ($t_{1/2}$ = 36.9 years), ^{152}Eu ($t_{1/2}$ = 13.516 years), and ^{154}Eu ($t_{1/2}$ = of 8.593 years). The rest of the synthetic radio isotopes have half-lives in the order of a few years to some seconds [85]. The isotopes lighter than ^{153}Eu mostly undergo electron capture decays producing Sm, whereas the heavier ones tend to have beta-minus decays producing Gd.

Eu^{3+} compounds are usually accompanied by six to nine ligands with oxygen atoms, fofr example, water. Europium reacts with all halogens and produces trivalent EuF_3 (white), $EuCl_3$ (yellow), $EuBr_3$ (gray), and EuI_3 (colorless) as well as the divalent EuF_2 (yellow-green), $EuCl_2$, $EuBr_2$ (colorless), and EuI_2 (green). The compounds of Eu^{3+} and the oxygen group elements are all stable, but Eu^{2+} only forms stable compounds with S, Se, and Te. Eu has three known oxides of EuO, Eu_2O_3, and Eu_3O_4 (where both the 2+ and 3+ oxidation states coexist) [86]. Trace amounts of Eu^{2+} have been reported to act as activators of the fluorescence of some mineral fluorite (CaF_2) samples [87-89].

Europium can be found in almost all ores containing rare elements but becasue of the difficulties in separating the elements, it was not until the late 1800s that the element was isolated in its pure form. William Crookes observed the phosphorescent spectra of the rare elements, and the spectral lines were later attributed to europium [90]. The discovery of europium-doped yttrium orthovanadate, which is a red phosphor, in the early 1960s and its application to solve the problem with the red color in color televisions, which boosted the image quality, increased the need for the element

[91]. The europium content of monazite (0.05 wt%) is very low, but fortunately the bastnaesite deposits discovered have unusually high amounts of the element (0.1 wt%). Another mineral containing europium is loparite which is the largest source for Lns produced in Russia [15].

The ion-exchange technology developed by Frank Spedding, in the mid-1950s had a great influence on the scales of producing rare earths commercially [91-93]. To produce a concentrate of Lns the ores are roasted, prior to acidic and basic leaching, which is followed by the precipitation of the Ce content if present. Next solvent extractions or ion exchange elution steps are used to enrich europium, and next the ionic species are reduced using zinc, zinc/amalgam, electrolysis or other methods which can reduce Eu^{3+} into Eu^{2+}europium(II), which can be precipitated in the form of carbonate or is co-precipitated with barium sulfate. Europium metal can also be prepared through the electrolysis of molten $EuCl_3$ and NaCl (or $CaCl_2$) in a graphite cell that also acts as the cathode [94-98].

GADOLINIUM

The name gadolinium comes from a mineral gadolin, which was discovered by a Johan Gadolin [43]. The element however was discovered in 1880, by a Swiss chemist who noticed unanticipated spectral lines in the spectrum of gadolinite and cerite samples. Regardless of the names of the minerals, cerite contains more gadolinium and hence it was used for the separation of this new element by Jean Charles Galissard de Marignac and the resulting oxide was proven to be that of a new element. The separation of the oxide was however the result of the works of a French chemist named Paul Émile Lecoq de Boisbaudran and was done in 1886.

Naturally occurring Gd samples contain 6 stable isotopes of ^{154}Gd, ^{155}Gd, ^{156}Gd, ^{157}Gd, ^{158}Gd and ^{160}Gd, in addition to a radioisotope, ^{152}Gd, among which ^{158}Gd has the highest abundance (24.84 wt%) [99]. 28 other synthetic radioisotopes have also been reported for the element, including ^{152}Gd ($t_{1/2}$= 1.08 × 1014 years), ^{150}Gd ($t_{1/2}$= 1.79 × 10^6 years). The remaining isotopes have $t_{1/2}$ values in the range of some decades to a few seconds. As a rule of thumb, isotopes that are lighter than ^{158}Gd, primarily undergo electron capture reactions forming isotopes of Eu, while those heavier decay β, producing terbium isotopes.

Being the 64th element in the periodic table, gadolinium (Gd) looks the same as the other lanthanides discussed so far in terms of color and shine, but has rather unconventional properties in combination with other metals or alloys, which is reflected by the grave improvements it causes in iron, chromium and the related alloys in terms of temperature resistance. Such effects are observed even when negligible amounts of the element

(1 wt%) are present in the formulation of the alloys. Besides, because of its high neutron absorption, elemental Gd or its salts are used in shielding devices in neutron radiography and in nuclear usages.

The trivalent cation Gd^{3+} is fluorescent and hence its salts are used as green phosphors in a range of applications. An anomaly in the behavior of this element with respect to its fellow elements is its stability in the atmosphere in the absence of humidity. There is no need to mention that moist air leads to the formation of Gd_2O_3 scales on the surface.

$$4Gd + 3O_2 \rightarrow 2Gd_2O_3$$

Although the behavior of gadolinium in air is different from those of the formerly discussed elements, it has identical reactions in cold and hot water and dilute sulfuric acid, and in the last case, it forms $[Gd(H_2O)_9]^{3+}$ [100]. The same is true for its reaction with halides. All three halides except for the yellow triiodide are white. The trivalent sate also follows the common Ln tendency of forming complexes with high coordination numbers, and an example is its reaction with 1,4,7,10-tetraazacyclododecane-1,4,7,10-tetraacetic acid (DOTA), the salts of which are used in MRI applications. Elemental gadolinium has strong reducing properties and reduces several oxides of different metals to their elemental form. The divalent state can be prepared through the thermal treatment of a mixture of the trivalent cation and the metallic form under an inert atmosphere in a tantalum crucible. A rather unconventional gadolinium compound is the sesquichloride (Gd_2Cl_3), which can yield graphite-like layers of GdCl through annealing at 800°C [101].

The major minerals used for producing the element, which is as abundant as 6.2 ppm in the earth's crust, are monazite and bastnaesite, and as already described the major deposits of these minerals are located in China, USA, Brazil, Sri Lanka, India, and Australia. The overall process for producing gadolinium includes converting the insoluble oxides of Lns into the soluble chlorides and sulfates using hydrochloric or sulfuric acids, and adjusting the pH of the acidic filtrates to 3–4, to precipitate thorium hydroxide. Next, the thorium-free solution is treated with ammonium oxalate to yield the insoluble oxalates salt of the remaining elements, which are next converted to oxide salts as mentioned earlier. Then cerium oxide, which is the only insoluble salts among the rest of the oxides in an HNO_3 environment, is separated, and the remaining solution next is treated as described before, to yield an SGE. For further separations, preparative ion-exchange procedures are used in combination with the application of suitable complexing agents. Pure Gd can be achieved through heating its oxides at 1450°C in an inert atmosphere, or by reducing molten $GdCl_3$ with an appropriate reducing agents below the melting point of the element [43].

TERBIUM

As in the case of lanthanum and praseodymium, terbium was discovered by C. G. Mosander of Sweden in 1843. He first noticed the presence of the element as an impurity in Y_2O_3, but the element could not be isolated until the development of ion-exchange procedures [102]. During his explorations, Mosander first could fractionate yttria into three compounds, among which what became known as terbia was the pink fraction. This compound was later found to get its color from another element that is called erbium today [102].

Because naturally occurring terbium is composed of its only stable isotope (i.e., ^{159}Tb), it is called mononuclidic or monoisotopic. Thirty-six synthetic radioisotopes have been characterized for the element, the heaviest and lightest being ^{171}Tb and ^{135}Tb, respectively [42]. The most stable synthetic radioisotopes of terbium are ^{158}Tb ($t_{1/2}$ = 180 years) and ^{157}Tb ($t_{1/2}$ = 71 years). All of the remaining radioactive isotopes have half-lives ranging from less than a quarter of a year to less than half a minute. The primary decay path before the most abundant stable isotope, that is, ^{159}Tb, is electron capture, which results in the production of gadolinium isotopes, which next undergoes a β- decay, producing dysprosium isotopes [42].

The properties of the element are rather similar to its neighboring element gadolinium, and hence rather different from the lighter Lns [14]. Terbium, which is the 65th element in the periodic table, has two major allotropes, which transform to one another below or above 1289°C [23]. The fluorescence of the Tb^{3+} cation is a bright lemon-yellow in color. Apart from the common 3+ oxidation state as in Tb_2O_3, the element also has a 4+ state, examples of which are seen in TbO_2 and TbF_4 [24]. Terbium has also been reported to readily burn to form mixed Tb^{3+} and Tb^{4+} oxides according to the reaction below:

$$8\,Tb + 7\,O_2 \rightarrow 2\,Tb_4O_7$$

Its reactions with water, halogens and sulfuric acid are the same as the already discussed lanthanides and can simply be summarized as below:

$$2\,Tb + 6\,H_2O \rightarrow 2\,Tb(OH)_3 + 3\,H_2\uparrow$$

$$2\,Tb + 3\,X_2 \rightarrow 2\,TbX_3\;(X = F, Cl, Br, I)$$

$$2\,Tb(s) + 3\,H_2SO_4 \rightarrow 2\,Tb^{3+} + 3\,SO_4^{2-} + 3\,H_2\uparrow$$

This element has also been reported to form Tb^{2+} and Tb^{3+} compounds like TbH_2, TbH_3, TbB_2, Tb_2S_3, TbSe, TbTe, and TbN through reacting with nitrogen, carbon, sulfur, phosphorus, boron, selenium, silicon, and arsenic at high temperatures [24]. The formation of the Tb^{2+} compounds is possible through the co-annealing of the elemental and 3+ terbium compounds under the conditions similar to those of gadolinium. A further similarity to gadolinium is the formation of sesquichloride (Tb_2Cl_3), which is reducible to TbCl under identical conditions as in the case of Gd [101]. Examples of other compounds are $TbCl_3$, $TbBr_3$, TbI_3, TbF_3, and TbF_4 [103].

Terbium concentration in the earth's crust is 1.2 ppm, and the element is found in minerals including cerite, gadolinite, bastnaesite, monazite, xenotime, and euxenite. The method of extraction of the element is based on the same procedure as mentioned in the case of the previous lanthanides, and after the separation of insoluble cerium using HNO_3, terbium is isolated as a double salt with ammonium nitrate [24]. The best separation technique is the application of ion-exchange techniques, in which course the ions are exchanged with hydrogen, ammonium, or cupric ions present in a suitable ion-exchange resin, and next selectively eluted by suitable chelates. Metallic terbium is prepared through the conventional reduction of its dry chloride or fluoride salts with calcium metal, which is next removed from the final product through vacuum or zone remelting, distillation, and amalgamation [24].

DYSPROSIUM

Dysprosium was isolated by Paul Émile Lecoq de Boisbaudran in 1886, from holmium oxide, through a process based on dissolving dysprosium oxide in acid and then adding NH_3 to increase the pH and hence precipitate its hydroxide. It took the researchers more than 30 attempts to isolate the element, and hence the element was named based on the Greek dysprositos (δυσπρόσιτοζ), which means difficult to get. Obtaining the pure form of the element, however, needed to wait until the 1950s and the introduction of ion-exchange techniques [55].

Naturally occurring dysprosium is composed of the stable isotopes ^{156}Dy, ^{158}Dy, ^{160}Dy, ^{161}Dy, ^{162}Dy, ^{163}Dy, and ^{164}Dy, among which only ^{156}Dy decays with a $t_{1/2}$ of 1×10^{18} years. ^{164}Dy and ^{162}Dy are the two most abundant isotopes accounting for 54% of the total weight of the samples [42]. About 29 synthetic radioisotopes with the mass numbers ranging from 138 to 173 have been reported for the element. The most stable ones are ^{154}Dy ($t_{1/2} = 3 \times 106$ years), and ^{159}Dy ($t_{1/2} = 144.4$ days). Roughly speaking, the heavier and lighter isotopes primarily undergo β+ and β− decays, respectively, but this is not always true [42].

The overall properties of the element are identical to its neighboring ones [48]. The magnetic properties of dysprosium and holmium are the highest among the lanthanides, especially at low temperatures [104]. Dysprosium metal reacts with O_2, water, and halogens in the same way as the other elements of the lanthanide family. Its trifluoride, triiodide, and tribromide are green, whereas the trichloride is white. The sulfate salt produced through its dissolution in dilute sulfuric acid is remarkably paramagnetic. Examples of other compounds of the element where it has 3+ or 2+ oxidation states include DyN, DyP, DyH_2, and DyH_3; DyS, DyS_2, Dy_2S_3, and Dy_5S_7; DyB_2, DyB_4, DyB_6, and DyB_{12}; and Dy_3C, Dy_2C_3, $Dy_2(CO_3)_3$, and $Dy_2(SO_4)_3$ [24]. The majority of dysprosium compounds, with exceptions like $Dy_2(CO_3)_3{\bullet}4H_2O$ and $Dy_2(C_2O_4)_3{\bullet}10H_2O$ are soluble in water [105].

Dysprosium (5.2 ppm in the earth's crust) [24] is not found in elemental form in nature, but its compounds in combination with other lanthanides are found in xenotime, fergusonite, gadolinite, euxenite, polycrase, blomstrandine, monazite, and bastnaesite. The major sources of the element are located in China [106] and Australia [107]. It has been reported that the ores with high yttrium contents contain dysprosium as the most abundant heavy lanthanide with 7%–8% of the element [102,108].

The process for the isolation of the element is actually the process for the purification of yttrium and starts with monazite, which is a mixture of various phosphates. The unwanted elements are removed by magnets or through flotation. The other procedures for the preparation of pure dysprosium are based on ion-exchange processes or thermal reactions of DyF_3, or $DyCl_3$ with calcium or lithium metals under an inert atmosphere as shown below:

$$3\,Ca + 2\,DyF_3 \rightarrow 2\,Dy + 3\,CaF_2$$

$$3\,Li + DyCl_3 \rightarrow Dy + 3\,LiCl$$

During the process, the by-products and molten dysprosium are differentiated based on their density differences.

HOLMIUM

Like many other Lns, the discovery of holmium was based on noticing unprecedented spectral lines in samples. This was done by Marc Delafontaine and Jacques-Louis Soret in 1878 and also during the independent investigations of T. Cleve while he was working on erbium oxide. He used the method of C. G. Mosander to remove the known contaminants from erbia and yielded two new brown and green products, which he named holmia (after the Latin name for Stockholm) and thulia respectively. Later works showed that what was called holmia was holmium oxide and the other one was thulium oxide [54]. Naturally occurring holmium contains only ^{165}Ho, which is a stable isotope, whereas some synthetic radio isotopes including ^{163}Ho ($t_{1/2}$ = 4570 years) have been reported. The rest of the radioisotopes have very short have half-lives, in the order of some days to a few hours.

Holmium is paramagnetic at room temperature and ferromagnetic at temperatures below 19 K, has a very high magnetic moment of 10.6 μB as well as other unconventional magnetic properties, and forms highly magnetic compounds with yttrium [102,109]. Like its close neighbors, Ho (atomic number = 67) is very resistant against corrosion in dry air or ambient temperature, whereas moist air or higher temperatures oxidize it to a yellowish compound, which tarnishes the metallic, bright silvery element. This oxide changes color based on the lighting and looks orange red under trichromatic light [110].

Ho also has identical reactions with water, halogens, and dilute sulfuric acid solutions, and its most stable oxidation state is 3+ [54]. Similar to the other rare earths, Ho (1.3 ppm in the earth's crust, 1 ppm in soil, and 400 parts per quadrillion [ppq] in seawater) occurs in combination with other lanthanide compounds in gadolinite, monazite, and lanthanide minerals, which are abundant in China, United States, Brazil, India, Sri Lanka, and Australia [17,54]. This element is commercially obtained through ion-exchange processes or the thermal reduction of its trichloride or trifluoride with metallic calcium, under an inert atmosphere [23,24].

ERBIUM

The element was discovered by C. G. Mosander in 1843, who could fractionate "yttria," which was obtained from gadolinite, to three parts. He named the three fractions yttria, erbia, and terbia. The last two elements were initially confused. In 1905, Georges Urbain and Charles James could prepare pure Er_2O_3, but the pure metallic form was prepared in 1934, through the reduction of the chloride salt of Er(III) by Klemm and Bommer, who used potassium vapor to this end [54]. The stable isotopes found in the natural erbium samples are ^{162}Er, ^{164}Er, ^{166}Er, ^{167}Er, ^{168}Er, and ^{170}Er, among which ^{166}Er has the highest concentration (33.50 wt%). Yet around 29 synthetic isotopes have been reported for erbium, including ^{169}Er ($t_{1/2}$ = 9.4 days), which is the most stable among all, ^{172}Er ($t_{1/2}$= 49.3 hours), ^{160}Er ($t_{1/2}$ = 28.58 hours), ^{165}Er ($t_{1/2}$ = 10.36 hours), and ^{171}Er ($t_{1/2}$ = 7.516 hours), as well as some other isotopes with very short half-lives [111]. The primary decay for the isotopes lighter than ^{166}Er is electron capture leading to the formation of holmium isotopes, whereas heavier isotopes undergo beta decays, in which case the products are isotopes of thulium [111]. Er has an atomic number of 68 and shares physical properties with those of the other Ln, but like its close neighbors and opposite to the lighter lanthanides, it is stable in air in the absence of humidity, but reacts with moist air and water, forming rose compounds. The element has not been proven as a biologically effective agent but has been found to have metabolic influences [54].

In terms of magnetic properties, Er has ferromagnetic properties below 19 K, is antiferromagnetic from 19 to 80 K, and has paramagnetic behaviors at temperatures higher than 80 K [111]. Except for the pink color of its trifluoride, the violet color of the other trihalides, and the red color of the solutions of Er in dilute sulfuric acid, its reactions with water, oxygen, halogens, and acid solutions are identical to the previous descriptions. This element is present in the earth's crust at about 2.8 ppm, but like the other rare earths, erbium is never found in its elemental form in nature [24]. Minerals containing the element are xenotime and euxenite, and erbium is extracted from the ores through a procedure that is identical to those mentioned in the case of other lanthanides as long as the separation of the insoluble cerium oxide using HNO_3, which dissolves the rest of the elements. The solution from that cerium oxide is filtered next and treated with magnesium nitrate, giving a crystallized mixture of double salts of rare-earth metals, which are then used in ion exchange, and subsequent elution with ligands [112]. A further conventional method involves the reduction of the oxide or halides through heating in the presence of a reducing agent such as calcium or lithium vapor at 1450°C, under an argon atmosphere [24].

THULIUM

P. T. Cleve discovered thulium in 1879 as he was searching for impurities in rare-earth oxides. Once he had removed Er_2O_3, he came across what looked like two unknown brown and green oxides, which he called holmia and thulia, respectively. This element was initially shown with the symbol "Tu" but is now illustrated by "Tm" [54]. Pure thulium was rather impossible to achieve, because of the scarcity of the element. Consequently, the studies on the element were mainly spectroscopic in nature and were based on the increased intensity of the two characteristic absorption bands, which was due to stepwise removal of erbium. The first report on obtaining pure thulium was made by Charles James in 1911. He used the technique he called bromate fractional crystallization after 15,000 purification operations [113]. Unlike the majority of Lns, however, the commercial supply of the element had to wait until the late 1950s and was based on the famous ion-exchange technique [35,36].

The isotopes of thulium cover all the range from ^{146}Tm to ^{177}Tm. The most stable and abundant one of these isotopes is ^{169}Tm. Isotopes lighter than ^{169}Tm are known to undergo electron capture reactions, transforming Tm to erbium, whereas the heavier ones form ytterbium after beta emissions [48]. ^{169}Tm is considered stable even given the predictions that it should decay α with a very long half-life; the rest of the isotopes are unstable. The most stable isotopes among these are ^{171}Tm ($t_{1/2}$ = 1.92 years) and ^{170}Tm ($t_{1/2}$ = 128.6 days), and the rest of the isotopes have $t_{1/2}$ values ranging from some minutes to seconds.

In terms of chemical properties, thulium, which is the 69th element in the periodic table, is mostly like the other lanthanides, being a silvery, soft, and ductile solid metal (with tetragonal α-Tm and the more stable hexagonal β-Tm allotropes) that slowly tarnishes in the air, with the most common oxidation state of 3+ , seen in its oxide, halides, and other compounds. Liquid thulium is reported to be very volatile [23,114]. Tm behaves ferromagnetically below 32 K, antiferromagnetically in the range of 32–56 K, and paramagnetically above 56 K [115].

Thulium reacts with water, halogens (producing yellow and white products), sulfuric acid, and moist air. The element also forms different compounds with both metals and nonmetals. Examples include TmN, TmS, TmC_2, Tm_2C_3, TmH_2, TmH_3, $TmSi_2$, $TmGe_3$, TmB_4, TmB_6 and TmB_{12}, in some of which the element adopts the less stable 2+ oxidation states in addition to the conventional 3+ state [24]. Tm^{2+} compounds are reddish-purple and can be prepared through reducing Tm^{3+}, but they are not stable [116]. $TmCl_2$ has vigorous reactions with H_2O, producing hydrogen gas and $Tm(OH)_3$ with a fading red color [117].

The reactions between thulium and HCl or HNO_3 result in the formation of $TmCl_3$ and $Tm(NO_3)_3$ and hydrogen gas. Thulium is commonly found in ores containing yttrium and gadolinium (e.g., gadolinite) [118], but it is also present in monazite, xenotime, and euxenite, which are abundant in Australia, Brazil, Greenland, India, Tanzania, and the United States. Its concentrations in the earth's crust and seawater are 50 ppb and 250 parts per quadrillion, respectively. The metallic form can be prepared by reducing lanthanum oxide in the presence of calcium as in the case of the other Lns.

YTTERBIUM

Jean Charles Galissard de Marignac discovered ytterbium in 1878 when working with gadolinite. He named the element after the Swedish village called Ytterby, near which the element was discovered [54,119]. Ytterbium is made up of ^{168}Yb, ^{170}Yb, ^{171}Yb, ^{172}Yb, ^{173}Yb, ^{174}Yb, and ^{176}Yb, among which ^{174}Yb is the most abundant isotope (31.8 wt%). This element also has 27 radioisotopes, including the rather stable ^{169}Yb ($t_{1/2}$ = 32.0 days), ^{175}Yb ($t_{1/2}$ = 4.18 days), and ^{166}Yb ($t_{1/2}$ = 56.7 hours), as well as the less stable ones with half-lives of less than some hours to a few minutes [111]. The decay of the isotopes heavier than ^{174}Yb takes place through electron capture reactions leading to the formation of thulium isotopes, whereas the heavier ones decay through emitting β and producing lutetium isotopes.

Three allotropes (α, β, and γ) have been reported for the element in the temperature range of −13°C to 795°C [120]. The α structure, which is stable at lower temperatures, is hexagonal, the β face-centered cubic crystal forms at ambient temperature, and the γ form, which has body-centered cubic crystalline structure, is formed at higher temperatures [23]. These allotropes have different electrical conductivities based on the temperature and pressure [23,54]. The magnetic behavior of the element is different from those of the other lanthanides, and ytterbium is paramagnetic at temperatures above 1.0 K [115].

Because ytterbium is the last f-block element, it has a comparatively more stable 2+ oxidation state, but it is ineluctable to state that its most stable oxidation state is 3+ . This is also observed in the case of samarium, thulium, and europium. The electron configuration of the 2+ oxidation state is $4f^{14}$, and this is the reason behind its rather unusual stability. This divalent ion has a green-yellow color, is a strong reducing agent, and reacts with water, giving off H_2 gas, to form colorless ytterbium(III) ion. The majority of ytterbium compounds, however, still belong to the 3+ oxidation state. As in the case of Eu, Sm, and Tm, the ytterbium trihalides can be changed to dihalides using pure ytterbium in the presence of a reducing agent such as zinc. Elemental ytterbium is no different from its other fellow elements, in terms of physical properties, as well as in terms of reactions with mineral acids, halogens, air, and water (the element reacts with water, producing hydrogen gas). Ytterbium is slowly oxidized in the air, and this reaction is faster for fine ytterbium particles [121]. Ytterbium(III) chloride has been reported as a catalyst in the Aldol [122] and Diels–Alder reactions [123]. YbI_2 can also be used as a reducing agent in coupling reactions [124]. YbF_3 has been reported as an inert, nontoxic material for filling teeth and it continuously releases beneficial fluoride ions [119].

Ytterbium, which is present in the earth's crust at 3 ppm, is an ingredient of ores like monazite (0.03 wt%), as well as in euxenite and xenotime, which are found in China, the United States, Brazil, India, Sri Lanka, and Australia [54]. Microscopic traces of this element have been used as dopants in the Yb:YAG lasers [125]. Because of the difficulty of separating the element from other Lns, the process is very time consuming. Initially, the ores are dissolved using inorganic acids and are then subjected to separation by ion-exchange routines [126,127]. Another separation procedure involves reduction by a sodium amalgam, where a buffered acidic solution of the trivalent cation is subjected to a molten sodium–mercury alloy, which leads to the formation of the reduced form of the cation of the element [24].

LUTETIUM

In 1907, Georges Urbain, Baron Carl Auer von Welsbach, and Charles James independently discovered lutetium as an impurity in Ytterbia [128,129]. Natural lutetium samples contain only two isotopes of the stable

^{175}Lu and radioactive ^{176}Lu ($t_{1/2} = 3.78 \times 10^{10}$ years) that undergoes beta decays [119]. The element also has 32 synthetic radioisotopes from ^{150}Lu to ^{184}Lu, among which the most stable ones are ^{174}Lu ($t_{1/2}$ = 3.31 years) and ^{173}Lu ($t_{1/2}$ = 1.37 years), and the rest have half-lives that are less than some days to some minutes [42]. The conventional decays among the isotopes lighter than ^{175}Lu are actually electron capture (and some α emissions), whereas the heavier ones have been reported to decay alpha and positron producing hafnium isotopes.

The physical appearance and other physical properties of the element are identical to those of its neighbors. As usual, the common oxidation state for the element is the colorless 3+ , which is present in many of the salts of the element. The soluble salts of lutetium $Lu(NO_3)_3$, $Lu_2(SO_4)_3$, and $Lu(CH_3COO)_3$ form hydrates on crystallization. The other salts like Lu_2O_3, $Lu(OH)_3$, LuF_3, $Lu_2(CO_3)_3$, $LuPO_4$, and $Lu_2(C_2O_4)_3$ do not dissolve in aqueous media [24].

Lutetium metal is rather unstable in air and like the other lanthanides starts burning at 150°C. The oxide of lutetium formed this way is a well-known water and carbon dioxide sorbent [114]. The reactions between the element and water, halogens, and sulfuric acids are also similar to those of the other Lns. Lutetium has an abundance of 0.5 ppm in the earth's crust but it is found in minerals like monazite at very low concentrations (0.0001 wt%) [114]. The element is very difficult to isolate from its minerals based on ion-exchange procedures as well as the reduction of its chlorides in the presence of calcium under heat and inert atmospheres [54].

References

[1] The 1985 "Red Book" (IUPAC Nomenclature of Inorganic Chemistry), p. 45.
[2] Bünzli JCG. In: Lanthanides, Kirk-Othmer encyclopedia of chemical technology. Wiley Online Library; 2013. p. 1–43.
[3] Castor B, Hedrick JB. In: Kogel JE, Trivedi NC, Barker JM, editors. Rare earth elements, in industrial minerals and rocks. Society for Mining, Metallurgy and Exploration; 2006. p. 769–92.
[4] [Online] available at: http://www.statista.com/statistics/277268/rare-earth-reserves-by-country/; The Statistics Portal; Statistics and Studies from more than 18,000 Sources.
[5] [Online] available at: http://www.metal-pages.com/.(released 31 December 2014; MP-Metal Pages).
[6] Kaltsoyannis N, Scott P. The f elements. Oxford: Oxford Science Publications; 1999.
[7] Cotton S. Lanthanide and Actinide Chemistry. Chichester: John Wiley & Sons Ltd; 2006.
[8] Bünzli JCG, Milicic-Tang A. In: Gschneidner KA Jr, Eyring L, editors. Hand book on the physics and chemistry of rare earths, 21. Amsterdam: Elsevier Science B.V; 1995. p. 306–66. [chapter 145].
[9] Bünzli JCG. Review: Lanthanide coordination chemistry: from old concepts to coordination polymers. J Coord Chem 2014;67:3706–33.
[10] Petrucci HH. General chemistry: principles & modern applications. New Jersey: Macmillan Publishing Company; 2007.
[11] Moeller T. The chemistry of the Lanthanides. New York: Reinhold Publishing Corporation; 1963.

[12] Cotton S. Lanthanides and Actinides. London: Macmillan Education Ltd; 1991.
[13] [Online] available at: http://www.webelements.com/lanthanum/; "Chemical reactions of Lanthanum". Webelements.
[14] [Online] "Rare-Earth Metal Long Term Air Exposure Test". available at: http://www.elementsales.com/re_exp/.
[15] Hedrick JB, Sinha SP, Kosynkin VD. Loparite - a rare-earth ore (Ce,Na,Sr,Ca)(Ti,Nb,Ta,Fe^{3+})O_3. J Alloy Compound 1997;250:467–70.
[16] [Online] "Visual Elements: Cerium". London: Royal Society of Chemistry; 1999–2012. Available at: http://www.rsc.org/periodic-table/element/58/cerium.
[17] Emsley J. Nature's building blocks: an A-Z guide to the elements. Oxford University Press; 2011. p. 120–125, ISBN 978-0-19-960563-7.
[18] Gray T. The elements. Black Dog & Leventhal Pub; 2010. ISBN 1579128955.
[19] Jackson, M. Periodic Table Advanced. Barcharts Incorporated; 2002, ISBN 1572225424.
[20] [Online] available at: NuDat 2.6. Nndc.bnl.gov.
[21] Koskimaki DC, Gschneidner KA, Panousis NT. Preparation of single phase β and α cerium samples for low temperature measurements. J Cryst Growth 1974;22:225–9.
[22] Stassis C, Gould T, McMasters O, Gschneidner K, Nicklow R. Lattice and spin dynamics of γ-Ce. Phys Rev B 1979;19:5746–53.
[23] Hammond CR. The Elements, in: Handbook of chemistry and physics. 81st ed. CRC press; 2000. ISBN 0-8493-0481-4.
[24] Patnaik P. Handbook of inorganic chemical compounds. McGraw-Hill; 2003. p. 199–200. ISBN 0-07-049439-8.
[25] [Online] available at: Cerium dioxide. nanopartikel.info.
[26] [Online] available at: www.webelements.com; "Chemical reactions of Cerium".
[27] Kreh RP, Spotnitz RM, Lundquist JT. Mediated electrochemical synthesis of aromatic aldehydes, ketones, and quinones using ceric methanesulfonate. J Org Chem 1989;54:1526–31.
[28] [Online] Cerium. Minor Metals in the Periodic Table. Mmta.co.uk (2012-08-30).
[29] Rare Earth Elements 101, IAMGOLD Corporation, April 2012, p. 5–7.
[30] [Online] "Chemical reactions of Praseodymium". Webelements. Available at: http://www.webelements.com/praseodymium/chemistry.html.
[31] NNDC contributors. Sonzogni AA (Database Manager), editor. "Chart of Nuclides". Upton (NY): National Nuclear Data Center, Brookhaven National Laboratory; 2008.
[32] [Online] available at: http://www.nyu.edu/cgi-bin/cgiwrap/aj39/NMRmap.cgi.
[33] Gschneidner KA, Eyring L. Handbook on the physics and chemistry of rare earths. Amsterdam: North Holland Publishing Co; 1978.
[34] [Online] "Chemical reactions of Neodymium". Webelements. Available at: http://www.webelements.com/neodymium/chemistry.html.
[35] Hedrick, F., James, B. Rare earths: chapter in Mineral commodity summaries 2010, U.S. Geological Survey; 2010, p. 128–129.
[36] Hedrick, J.B. Rare earths - The lanthanides, yttrium, and scandium: chapter in Minerals Yearbook 1990, vol. 1, U.S. Geological Survey; 1990, p. 903–922.
[37] Beaudry BJ, Gschneidner Jr. KA, Preparation and basic properties of the rare earth metals. In: Gschneidner Jr E, editor. Handbook on the physics and chemistry of rare earths volume 1 metals. North Holland, New York; 1978, p. 173–232.
[38] Weeks ME, Leicester HM. Discovery of the elements. 7th ed. Easton, Pennsylvania. J Chem Educat 1968;896.
[39] Marinsky JA, Glendenin LE, Coryell CD. The chemical identification of radioisotopes of neodymium and element 61. J Am Chem Soc 1947;69:2781–5.
[40] Weigel F. Darstellung von metallischen promethium. Angew Chem 1963;75:451–1451.
[41] Wheelwright EJ. Preparation of promethium-147 metal and determination of the density and melting points. J Phys Chem 1969;78:2867–71.

[42] Audi G, Wapstra AH, Thibault C, Blachot J, Bersillon O. The NUBASE evaluation of nuclear and decay properties. Nucl Phys A 2003;729:3–128.
[43] Greenwood NN, Earnshaw A, Chemistry of the elements. 2nd ed. Butterworth-Heinemann; 1997, p. 1233, ISBN 0080379419.
[44] Cotton FA, Wilkinson G. Advanced inorganic chemistry. 5th ed. New York: Wiley-Interscience; 1988. p. 776–955, ISBN 0-471-84997-9.
[45] Lavrukhina AK, Pozdnyakov AA, The oxalate of the element ($Pm_2(C_2O_4)_3 \cdot 10\ H_2O$) 1966, p. 120–123.
[46] Chikalla TD, McNeilly CE, Roberts FP. Polymorphic Modifications of Pm_2O_3. J Am Chem Soc 1972;55:428.
[47] Attrep M, Kuroda PK. Promethium in pitchblende. J Inorg Nucl Chem 1968;30:699–703.
[48] Holden NE. Table of the isotopes. In: Lide DR, editor. CRC Handbook of chemistry and physics. 86th ed. Boca Raton (FL): CRC Press; 2005. ISBN 0-8493-0486-5.
[49] Lock JM. The magnetic susceptibilities of Lanthanum, Cerium, Praseodymium, Neodymium and Samarium, from 1.5 K to 300 K. Procee Phys Soc Seri B 1957;70:566–76.
[50] Huray P, Nave S, Haire R. Magnetism of the heavy 5f elements. J Less Common Metal. 1983;93:293–300.
[51] Okazaki T, Suenaga K, Hirahara K, Bandow S, Iijima S, Shinohara H. Electronic and geometric structures of metallofullerene peapods. Physica B 2002;323:97–9.
[52] Chen X, Roth G. Superconductivity at 8 K in samarium-doped C60. Phys Rev B 1995;52:15534–6.
[53] Wu G, Xie YL, Chen H, Zhong M, Liu RH, Shi BC, Li QJ, Wang XF, Wu T, Yan YJ, Ying JJ, Chen XH. Superconductivity at 56 K in Samarium-doped SrFeAsF. J Phys Condens Matter 2008;21:142203.
[54] Emsley J. Samarium. Nature's building blocks: an A–Z guide to the elements. Oxford, England, UK: Oxford University Press; 2001. ISBN 0-19-850340-7.
[55] [Online] "Chemical reactions of Samarium". Webelements. Available at: http://www.webelements.com/samarium/chemistry.html.
[56] Gouteron J, Michel D, Lejus AM, Zarembowitch J. Raman spectra of lanthanide sesquioxide single crystals: Correlation between A and B-type structures. J Solid State Chem 1981;38:288–96.
[57] Taylor D. Thermal expansion data: III. Sesquioxides, M_2O_3, with the corundum and the A-, B- and C-M_2O_3 structures. Br Ceram Trans J 1984;83:92–8.
[58] Leger J, Yacoubi N, Loriers J. Synthesis of rare earth monoxides. J Solid State Chem 1981;36:261–70.
[59] Beaurepaire E, editor. Magnetism: a synchrotron radiation approach. Springer; 2006. p. 393. ISBN 3-540-33241-3.
[60] Greis O. Über neue Verbindungen im system SmF_2_SmF_3. J. Solid State Chem 1978;24:227–32.
[61] Baernighausen H, Haschke JM. Compositions and crystal structures of the intermediate phases in the samarium-bromine system. Inorg Chem 1978;17:18.
[62] Nickerson J, White R, Lee K, Bachmann R, Geballe T, Hull, G. Physical Properties of SmB6. Phys Rev B, 1971, 3, 2030-&.
[63] Solovyev GI, Spear KE. Phase behavior in the Sm-B system. J Am Ceram Soc 1972;55: 475–9.
[64] Cannon J, Cannon D, Tracyhall H. High pressure syntheses of SmB_2 and GdB_{12}. J Less Common Metal 1977;56:83–90.
[65] Nyhus P, Cooper S, Fisk Z, Sarrao J. Light scattering from gap excitations and bound states in SmB6. Phys Rev B 1995;52:R14308.
[66] Sera M, Kobayashi S, Hiroi M, Kobayashi N, Kunii S. Thermal conductivity of RB6 (R = Ce, Pr, Nd, Sm, Gd) single crystals. Phys Rev B 1996;54. R5207.
[67] Botimer J, Kim T, Grant F, Jing X. Robust surface hall effect and nonlocal transport in SmB6: indication for an ideal topological insulator; 2012. arXiv:1211.6769 [cond-mat.str-el].

[68] Zhang XH, Butch NP, Syers P, Ziemak S, Greene RL, Paglione J. Hybridization, correlation, and in-gap states in the Kondo insulator SmB6. Phys Rev 2013;3:011011.
[69] Cagliyan Kurdak W, Allen KS, Kim DJ, Fisk Z. Discovery of the first topological Kondo insulator: samarium hexaboride; 2012. arXiv:1211.5104 [cond-mat.str-el].
[70] Spedding FH, Gschneidner K, Daane AH. The crystal structures of some of the rare earth carbides. J Am Chem Soc 1958;80:4499–503.
[71] Meng J, Ren Y. Studies on the electrical properties of rare earth monophosphides. J Solid State Chem 1991;95:346–51.
[72] Beeken R, Schweitzer J. Intermediate valence in alloys of SmSe with SmAs. Phys Rev B 1981;23:3620–6.
[73] Gladyshevskii EI, Kripyakevich PI. Monosilicides of rare earth metals and their crystal structures. J Struct Chem 1965;5:789.
[74] Smith GS, Tharp AG, Johnson W. Rare earth–germanium and –silicon compounds at 5:4 and 5:3 compositions. Acta Crystallogr 1967;22:940–3.
[75] Yarembash EI, Tyurin EG, Reshchikova AA, Karabekov A, Klinaeva NN. Inorg Mater 1971;7:661–5.
[76] Evans WJ, Hughes LA, Hanusa TP. Synthesis and x-ray crystal structure of bis(pent amethylcyclopentadienyl) complexes of samarium and europium: (C5Me5)2Sm and (C5Me5)2Eu. Organometallics 1986;5:1285–91.
[77] [Online] Chemistry in Its Element – Samarium, Royal Society of Chemistry. Available at: http://www.rsc.org/chemistryworld/podcast/Interactive_Periodic_Table_Transcripts/Samarium.asp.
[78] Holleman AF, Wiberg E. Inorganic Chemistry. San Diego: Academic Press; 2001. ISBN 0-12-352651-5.
[79] Johansson B, Rosengren A. Generalized phase diagram for the rare-earth elements: Calculations and correlations of bulk properties. Phys Rev B 1975;11:2836–57.
[80] Debessai M, Matsuoka T, Hamlin J, Schilling J, Shimizu K. Pressure-Induced Superconducting State of Europium Metal at Low Temperatures. Phys Rev Lett 2009;102:197002.
[81] [Online] Chemical reactions of Europium. Webelements. Available at: http://www.webelements.com/europium/chemistry.html.
[82] Cooley RA, Yost DM, Stone HW. Europium(II) Salts. Inorganic Syntheses. Inorg Synth 1946;2:69–73.
[83] Sinha SP. Scientific Affairs Division, North Atlantic Treaty Organization, The Europium anomaly. Systematics and the properties of the lanthanides. 1983;550–3. ISBN 978-90-277-1613-2.
[84] Belli P, Bernabei R, Cappella F, Cerulli R, Dai CJ, Danevich FA, d'Angelo A, Incicchitti A, Kobychev VV, Nagorny SS, Nisi S, Nozzoli F, Prosperi D, Tretyak VI, Yurchenko SS. Search for α decay of natural europium. Nucl Phys A 2007;789:15–29.
[85] Nucleonica. Nucleonica: Universal Nuclide Chart. Nucleonica: Universal Nuclide Chart. Nucleonica. Archer RD, Mitchell WN, Mazelsky R. Europium(II) Sulfide. Inorg Synth 1967;10:77–9. (2007-2011).
[86] Bill H, Calas G. Color centers, associated rare-earth ions and the origin of coloration in natural fluorites. Phys Chem Miner 1978;3:117–31.
[87] Valeur B, Berberan-Santos MN. A brief history of fluorescence and phosphorescence before the emergence of quantum theory. J Chem Educat 2011;88:731–8.
[88] Sidike A, Kusachi I, Yamashita N. Natural fluorite emitting yellow fluorescence under UV light. Phys Chem Mineral 2003;30:478–85.
[89] Przibram K. Fluorescence of Fluorite and the Bivalent Europium Ion. Nature 1935;135:100–1100.
[90] Srivastava AM, Ronda CR. Phosphors. Electrochem Soc Interface 2003;12:48–51.
[91] Spedding FH. Large-scale separation of rare-earth salts and the preparation of the pure metals. Discuss. Faraday Soc 1949;7:214–31.

[92] Corbett JD. Frank Harold Spedding. Biography. Memoirs, Natl Acad Sci 1986;80:106.
[93] Gupta CK, Krishnamurthy N. Extractive metallurgy of rare earths. Int Mater Rev 1992;37:197–248.
[94] Morais C, Ciminelli VST. Recovery of europium by chemical reduction of a commercial solution of europium and gadolinium chlorides. Hydrometallurgy 2001;60:247–53.
[95] Neikov OD, Naboychenko S, Gopienko VG, Frishberg IV. Handbook of non-ferrous metal powders: technologies and applications. 2009;505. ISBN 978-1-85617-422-0.
[96] Drewa LJ, Qingrunb M, Weijun S. The Bayan Obo iron-rare-earth-niobium deposits, Inner Mongolia. China. Lithos 1990;26:43–65.
[97] Yang XM, Le Bas MJ. Chemical compositions of carbonate minerals from Bayan Obo, Inner Mongolia, China: implications for petrogenesis. Lithos 2004;72:97–116.
[98] Wu C. Bayan Obo Controversy: Carbonatites versus Iron Oxide-Cu-Au-(REE-U). Resour Geol 2007;58:348–54.
[99] Danevich FA, Kobychev VV, Ponkratenko OA, Tretyak VI, Zdesenko GY. Quest for double beta decay of 160Gd and Ce isotopes. Nucl Phys A 2001;694:375–91.
[100] [Online] Chemical reactions of Gadolinium. Webelements. Available at: http://www.webelements.com/gadolinium/chemistry.html.
[101] Cotton, Advanced inorganic chemistry. 6th ed. Wiley-India; 2007, p 1128, ISBN 81-265-1338-1.
[102] Gupta CK. Nagaiyar Krishnamurthy, Extractive metallurgy of rare earths. CRC Press; 2004. p. 5. ISBN 0-415-33340-7.
[103] Rau JV, Chilingarov NS, Leskiv MS, Sukhoverkhov' VF, Rossi AV, Sidorov LN. Transition and rare earth metal fluorides as thermal sources of atomic and molecular fluorine, 2001.
[104] Krebs RE. Dysprosium. The history and use of our earth's chemical elements. Greenwood Press; 1998. p. 234–235, ISBN 0-313-30123-9.
[105] Perry DL. Handbook of inorganic compounds. CRC Press; 1995. p. 152–154, ISBN 0-8493-8671-3.
[106] Bradsher K. Earth-friendly elements, mined destructively. The New York Times; 2009.
[107] Brann, M. Halls Creek turning into a hub for rare earths; 2011.
[108] Naumov AV. Review of the world market of rare-earth metals. Russ J Non-Ferr Met 2008;49:14–22.
[109] Jiles D. Introduction to magnetism and magnetic materials. CRC Press; 1998. p. 228, ISBN 0-412-79860-3.
[110] Su YG, Li GS, Chen XB, Liu JJ, Li LP. Hydrothermal Synthesis of GdVO4:Ho^{3+} Nanorods with a Novel White-light Emission. Chem Lett 2008;37:762–3.
[111] Georges A, Bersillon O, Blachot J, Wapstra AH. The NUBASE evaluation of nuclear and decay properties. Nucl Phys A 2003;729:3–128.
[112] Early paper on the use of displacement ion-exchange chromatography to separate rare earth, Spedding FH, Powell JE. A practical separation of yttrium group rare earths from gadolinite by ion-exchange. Chem Eng Prog 1954;50:7–15.
[113] James C. Thulium I. J Am Chem Soc 1911;33:1332–44.
[114] Krebs RE. The history and use of our earth's chemical elements: a reference guide. Greenwood Publishing Group; 2006. p 299, ISBN 978-0-313-33438-2.
[115] Jackson M. Magnetism of rare earth. The IRM quarterly 2000;10:1.
[116] Eagleson M. Concise Encyclopedia Chemistry. Walter de Gruyter; 1994. p. 1105, ISBN 978-3-11-011451-5.
[117] Emeléus HJ, Sharpe AG. Advances in inorganic chemistry and radiochemistry. Academic Press; 1977. ISBN 978-0-08-057869-9.
[118] Walker P, Tarn WH. CRC Handbook of metal etchants. CRC Press; 2010. p. 1241. ISBN 978-1-4398-2253-1.

[119] Enghag, P. Encyclopedia of the elements: technical data, history, processing, applications. John Wiley & Sons, ISBN 978-3-527-30666-4, **2004**, p 448.
[120] Bucher E, Schmidt P, Jayaraman A, Andres K, Maita J, Nassau K, Dernier P. New first-order phase transition in high-purity ytterbium metal. Phys Rev B 1970;2:3911–7.
[121] Koch EC, Weiser V, Roth E, Knapp S, Kelzenberg S. Combustion of Ytterbium metal. Propell Explos Pyrot 2012;37:9–11.
[122] Lou S, Westbrook JA, Schaus SE. Decarboxylative Aldol Reactions of Allyl β-Keto Esters via heterobimetallic catalysis. J Am Chem Soc 2004;126:11440–1.
[123] Fang X, Watkin JG, Warner BP. Ytterbium trichloride-catalyzed allylation of aldehydes with allyltrimethylsilane. Tetrahedron Lett 2000;41:447–9.
[124] Girard P, Namy JL, Kagan HB. Divalent lanthanide derivatives in organic synthesis. 1. Mild preparation of samarium iodide and ytterbium iodide and their use as reducing or coupling agents. J Am Chem Soc 1980;102:2693–8.
[125] Lacovara P, Choi HK, Wang CA, Aggarwal RL, Fan TY. Room-temperature diode-pumped Yb:YAG laser. Optics Lett 1991;16:1089–91.
[126] Gelis VM, Chuveleva EA, Firsova LA, Kozlitin EA, Barabanov IR. Optimization of Separation of Ytterbium and Lutetium by Displacement Complexing Chromatography. Russ. J Appl Chem 2005;78:1420.
[127] Hubicka H, Drobek D. Anion-exchange method for separation of ytterbium from holmium and erbium. Hydrometallurgy 1997;47:127–36.
[128] James C. Separation of Rare Earth Elements. National Historic Chemical Landmarks. American Chemical Society; 1999.
[129] Urbain G. Lutetium und Neoytterbium oder Cassiopeium und Aldebaranium Erwiderung auf den Artikel des Herrn Auer v. Welsbach. Monatsh Chem 1909;31:1.

CHAPTER

2

Applications of the Lanthanide Series in Human Life

OUTLINE

As already mentioned, rare earth elements (the majority of which are lanthanides) are not as rare as implied by their names, and these elements can be found in many of the devices and materials human beings deal with every day. Many lanthanides are used in batteries, capacitors, magnets, and superconductors [1–3] that are widely used in everyday electronic gadgets. Other examples include the use of the elements in

Lanthanides Series Determination by Various Analytical Methods. http://dx.doi.org/10.1016/B978-0-12-804704-0.00002-5

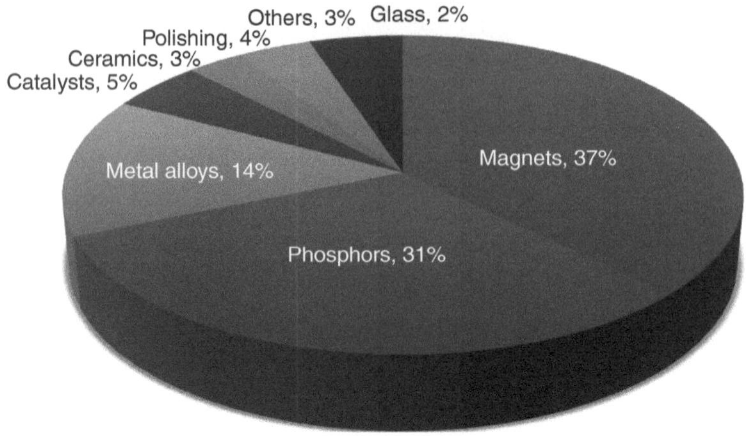

FIGURE 2.1 Applications of rare earth elements in terms of economic value.

phosphors and luminescence applications, metal alloys, catalysts, as well as glass polishing and ceramics [4]. The areas of magnets and phosphors are clearly the major sectors, consuming the majority of lanthanides produced, and the growth and development of technological applications is increasing the demand for lanthanides in these sectors.

Figure 2.1 illustrates the share of economic values of the different applications of lanthanides [4,5], among which use in magnets and as phosphors constitute the highest value (about 70%). The other applications include areas related to metal alloys, batteries, catalysts, glass and polishing, and ceramics.

A rough overview of the applications of these elements is summarized in Table 2.1, and this chapter shall further focus on the applications of these elements in some detail.

One classification of lanthanides, which were basically introduced in the previous chapter, would involve dividing them into the light lanthanides (LLs), including the elements from lanthanum to samarium (their oxides also known as the light rare earth oxides [LEROs]) and the heavy lanthanides (HLs), ranging from europium to lutetium (their oxides also known as heavy rare earth oxides [HEROs]). Another ground for their classification could be forming a group of the five most important lanthanides, namely, Nd, Pr, Dy, Tb, and Eu, or the so-called critical rare earths oxides (CREOs). These elements are also referred to as the big 5 [5].

An interesting illustration of the applications of lanthanides in everyday use could be the figure of the Toyota Prius hybrid (Fig. 2.2), which lists the utilization of lanthanides in the different parts of the vehicle [7].

The former lanthanides' use began in the 1960s with the advent of color televisions using Y, Eu, Gd, and Tb as phosphors. The catalysts used in the automotive industry, such as cerium compounds, create another major

TABLE 2.1 Application of Lanthanides in Human Life and Industries [6]

Industry	Ln	Applications
Automotive	Ce, La, Nd, Tb, Dy, Pr, Eu, Yb	• Catalysts for pollution control • Catalytic converter • Catalyst substrate • Rechargeable batteries • Fuel cells • Colored plastics
Ceramics	Nd, Eu, Gd, Lu, Dy, Pr, Ce, La, Er	• Capacitors • Oxygen sensors • Structural ceramics for bearings • Colorants • Jet engine coatings • Scintillators • Investment molds • Refractories • Pigments
Catalysts	Nd, Pr, Ce, La, Lu	• Petroleum refining • Catalytic convertor • Fuel additives • Chemical processing
Defense	Nd, Eu, Tb, Dy, Sm, Lu, Pr, La, Ho	• Satellite communications • Guidance systems • Aircraft structures • Shielding • Fly-by-wire • Smart missiles • Radars
Electronics	Ce, La, Pr, Nd, Tb	• Capacitors • Cathodes • Electrodes • Semiconductors • Thermistors • Traveling wave tubes (TWTs) • Radio frequency circulators and toroid • NimH battery • Fuel cell • Data storage (CDs and DVDs)
Glass and polishing	Ce, La, Pr, Nd, Gd, Er, Ho, Sm	• Polishing compounds • Pigments and coatings • UV resistance glass • Photo-optical glass • Decolorizing • Colorizing • Coloring glass • Increase refraction • Decrease dispersion

(*Continued*)

TABLE 2.1 Application of Lanthanides in Human Life and Industries [6] (*cont.*)

Industry	Ln	Applications
		• Radiation stabilization • Absorber • Expensive camera lenses
Magnetics	Nd, Tb Dy, Pr, Sm, Gd, Ho	• Speakers and headphones • Samarium-cobalt magnets • Linear motors • Antilock braking systems • Tape and disk drives • Gauges • Electric motors • Pumps • Ignition
Metallurgy	Nd, La Ce, Pr, Yb, Gd, Er	• Alloying agents in aluminum • Magnesium, iron, nickel, and steel alloys • Super-alloys • Pyrophoric alloys • Lighter flints • Armaments
Pharmaceutics	Gd, Lu, Ce, La	• Pharmaceuticals (Fosrenol, texaphyrin) • Contrast agents • Magnetic resonance imaging (MRI) • Positron emission tomography (PET) • Radioisotope • Tracers and emitters
Phosphorous	Nd, Eu, Tb, Er, Gd, Lu, Pr, Ce, Tm, La, Sm, Ho, Yb	• Display phosphors (CRT, LED, LCD) • Color TV • Lasers • Cathode-ray tubes (CRTs) • Fluorescent lighting • X-ray intensifying screens • Trichromatic fluorescent lamps • Mercury lamps • Carbon arc lamps • Gas mantles • Auto headlamps • Long-glow phosphors

area of application of lanthanides. Another important example is the use of lanthanum in fluid cracking catalysts (FCCs).

Although the worldwide demand for REEs has had a general upward trend for many decades, REEs' economics are as complex as their chemistry. Because individual REEs occur together in most deposits in fixed ratios, the economic problem is how to balance the production against market demand. To resolve this complication, the natural abundance of

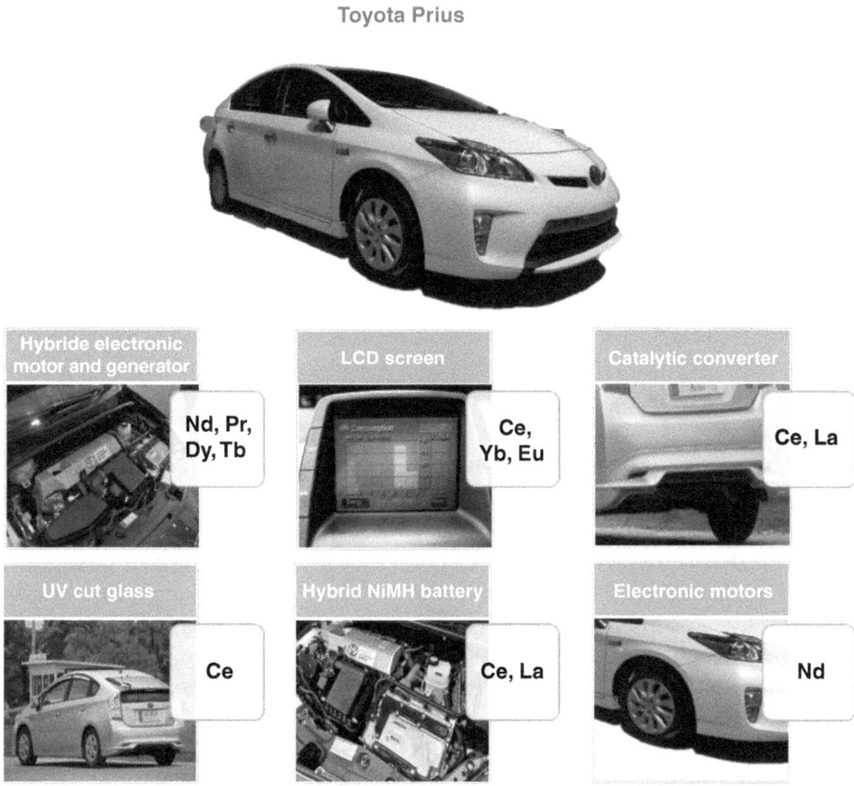

FIGURE 2.2 List of lanthanides used in the manufacturing of a Toyota Prius hybrid vehicle. *The images are provided by Shutterstock (Shutterstock.com) with the following credit attributions: Image 99514661, Deva Studio; Image 88629427, Ahmad Faizal Yahya; Image 333753818, nitinut380; Image 164440673, Christopher Halloran; Image 308116397, A. Aleksandravicius.*

each lanthanide in the ores should be clear to evaluate the uncertainties about the price and market. This has become more difficult by the fact that separation and purification costs are varied for different lanthanides, and this can be influenced by the market trends for each element, which depends on the changes in the technologies.

China produces and exports about 90% of the lanthanides used worldwide and, naturally, controls the prices. Some raw lanthanides are also provided by the United States and Australia [8,9], which are processed in different countries, for example, France and Japan. It is interesting that neither France nor Japan possesses natural deposits of the element but are known as the major producers of the elements and their compounds [6,10].

INDUSTRIAL APPLICATIONS

Regarding the widespread applications of the elements in different industries, some detailed information shall follow on the specific applications of each element. There is no need to mention that many of these applications may be similar, but they have been included in each case for the sake of clarity and comprehensiveness.

Lanthanum

Up to 25% to 45% of lanthanum compounds are used in Misch Metal alloy. Misch Metal is a phosphoric alloy that is used in used in cigarette lighters because of its good sparking properties [6,11].

Lanthanum has applications in different fields. An example would be the use of lanthanum compounds in carbon arc lamps, in which the electricity passes through the lamp electrode. The electrode is composed of carbon and traces of other doping materials, which increase the electrical resistance of the electrode. This way, the electrode is heated on passage of electricity and emits a bright white light, the color of which can be altered based on the doping materials, which can be chosen from lanthanum compounds such as LaF_3 and La_2O_3, which also act as phosphors [6,11]. Lanthanum oxides and borides are also used in hot cathode materials in electronic vacuum tubes, because of their high electron-emitting property [12].

$La(Ni_{3.6}Mn_{0.4}Al_{0.3}Co_{0.7})$, which is an intermetallic compound of lanthanum, nickel, manganese, aluminum, and cobalt with an AB5 structure, is used as the anodic material in nickel-metal hydride batteries [13,14], which are used in many hybrid cars. The amount of lanthanum in such batteries is rather high and, for example, in the case of a Toyota Prius, almost 10 to 15 kg of lanthanum is present in the car [14]. This amount is expected to increase because studies have revealed that the fuel mileage of such vehicles is proportional to the amount of the lanthanum compounds present in the battery. Another application of lanthanides is in H_2 sponge alloys used for storing high amounts of H_2 per volume (up to 400 times their own volume) and can, therefore, act as energy conservation systems [11,15].

Lanthanum fluoride (LaF_3) is an essential component of a heavy fluoride glass named ZBLAN, which was invented by a French team in 1974. ZBLAN is composed of zirconium, barium, lanthanum, aluminum, and sodium ("ZBLAN" stands for the first letters of each element). One of the new applications of ZBLAN, which is also known as lanthanum glass, is in the preparation of optical fibers. Optical fibers are wirelike materials made of glass, used to carry light, in the same way a copper wire carries electricity. These fibers are becoming more and more popular as a means for carrying audio, video, and digital messages, and have even replaced conventional copper wires in many cases. ZBLAN optical fibers, as opposed to conventional optical fibers, enjoy the advantage of carrying the messages more than 100 times faster and with higher quality. This has been attributed to the excellent transmittance of infrared (IR) spectra by the ZBLAN fibers [16].

La_2O_3 has also found applications in the optical glass formulations, as an agent to improve its resistance against alkaline media. Such glass is used in special optical glass like that used for filtering IR rays, or in camera and telescope lenses [17]. Another example may be the use of $LaBr_3$ and $LaCl_3$ salts as doping agents for the cerium samples used in high-yield inorganic scintillators with high energy resolution, and fast responses. Such modified scintillators have proven to be effective neutron or gamma detectors [18].

As an additive, lanthanum containing steel formulations has proven to possess improved malleability, impact resistance, and ductility [11]. In the case of tungsten welding electrodes, the presence of small amounts of lanthanum oxide modifies the function of the electrodes the same as radioactive thorium does [19,20].

The element has also proven to possess anti-phosphate properties in pools. This is good for the removal of phosphates that feed algae. The same effect has been observed in the case of bentonite modified with La [21].

Other applications of such elements include the use of its compounds (oxides, chlorides, etc.) as promoters [22], radiometric dating techniques (lanthanum-barium), phosphor lamp coatings (LaF_3), as a doping agent in the crystal membranes electrodes for fluoride ion (LaF_3 and EuF_3) [23,24].

Cerium

Cerium is used in a variety of fields, ranging from catalysts, fuel additives, glasses, and enamel dopants to glass polishing powders and phosphors used in screens and lamps. In the automotive field, Ce_2O_3 is used in catalytic converters for oxidizing CO and NO_x present in exhaust gases [25], or in diesel fuels, where compounds of the element are used as additives for increasing the efficiency of the fuel [26]. Like any other widely used material, an issue under debate about cerium is the health effects it can have, especially in the form of cerium oxide in exhaust gases [27,28].

The oxide of Ce^{3+} has also been used in the catalysts used in self-cleaning ovens and those used for petroleum cracking in refining industries [29]. As in the case of lanthanum, cerium compounds have many applications in the production and treatment of glass, both as additives and colorants. As examples of the latter case, CeO_2 and TiO_2 have been found to create a golden yellow color in glass, in addition to boosting its UV absorption. CeO_2 is further used as a fine polishing compound in the treatment of optical elements made of glass. Another application of cerium oxide is in enamel formulations, where it is used to make enamel opaque [29]. Mixtures of CeO_2 with the oxides of lanthanum, yttrium, thorium, and magnesium are used in Welsbach mantle [30]. The oxide of the element has also been reportedly used in solid electrolytes suited for intermediate-temperature solid oxide fuel cells [31]. The stability of long-life Ce phosphors in TV screens has led to tremendous increases in the application of the element. The presence of these stable compounds on the inner walls of TV screens prevents the gradual darkening due to chronic electronic bombardment [29].

Cerium is also present in Misch Metal [32] and in the formulation of various aluminum and iron alloys. This leads to increased corrosion resistance of these alloys. Additionally, the great affinity of the element toward sulfur and oxygen helps reduce sulfides and oxides present in steel. The precipitates of the cerium additives and these elements have been found to increase the hardness of stainless steel, as a result of preventing the graphitization phenomenon. Trace amounts of cerium and zirconium in magnesium alloys has also proven to modify the alloys in terms of casting and heat resistance [29,33]. Like lanthanum, cerium alloys are also used in permanent magnets and tungsten arc welding electrodes.

In the area of pharmacy, cerium oxalate is used in treating vomiting and nausea. Ce^{2+} sulfate and ammonium nitrate have been reported to act as oxidizing agents and have been used in quantitative analysis routines [11,22]. Further applications of the element include its use in zinc-cerium redox flow batteries [34]. These batteries function based on the generation of electricity in the presence of Ce^{4+} during the charge phase, or reduction reactions through the oxidation of zinc, which produces about 2.43 V [35]. A major component of these batteries is the electrolyte (i.e., methanesulfonic acid), which easily dissolves cerium compounds [36]. Ce^{4+} is produced through electrical driving forces that are used in the electrosynthesis of naphthoquinone [37].

Praseodymium

In combination with neodymium, praseodymium has been used in very strong and durable magnets [38], while its mixtures with magnesium forms highly mechanically resistant alloys that are used in parts of aircraft engines [39]. Like the previous two elements, praseodymium is also present in rather high concentrations (of up to 5%) in Misch Metal formulations. Mixtures of PrF_3 and the fluorides of other lanthanides have reportedly been used in the core of carbon arc and projector lights.

Praseodymium is also used to induce color to glass [11]. The glass formulations contains the presence of the mixtures of neodymium and praseodymium salts, also known as didymium glass, is used in welders' and glass blowers' goggles [11]. Crystals of praseodymium silicate create light pulses down to a few hundred meters per second and are hence used in devices where such pulses are required [40]. Nickel alloys containing praseodymium ($PrNi_5$) have proven to have strong magneto-caloric effects, used for reaching very low temperatures of down to about one thousandth of a degree of absolute zero [33]. The element has also been reported as a dopant in fluoride glass, for use in single-mode fiber-optical amplifiers [41] and its oxide, have been used in solid solutions with ceria, or ceria-zirconia for oxidation catalysts [42].

Neodymium

Like the majority of lanthanides, neodymium was initially used for coloring glass, which are known as Alexandrite or neodymium glass. They can change the color of the glass from reddish-purple to yellow under sunlight and artificial lighting and further to blue under white fluorescent or green under trichromatic lights [43]. The color change is attributed to the sharp absorption bands of neodymium [44], which leads to the manifestation of distinct colors based on the lighting. The sharp bands of the element make the Nd-doped glass suitable for astronomical applications, which require the calibration of spectral lines. Neodymium is also used to reduce the greenish touch given to normal glass by the iron contaminants. Another previously mentioned application of the element is its use in combination with praseodymium, in didymium glass for making welders' and glass-blowers' goggles. This glass is further used in color-enhancing filters for indoor photography, or in incandescent light bulbs, where the neodymium filters yellow light, producing whiter light. The glass is used in automobile rear-view mirrors as well, which reduce the glare at night.

Further applications of the element are in the neodymium-iron-boron magnets used in wind turbine generators, or electric motors in hybrid "HEV" and electric vehicles. Nd-containing magnets also include NdFeB systems used in magnetic resonance imaging (MRI) or $Nd_2Fe_{14}B$ permanent magnets. The element is also used in the neodymium yttrium aluminum garnets (Nd:YAG) used in commercial and military lasers.

Promethium

None of the promethium isotopes, except for ^{147}Pm, can be found outside laboratories [33]. This stable isotope (i.e., ^{147}Pm), which does not have any gamma emissions, is commonly supplied as the oxide or chloride, in milligram quantities. This isotope majorly undergoes beta radiation that is absorbed by air molecules or other martial nearby, making it safe to use.

Moreover, this element is used in signal lights, with beta-absorbing phosphors [45]. ^{147}Pm does not damage phosphors like the other sources of beta radiation, extending the life of the devices where it is used [46] in addition to being safe and hence preferred over other more dangerous sources [47].

This isotope is also used in long-life atomic batteries, where small Pm pieces are sandwiched within a semiconductor matrix to convert their beta emissions into electricity [48]. Another application of the isotopes of the element is in devices used for the measurement of thickness of materials. This is done through the measurement of radiation passing through the samples [33]. The element is also used in the defense industry, where its luminosity is used for range-marking in targeting sights of shoulder-fired missiles, or in watch hands and dials to make them luminous in the dark. Pm is also used in compact fluorescent lamps with high energy efficiency as the starter switch.

Samarium

One of the prominent applications of the element is in samarium-cobalt magnets with high-temperature stability (up to 700°C), where it is used as $SmCo_5$ or Sm_2Co_{17}. The ^{149}Sm isotope has strong neutron-absorbing properties and is hence used in the composition of control rods in nuclear reactors. Further applications of the element include catalyst promotion and radioactive dating as well as in X-ray lasers.

In the area of catalysis, Sm^{3+} triflate ($Sm(OTf)_3$ is an important Lewis acid used in halogen-promoted Friedel–Crafts catalysts [49]; in addition, SmI_2 is a conventional reducing and coupling agent for desulfunylation and annulation reactions as well as in Danishefsky, Kuwajima, Mukaiyama, and Holton Taxol total reactions; strychnine total syntheses; and Barbier reaction [50]. Stable Sm cations are added to ceramics and glasses, for enhancing the infrared absorption, and like many other lanthanides, the element is also used in low concentrations in Misch Metal [11,33].

CaF_2 crystals doped with samarium have been used as the active medium for solid-state lasers, which emitted pulses of red light (708.5 nm) [51,52]. Other samarium-based lasers include the saturated X-ray system, reportedly providing 50-picosecond pulses at 7.3 and 6.8 nm, which are applicable in holographic, deflectometric, interferometric, radiographic, and microscopic uses [53]. This element has also been used in pressure sensors, which work based on the electrical resistivity of its monochalcogenides [54]. Heating Samarium monosulfide (SmS) up to 150°C creates electric voltage that has been used in thermoelectric power converters [55]. The age and origin of rocks and meteorites is determined based on the proportion of ^{147}Sm, ^{144}Nd, and ^{143}Nd isotopes in their compositions [11]. Sm^{3+} is used in LEDs offering high luminous efficacy [56], and in very strong magnets used in miniaturized appliances like tape players, computer disk drives, and headphones.

Europium

An application of the complexes of the element (e.g., $Eu(fod)_3$) is as shift reagents in nuclear magnetic resonance. This has, however, been banished because of the advent of cheaper superconducting magnets spectroscopy. Chiral shift reagents, such as $Eu(hfc)_3$, are still used to determine enantiomeric purity [57–61]. Other applications of Eu in the form of 2+ and 3+ cations, are rather specialized and based on its phosphorescence properties. Similar to a majority of lanthanides, europium was used as a phosphor in color CRT (at concentrations of 0.5–1 g) and is currently used in LED televisions, scoreboards, and billboards. Eu^{3+} has the capability to produce a range of colors including reds, greens, and blues. The element has two red and blue phosphors, which even can be used to produce white color in combination with yellow/green terbium phosphors, as in helical fluorescent light bulbs. Europium has also been used in the preparation of fluorescent glass [62], or as an activator for yttrium phosphors. Europium is also present in some laser sources and electronic devices as a dopant [63].

Another application of Eu is in anticounterfeiting marks on the euro as well as many other bank notes. This application is possible as a result of the red fluorescence produced on the application of Eu^{3+}-β-diketone complexes, in printing the bank notes. Europium-doped strontium aluminate has also been reported as a phosphor with persistent elimination afterglow, which acts like copper-doped zinc sulfide [62].

Gadolinium

Gadolinium is not used in large-scale applications and has rather found a range of specialized uses in different fields as in the case of europium. The high neutron cross-section of element as compared to its family members (259,000 barns for ^{157}Gd) has made it interesting in neutron radiotherapy and radiographic applications. Owing to this property, Gd is also used in emergency shut-down systems in nuclear reactors [64]. The addition of low amounts of gadolinium (about 1 wt%) in iron, chromium, and other related alloys has also proven to improve the properties in terms of workability, mechanical resistance, and resistance to oxidation at high temperatures.

Possessing paramagnetic properties, gadolinium and other similar lanthanide ions with different movement speeds in a magnetic field are used in MRI. In these applications, organic complexes of the lanthanide ions are intravenously injected into the patient to boost the contrast of the MRI images. This effect is attributed to the accumulation of the ions in abnormal tissues, which increases the contrast of their image [65–67]. This effect has been reported to further boost up to about 40 times when using nanotubes packed with gadolinium [68]. Being a phosphor like the majority of lanthanides, gadolinium compounds (e.g., gadolinium oxysulfide [Gd_2O_2S])

are also used in the detection systems in other imaging methods, such as X-ray devices, where they emit green light at 540 nm (in the presence of Tb^{3+}) on receiving the X-ray signals. Tb^{3+} species are used because of their image-enhancing qualities. In these detection systems, almost 20% of the signal is converted to light as a result of the energy conversion of Gd, which is 20%. The use of gadolinium oxyorthosilicate (Gd_2SiO_5) single crystals in scintillation devices in medical imaging has also been reported [69].

The ^{153}Gd isotope, which is a product of the fission reaction in nuclear reactors, is also used in many quality assurance applications, including spectral calibration of photon sources, which help ensure whether the nuclear imaging instruments are working correctly. Another application of this isotope is in gamma ray sources used in X-ray absorption, in bone density gauges, as well as in some portable X-ray imaging systems.

Like the other lanthanides discussed so far, gadolinium compounds have been used as green phosphors in color TV tubes. Another application of the element in gadolinium yttrium ($Gd{:}Y_3Al_5O_{12}$), as well as gadolinium and gadolinium gallium garnets (GGGs, Gd_3Ga5O_{12}), which were used in computer bubble memory. Other examples of the uses of the compounds containing the element include microwave applications, use in fabrication of optical components, as electrolytes in solid oxide fuel cells (SOFCs), or as doping agents in mixtures containing cerium oxide or the like, leading to the high ionic conductivity and low operating temperatures desirable for fuel cells.

Terbium

There have been reports on the application of terbium as a doping agent in calcium fluoride and tungstate, as well as strontium molybdate, which are used in solid-state devices with ZrO_2 [11]. The element is also used in alloys (like the iron-based Terfenol-D) that are applied in naval sonar systems, sensors, in sound bug devices, and in magnetomechanical devices [70].

Terbium oxide is another green phosphor that is used in fluorescent lamps and color TV tubes, flat displays, trichromatic fluorescent bulbs, as well as X-ray intensifying screens [11,33]. Another terbium-containing compound used in solid-state devices is sodium terbium borate. The fluorescence of terbium also allows for its use in biochemical applications. Combinations of the green, blue, and red phosphors of terbium and di- and trivalent europium are used in the trichromatic lighting technology. Terbium-iron-cobalt mixtures are found in CDs and DVDs and silica glass and yttrium aluminum garnet doped with terbium (SiO2:Tb and Tb:YAG), which are used in fiber-optic temperature sensors. The element has also been used as an additive in neodymium-iron-boron (NdFeB) magnets in hybrid and electric vehicle motors [6,11].

Dysprosium

Mixtures of dysprosium, vanadium, and other elements are used in laser materials and Dy-Cd chalcogenides, which act as IR sources [65]. The element is also used in dysprosium-oxide nickel cermet in neutron-absorbing control rods in nuclear reactors. This application is due to the high neutron absorption of the element [71].

The magnetic character of dysprosium and its compounds has also changed them to good candidates for use in different data storage devices, as well as in Nd-Fe-B magnets. In these magnets, dysprosium can constitute up to 6% of the neodymium content [72], which leads to improved magnetic properties for uses in drive motors for electric vehicles. The presence of Dy also improves the corrosion resistance of the magnets [73].

Similar to Tb, Dy is a component of the iron-based Terfenol-D, which is highly magnetic in room temperature, and hence is suitable for use in transducers, and wide-band mechanical resonators as well as highly precise liquid fuel injectors [74,75].

Dysprosium is also used in dosimeters applied to the measurement of ionizing radiations. In these devices crystals of calcium sulfate or fluoride are doped with the Dy compounds, which become excited and luminescent upon being subject to radiation. The signals produced consequently are measured and used as an indication of the extent of exposure to radiation [33]. The iodide and bromide salts of the element have been used in high-intensity lamps, where they dissociate around the hot center of the lamp releasing atomized dysprosium and emit green and red light [33].

Furthermore, dysprosium oxide and fluoride nanofibers have been used to mechanically reinforce other materials, in addition to their catalytic applications [76]. The use of paramagnetic crystals of Dy like dysprosium gallium, dysprosium-aluminum, and dysprosium-iron garnets in adiabatic demagnetization refrigerators has also been reported [77].

Dysprosium phosphide (DyP), dysprosium arsenide (DyAs), and dysprosium scandium oxide ($DyScO_3$) have been reported as semiconductors in laser diodes and photo-optic applications, and dysprosium iodide (DyI_3) has applications as a heat and light stabilizer for nylon fabrics [11,33].

Holmium

Owing to the fact that holmium has the highest magnetic strength among all elements, it is used in devices for creating very strong artificial magnetic fields in different applications. Besides, holmium is used in the control systems of nuclear reactors [33]. Another application of Ho is in yttrium-iron-garnet and yttrium-lanthanum-fluoride lasers. These lasers have several applications in medicine, as well as space-based laser distance and ranging (LADAR) systems; infrared countermeasure (IRCM) for confusing shoulder-launched, infrared "heat-seeking" missiles; space-based high-pulse energy applications; and the detection systems for hidden remote targets based on the high-resolution Doppler technique [22]. One more application of holmium is as a yellow to red colorant for cubic zirconia and glass, the former being used as a calibration standard for optical spectrophotometers [78]. The radioactive but long-lived Ho-166m1 is used in the calibration of gamma ray spectrometers [79].

Erbium

Erbium has a rather wide spectrum of everyday applications ranging from photographic filters to metallurgical additives. The same as other high neutron-absorbing lanthanides, it is used in nuclear technologies, too [33]. In metallurgy, it is used as an additive in vanadium as an alloy to increase their mechanical resistance [11]. Its pink color leads to its application in glass, cubic zirconium, and porcelain coloring, and the glass colored this way is used in sunglasses and fake jewelry.

Silica-glass optical fibers doped with erbium have widespread use in optical communications [80] and laser applications. Erbium is usually co-doped with glass modifiers like aluminum or phosphorus for increased efficiency. For example, Er and Yb co-doped optical fibers are used in waveguide amplifiers [33]. Additionally, erbium-nickel alloys (Er_3Ni) have been used in cryogenic coolers because of their heat capacity [81]. Er is further used in a range of medical applications where Er^{3+} emission at 2940 nm, which is well absorbed by water, is applied for laser surgery, and the production of steam for enamel ablation in dentistry.

Thulium

In the area of solid-state lasers, holmium-chromium-thulium garnet (Ho:Cr:Tm:YAG) has been found to produce high-efficiency lasers at 2097 nm. These lasers are applicable in the medical, astronomical, and military fields. Thulium-doped YAG, on the other hand, produces lasers in the range of 1930 and 2040 nm [82]. The strength and wavelength of these lasers changes them to useful tools for surgical applications [83].

If thulium is bombarded in nuclear reactors, it can be used as a radiation source in portable X-ray devices [22]. Such devices can be applied to the imaging and cancer treatment via brachytherapy. The latter application requires the use of ^{170}Tm isotope [84], which has a $t_{1/2}$ of 128.6 days and five major emission lines of 7.4, 51.354, 52.389, 59.4, and 84.253 keV with similar intensities. The sources using this isotope have been used for industrial radiography [85].

Examples of other applications of thulium are high-temperature superconductors (like yttrium) as well as in ferrites, ceramic magnetic materials, and microwave instruments [22]. Tm is used in arc lighting (like scandium) to produce green emission lines that are not emitted by other elements, too. Additionally, because of the blue fluorescence properties of Tm upon excitation with UV light, the element is used for anticounterfeiting markings in euro banknotes [86]. This property is also used in Tm-doped calcium sulfate that is used in personal dosimeters for radiation monitoring [33].

Ytterbium

The ^{169}Yb isotope ($t_{1/2}$ = 32 days), which is formed through the neutron activation of ytterbium, together with ^{175}Yb ($t_{1/2}$ = 4.2 days), is used as a radiation source in portable X-ray instruments [87]. A rather specific application of Ytterbium is its use in high precision and very stable atomic clocks [88]. Yb is used as a doping agent to improve physical properties in alloys used in dental treatments. One effect of the presence of the element in alloys is the refinement of grains and increased strength (e.g., in stainless steel samples) [11,33].

Yb^{3+} is also a doping agent in laser-producing substrates, especially in long-life solid-state lasers [89], radiating short pulses in the range of 1.06–1.12 μm [90]. Low concentrations of Yb are used in laser materials, because high concentrations of the element have been reported to cause photo darkening [91]. A lot of successful work has been performed in the area of the power scaling lasers and amplifiers using optical fibers doped with Yb, which have advantages including larger mode field diameters (MFDs), which reduce the influence of nonlinear effects like stimulated Brillion scattering (SBS) and stimulated Raman scattering (SRS). Further applications of the compounds of the element include their use in stress gauges for monitoring ground deformations caused by earthquakes and explosions [22], as well as a substitute for magnesium in high density pyrotechnic payloads [92].

Lutetium

Because the element is both scarce and expensive, it is not used in many commercial applications. Some of its uses are in petroleum cracking, alkylation, hydrogenation, and polymerization catalysts. Lu is also used in lutetium-aluminum-garnet ($Al_5Lu_3O_{12}$), which has found application in high–refractive index immersion lithography [93], or in small amounts in gadolinium gallium garnet (GGG). Lutetium oxyorthosilicate (LSO) is also used in doping cerium for use in positron emission tomography (PET) detectors [94,95] as well as in LED light bulbs, where it acts as a phosphor [96].

The radioactive isotopes of the element like ^{176}Lu or ^{177}Lu are used in pure β-sources [97] and for targeted radionuclide therapy of neuroendocrine tumors [98]. A use of lutetium tantalate ($LuTaO_4$), which is a dense white material (with a density of 9.81 g/cm^3, which is second only to thorium [density 10 g/cm^3]) [99], is a host in X-ray phosphors [100].

AGRICULTURAL APPLICATIONS OF LANTHANIDES

To increase the production of livestock needed to support the population worldwide, safe and affordable additives are required, and rare earths are considered as promising compounds in this area. Rare earth elements have reportedly caused improved weight gain in almost all categories of farming animals as well as improvements in milk and egg production among dairy cows and laying hens, in addition to improved rate of survival of fish and egg hatching among fish and shrimps. The additives have included La, Ce, Pr, and Nd [101].

The use of low concentrations of lanthanides as food additives has been common in China for a long time. The results have indicated that these can boost the growth and productivity of several crops, and the mechanisms of these improvements have been studied, yielding a great deal of information on the issue [101]. An overall conclusion on the health implications of using these elements in agriculture might be that this can be considered safe and will not lead to serious damages to the environment.

BIOLOGICAL APPLICATIONS OF LANTHANIDES

In light of the conclusions above, because the presence of rare earths has not revealed to be essentially harmful, and these elements can be used in agriculture, with more confidence than many transition and other heavy metals. However, in case biological influences has been observed for lanthanides, they have been attributed to their similarity to calcium in terms of radii and higher charge densities that lead to their high affinity for the Ca^{2+} sites in biological molecules and stronger binding to water molecules [102–104].

In addition to the physiological effects of the lanthanide ions, such as blocking the voltage and the receptor-operated calcium channels, inhibiting the skeletal, smooth, and cardiac muscle contraction [105], the ions can also substitute Ca^{2+} in proteins (as well as Mg^{2+}, Fe^{3+}, and Mn^{2+}), inhibiting calcium-related enzymes leading to stimulatory or inhibitory effects [3,106].

Further to the above mentioned, lanthanum is a specific GABA receptor, acting as zinc in its modulatory site, but being a positive allosteric modulator, as opposed to zinc, which is a negative one. As a result, La^{3+} increases the open channel time and decreases the desensitization [107]. Some cerium is found in tobacco, barley, and the wood of beech trees, but trace amounts of the element are also present in the food chain. Cerium (or other lanthanides) acts as cofactors for the methanol dehydrogenase of the methanotrophic bacterium *Methylacidiphilum fumariolicum* [33,108]. Soluble Sm^{3+} salts have been reported to be mildly toxic if ingested. It causes skin and eye irritation on exposure. Gadolinium compounds have been in various ion channel electrophysiology experiments to block sodium leak channels and stretch activated ion channels [110]. Holmium salts have been found to have stimulating properties in metabolism [11,111].

CLINICAL AND PHARMACEUTICAL APPLICATIONS OF LANTHANIDES

Lanthanides have shown antimicrobial, anticoagulant, cytotoxic, and phosphate-binding influences because of their capability to replace Ca^{2+}, which is the basis for some of the pharmacologic applications of the elements. These elements can accumulate in tumors providing various alpha-, beta-, or gamma-emitting radiations, as well as paramagnetic properties, which gives them the potential for cancer diagnosis and therapy. Initial efforts for using lanthanides as pharmaceuticals had not been very successful, but this changed in the 20th century [103,112,113].

The older uses of the elements were as antiemetic, antimicrobial, and anticoagulation agents, but they were withdrawn from the market because of radioactive residues as well as the introduction of less toxic, more affordable, and effective drugs. Currently, the elements are used in more specific drugs, such as the use of lanthanum carbonate (named Fosrenol) for absorbing excess phosphate in end-stage renal failure [114] or the application of cerium nitrate for suppressing the elevation of TNF-α levels through increasing IL-6 levels [115–117].

The use of neodymium magnets in magnetic braces and bone repair [118] has been reported, too. Further, the use of radioactive ^{153}Sm in the drug lexidronam, which is used for killing lung, breast, and prostate cancer osteosarcoma cells, can be mentioned. Free Gd^{3+} ions have been found to show toxicity among some mammals, while this effect is greatly decreased in chelated forms, because of the fact that they leave the body through the kidneys before affecting the tissues. Some reports also mentioned that Gd ions injected intravenously to patients, for increasing the contrast in MRI, led to a rare nodular inflammatory disease, namely, nephrogenic systemic fibrosis among some patients with renal failure experience. Erbium has no serious biological effects, but its salts can stimulate metabolism [33]. Complexes of lanthanides with texaphyrins, which are actually extended porphyrins, have been used in clinical tests for the treatment of brain metastases [102]. An additional example of the application of lanthanides in clinical materials or devices is the case of lutetium motexafin, which is used as a texaphyrin photosensitizer for photodynamic therapy to treat skin conditions and superficial cancers [119,120].

References

[1] Goldman JA. The U.S. rare earth industry: its growth and decline. J Policy Hist 2014;26:299–304.

[2] Graedel D. Uncovering the end uses of the rare earth elements. Sci Total Environ 2013;781–4:461–2.

[3] Sastri VS, Bünzli JC, Ramachandra Rao V, Rayudu GVS, Perumareddi JR. Modern aspect of rare earth and their complexes. Elsevier; 2003. ISBN: 978-0-444-51010-5.

[4] [Online] available at: http://www.frontierrareearths.com/rare-earths-101/.

[5] [Online] IMCOA (2012); available at: http://www.frontierrareearths.com/rare-earth-market/.

[6] Castor SB, Hedrick JB. Rare earth elements. In: Kogel JE, Trivedi NC, Barker JM, editors. Industrial minerals and rocks. Society for Mining, Metallurgy and Exploration; 2006. p. 769–92.

[7] [Online] available at: http://www.earthexplorer.com/2009-11/Rare_Earth.asp; Rare earth exploration techniques for rare earth element deposits; by Virginia Heffernan.

[8] Massari R. Rare earth elements as critical raw materials: Focus on international markets and future strategies. Res Policy 2013;38:36–43.

[9] Binnemans J. Perspectives for the recovery of rare earths from end-of life fluorescent lamps. J Rare Earth 2013;32:159–212.

[10] Haxel GB, Hedrick JB, Orris GJ. Rare earth elements – critical resources for high technology. USGS fact sheet 2002.

[11] Hammond CR. The elements in handbook of chemistry and physics. 81st ed. CRC press; 2000. ISBN 0-8493-0481-4.

[12] Sommerville, JD, King, LB. Effect of cathode position on Hall-effect thruster performance and cathode coupling voltage. 43rd AIAA/ASME/SAE/ASEE joint propulsion conference & exhibit, 8–11 July 2007, Cincinnati, OH. Retrieved 2009-06-06.

[13] Tliha M, Mathlouthi H, Lamloumi J, Percheronguegan A. AB5-type hydrogen storage alloy used as anodic materials in Ni-MH batteries. J Alloy Comp 2007;436:221–5.

[14] Bauerlein P, Antonius C, Loffler J, Kumpers J. Progress in high-power nickel–metal hydride batteries. J Power Source. 2008;176:547–54.

[15] Uchida H. Hydrogen solubility in rare earth based hydrogen storage alloys. Int J Hydrogen Energ 1999;24:871–7.

[16] Harrington JA. Infrared fiber optics. SPIE press, Rutgers University; 2004.

[17] Kim K, Shim KB. The effect of lanthanum on the fabrication of ZrB2–ZrC composites by spark plasma sintering. Mater Character 2003;50:31–7.

[18] [Online] BrilLanCe-NxGen. Retrieved 2009-06-06.

[19] Howard BC. Arc welding automation. CRC Press; 1995. p. 139, ISBN 0-8247-9645-4.

[20] Larry J. Types of tungsten. Welding: principles and applications. Clifton Park, N.Y: Thomson/Delmar Learning; 2003. p. 350, ISBN 978-1-4018-1046-7.

[21] Haghseresht F, Wang S, Do DD. A novel lanthanum-modified bentonite, Phoslock, for phosphate removal from wastewaters. Appl Clay Sci 2009;46:369–75.

[22] Gupta CK. Nagaiyar Krishnamurthy, Extractive metallurgy of rare earths. CRC Press; 2004. p. 441, ISBN 0-415-33340-7.

[23] Nakai S, Masuda A, Lehmann B. La-Ba dating of bastnaesite. Am Mineralogist 1988;73:1111–3.

[24] Patnaik P. Handbook of inorganic chemical compounds. McGraw-Hill; 2003. p. 199–200, ISBN 0-07-049439-8.

[25] Bleiwas, DI. Potential for recovery of cerium contained in automotive catalytic converters. Reston, Va.: U.S. Department of the Interior, U.S. Geological Survey, 2013.

[26] Exploring nano-sized fuel additives EPA scientists examine nanoparticle impacts on vehicle emissions and air pollution. United States Environmental Protection Agency. April 2001.

[27] Nanoparticles used as additives in diesel fuels can travel from lungs to liver. Science Daily. November 18, 2011.
[28] Park B, Donaldson K, Duffin R, Tran L, Kelly F, Mudway I, Morin JP, Guest R, Jenkinson P, Samaras Z, Giannouli M, Kouridis H, Martin P. Hazard and risk assessment of a nanoparticulate cerium oxide-based diesel fuel additive—a case study. Inhal Toxicol 2008;20:547–66.
[29] Trovarelli A. Catalysis by ceria and related materials. Imperial College Press; 2002. p. 6–11, ISBN 978-1-86094-299-0.
[30] [Online] Cerium dioxide.nanoparticle.info; available at: http://www.nanopartikel.info/en/nanoinfo/materials/cerium-dioxide/exposure.
[31] [Online] Hydrogen production from solar thermochemical water splitting cycles. available at: solarpaces.org.
[32] Gray T. The elements. Black Dog & Leventhal Pub; 2010. ISBN 1579128955.
[33] Emsley J. Nature's building blocks: An A-Z guide to the elements. Oxford University Press; 2011. p. 120–125, ISBN 978-0-19-960563-7.
[34] Xie Z, Liu Q, Chang Z, Zhang X. The developments and challenges of cerium half-cell in zinc–cerium redox flow battery for energy storage. Electrochim Acta 2013;90:695–704.
[35] Clarke, RL et al. U.S. Patent 20,040,202,925, Redox system includes cerium ions complexed by an organic acid or a chelating agent and a second element or compound that reduces hydrogen overpotential, 2004.
[36] Kreh, Robert P, Spotnitz, Robert M, Lundquist, Joseph T. Mediated electrochemical synthesis of aromatic aldehydes, ketones, and quinones using ceric methanesulfonate. The J Org Chem 1989;54:1526–31.
[37] Harrison S, Théorêt A. The electrosynthesis of naphthoquinone, and tetrahydroanthraquinone. J New Mater Electrochem Sys 1999;2:1–9.
[38] [Online] Rare earth elements 101, IAMGOLD Corporation, 2012, 5–7; available at: http://www.iamgold.com/files/REE101_April_2012.pdf.
[39] Rokhlin LL. Magnesium alloys containing rare earth metals: structure and properties. CRC Press; 2003. ISBN 0-415-28414-7.
[40] [Online] ANU team stops light in quantum leap; available at: https://news-test.weblogs.anu.edu.au/2005/08/29/anu-team-stops-light-in-quantum-leap/.
[41] Jha A, Naftaly M, Jordery S, Samson BN, Taylor ER, Hewak D, Payne DN, Poulain M, Zhang G. Design and fabrication of Pr3 + -doped fluoride glass optical fibers for efficient 1.3 mu m amplifiers. Pure Appl Optic J Eur Optic Soc A 1995;4:417.
[42] Borchert Y, Sonstrom P, Wilhelm M, Borchert H, Baumer M. Nanostructured Praseodymium Oxide: Preparation, Structure, and Catalytic Properties. J Phys Chem C 2008;112:3054–63.
[43] [Online] Chameleon Glass Changes Color. Available at: http://web.archive.org/web/20080403165916/http://coloradosprings.yourhub.com/CrippleCreekTellerCounty/Stories/Arts/Story~443258.aspx.
[44] Bray C. Dictionary of glass: materials and techniques. University of Pennsylvania Press; 2001. p. 102. ISBN 0-8122-3619-X.
[45] Hammond CR. Promethium in the elements. In: Haynes William M, editor. CRC Handbook of chemistry and physics. 92nd ed. CRC Press; 2011. p. 4–28. ISBN 1439855110.
[46] Tykva R, Berg D. Man-made and natural radioactivity in environmental pollution and radiochronology. Springer; 2004. p. 78, ISBN 1-4020-1860-6.
[47] Deeter, DP. Disease and the environment. Government Printing Office, 1993, 187.
[48] Flicker H, Loferski JJ, Elleman TS. Construction of a promethium-147 atomic battery. IEEE Trans Electron Device 1964;11:2–8.
[49] Hajra S, Maji B, Bar S. Samarium Triflate-Catalyzed Halogen-Promoted Friedel-Crafts Alkylation with Alkenes. Org Lett 2007;9:2783–6.

[50] Cotton. Advanced inorganic chemistry. 6th ed. Wiley-India; 2007. p. 1128, ISBN 81-265-1338-1.
[51] Bud, R., Gummett, P. Cold war, hot science: applied research in Britain's defense laboratories, 1945–1990, NMSI Trading Ltd, 2002, 268, ISBN 1-900747-47-2.
[52] Sorokin PP. Contributions of IBM to laser science 1960 to the Present. IBM J Res Develop. 1979;23:476–89.
[53] Zhang J. A saturated X-ray laser beam at 7 nanometers. Science 1997;276:1097–100.
[54] Elmegreen BG, Krusin-elbaum L, Liu XH, Martyna GJ, Muser M, Newns DM. Piezo-driven non-volatile memory cell with hysteretic resistance US patent application number 12/234100, Filling date: 09/19/2008.
[55] Kaminskii VV, Solov'ev SM, Golubkov AV. Electromotive force generation in homogeneously heated semiconducting samarium monosulfide. Tech Phys Lett 2002;28:229–31.
[56] Baur F, Katelnikovas A, Sazirnakovas S, Jüstel T. Synthesis and optical properties of Li3Ba2La3(MoO4)8:Sm3 +. Z Naturforsch 2014;69:183–92.
[57] Richards, S., Hollerton, J. Essential practical NMR for organic chemistry. (2011-02-15), ISBN 978-0-470-71092-0.
[58] Pavia, DL, Lampman, GM. Introduction to spectroscopy. 2009, ISBN 978-0-495-11478-9.
[59] Wenzel, Thomas, J. Discrimination of chiral compounds using NMR spectroscopy, 2007, ISBN 978-0-471-76352-9.
[60] Cotton, S. Lanthanide and actinide chemistry, 2006, ISBN 978-0-470-01006-8.
[61] Gschneidner, KA, Bünzli, JC, Pecharsky, VK. Handbook on the physics and chemistry of rare Earths. (2005-09-26), ISBN 978-0-444-52028-9.
[62] Lakshmanan, A. Persistent afterglow phosphors. Luminescence and display phosphors: phenomena and applications. 2008, ISBN 978-1-60456-018-3.
[63] Bamfield, P. Inorganic phosphors. Chromic phenomena: technological applications of colour chemistry. 2001, p. 159–171, ISBN 978-0-85404-474-0.
[64] Greenwood NN, Earnshaw A. Chemistry of the elements. 2nd ed. Butterworth-Heinemann; 1997. ISBN 0080379419.
[65] Lide DR, editor. CRC handbook of chemistry and physics. 86th ed. Boca Raton (FL): CRC Press; 2005. p. 4.122, ISBN 0-8493-0486-5.
[66] Gary L. MRI in clinical practice. Springer; 2006. p. 13–30, ISBN 1-84628-161-X.
[67] Raymond KN, Pierre VC. Next generation, high relaxivity gadolinium MRI agents. Bioconjug Chem 2005;16:3–8.
[68] [Online] Wendler, R. Magnets guide stem cells to damaged hearts. Texas Medical Center, 2009; available at: http://web.archive.org/web/20110728091851/http://www.texasmedicalcenter.org/root/en/TMCServices/News/2009/12-01/Magnets+Guide+Stem+Cells+to+Damaged+Hearts.htm.
[69] Ryzhikov VD, Grinev BV, Pirogov EN, Onyshchenko GM, Ivanov AI, Bondar VG, Katrunov KA, Kostyukevich SA. Use of gadolinium oxyorthosilicate scintillators in x-ray radiometers. Optic Eng 2005;44:016403–6.
[70] Rodriguez C, Rodriguez M, Orue I, Vilas J, Barandiaran J, Gubieda M, Leon L. New elastomer–Terfenol-D magnetostrictive composites. Sens Actuators A Phys 2009;149:251–4.
[71] Amit S, Sharma BP. Development of Dysprosium Titanate Based Ceramics. J Am Ceram Soc 2005;88:1064–6.
[72] Shi FX, Shi Y, Jiles DC. Modeling of magnetic properties of heat treated Dy-doped NdFeBparticles bonded in isotropic and anisotropic arrangements. IEEE Trans Mag 1998;34:1291–3.
[73] Yu LQ, Wen Y, Yan M. Effects of Dy and Nb on the magnetic properties and corrosion resistance of sintered NdFeB. J Mag Mag Mater 2004;283:353–6.
[74] Kellogg R, Flatau A. Wide band tunable mechanical resonator employing the ΔE effect of terfenol-D. J Intell Mater Sys Struct 2004;15:355–68.

[75] Leavitt W. Take Terfenol-D and call me. Fleet Owner (RODI Power Systems Inc) 2000;95:97.
[76] Hoffman MM, Young JS, Fulton JL. Unusual dysprosium ceramic nano-fiber growth in a supercritical aqueous solution. J Mater Sci 2000;35:4177–83.
[77] Milward, S, Harrison, S, Allen, RS, Hepburn, I, Brockley-Blatt, C. Design, manufacture and test of an adiabatic demagnetization refrigerator magnet for use in space. 2004, available at: http://www.ucl.ac.uk/mssl/cryogenics/documents/5LH01.pdf.
[78] MacDonald RP. Uses for a holmium oxide filter in spectrophotometry. Clinic Chem 1964;10:1117–20.
[79] Yuan MC, Lee JH, Hwang WS. The absolute counting of 166mHo, 58Co and 88Y. Appl Rad Isotop 2002;56:424.
[80] Becker PC, Olsson NA, Simpson JR. Erbium-doped fiber amplifiers fundamentals and technology. San Diego: Academic Press; 1999. ISBN 978-0-12-084590-3.
[81] Ackermann RA. Cryogenic regenerative heat exchangers. Springer; 1997. p. 58, ISBN 978-0-306-45449-3.
[82] Koechner W. Solid-state laser engineering. Springer; 2006. p. 49, ISBN 0-387-29094-X.
[83] Duarte FJ. Tunable laser applications. CRC Press; 2008. p. 214, ISBN 1-4200-6009-0.
[84] Krishnamurthy D, Weinberg V, Cunha JAM, Hsu IC, Pouliot J. Comparison of high–dose rate prostate brachytherapy dose distributions with iridium-192, ytterbium-169, and thulium-170 sources. Brachytherapy 2011;10:461–5.
[85] Raj, B, Venkataraman, B. Practical radiography, 2004, ISBN 978-1-84265-188-9.
[86] Wardle, B. Principles and applications of photochemistry. (2009-11-06), p. 75, ISBN 978-0-470-71013-5.
[87] Halmshaw R. Industrial radiology: theory and practice. Springer; 1995. p. 168–169, ISBN 0-412-62780-9.
[88] [Online] NIST ytterbium atomic clocks set record for stability; available at: http://www.nist.gov/pml/div688/clock-082213.cfm.
[89] Ostby, EP. Photonic whispering-gallery resonations in new environments. Dissertation (PhD), California institute of technology, 2009. Available at: http://thesis.library.caltech.edu/2284/.
[90] Grukh DA, Bogatyrev VA, Sysolyatin AA, Paramonov VM, Kurkov AS, Dianov EM. Broadband radiation source based on an ytterbium-doped fibre with fibre-length-distributed pumping. Quantum Electron. 2004;34:247–8.
[91] Koponen JJ, Söderlund MJ, Hoffman HJ, Tammela SKT. Measuring photodarkening from single-mode ytterbium doped silica fibers. Opt Express 2006;14(24):11539–44.
[92] Koch EC. Hahma, A. Metal-fluorocarbon pyrolants. XIV: high density-high performance decoy flare compositions based on ytterbium/polytetrafluoroethylene/Viton®. Z Anorg Allgem Chem 2012;638:721.
[93] Wei Y, Brainard RL. Advanced processes for 193-NM immersion lithography. SPIE Press; 2009. p. 12, ISBN 0-8194-7557-2.
[94] Wahl RL. Instrumentation. Principles and practice of positron emission tomography. Philadelphia: Lippincott: Williams and Wilkins; 2002. p. 51.
[95] Daghighian F, Shenderov P, Pentlow KS, Graham MC, Eshaghian B, Melcher CL, Schweitzer JS. Evaluation of cerium doped lutetium oxyorthosilicate (LSO)scintillation crystals for PET. Nucl Sci 1993;40:1045–7.
[96] Martine Simard-Normandin A19 LED bulbs: What's under the frosting?, EE Times (July 18), 2011, p. 44–45,
[97] Gargaud M, Martin H, Claeys P. Lectures in astrobiology. Springer; 2007. p. 51, ISBN 3-540-33692-3.
[98] Helmut S. Metal complexes in tumor diagnosis and as anticancer agents. CRC Press; 2004. p. 98, ISBN 0-8247-5494-8.

[99] Blasse G, Dirksen G, Brixner L, Crawford M. Luminescence of materials based on LuTaO4. J Alloy Comp 1994;209:1–2.
[100] Shigeo S. Phosphor handbook. CRC Press; 1998. p. 846, ISBN 0-8493-7560-6.
[101] Tyler G. Rare earth elements in soil and plant systems – A review. Plant Soil 2004;267:191–206.
[102] Fricker SP. The therapeutic application of lanthanides. Chem Soc Rev 2006;35:524–33.
[103] Wang K, Cheng Y, Yang X, Li R. Cell responses to lanthanides and potential pharmacological actions of lanthanides. Met Ions Biol Syst 2003;40:707–51.
[104] Evans CH. Interesting and useful biochemical properties of lanthanides. Trends Biochem Sci 1983;8:445–9.
[105] Evans CH. Calcium in drug actions. Handbook of experimental pharmacology. Berlin: Springer-Verlag; 1988.
[106] Goecke F, Jerez CG, Zachleder V, Figueroa FL, Bišová K, ezanka T, Vítová M. Use of lanthanides to alleviate the effects of metal ion-deficiency in *Desmodesmus quadricauda* (Sphaeropleales, Chlorophyta). Front Microbiol 2015;6:2–12.
[107] Boldyreva AA. Lanthanum potentiates GABA-activated currents in rat pyramidal neurons of CA1 hippocampal field. Bull Exp Biol Med 2005;140:403–5.
[108] Pol A, Barends TRM, Dietl A, Khadem AF, Eygensteyn J, Jetten MSM, Op den Camp HJ. M. Rare earth metals are essential for methanotrophic life in volcanic mudpots. Environ Microbiol 2013;16:255–64.
[109] Fitriyanto NA, Fushimi M, Matsunaga M, Pertiwiningrum A, Iwama T, Kawai K. Molecular structure and gene analysis of Ce3+-induced methanol dehydrogenase of Bradyrhizobium sp. MAFF211645. J Biosci Bioeng 2011;111:613–7.
[110] Yeung EW, Allen DG. Stretch-activated channels in stretch-induced muscle damage: role in muscular dystrophy. Clin Exp Pharmacol Physiol 2004;31:551–6.
[111] Suzuki YS, Momose Y, Higashi N, Shigematsu A, Park KB, Kim YM, Kim JR, Ryu JM. Biodistribution and kinetics of holmium-166-chitosan complex (DW-166HC) in rats and mice. J Nucl Med 1998;39(12):2161–6.
[112] Evans CH. Biochemistry of the lanthanides. New York and London: Plenum Press; 1990.
[113] Damment SJP, Totten W. The pharmacology of a new phosphate binder, lanthanum carbon-ate. Nephrol Dial Transplant 2004;19(Suppl. 1):i19–24.
[114] FDA approves Fosrenol (R) in end-stage renal disease (ESRD) patients. 28 October 2004. Available at: http://www.medilexicon.com/drugs/fosrenol_lanthanum_carbonate.php.
[115] Sparkes BG. Immunological responses to thermal injury. Burns 1997;23:106–13.
[116] Deveci M, Eski M, Sengezer M, Kisa U. Effects of cerium nitrate bathing and prompt burn wound excision on IL-6 and TNF-a levels in burned rats. Burns 2000;26:41–5.
[117] Garner JP, Heppell PS. The use of Flammacerium in British Burns Units. Burns 2005;31:379–82.
[118] Swain, F. How to remove a finger with two super magnets. Seed Media Group LLC, (March 6, 2009); available at: http://scienceblogs.com/sciencepunk/2009/03/06/how-to-remove-a-finger-with-tw/.
[119] O'Connor AE, Gallagher WM, Byrne AT. Porphyrin and nonporphyrin photosensitizers in oncology: Preclinical and clinical advances in photodynamic therapy. Photochem Photobiol 2009;85:1053–74.
[120] Patel H, Mick R, Finlay JC, Zhu TC, Rickter E, Cengel KA, Malkowicz SB, Hahn SM, Busch TM. Motexafin lutetium-photodynamic therapy of prostate cancer: Short and long term effects on PSA. Clin Cancer Res Am Assoc Cancer Res 2008;14:4869–76.

CHAPTER

3

Importance of Measuring Lanthanides

OUTLINE

WHY MEASURE LANTHANIDES?

For a long time, the presence of lanthanides in samples was taken for granted. The widespread applications of the elements in various areas have however shifted this paradigm. This has been highlighted by the new discoveries of the potential or actual biological roles the elements can have, with a focus on the various influences on humans.

The major source of lanthanide emissions into the environment is industrial effluents, taking into account the fact that the elements are present in alloys, glass formulations, superconductors used in various devices, magnets, and fluorescence- and laser-producing materials. The rate of consumption of lanthanides in the various applications that have been mentioned in some detail in the previous chapter is rapidly increasing. Further to the scattering of residual amounts of the elements during or after their production procedures, the elements find their way into the air (in the form of particulates), water, and soil during or after their use in

Lanthanides Series Determination by Various Analytical Methods. http://dx.doi.org/10.1016/B978-0-12-804704-0.00003-7

the new applications [1-5]. Agricultural applications of the elements, on the other hand, which was intended to improve the yield and quality in plants and crops, and in animal husbandry, has led to large-scale applications, and consequently emissions of these elements into nature [6-10]. It would be ineluctable to state that the production and purification cycles of lanthanide elements require consuming large amounts of water and chemicals, which can carry them into the environment.

It is worthy to note that although these elements released into the environment in scales that can be considered as negligible compared to their natural occurrence, the nature of the compounds released is much different from those naturally present in the environment [11,12]. This special nature of the released compounds, especially in terms of considerably higher solubility and reactivity, increases the chances that the elements would enter biological systems [13,14]. This raises concerns on interference with the balance of the biogeochemical cycle of lanthanides. Additionally, working with these elements mandate costly measures for ensuring safety of the environment, as well as of the workers dealing with their preparation or application, especially given the radioactive properties of many lanthanide isotopes. The importance of measuring these elements is increased by concerns over reports on the environmental situation and the medical conditions of the people living in the vicinity of lanthanide mines [14-17].

Based on the applications discussed for these elements, the improvements they can bring about in different fields of technology and human life (Fig. 3.1), and in light of the possible adverse effects they can cause to the humans and nature beginning from their extraction to their various applications, determination of these elements has assumed great importance, at least in the modern world.

LANTHANIDES AS SOIL AND WATER POLLUTANTS

Based on the discussions so far, and given that the abundance of lanthanides in the earth's crust exceeds 100 ppm, it is well known that naturally occurring compounds of the elements cannot infiltrate into natural waters because of their insolubility [18]. This has been established by some reports, on the leaching of rare earths into groundwater sources, revealing that the risk of pollution through such mechanisms is rather low [11, 19-21]. Yet there have been reports on the severe contamination of groundwater and plant life by lanthanides in the vicinity of lanthanide mines, with ionic lanthanide deposits or, in case of long-term-application, sewage sludge soils.

A study conducted on more than 480 soil samples from different areas in China, for example, revealed that on average La, Ce, Nd, Sm, and Eu, which

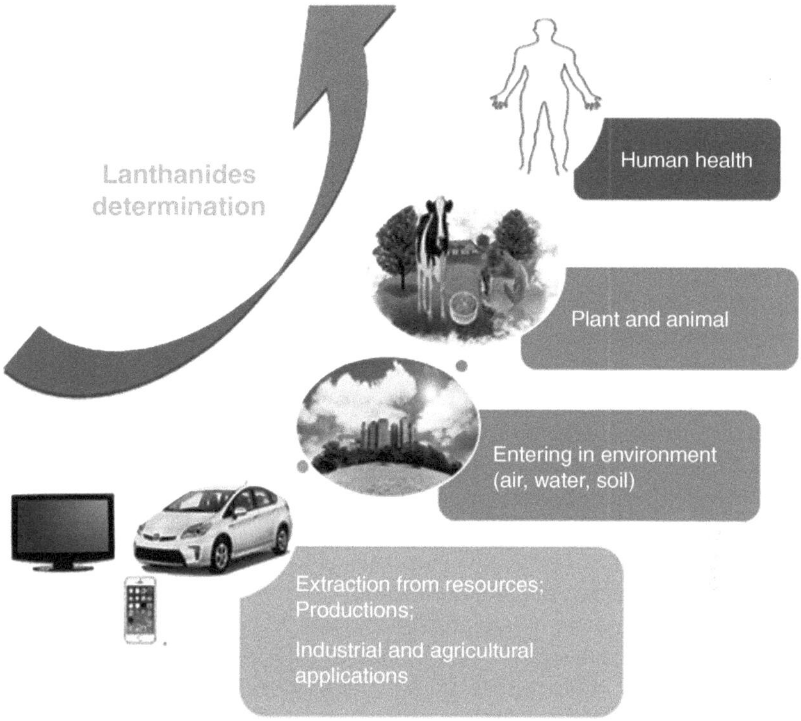

FIGURE 3.1 Application of lanthanides can improve human life or contribute to treatments for human health.

account for almost 90% of the total content of the rare earths, were present in amounts of 41, 74, 7, 28, and 6 ppm [21] in the sample. These values exceed the anticipated values of 39, 60, and 8.2 ppm, which were previously reported as the average amounts from the first three first elements in the earth's crust. One should pay close attention to the fact that the values vary in different soil samples [21]. As another example, a study on 1790 samples by Xiong et al. [22] that focused on the rare earth content of the plants grown in Chinese soils showed that the total concentration of the elements in plant tissues, though negligible in some cases, even reached the very high value of 208 ppm. The highest amounts of lanthanides were found in red (Ferralic Cambisol) and paddy (Hydrgric Anthrosols) soils, which reached up to 18.8 and 17.1 ppm, respectively [22,23]. The researchers also concluded that this content depended on the physicochemical properties of the soil. For example, plants grown in acidic soils were found to have higher rare earth contents as compared to those in calcareous soils, and the lowest rare earth contents were observed in black (Haplic Phaeozems) soils, chernozem (Haplic chernozem) soils, dark brown (Eutric Cambisol) soils, gray sand (Cumulic-calcaric Regosol) soils, and Shajiang (Gleyic Cambisol) soils [22,23].

Apart from the widespread use of lanthanides in different areas during recent years, other factors can influence the distribution of these elements in different samples, and much research effort has hence been directed to unveiling the related phenomenon [24-26]. As mentioned earlier, the physicochemical characteristics of soils including the pH, concentration of ions, etc. are of utmost importance in this regard. Another important factor to be considered is the type of plants growing in the soils, since the properties of soils can be altered by the changes in the chemistry of plant roots during their growth stages. For example, plants release H^+ to the soil while trying to absorb the nutrients, which consequently changes the properties of the soil [23].

Studies on the release of some lanthanides under different pH values and redox potentials showed that, on average, these elements release more lanthanides with decreasing pH and redox potentials. Nonetheless, the extent of this phenomenon is different in the case of different ions. (Ce release showed higher sensitivity to the variations of the redox potential as compared to La, Gd, and Y, while the release of La, Ce, Gd, and Y reacted more to the lowering of redox potentials at lower pH values, which was followed by changes in the species too [27].) Studies have further revealed a considerable negative correlation between the amount of lanthanides present in plant tissues and the soil pH from 6 to 10 and a positive one with the increase in the content of clay and organic matter in the soils [22]. Further studies have shown that reducing conditions and low pH leads to the dissociation of Fe–Mn oxyhydroxides, which in turn release rare earths [21].

In a research on the influence of soil amendments on the concentration of rare earths, which was conducted on Australian soil samples, Diatloff et al. [28] discovered that extraction solutions of 10 acid soils were from 3.9 to 4.9, whereas the addition of calcium carbonate and sulfate to the samples led to increased and decreased pH, respectively. As a result, the concentration of the lanthanides in the solutions were decreased and increased correspondingly.

One of the phenomena that greatly influence the mobility, solubility, and reactivity of lanthanides is the formation of complexes between cations and chelating agents [29,30]. It is well established that the toxicity and bioavailability of lanthanides mostly relate to their free-ionic forms [31,32], which implies the importance of the type and nature of the species and ligands interacting with these ions in natural water samples. Interactions with organic and inorganic species greatly influence the biological availability of lanthanides in soil, where these elements can influence the microorganisms [33,34]. As an example, low concentrations of lanthanides have been found to slightly stimulate the soil bacteria (e.g., in the nitrification process) [35,36], while higher concentrations have been found to do the reverse [33]. It is possible that microorganisms that can actively accumulate rare earths may exist in nature but so far no proof has been found in this regard. The lanthanide–bacteria interactions are varied, but the most basic and well-known interactions, which certainly leads to further consequences, is the strong binding of the trivalent lanthanides with microorganisms [18]. The diffusion of the elements inside the cell membranes of species like algae and yeast, on the other hand, has been considered not very probable [18]. There are, on the other hand, reports on the uptake of nutrients and various inorganic elements by oligotrophic microorganisms to support their growth [37].

In the last 15 years, numerous studies have shown that microorganisms (bacteria, yeast, and fungi) may interact with the ionic forms of heavy metals or radionuclides. The interactions cover a wide range of possibilities, like biosorption and bioabsorption, bioaccumulation, resistance, detoxification mechanisms, and direct or indirect utilization [38]. The studies on the effects of La on actinomycetes, bacteria, and fungi have proven the element to be toxic to these microorganisms [31], but the extent of this influence has been found to depend on the nature of the microorganism; for example, the fungi in soil have been found to be more resilient against heavy metals as compared to the bacteria [39].

Adding low doses of lanthanides to the soil has been subject to some debate. A number of Chinese research studies claim that at such doses, lanthanides are beneficial for a wide range of crops [8,40-42], whereas others mention that the elements tend to be very toxic to the microorganisms in the soil [43,44]. In the case of high dose of the elements, on the other hand, there is unanimity because all concur that such doses of lanthanides leave severe harmful effects on the microbial biomass in soil [45], the nitrogen cycle [36,46], evolution of CO_2, and enzymatic catalysis of the cycles of elements such as C, N, P, and S [43,47]. This feedback has raised concerns even on the application of low doses of lanthanides for agricultural purposes, which is rather common in China [14,17,48]. The influence of lanthanides on the activity of microbial and plant enzymes in the soil has been further studied, and the results have revealed that lanthanum and cerium decrease the acid phosphatase activity in Chinese red soil samples [47, 49], whereas in the case of yellow soils, a reverse phenomenon was observed. The latter condition was attributed to the different properties of soil, for example, their pH both before and after the addition of the lanthanides [47,50].

It is also important to note that soil enzymes form a part of the soil matrix and that the activities of these enzymes commonly correlate with microbial parameters. Soils contain assays of enzymatic species ranging from biotic to abiotic enzymatic forms [51]. These enzymes, produced by microorganisms and plants, act as biological catalysts for important reactions. Also, it produces essential compounds for both soil microorganisms and plants. In fact, the enzymatic activity of soils under are further dependent on conditions like moisture, pH, temperature, and substrate concentration, and may vary under stress, as when soil is contaminated by heavy metals [52]. Soil enzymes are vital for catalyzing countless reactions necessary for life processes of microorganisms in soils, decomposition of organic residues, cycling of nutrients, and formation of organic matter and soil structure [53].

In light of what was mentioned above, and the intricate interactions and relations observed between soil enzymes and the lanthanides, soil enzymes have also been used as indicators for the analysis of lanthanides in samples [54,55]. However, the complicated relations between the species has led to some studies on the nature of the reactions [34,47,56]. As an example, studies on the activity of the dehydrogenase enzyme have shown that its activity diminishes in the presence of metals but gradually build up over time in case the metal is lanthanum [34].

EFFECTS OF LANTHANIDES ON PLANTS

Some experimental data indicate that rather large amounts of lanthanides can be absorbed by a few plants, which is the reason behind the low concentration of these elements in humans and animal tissues and organs. The agricultural applications of lanthanides in countries like China have also attracted some focus toward the behavior of these elements in the soil. As already mentioned, the insolubility of many salts of these elements, especially those present in the naturally occurring minerals, has led to the increased concentration of lanthanide ions in water, where they have a better chance of finding their way into living organisms like plants as complexes with different anions and ligands (like fulvic and humic acids ([57]. Research has also confirmed that the pH, reduction and oxidation (redox) potential, salinity, and size distribution and organic content of the soil has a great influence on the claimed fertilizing effects of these elements.

Some have reported the natural concentration of lanthanides in Chinese soil samples to range from 0.01% to 0.02%, whereas in the case of farms, where the elements are used for fertilizing purposes, these values exceed the natural value ranging from 68 to about 630 ppm (181 ppm on average) [21,58]. Knowledge on the amounts of the elements is, however, an uncertain indicator for estimating how much plants can uptake. Some studies, on the other hand, have shown correlations between lanthanide content of rice, corn, and wheat and the portion of water-soluble, exchangeable, and carbonate-bound lanthanide species present in the environment [14,59,60].

Although lanthanides have long been considered as elements that can improve the nutritional value and rate of growth of crops [40, 61], a range of studies reject this claim, considering them as chemicals that at best have such influences during limited stages of the life cycle of the plants [40,61,62].

The analyses of the lanthanide content of plant tissues have been reportedly performed through instrumental neutron activation analysis (INAA) [63,64], which can be used for the detection of La, Ce, Nd, Sm, Eu, Tb, Yb, and Lu, or through the rather recent high-resolution inductively coupled plasma mass spectrometry (HR-ICP-MS) [65]. Some studies have reported the amount of lanthanides present in plant tissues to range from 4 to 168 mg g^{-1} [11,58].

Obviously, the more abundant the lanthanides are, the more they are taken up by plants, but the soil pH, its redox potential, and its clay content affects the extent of this uptake, and hence some studies have been directed toward such factors. For instance, the role of organic acids released at the root–soil interfaces, and the amount of the elements absorbed by plants, have been investigated [66].

More research on the details and distribution of the elements throughout plant tissues have indicated that Ce^{3+} has a higher tendency to accumulate in plant leaves than in its stems and roots. This behavior, however, is different with different lanthanides and plants [65,67]. For instance, roots of maize and mungbean grown in synthetic solutions have been reported to accumulate lanthanum at about 20 to 150 times more than the small branches [68], and a different accumulation pattern has been reported in the case of plants like wheat and rice or citrus trees [11]. Other instances include pteridophytes with tendencies to absorb La and Ce or genera Dryopteris, Asplenium, Adiantum, and Dicranopteris [62,69-73]. Several research studies have indicated that the presence of chelating agents leads to reduced lanthanide uptake in fern but had no influence on that of non-accumulating plants [74].

The metabolic influences of the lanthanides present in plant tissues are not very clear but there are, as already mentioned, reports on the stimulatory effects these elements induce on the process of absorbing and using other nutrients like nitrogen, phosphorus, and potassium from soil. These effects were found to be influenced by the mechanism through which lanthanides enter plant tissues [75-77]. Further influences of the lanthanides on plant metabolism include stabilizing the cytoplasmic membranes, improving the influence of plant hormones, increasing the nitrogen fixation capability, and reducing water loss [78,79]. Other metabolic influences of lanthanides include the inhibition of the formation of OH^- by Ce^{3+} and La^{3+}, which greatly reduces the peroxidation of the membrane lipids [80]; or boosting the cellular stability against osmotic pressure because of the presence of different lanthanides [81-86]. The intracellular release of H^+ or diffusion of biomolecules into the cells has also been found to be affected in the presence of lanthanum ions [82,83,87].

Lanthanides have also been reported to fortify the cellular membranes through ion-exchange reactions where the calcium ions are replaced with lanthanides [88], but high concentrations of the ions could damage the cell membranes and lead to increased cell permeability [61,89]. At concentrations as low as 0.001 to 1.0 ppm, they have been found to show hormonelike effects in plants [84]. These elements have also proved to influence the hormones in plants [90,91]; for instance, lanthanides have been found to boost the formation of α-amylase in wheat seed, increase the activity of glutamic oxaloacetic transaminase and its reactions [84,92-95], enhance the activity of superoxide dismutase, catalase, and peroxidase, or that of deoxidase in soybean [96,97] or act as regulators for certain hormones [92], or even decrease the activity of enzymes [93,94]. Other examples include the decreased activity of the sucrose-transform enzyme upon being sprayed with lanthanide solutions, or increased nitrate deoxidase activity in soybean leaves [97,98], or the same behavior in the case of nitrate reductase in cotton leaves or increased activity of other enzymes like superoxide dismutase, catalase, and peroxidase [99-101]. There are also reports on the influence of La and Ce trivalent cations on promoting the enzyme concentration. The suggested mechanism is based on the substitution of the two cations with Mg^{2+} leading to the enhanced electron transfer behaviors of PS II [102]. Although similar behaviors have been observed in the case of Nd^{3+} [103], some studies have shown that increasing the concentrations of lanthanides could reverse the effects on chlorophyl content [104].

The presence of lanthanides in plant tissues has also been found to increase their water retention and hence drought tolerance [99,105-107]. The interesting point is that an increased resistance (of about 20%) has been observed without increasing the water use of the plants [99]. There are also some rather inconsistent reports on the influence of lanthanides on plant growth, claiming both inhibitory and stimulatory effects for lanthanides. La^{3+} and Nd^{3+} cations, for instance, have been reported to inhibit the elongation of oat and barley roots [99], or the same effect has been observed in corn roots upon administering La^{3+} and Ce^{3+} cations [108]. Some have reported threshold concentrations for specific plants above which lanthanides act as poison for the plants. The concentrations are however influenced by the addition of proper chelating agents [109].

Further effects of lanthanides include increases in the sugar, vitamin C, and protein contents of sugarcane, grape and apple, and soybean, respectively [58, 77, 90], and promoting the growth of plants. For example, the presence of lanthanides has also showed to promote the growth as well as starch and ascorbic acid content of potatoes [110]. The effect of the growth medium and the method of administering lanthanides have also been found to have a significant role in the impact of lanthanides on plant growth [111], but apart from all considerations, the mere presence of such elements has confirmed to be beneficial [112-127]. It is believed that they influence the metabolism of nutrients.

TOXICOLOGICAL AND METABOLIC EFFECTS OF LANTHANIDES

The extensive industrial and agricultural usages of lanthanides shows an incremental trend, which explains why detailed information on the metabolic and toxicological influences these elements can have is required. They are known to be naturally present in air, groundwater, and soil and hence find their way into plant, animal, and human tissues, which increases the probability of metabolic effects. Figure 3.2 lists some of the known and important metabolic aspects of lanthanides in living organisms [128]. These effects can be positive (enhancing) or negative (inhibiting) depending on the type of the compound and its concentration.

In the human body, the elements are distributed in different organs. They mostly occur in the liver and skeleton, as well as the spleen, lungs, and endocrine glands to a lesser degree. The lighter lanthanides (that are also more basic) prefer the liver, whereas the heavier (more acidic) ones target the skeletal system. The different distribution patterns can be

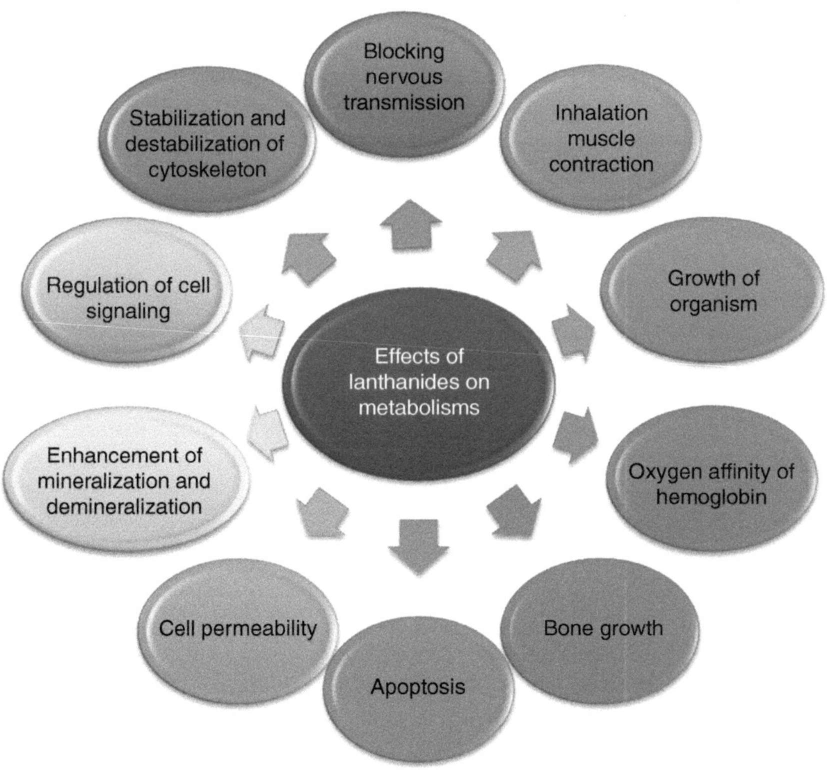

FIGURE 3.2 Lanthanides' effects on metabolism.

attributed to the different ionic radii, solubility, and the stability of the chelates as well as the basic character of the ions that leads to dissimilar distributions of the elements [129]. The majority of lanthanides stored in the liver have been reported to leave it in 4 days, but the rest need 10–40 days for completely leaving the organ. The trend is different with bones, and the time required for it is in the scale of about 2–3 years [130]. The majority of the elements leave the body through the urinary system, especially as complexed ions. The complexation reduces the possibility of interactions (like ion-exchange reactions) and the absorption of the elements in the body [129-131].

It has been found that the metabolic behavior of the elements is very much dependent on the method of administration. Another important consideration is that the study of the metabolism of lanthanides is complicated because of their tendency to form insoluble hydroxides at the pH values equal to those in the body. Studies have shown that the elements form so-called radio-colloids before their precipitation (forming insoluble phosphates and carbonates) or forming complexes with the biomolecules (ligands) present in the body [129]. The formation of precipitates during IV injection of lanthanides, however, has not been observed [132], which could be attributed to presence of physiologic ligands like albumin in blood plasma. This can lead to a parallel rivaling reaction that reduces the chances of precipitation. There is no need to mention that even in the presence of ligands, the precipitation can be observed above certain concentrations of the cation.

Ingested lanthanides reportedly accumulate in kidneys, liver, spleen, skeleton, and teeth [129,133-140]. It has been found that these elements mainly remain in the tissues or in organs they are injected to. If lanthanides are not absorbed by other organs, there is a high chance that they would be stored in the skeletal system. This phenomenon is under debate for lanthanides not absorbed, due to inconsistent data. Also, it indicates that the elements are finally transferred to the gastrointestinal tract [130-140], and if not absorbed by the bones, leave the body through the intestines. In particular, injected lanthanides are mainly reported to leave the body through the kidneys [130-140].

Toxicological studies on the effects of lanthanides on microorganisms have shown that the presence of these elements affect the growth of bacteria, fungi, and yeast [141-144]. The toxicity of lanthanides has also been reported to vary based on the compounds used as well as the way they are introduced to the subjects. Ingestion of the elements has proven to be of little effectiveness, in terms of absorption and accumulation, because the gastrointestinal tracts have shown a poor capability for absorbing them [129,130,142,143]. Consequently, the LD_{50} for them is in the order of several grams/kilogram of body weight. The most lethal method of usage, on the other hand, has been found to be injection, especially intravenous.

Ingesting daily doses of lanthanum (up to 3 g per day), or drinking water containing up to 2 ppm of lanthanides, have been reported to be safe and not cause health problems [129-133,145-148], and hence they are considered as low-toxicity species according to the Hodge-Sterner classification system [129].

Symptoms of receiving lethal doses include calcification, hypotension, hypoglycemia, and increased clotting times and may lead to death by heart or lung failure [145-152]. Symptoms due to the elements, on the other hand, have been reported to be internal peritonitis, adhesions, ascites and eye irritation, and even ulceration. They have also shown anticoagulation properties due to interference with clotting reactions based on Ca^{2+} ions [153]. Chronic exposure with air polluted with lanthanides could lead to emphysema, pneumonitis, bronchitis, and pulmonary fibrosis or pulmonary reticulosarcoma [154]. Table 3.1 gives an overview of the common lanthanide compounds and their toxicities [155-158], which are very much dependent on the chemical sources.

The toxicity of the elements can be varied using different ligands or other compounds like drug-metabolizing enzymes, adenosine triphosphate, adenosine, and tryptophan, which have been found to lower the toxicity of these elements. It is worthy to note that some compounds (e.g., dexamethasone and glucocorticoid) have reportedly caused increased toxicities. The carcinogenicity of these elements, alternatively, has not been proven experimentally, in the face of older reports [129,130,136,137].

EFFECTS OF LANTHANIDES ON THE INTERNAL ORGANS

Lanthanides deposited in the bones have not been reported to show toxic effects. Lanthanide ions, as functional mimics of Ca(II) ions, have been shown to affect the bone remodeling cycle and has potential for the treatment of bone density disorders, such as osteoporosis [159]. Yet, high doses of the elements have been reported to cause bone mineralization defects in rats [160,161]. Although lanthanides have not shown any effects on healthy skin or, abraded rabbit skin, they showed serious interactions with the elements, leading to epilation and scar formation [130]. Their injection has also reportedly caused granulomas and local calcification, with mild fibrosis, as well as soft-tissue calcification.

The effect of inhalation of lanthanides on lungs has been investigated, revealing that exposure to such atmospheres (e.g., smoke of cored carbon arc lamps) can lead to chronic pneumoconiosis. This condition is also dependent on the nature and dose of the compounds as well as the exposure duration. Others have suggested that conditions like interstitial disorder, emphysema, and severe obstructive impairment might be caused

TABLE 3.1 An Overview of the Common Lanthanide Compounds and Their Toxicities

Administered substance	Animal	LD_{50} (mg/kg BW)	Reference
La^{3+} -acetate	Rat	10000	[155]
La^{3+} -ammonium nitrate	Rat	3400	[155]
$LaCl_3$	Rat	4200	[155]
$LaCl_3$	Mouse male	2354	[156]
$La(NO_3)_3$	Rat	4500	[155]
$La_2(SO_4)_3$	Rat	>5000	[155]
La_2O_3	Rat	>10000	[155]
$CeCl_3$	Mouse male	1959	[156]
$CeCl_3$	Rat	2110	[157]
$CeCl_3$	Mouse	5277	[157]
$Ce(NO_3)_3$	Rat female	4200	[158]
CeO_2	Rat	>5000	[157]
$PrCl_3$	Mouse male	4500 (4054–4995)	[130]
$Pr(NO_3)_3$	Rat female	3500	[158]
Pr_6O_{11}	Rat	>1000	[156]
$NdCl_3$	Mouse male	5250 (4730–5830)	[130]
$Nd(NO_3)_3$	Rat female	2750	[158]
Nd_2O_3	Rat	>1000	[157]
$SmCl_3$	Mouse male	>2000	[130]
$Sm(NO_3)_3$	Rat female	2900	[157]
Sm_2O_3	Rat	>1000	[157]
$EuCl_3$	Mouse	5000	[157]
$Eu(NO_3)_3$	Rat female	>5000	[158]
Eu_2O_3	Rat	>1000	[157]
$GdCl_3$	Mouse male	>2000	[130]
$Gd(NO_3)_3$	Rat female	>5000	[158]
Gd_2O_3	Rat	>1000	[157]
$TbCl_3$	Mouse male	5100 (5049–5130)	[130]
$TbCl_3$	Mouse male	2652	[156]
$Tb(NO_3)_3$	Rat female	>5000	[158]
Tb_4O_7	Rat	>1000	[157]

TABLE 3.1 An Overview of the Common Lanthanide Compounds and Their Toxicities (*cont.*)

Administered substance	Animal	LD_{50} (mg/kg BW)	Reference
$DyCl_3$	Mouse male	7650 (7150–8158)	[130]
$Dy(NO_3)_3$	Rat female	3100	[158]
Dy_2O_3	Rat	>1000	[157]
$HoCl_3$	Mouse male	7200 (6667–7776)	[130]
$HoCl_3$	Mouse male	2568	[156]
$Ho(NO_3)_3$	Rat female	3000	[158]
Ho_2O_3	Rat	>1000	[157]
$ErCl_3$	Mouse male	6200 (5390–7140)	[130]
Er_2O_3	Rat	>1000	[156]
$TmCl_3$	Mouse male	6250 (5390–7140)	[130]
Tm_2O_3	Rat	>1000	[157]
$YbCl_3$	Mouse male	6700 (6374.9–7041.7)	[130]
$Yb(NO_3)_3$	Rat female	3100	[158]
$LuCl_3$	Mouse male	7100 (6630–7590)	[130]
Lu_2O_3	Rat	>1000	[157]

as a result of exposure to lanthanide containing dust [162,163], but the elements are still considered to have mild toxic potentials in comparison to other fibrogenic dusts like quartz or silica containing dust.

In the case of liver, which the light, basic lanthanides tend to target, studies have shown that handing them out through injection can cause temporary changes in the activities of GPT and GOT enzymes. Some reports mention the development of hepatic necrosis as a result of injecting lanthanide solutions, which could fortunately be prevented by certain drugs [164-166]. Another condition reported as a result of receiving them through injection is the so-called rare-earth fatty liver [129].

The conditions attributed to the influence of lanthanides on the nervous system and the brain is limited. This could be due to the existence of the blood–brain barrier, which hinders them from reaching the central nervous system [129]. Yet examples of such influences include the analgesic effects of La^{3+} on the brain, observed through their direct injection of in a solution into brain tissue. This revealed that they act in the same way as other opiates, under the conditions. For example, injection of gadolinium chloride or its EDTA or DTPA complexes has been found to drop the motor function and epileptoid fits.

EXAMPLES OF TOXICOLOGICAL OR METABOLISM EFFECT OF LANTHANIDE MEMBERS

The following discussion attempts to provide some examples of the effects of the specific elements. The toxicity of lanthanum is rather low, but its injection has been reported to cause hyperglycemia, low blood pressure, degeneration of the spleen, and hepatic alterations. Exposure to the oxides of element in air leads to pneumoconiosis [167,168].

Cerium has the same toxicity as lanthanum. The element readily ignites in air as mentioned in previous chapters, and the fumes produced can be dangerous. Workers whose skins have been exposed to cerium complained of itching and showed heat sensitivity and skin lesions. Ingestion of cerium does not show toxicity, but injection of high doses of the element to animals reportedly caused death due to cardiovascular collapse [169]. Cerium can cause more danger to aquatic organisms by damaging cell membranes. Cerium(IV) oxide is a strong oxidizing agent at high temperatures and will react with combustible organic materials. Although cerium is not radioactive, the impure commercial grade may contain traces of thorium, which is weakly radioactive.

The toxicity of praseodymium is similar to the previous lanthanides. Its solutions show mild toxicity if ingested. Although its insoluble salts do not show toxic effects, they irritate the skin and eyes if the tissues are directly exposed. Workers chronically exposed to air containing the element suffer from lung embolism. The element is also a threat to the liver.

Because of the very low amount of neodymium in the human body, it normally has no biological effects, but in case one is expressed to high concentrations of the element, some parts of the body can be influenced. Its dust and salts can irritate the eyes, yet its oral usage does not create toxic effects. Neodymium metal dust can ignite and may even explode. Inhaling the dust can lead to lung embolisms [169].

Promethium in normal concentrations does not show any biological effects. The 147 isotope that emits X-rays is dangerous for all life forms and should be handled with protective clothing [170,171].

Samarium is a slightly toxic element and does not have serious biological effects in human body. Samarium salts are claimed to have stimulatory effects on the metabolic process, but it is suspected that this may be caused by other lanthanides that accompany it. The element mostly accumulates in the liver and kidneys, and some 8 μg/L of the element is present in in the blood. Samarium is not considerably absorbed by many plants, and consequently it is not very abundant in human diet, but a few plants and vegetables may contain up to 1 ppm of the element. Ingestion of up to 0.05% of its salts does not produce serious problems because it is marginally absorbed into the bloodstream and the rest is excreted [169, 172].

Europium is no different from the other elements discussed so far, in terms of toxicity. The LD_{50} of its chloride, nitrate, and oxide are around

550, 320, and 5000 mg/kg, respectively [158,173]. The metallic powder can ignite and explode, which can be dangerous [174].

Gadolinium ion is often reported as highly toxic, but in chelated form it is used as an MRI contrast agent [175]. The toxicity of the free ion (LD_{50} = 100–200 mg/kg) is due to its interference with the process of calcium-ion channels [176]. The stronger complexes of the ion show relatively less toxicities [177-180]. Gadolinium compounds have been effectively used on patients with renal impairment, but if the condition is serious (requiring dialysis), the patients are at the risk of rare conditions like nephrogenic systemic fibrosis [180] or nephrogenic fibrosing dermopathy [181].

Although terbium and dysprosium are not known for any serious biological role, except for the danger of the ignition of their powders [182-184], the soluble chloride and nitrate of holmium are mildly toxic when orally administered, whereas the insoluble salts are not. Considering the toxicity of the element to mice, it is estimated that ingesting about 500 g of it should be fatal to a human [169]. Inhaling, ingesting, or injecting large amounts of holmium salts can cause severe damage, but at low doses it has a low acute toxicity [185]. Erbium shows similar properties as the previous elements [186-188].

Thulium is present in the human body at very low concentrations and is not known to have a major biological role. Small amounts of the element, however, are known for stimulating metabolism. Its soluble salts are mildly toxic, but the insoluble ones are not [169]. Injecting thulium can lead to liver and spleen degeneration and can further lead to fluctuations in the hemoglobin. The element accumulates in the liver, kidneys, and bones in the human body [189]. Radioactive thulium isotopes can cause radiation poisoning [190].

Ytterbium compounds are treated as highly toxic, although studies show their danger to be minimal. The compounds are known for irritation to the human skin and eyes, and some might be teratogenic [191]. Metallic ytterbium dust can spontaneously combust, producing hazardous fumes [192].

The last lanthanide (Lutetium) is like its other family members and has low toxicity, but like the other lanthanides its compounds are handled with care: inhaling its compounds such as lutetium fluoride, oxide, etc. is known to be dangerous and its salts are known to irritate skin [190].

References

[1] Ichihashi H, Morita H, Tatsukawa R. Rare earth elements (REEs) in naturally grown plants in relation to their variation in soils. Environ Pollut 1992;76:157–62.

[2] Gonzalez V, Vignati DAL, Leyval C, Giamberini L. Environmental fate and ecotoxicity of lanthanides: Are they a uniform group beyond chemistry? Environ Int 2014;71:148–57.

[3] Kulaksız S, Bau M. Rare earth elements in the Rhine River, Germany: first case of anthropogenic lanthanum as a dissolved microcontaminant in the hydrosphere. Environ Int 2011;37:973–9.

[4] Kulaksız S, Bau M. Anthropogenic gadoliniumas a microcontaminant in tapwater used as drinking water in urban areas and megacities. Appl Geochem 2011;26:1877–85.
[5] Li J, Hong M, Yin X, Liu J. Effects of the accumulation of the rare earth elements on soil macrofauna community. J Rare Earths 2010;28:957–64.
[6] Diatloff E. Foliar application of rare earth elements to maize and mungbean. Aust. J Exp Agr 1999;39:189–94.
[7] Maheswaran J, Meehan B, Reddy N, Peverill K, Buckingham S. Impact of rare earth elements on plant physiology and productivity. Rural Ind Res Dev Corp 2001;1:1–40.
[8] Xiong BK. Application of rare earths in Chinese agriculture and their perspectives of development. In Proceeding of the rare earths in agriculture seminar, 20 September 1995, 5–9, Canberra, ACT Australia.
[9] Zhang FS, Yamasaki S, Kimura K. Rare earth element content in various waste ashes and the potential risk to Japanese soils. Environ Int 2001;27:393–8.
[10] Wang HY. Special topic in agricultural use of rare earths in China (in Chinese). Chinese Rare Earth. 1998;19:68–72.
[11] Tyler G. Rare earth elements in soil and plant systems - a review. Plant Soil 2004;267: 191–206.
[12] Anawar HM, Freitas MC, Canha N, Dionisio I, Dung HM, Galinha C, et al. Assessment of bioaccumulation of REEs by plant species in a mining area by INAA. J Radioanal Nucl Chem 2012;294:377–81.
[13] Zhang S, Shan XQ. Speciation of rare earth elements in soil and accumulation by wheat with rare earth fertilizer application. Environ Pollut 2001;112:395–405.
[14] Binggan W, Yonghua L, Hairong L, Jiangping Y, Bixiong Y, Tao L. Rare earth elements in human hair from a mining area of China. Ecotoxicol Environ Saf 2013;96: 118–23.
[15] Schuler D, Buchert M, Liu R, Dittrich S, Merz C. Study on rare earths and their recycling. Final report for the Greens/EFA Group in the European Parliament, Öko-Institut e.V., Darmstadt, January, 2011.
[16] Massari S, Ruberti M. Rare earth elements as critical raw materials: focus on international markets and future strategies. Res Policy 2013;38:36–43.
[17] Liang T, Li K, Wang L. State of rare earth elements in different environmental components in mining areas of China. 2014;186:1499–513.
[18] Sastri VS, Bünzli JC, Ramachandra Rao V, Rayudu GVS, Perumareddi JR. Modern aspect of rare earth and their complexes. Elsevier; 2003. ISBN: 978-0-444-51010-5.
[19] Zhang JZ, Zhu WM, Zhang L. Accumulation, distribution and migration of ^{141}Ce and ^{147}Nd in soils of China. In: Xu GX, Xia JM, Yu ZS, Yan CH, editors. Proceeding of the third international conference on rare earth development, application, Baotou, Inne Mangolia, China, August, 1995, p. 21–25, vol. II, Beijing, China.
[20] Stockes J, Meehan B, Fowless B, Peverill A. The bioavailability of applied lanthanum in 7 different soils from Victoria, Australia. In: Yu ZS, Yan CH, Xu GY, Niu JK, Chen ZH, editors. Proceeding of the fourth international conference on rare earth development, application., Part I, p., 255–261, Beijing, China, June 16–18, 2001, Beijing: Metallurgical Industry Press; 2001.
[21] Hu Z, Haneklaus S, Sparovek G, Schnug E. Rare earth elements in soils. Commun Soil Sci Plant Anal 2006;37:1381–420.
[22] Xiong BK, Chen P, Guo BS, Zheng W. Rare earth element research and applications in Chinese agriculture and forest (in Chinese). Beijing, China: Metallurgical Industry Press; 2000. p. 1–151.
[23] El-Ramady HRH. A contribution on the bio-actions of rare earth elements in the soil-plant environment; Braunschweig, Techn Univ, Diss, 2008, Online-Resource: http://nbn-resolving.de/urn:nbn:de:gbv:084-20426.
[24] Wyttenbach A, Furrer V, Schleppi P, Tobler L. Rare earth elements in soil and in soil-grown plants. Plant Soil 1998;199:267–73.

[25] Wyttenbach A, Tobler L, Schleppi P, Furrer V. Variation of the rare earth element concentrations in the soil, soil extract and in individual plants from the same site. J Radioanal Nuclear Chem 1998;231:101–6.
[26] Lu A, Zhang S, Shan X, Wang S, Wang Z. Application of microwave extraction for the evaluation of bioavailability of rare earth elements in soils. Chemosphere 2003;53: 1067–75.
[27] Cao X, Chen Y, Wang X, Deng X. Effects of redox potential and pH value on the release of rare earth elements from soil. Chemosphere 2001;44:655–61.
[28] Diatloff E, Asher CJ, Smith FW. Concentration of rare earth elements in some Australian soils. Aust J, Soil Res 1996;34:735–47.
[29] Ding S, Liang T, Zhang C, Yan J, Zhang Z. Accumulation and fractionation of rare earth elements (REEs) in wheat: controlled by phosphate precipitation, cell wall absorbtion and solution complexation. J Exp Bot 2005;56:2765–75.
[30] Shan X, Lian J, Wen B. Effect of organic acids on adsorption and desorption of rare earth elements. Chemosphere 2002;47:701–10.
[31] Wang J, Chunji N, Kuiyue Y, Jiazuan N. Formation equilibria of ternary metal complexes with citric acid and glutamine (alanine) in aqueous solution. J Rare Earth 2004;22:183–6.
[32] Wen B, Liu Y, Hu X, Shan X. Effect of earthworms (*Eisenia fetida*) on the fractionation and bioavailability of rare earth elements in nine Chinese soils. Chemosphere 2006;63:1179–86.
[33] Chu H, Zunhua W, Xie Z, Jianguo Z, Zhengao ZL, Cao Z. Effect of lanthanum on major microbial populations in red soil. Pedosphere 2001;11:73–6.
[34] Chu HY, Zhu JG, Xie ZB, Zhang HY, Cao ZH, Li ZG. Effects of lanthanum on dehydrogenase activity and carbon dioxide evolution in a Haplic Acrisol Aust Soil Res 2003;41:731–9.
[35] Chen XH, Zhao B. Arbuscular mycorrhizal fungi mediated uptake of lanthanum in Chinese milk vetch (Astragalus sinicus L.). Chemosphere 2007;68:1548–55.
[36] Zhu JG, Chu HY, Xie ZB, Yagi K. Effects of lanthanum on nitrification and ammonification in three Chinese soils. Nutr Cycl Agroecosys 2002;63:309–14.
[37] Kamijo M, Suzuki T, Kawai K, Fujii T, Murase H. Ytterbium-decreasing *Streptomyces* sp. and its naphthoquinone-pigment production in the presence of rare-earth elements. J Biosci Bioengin 1999;87:340–3.
[38] Tsuruta T. Accumulation of rare earth elements in various microorganisms. J Rare Earth 2007;25:526–32.
[39] Rajapaksha RMCP, Tobor-Kapłon MA, Baath E. Metal Toxicity Affects Fungal and Bacterial Activities in Soil Differently. Appl Environ Microbiol 2004;70:2966–73.
[40] Thomas PJ, Carpenter D, Boutin C, Allison JE. Rare earth elements (REEs): effects on germination and growth of selected crop and native plant species. Chemosphere 2014;96:57–66.
[41] Xie Z, Zhu JG, Chu HY, Zhang YL, Ma HL, Cao ZH. Effect of lanthanum on rice production, nutrient uptake and distribution. J Plant Nutr 2002;25:2315–31.
[42] Zhou F, Fu C, Jianhua C, Lijie P, Buzhuo P. Effect of exogenous rare earths on microbial characteristics in paddy soil. J Rare Earth 2004;22:296–300.
[43] Chu HY, Zhu JG, Xie ZB, Zeng Q, Li ZG, Cao ZH. Availability and toxicity of exogenous lanthanum in a Haplic acrisols. Geoderma 2003;115:121–8.
[44] Tang X, Sun Y, Xia M, Wen C, Zhang Z. Ecological effects of low dosage mixed rare earth elements accumulation on major soil microbial groups in a yellow cinnamon soil. Ying Yong Sheng Tai Xue Bao 2004;15:2137–41.
[45] Chu H, Zhu J, Xie Z, Cao Z, Li Z, Zeng Q. Effects of lanthanum on microbial biomass carbon and nitrogen in red soil. J Rare Earth 2001;19:63–6.
[46] Xu X, Wang Z. Effects of lanthanum and mixture of rare earths on ammonium oxidation and mineralization of nitrogen in soil. Eur J Soil Sci 2001;52:323–9.

[47] Xu D, Guangshen L, Jie X, Weiping L. Effects of lanthanum and cerium on acid phosphatase activities in two soils J Rare Earth 2004;22:725–8.
[48] Chu H, Zhu J, Lin X, Yin R, Xie Z, Cao Z, Fujii T. Short-term decomposition of 14 C-labelled glucose in a fluvo-aquic soil as affected by lanthanum amendment. Biol Fert Soils 2007;43:811–4.
[49] Pang X, Li D, Peng A. Application of rare-earth elements in the agriculture of China and its environmental behavior in soil. J Soils Sediments 2001;1:124–9.
[50] Dick RP, Sandor JA, Eash NS. Soil enzyme activities after 1500 years of terrace agriculture in the Colca Valley. Peru Agric Ecosys Environ 1994;50:123–31.
[51] Nannipieri P. The potential use of soil enzymes as indicators of productivity, sustainability and pollution. In: Pankhurst CE, Doube BM, Gupta VVSR, Grace PR, editors. Soil Biota Management in Sustainable Farming Systems. Melbourne: CSIRO; 1994. p. 238–44.
[52] Dick RP. Soil enzyme activities as integrative indicators of soil health. In: Pankhurst CE, Doube BM, Gupta VVSR, editors. Biological Indicators of Soil Health. CAB International; 1997. p. 121–56.
[53] Balota EL, Kanashiro M, Filho AC, Andrade DS, Dick RP. Soil enzyme activities under long-term tillage and crop rotation systems in subtropical agro-ecosystems. Braz J Microbiol 2004;35:300–6.
[54] Banerjee MR, Burton DL, Depoe S. Impact of sewage sludge application on soil biological characteristics. Agric Ecosyst Environ 1997;66:241–9.
[55] Chu H, Zhu J, Xie Z, Li Z, Cao Z, Zeng Q, Lin X. Effect of lanthanum on hydrolytic enzyme activities in red soil. J Rare Earth 2002;20:158–60.
[56] Chu HY, Zhu JG, Xie ZB, Li ZG, Cao ZH. Effects of lanthanum on the microflora of red soil. Environ Sci 2000;21:28–31.
[57] Pang X, Li D, Peng A. Application of rare earth elements in the agriculture of China and its environmental behaviour in soil. Environ Sci Pollut Res Int 2002;9:143–8.
[58] Zeng Q, Zhu JG, Cheng HL, Xie ZB, Chu HY. Phytotoxicity of lanthanum in rice in haplic acrisols and cambisols. Ecotoxicol Environ Saf 2006;64:226–33.
[59] Li F, Shan X, Zhang T, Zhang S. Evaluation of plant availability of rare earth elements in soils by chemical fractionation and multiple regression analysis. Environ Pollut 1998;102:269–77.
[60] von Tucher S, Schmidhalter U. Lanthanum uptake from soil and nutrient solution and its effects on plant growth. J Plant Nutr Soil Sci 2005;168:574–80.
[61] Chang J. Effects of lanthanum on the permeability of root plasmalemma and the absorption and accumulation of nutrients in rice and wheat. Plant Physiol Commun 1991;27:17–21.
[62] Wang Z, Liu D, Lu P, Wang C. Accumulation of rare earth elements in corn after agricultural application. J Environ Qual 2001;30:37–45.
[63] Ni BF, Tian WZ, Nie HN, Wang PS, He GK. Study on air pollution in Beijing's major industrial areas using multielements in biomonitors and NAA techniques. Biol Trace Elem Res 1999;267–72.
[64] Wang YQ, Sun JX, Chen HM, Guo FQ. Determination of the contents and distribution characteristics of REE in natural plants by NAA. J Radioanal Nucl Chem 1997;219:99–103.
[65] Xu X, Zhu W, Wang Z, Witkamp G. Accumulation of rare earth elements in maize plants (Zea mays L.) after application of mixtures of rare earth elements and lanthanum. Plant Soil 2003;252:267–77.
[66] Han F, Shan X, Zhang J, Xie Y, Pei Z, Zhang S, Zhu Y, Wen B. Organic acids promote the uptake of lanthanum by barley roots. New Phytologist 2005;165:481–92.
[67] Wyttenbach A, Schleppi P, Bucher J, Furrer V, Tobler L. The accumulation of the rare earth elements and of scandium in successive needle age of Norway spruce. Biol Trace Elem Res 1994;41:13–22.

[68] Diatloff E, Smith FW, Asher CJ. Rare earth elements and plant growth. II. Response of corn and mungbean to low concentrations of lanthanum in dilute, continuously flowing nutrient solutions. J Plant Nutr 1995;18:1977–89.
[69] Ozaki T, Enomoto S, Minai Y, Ambe S, Makide Y. A survey of trace elements in pteridophytes. Biol Trace Elem Res 2000;74:259–73.
[70] Lao X, Liu C, Yang S, Li X, Ma Y, Wang W. Rare earth absorption and distribution of maize plant. Chin Bull Bot 1996;13:59–61.
[71] Liu P, Li Y, Ju T, Zhang M. Studies on the absorption and distribution of rare earth elements in soybean. Acta Agric Boreali-occidentalis Sinica 1997;6:58–60.
[72] Liu S, Wang L, Zhang S. Effect of long-term foliage-dressing rare earth elements on their distribution, accumulation and transportation in soil spring wheat system. Chin J Appl Ecol 1997;8:55–8.
[73] Peng A, Wang Z. Research progress in environmental chemistry of rare earth elements. Chin J, Environ Chem 1995;3:22–32.
[74] Ozaki T, Ambe S, Enomoto S, Minai Y, Yoshida S, Makide Y. Multitracer study on the uptake mechanism of yttrium and rare earth elements by autumn by fern. Radiochim Acta 2002;90:303–7.
[75] Liu DW, Wang X, Chen ZW. Effects of rare earth elements and REE-binding proteins on physiological responses in plants. Protein and Peptide Lett 2012;19:198–202.
[76] Cunha MDLE, Nardi LVS, Muller IF. Biogeochemistry of REE elements and tetrad effect in the soil-plant system: a study on volcanic rock covers in southernmost Brazil. Anais da Academia Brasileira de Ciencias 2012;84:911–8.
[77] Wang Q. Technologies and effect of REEs enriched La on rice. Chinese Rare Earth 1995;16:67–76.
[78] Wen B, Yuan D, Shan X, Li F, Zhang SZ. The influence of rare earth element fertilizer application on the distribution and bioaccumulation of rare earth elements in plants under field conditions. Chem Spec Bioavailab 2001;13:39–48.
[79] Xing JM, Weng LP. Effect of Ca, La and Ce on staved doat of radish leaf. Rare Metal 1991;15:262–6.
[80] Zeng FA, Zhang Y, Zhang HT. The effects of La (III) on the peroxidation of membrane lipids in wheat seedling leaves under osmotic stress. Biol Trace Element Res 1999;69:145–50.
[81] Zheng HL, Zhao ZQ, Zhang CG, Feng JZ, Ke ZL, Su MJ. Changes in lipid peroxidation, the redox system and ATPase activities in plasma membranes of rice seeding roots caused by lanthanum chloride. Biometals 2000;13:137–63.
[82] Dong B, Wu ZM, Tan XK. Effects of $LaCl_3$ on physiology of cucumber with Ca deficiency. J Chinese Rare Earth Soc 1993;11:65–8.
[83] Qiao GL, Tang XK, Wu ZM. Study dynamics of $NdCl_3$ on K^+ exosmosis form corn roots. Acta Bot Sin 1993;35:286–90.
[84] Redling, K. Rare earth elements in agriculture with emphasis on animal husbandry. PhD. Thesis, München Uni., Faculty of Veterinary Medicine, 2006.
[85] Tian WX. Effect of REEs on seed germination, growth of seedling and yield of tuber of sugar beet. Bot Commun 1990;7:37–40.
[86] Shen H, Yan XL. Membrane permeability in roots of crotalaria seedlings as affected by low temperature and low phosphorus stress. J Plant Nutr 2002;25:1033–47.
[87] Ni JZ. The use of rare earth elements in agriculture and medicine. Bioinorganic Chemistry of rare earth elements. Beijing, China: Science Press; 1995. p. 13–55.
[88] Hu QH, Ye ZJ. Physiological effects of rare earth elements on plants. Chinese Plant Physiol Commun 1996;32:296–300.
[89] Wang X, Shan X, Zhang S. Distribution of rare earth elements among chloroplast compounds of hyperaccumulators Dicranopteris dichotoma. Anal Bioanal Chem 2003;376:913–7.

[90] Brown PH, Rathjien AH, Graham RD, Tribe DE. Rare earth elements in biological systems. In: Geschneider KA, Eyring Jr L, editors. Handbook on the physics and chemistry of rare earths, vol. 13. New York: Elsevier Science Publisher B. V; 1990. p. 423–53.
[91] Enyeart J, Xu JL, Enyeart JA. Dual actions of lanthanides on ACTH-inhibited leak K^+ channels. Am J Physiol Endocrinol Metab 2002;282:1255–66.
[92] Taiz L, Zeiger E. Plant Physiology. 2nd ed. Sunderland, MA, USA: Sinauer Associates; 1998. ISBN 0-87893-823-0.
[93] Allan AC, Rubery PH. Calcium deficiency and auxin transport in Cucurbita pep L. seedlings. Planta 1991;183:604–12.
[94] Bai BZ, Chen FY. Effect of REEs on some index of modality and physiology for sugar beet. Chinese J Sugar Beet 1989;11:10–2.
[95] Xu SJ, Harrison H, Chalupa W. The effects of rumminal bypass lysine and methionine on milk-yield and composition of lactating cows. J Dairy Sci 1998;81:1062–77.
[96] Hong FS, Wie ZG, Zhao GW. Mechanism of effect of lanthanum nitrate on vigor of aged rice seeds. J Rare Earth 2000;18:219–23.
[97] Chen, QP. Effect of REEs on yield of soybean. In: Chief Office of Helongjiang Farm, editors, Literature compile for application of REEs in agriculture (in, Chinese)., Helongjiang, China: Metallurgical Industry Press. 1991, p. 76–85.
[98] Wang CD, Ye S, Qi H, Zhao G. Effects of spraying rare earths on contents of rare earth elements and effective components in tea. J Agric Food Chem 2003;51:6731–5.
[99] Hu Z, Richter Y, Sparovek G, Schnug E. Physiological and biochemical effects of rare earth elements on plants and their agricultural significance: a review. J Plant Nutr 2004;27:183–220.
[100] Liu C, Hong F, Zheng L, Tang P, Wang Z. Effects of rare earth elements on vigor enhancement of aged spinach seeds. J Rare Earth 2004;22:547–51.
[101] Bai S, Deng XM. Effect of REEs on anti-blight of cotton. J South West Agric Univ 1995;17:28–31.
[102] Hong FS, Wie ZG, Zhao GW. Mechanism of lanthanum effect on the chlorophyll of spinach. J Science in China Series C 2002;45:166.
[103] Liu C, Hong F, Wang L, Zheng L. Effect of Nd^{3+} on photosynthesis of spinach. J Rare Earth 2004;22:306–10.
[104] Fashui H, Ling W, Xiangxuan M, Zheng W, Guiwen Z. The effect of cerium (III) on the chlorophyl formation in spinach. Biolol Trace Element Res 2002;89:263–76.
[105] Wen QK, Lai ZS, Fang B. Study the effect of REEs on drought tolerance in corn. Chinese Rare Earth 1992;13:64–7.
[106] Yu YH, Liu QY. Distribution cumulative of REEs in sugarcane plant and effect of REEs on adversity tolerance in sugarcane. J South China Agric Univ 1992;13:47–54.
[107] Meehan B, Peverill K, Maheswaran J, Bidwell S. The application of rare earth elements in enhancement of crop production in Australia. In: Yu ZS, Yan CH, Xu GY, Niu JK, Chen ZH, editors. Proceedings of the fourth international conference on rare earth development and application. Part 1. Beijing, China: Metallurgical Industry Press: Beijing; 2001. p. 244–50. June 16-18.
[108] Diatloff E, Smith FW, Asher CJ. Rare earth elements and plant growth. I. Effects of lanthanum and cerium on root elongation of corn and mungbean. J Plant Nutr 1995;18:1963–76.
[109] Harper SM, Edwards DG, Kerven GL, Asher CJ. Effects of organic acid fractions extracted from Eucalyptus camaldulensis leaves on rootelongation of maize (*Zea mays* L.) in the presence and absence of aluminum. Plant Soil 1995;171:189–92.
[110] Chen FM, Zheng TJ. Study application of REEs in potatoes. J Potatoes 1990;4:224–7.
[111] Perverill K, Meehan B, Buckingham S, Maheswaran J. The role of application of rare earth elements in enhancement of crop and pasture production. Rare Earth 1997;31:69–85.
[112] Chang J, Zhu W, Zhang L, Xiong J, Zhang J, Hu Z. Study on environmental effects of rare earth elements. In: Second international symposium on trace elements and food chain, 15–17 November 1998, p. 24, Wuhan, China.

[113] Zhu YY. Effects of REEs with enriched La on wheat Nd uptake. Nucl Agron Bull 1992;13:176–8.
[114] Zhang C, Li Q, Zhang M, Zhang N, Li M. Effects of rare earth elements on growth and metabolism of medicinal plants. Acta Pharmaceut Sin B 2013;3:20–4.
[115] Hu Z, Zhu WM. Kinetics of ion absorption by rice and ion interaction. Soils 1994;25: 278–81.
[116] Sun T, Yu J, Mao J. Effect of rare earth elements on physiological properties of Jun Date (Jujuba Mill, Z.). In: Second international symposium on trace elements and food chain, 12–15. November 1998, p. 38, Wuhan, China.
[117] Diatloff E, Smith FW, Asher CJ. Rare earth elements and plant growth. III. Response of corn and mungbean to low concentrations of cerium in dilute, continuously flowing nutrient solutions. J Plant Nutr 1995;18:1991–2003.
[118] Diatloff E, Asher CJ, Smith FW. Effects of rare earth elements on the growth and mineral nutrition of plants. In: Rare earths in agriculture seminar, Australian Academy of Technologies, Engineering, 1995.
[119] Walid PA, Valiathan MS, Kamalam NV, Eapen JT, Vijayalakshmi S, Prabhu RK, Mahalingam TR. Effect of rare earth elements on growth and nutrition of coconut palm and root competition for these elements between the palm and Calotropis gigantea. J Plant Nutr 2000;23:329–38.
[120] Aidid SB. Rare earth element abundances and distribution patterns in plant materials. J Radioanal Nucl Chem 1994;183:351–8.
[121] Guo BS. Recent research advance of rare earth in the field of biology. Chinese Rare Earth 1999;20:64–8.
[122] Wang, S.M., Zheng, W. The developing technology of rare earth application on agriculture and biology field in China. In: Yu ZS, Yan CH, Xu GY, Niu JK, Chen ZH, editors. Proceedings of the fourth international conference on rare earth development, Application, Beijing, China, June 16–18. 2001, p. 237–240.
[123] Wang X, Shan X, Zhang S. Preliminary characterization of light-rare-earth-element-binding peptide of a natural perennial fern Dicranopteris dichotoma. Anal Bioanal Chem 2003;376:49–52.
[124] Wang X, Shan X, Zhang S. Effect of exogenous rare earth elements on fraction of heavy metals in soils and bioaccumulation by plants. Commun Soil Sci Plant Anal 2003;34:1573–88.
[125] Wang D, Wang C, Wie Z, Qi H, Zhao G. Effect of rare earth elements on peroxidase activity in tea shoots. J Sci Food Agric 2003;83:1109–13.
[126] Wang L, Bing H, Bai KZ, Li LB, Kuang TY. Photosynthetic characterization of the plant *Dicranopteris dichotoma* Bernh in a rare earth elements mine J, Int Plant Biol (Acta Bota. Sin.) 2005;47:1092–100.
[127] Hu X, Ding ZH, Chen Y, Wang X, Dai L. Bioaccumulation of lanthanum and cerium and their effects on the grown of wheat (*Triticum aestivum* L.) seedlings. Chemosphere 2002;48:621–9.
[128] Khalil S. Ion-selective electrode for lanthanum determination in standard samples. Anal Lett 2003;36:1335–49.
[129] Evans CH. Biochemistry of the lanthanides. Biochemistry of the Elements Series, vol. 8. New York and London: Plenum Press; 1990. ISBN 0-306-4317-9.
[130] Haley TJ. Toxicity. In: Gschneidner Jr KA, Eyring LR, editors. Handbook on the physics and chemistry of rare earths, vol. 4. Amsterdam, New York, Oxford: Elsevier, North-Holland; 1979. p. 553–79.
[131] Marang L, Reiller PE, Eidner S, Kumke MU, Benedetti MF. Combining spectroscopic and potentiometric approaches to characterize competitive binding to humic substances. Environ Sci Technol 2008;42:5094–8.
[132] Kanapilly GM. In vitro precipitation behavior of trivalent lanthanides. Health Phys 1980;39:343–6.

[133] Harrison TS, Scott LJ. Lanthanum carbonate. Drugs 2004;64:985–96.
[134] Marciniak JC, Baltrukiewicz Z. Transport of lanthanides in milk into suckling rats. The Quart J, Nucl Med 1996;40:351–8.
[135] Suzuki KT, Kobayashi E, Ito Y, Ozawa H, Suzuki E. Localization and health effects of lanthanum chloride instilled intratracheally into rats. Toxicology 1992;76:141–52.
[136] Hirano S, Kodama N, Shibata K, Suzuki KT. Distribution, localization and pulmonary effects of yttrium chloride following intratracheal instillation into the rat. Toxicol Appl Pharmacol 1990;104:301–11.
[137] Bulman RA. Metabolism and toxicity of the Lanthanides. In: Sigel A, Sigel H, editors. The Lanthanides and their interrelations with biosystems, vol. 40. New York: Marcel Dekker; 2003. p. 683–706.
[138] Hirano S, Suzuki KT. Exposure, metabolism and toxicity of rare earth and related compounds. Environ Health Perspect 1996;104:85–95.
[139] D'Haese PC, Spasovski GB, Sikole A, Hutchison A, Freemont TJ, Sulkova S, Swanepoel C, Pejanovic S, Djukanovic L, Balducci A, Coen G, Sulowicz W, Ferreira A, Torres A, Curic S, Popovic M, Dimkovic N, Debroe ME. A multicenter study on the effects of lanthanum carbonate (Fosrenol™) and calcium carbonate on renal bone disease in dialysis patients. Kidney Int 2003;63:73–8.
[140] Hutchison HAJ, Albaaj F. Lanthanum carbonate for the treatment of hyperphosphataemia in renal failure and dialysis patients. Expert Opin Pharmacother 2005;6:319–28.
[141] Haley T. Toxicity. In: Gscheidner Jr, KA, Eyring L, editors. Handbook of physics and chemistry of rare earths, vol. 4, North-Holland, Amsterdam, 1979, p. 553. (Chapter 40).
[142] Venugopal B, Luckey TD. Metal Toxicity in Mammals, vol. 2, Chemical toxicity of metals and metalloids. NY: Plenum Press; 1978. p. 101–173 (chapter 3).
[143] Arvela P. Toxicity of rare-earths. Prog Pharmacol 1979;2:71–114.
[144] Pol A, Barends TRM, Dietl A, Khadem AF, Eygensteyn J, Jetten MSM, Op den Camp HJM. Rare earth metals are essential for methanotrophic life in volcanic. Environ Microb 2014;16:255–64.
[145] Joy MS, Finn WF, Randomized. Double-Blind, Placebo-Controlled, Dose Titration, Phase III Study assessing the efficacy and tolerability of Lanthanum carbonate: a new phosphate binder for the treatment of hyperphosphatemia. AJKD Am J. Kidney Dis 2003;42:96–107.
[146] Locatelli F, D'Amico M, Pontoriero G. Lanthanum carbonate Shire. Drugs 2003;6: 688–95.
[147] Ritz E. Managing mineral balance in end-stage renal disease. Nephrol Dial Transplanant 2004;19:1–3.
[148] de Boer JLM, Verweij W, van der Velde-Koerts T, Mennes W. Levels of rare earth elements in Dutch drinking water and its sources. Determination by inductively coupled plasma mass spectrometry and toxicological implications. A pilot study. Water Res 1996;30:190–8.
[149] Haley TJ, Upham HC. Skin reaction to intradermal injection of rare earths. Nature 1963;200:271.
[150] von Lehmann B, Oberdisse E, Grajewski O, Arntz HR. Subcellular distribution of phospholipids during liver damage induced by rare earths. Arch Toxicol 1975;34:89–101.
[151] Rao VR, Rao KVB, Nagabhushanam R. In: Thompson MF, editor. An Indo-United States symposium on 'Bioactive Compounds from Marine Organisms' with emphasis on Indian Ocean. New Delhi: Oxford & IBH Publishing Co; 1991.
[152] Haley TJ, Raymond K, Komesu N, Upham HC. Toxicological and pharmacological effects of gadolinium and samarium chlorides. Br J Pharmacol 1961;17:526–32.
[153] Hunter RB, Walker W. Anticoagulant action of neodymium 3-sulpho-isonicotinate. Nature 1956;178:47–147.
[154] Husain MH, Dick JA, Kaplan YS. Rare earth pneumoconiosis. J Soc Occup Med 1980;30:15–9.

[155] Cochran KW, Daull J, Mazur M, DuBois KP. Acute toxicity of zirconium, columbium, strontium, lanthanum, cesium, tantalum and yttrium. AMA Arch Ind Hyg Occup Med 1950;1:637–50.
[156] Shimomura H, Tsutsumi S, Mizobuchi J, Takuma S, Nakagawa K, Asai Y. Experimental studies on acute toxicity of rare earth elements. Shika Gakuho 1980;80:1303–7.
[157] Rhodia, Rare earth based products. World Wide Web, http://www.rhodia-ec.com/site_ec_us/products/index_products.htm, 2005, Accessed 21. October 2005.
[158] Bruce DW, Hietbrink BE, DuBois KP. The acute mammalian toxicity of rare earth nitrates and oxides. Toxicol Appl Pharmacol 1963;5:750–9.
[159] Barta CA, Sachs-Barrable K, Jia J, Thompson KH, Wasan KM, Orvig C. Lanthanide containing compounds for therapeutic care in bone resorption disorders. Dalton Trans 2007;43:5019–30.
[160] Damment SJ, Shen V. Assessment of effects of lanthanum carbonate with and without phosphate supplementation on bone mineralization in uremic rats. Clin Nephrol 2005;63:127–37.
[161] Freemont A, Denton J. Effects of lanthanum carbonate and calcium carbonate on bone in patients with Chronic Kidney Disease. In National Kidney Foundation Meeting, 28. April - 2. May 2004, USA: Chicago, IL; 2004, Poster.
[162] Haley PJ. Pulmonary toxicity of stable and radioactive lanthanides. Health Phys 1991;61:809–20.
[163] Porru S, Placidi D, Quarta C, Sabbioni E, Pietra R, Fortaner S. The potential role of rare earths in the pathogenesis of interstitial lung disease: a case report of movie projectionist as investigated by neutron activation analysis. J Trace Element Med Biol 2000;14:232–6.
[164] Salonpaa P, Iscan M, Pasanen M, Arvela P, Pelkonen O, Raunio H. Cerium-induced strain dependent increase in Cyp2a-4/5 (cytochrom P 4502a-4/5) expression in the liver and kidneys of inbred mice. Biochem Pharmacol 1992;44:1269–74.
[165] Salas M, Tuchweber B. Prevention by steroids of cerium hepatotoxicity. Arch Toxicol 1976;35:115–25.
[166] Tuchweber B, Trost R, Salas M, Sieck R. Effect of praseodymium nitrate on hepatocytes and Kupffer cells in the rat. Can J Phys Pharmacol 1976;54:898–906.
[167] Dufresne A, Krier G, Muller J, Case B, Perrault G. Lanthanide particles in the lung of a printer. Sci Total Environ 1994;151:249–52.
[168] Waring PM, Watling RJ. Rare earth deposits in a deceased movie projectionist. A new case of rare earth pneumoconiosis. Med J Aust 1990;153:726–30.
[169] Emsley J. Nature's building blocks: an A-Z guide to the elements. Oxford University Press; 2011. p. 120–125.
[170] Operator, organizational, direct support, and general support maintenance manual: installation, operation, and checkout procedures for Joint-Services Interior Intrusion Detection System (J-SIIDS). Headquarters, Departments of the Army, Navy, and Air Force, 1991, p. 5.
[171] [Online] Hunt S, Associates L. Radioactive Material Safety Data Sheet, available at: http://ehsrms.uaa.alaska.edu/RSDSPm147.pdf.
[172] [Online] Facts About Samarium, Live science, available at: http://www.livescience.com/38162-samarium.html.
[173] Haley TJ, Komesu N, Colvin G, Koste L, Upham HC. Pharmacology and toxicology of europium chloride. J Pharm Sci 1965;54:643–5.
[174] [Online] Lenntech, B.V. Europium (Eu)-Chemical properties, health and environmental effects. Lenntech Periodic Table, 2011, available at: http://www.lenntech.com/periodic/elements/eu.htm.
[175] Murphy KJ, Brunberg JA, Cohan RH. Adverse reactions to gadolinium contrast media: A review of 36 cases. AJR. Am J Roentgenol 1996;167:847–9.
[176] Penfield JG, Reilly Jr RF. What nephrologists need to know about gadolinium. Nat Clin Pract Nephrol 2007;3:654–68.

[177] [Online] Questions and Answers on Magnetic resonance imaging. International Society for Magnetic Resonance in Medicine, available at: http://www.ismrm.org/special/EMEA2.pdf.
[178] [Online] Information on Gadolinium-Containing Contrast Agents. US Food and Drug Administration, Available at: http://www.fda.gov/Safety/MedWatch/SafetyInformation/SafetyAlertsforHumanMedicalProducts/ucm150564.htm.
[179] Gray T. The Elements, A visual exploration of every known atom in the universe. Black Dog & Leventhal Publishers; 2009.
[180] Thomsen HS, Morcos SK, Dawson P. Is there a causal relation between the administration of gadolinium-based contrast media and the development of nephrogenic systemic fibrosis (NSF). Clin Radiol 2006;61:905–6.
[181] Grobner T. Gadolinium- a specific trigger for the development of nephrogenic fibrosing dermopathy and nephrogenic systemic fibrosis. Nephrol Dial Transplant 2006;21:1104–8.
[182] Hammond CR. The Elements. In: Handbook of chemistry and physics. 81st ed. CRC press; 2000. ISBN 0-8493-0481-4.
[183] [Online] Dierks, S. Dysprosium. Material Safety Data Sheets. Electronic Space Products International, January 2003, available at: http://www.espi-metals.com/msds%27s/Dysprosium.htm.
[184] [Online] Dierks, S. Dysprosium Chloride. Material Safety Data Sheets. Electronic Space Products International, January 1995, available at: http://www.espi-metals.com/msds%27s/Dysprosium%20Chloride.htm.
[185] [Online] available at: http://nautilus.fis.uc.pt/st2.5/scenes-e/elem/e06740.html.
[186] Emsley J Erbium. Nature's building blocks: an A-Z guide to the elements. Oxford, England, UK: Oxford University Press; 2001. p 136–139, ISBN 0-19-850340-7.
[187] Haley TJ, Koste L, Komesu N, Efros M, Upham HC. Pharmacology and toxicology of dysprosium, holmium, and erbium chlorides. Toxicol Appl Pharm 1966;8:37–43.
[188] Haley TJ. Pharmacology and toxicology of the rare earth elements. J Pharm Sci 1965;54:663–70.
[189] Houk RS, Fassel VA, Reach GD, Svec HJ. Inductively coupled argon plasma for mass spectrometric determination of trace elements. Anal. Chem. 1980;52:2283–9.
[190] Krebs, R.E. The history and use of our earth's chemical elements: a reference guide, 2006, ISBN 978-0-313-33438-2.
[191] Gale TF. The embryotoxicity of ytterbium chloride in golden hamsters. Teratology 1975;11:289–95.
[192] Ivanov VG, Ivanov GV. High-temperature oxidation and spontaneous combustion of rare-earth metal powders. Comb Exp Shock Waves 1985;21:656–9.

CHAPTER

4

Overview of Lanthanides Series Determination

As previously discussed, lanthanides are of great interest because of their specific characteristics that can be used in several industries, material sciences, catalysis, medical diagnostic and therapeutic applications, geology, agriculture, and in the nuclear fuel cycle. Parallel to their valuable applications, lanthanides can cause negative effects on ecosystems and even threaten the living organisms. Because of their extensive usage, more and more lanthanide compounds enter the air, water, wastewater, and soil and consequently accumulate in living organisms, find their way into the food chain, and eventually may enter the human body.

In Chapters 1, 2, and 3, lanthanides' industrial and geochemical applications were introduced. Further, the biological effects of the elements on humans, animals, and plants were reviewed in some detail. In view of these facts, the monitoring and determination of trace amounts of lanthanides in various matrices is of increasing significance, and several techniques have hence been developed and used to this end.

Lanthanides can be found in a variety of samples (Fig. 4.1), of geological, environmental, industrial, biological and clinical, and food origins as individual, or mixtures of different, elements. The samples can also be divided into the two classes of solid and liquid (solutions).

Because of the various influences of each element, and considering the chemical and physical similarities shared in the family, the capability of analytical methods to selectively determine individual elements in mixtures of lanthanides is of crucial value. Additionally, fast determination and online monitoring of lanthanides by simple methods, especially in biology, is of great importance in today's world. Furthermore, determination of lanthanides without any time-consumed preliminary separation becomes a higher priority on the scientific agenda of the world's industrialized nations.

Few wet chemical methods, namely, volumetric and gravimetric procedures, are available to determine lanthanides in solutions. These methods, however, suffer from the intrinsic disadvantage of being unable

Lanthanides Series Determination by Various Analytical Methods. http://dx.doi.org/10.1016/B978-0-12-804704-0.00004-9

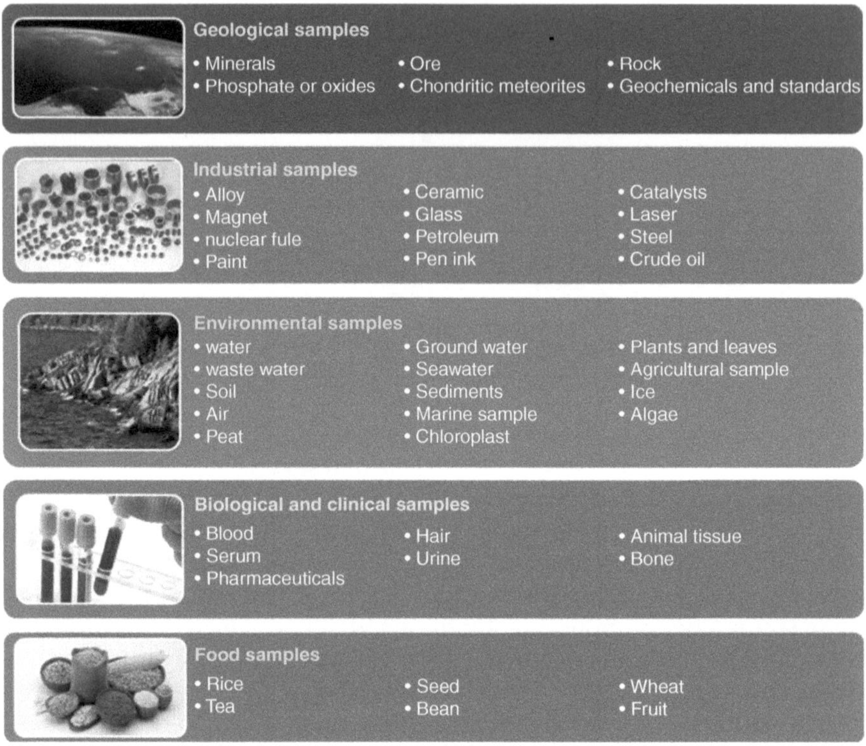

FIGURE 4.1 Various types of real samples containing lanthanides.

to selectively measure the elements, high detection limits, and low sensitivity, accuracy, and precision, which causes instrumental methods applied for the determination of individual lanthanides. Fig. 4.2 presents a schematic classification of the different instrumental techniques applied to the determination of lanthanides to date.

Historically speaking, rare earths (later called lanthanides) were not commonly used during the 1950s and 1960s, because of the nature of the industries, which had not entered the modern high-tech era yet. In those times, the chemistry and effects of lanthanides were often ignored, which might also be attributed to the scarcity of the elements. During this period, the analyses tended to measure the total content of lanthanides in samples without any efforts on the selective evaluation of each element. The analytical methods were, hence, gravimetric, colorimetric, or volumetric in nature. Of course, during the 1950s, paper chromatography was used for the separation and determination of light lanthanides.

Increased application of lanthanides in later years, however, not only led to emission of lanthanide species into the natural environment but also changed the nature of the lanthanide compounds found in nature

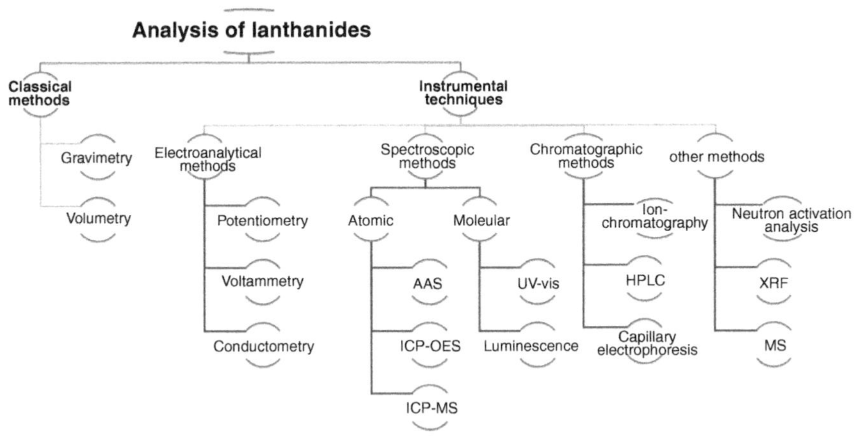

FIGURE 4.2 Various analytical techniques used for measurement of lanthanides.

in terms of increased solubility. Hence, the compounds found in nature were now more soluble compared with the naturally occurring lanthanide species, and the chances of the ions' finding their way into the living organisms and human body was greatly increased, and so did the need for the selective monitoring of the elements for the precise evaluation of their physiological effects. This has been of special importance, because of the similarities of the lanthanides with the biologically important Ca^{2+} ions, giving them the potential to replace the ion in biological systems and hence a great potential to disturb such systems (Chapters 1 and 3). This has further increased the importance of determination and quantification of lanthanides in trace amounts.

Hence, since the 1960s, more selective methods like more advanced analytical and chromatographic techniques have been used for analysis of the elements. The application of column and thin-layer chromatographic techniques, which is reflected by the large number of publications in this field, is an example of the beginning of the trend toward more selective specification and determination of lanthanides (Chapter 7). In general, liquid chromatographic techniques can be regarded as the most conventional techniques used for the separation of lanthanide samples. Yet the most efficient method for separation of lanthanides is the high-performance liquid chromatography (HPLC), which was first used to this end in the early 1970s, after its advent in the late 1960s. Other chromatographic procedures used for the separation of lanthanides include cation and anion exchange, ion-pair, and ion chromatographic techniques. Reports on the application of gas chromatography based on the use of complexes of lanthanides with dipivaloylmethane also exist. Further chromatography and electrophoresis techniques used for the determination of lanthanides include thermochromatography, extraction chromatography, supercritical fluid chromatography (SFC), zone electrophoresis, and isotachophoresis.

The late 1960s was also the time when ion-exchange resins were developed for separation and consequently more selective determination of the elements (Chapter 7). This mechanism also represented the onset of the application of methods for the purification of lanthanides, which in turn increased the production and industrial application of the element. This increased the chances of the appearance of these elements in the environment, because of the large-scale production and application of the elements, and consequently necessitated more accurate analysis of the elements in different samples.

As a consequence of the large-scale production of the elements during the past three to four decades, and the light of the unique properties of each element and its widespread applications in industry, agriculture, and medicine (Chapters 1 and 2 Chapters 1, 2), the importance of the analysis of individual elements was further highlighted.

Another indicator of the increased need for the determination of lanthanides is the trend of the application of more precise and selective techniques such as atomic absorption techniques (Chapter 6). The need for precise and accurate determination led to the use of graphite furnace atomic absorption (GFAA) and electrothermal atomization (ETA) methods, instead of flame atomization systems, during the 1970s and 1980s. AAS and its variations, however, proved to have low sensitivities in the analyses of lanthanide samples, especially in metallurgical samples and electronic materials, and hence interest in the technique declined in the 1990s. AAS was thus replaced with more modern and accurate techniques such as inductively coupled plasma optical emission spectrometry (ICP-OES), inductively coupled plasma mass spectrometry (ICP-MS), neutron activation analysis (NAA), and X-ray fluorescence (XRF) for the determination of lanthanides.

However, as already described, because of the similarly outer filled 4f orbital in all elements, and the consequent lanthanide contraction, which leads to minor differences in the ionic radii ranging from 1.06 Å in La^{3+} to 0.85 Å in Lu^{3+}, the physical and chemical properties of the element are similar. This makes the analysis of mixtures of the elements especially difficult because of numerous interferences and coincidences. The problem was worsened by the trace amounts of lanthanides in most samples and the presence of interfering ions (including other lanthanides). Consequently, ICP-OES and ICP-MS techniques were hyphenated with chromatographic separation (Chapter 6) to further boost the capability to analyze the species through preconcentration and elimination of interfering species.

This way, the selectivity and sensitivity of the detection procedures were greatly enhanced (Chapters 6 and 7). The most successful hyphenated technique is the HPLC-ICP-MS, with detection limits lower than those of HPLC-ICP-AES. Other measures, reportedly taken in the area of spectrometric analyses, include the use of chemical (e.g., halogenating) reagents for electrothermal vaporization and lowering the detection limits, as well as the selective volatilization of lanthanides from the matrix.

Another very sensitive technique used in the analysis of samples, when appropriate, is the isotope dilution mass spectrometry (IDMS) procedure (Chapter 7). This technique enjoys the advantages of high selectivity and elimination of spectral interference, which is a major problem in spectrometric methods. Other modern techniques used to this end are neutron activation analysis (NAA) and X-ray fluorescence (XRF). The former technique requires complicated instrumentation, and its analysis times are very long; however, it offers very high selectivities (low detection limits) in cases where it is applicable (as described in Chapter 7, the method does not show good selectivity with all lanthanides). The latter, on the other hand, requires little analysis times and is capable of very good qualitative and quantitative analysis of lanthanides, and it has found application in this area (Chapter 7). A further advantage of XRF is its nondestructive nature, which makes it a unique tool for analysis of solid samples. The method, however, suffers from the inability to differentiate some spectral overlaps (L lines of REE and the K lines of elements with medium atomic numbers), leading to limitations that sometime render the method incapable of the analysis of some lanthanide species.

Variations of XRF are the energy-dispersive and wavelength-dispersive x-ray fluorescence (EDXFR and WDXRF, respectively). The former suffers from low sensitivities, whereas the latter has more reasonable sensitivity values. A third variation of the method is the total reflectance XRF (TRXRF).

A general conclusion on the methods applied to the determination of lanthanides during the past decades could be that ICP-MS, ICP-OES, NAA, and XRF are the most conventional methods used to this end (Fig. 4.3).

One should note that as will be seen in detail later in Chapter 7, the different on-line and off-line separation and preconcentration techniques used to improve sensitivity and selectivity of the analyses increase the time and costs as well as use of chemical reagents and solvents and the chances of human-induced errors.

Other techniques that will be covered in the following chapters include molecular absorption spectrometry (MAS), spark source mass spectrometry (SSMS), particle-induced X-ray emission (PIX), direct current plasma atomic emission spectrometry (DCPAES), secondary ion mass spectrometry (SIMS), scanning electron microscopy (SEM), and electron probe microanalysis (EPM).

As can be seen in Fig. 4.3, another important class of analytical techniques used for the determination of lanthanides, as well as many other

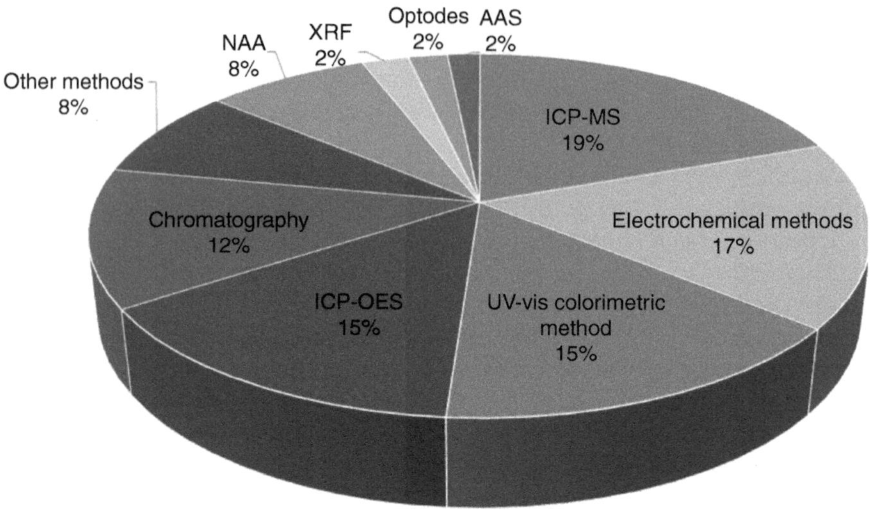

FIGURE 4.3 A statistical illustration of the share of various methods in the analyses of lanthanides.

analytes, include the electroanalytical techniques (Chapter 5). Electrochemical sensors, for instance, offer a wide range of advantages, including being portable, requiring very small analysis times, good detection limits, and low analysis costs. A rivaling technique for potentiometric sensors include the optical sensors. Voltammetric detection procedures also constitute another class of accurate electroanalytical techniques, which can be used for the direct analysis of electroactive lanthanides, or the indirect determination of the nonelectroactive species (Chapter 5).

A unique advantage of electrochemical techniques is their potential for application to online monitoring of analytes (e.g., lanthanides) during different processes. The increasing number of reports in the field, as well as the figures of merit the techniques enjoy, allows concluding that miniaturized and portable electrochemical devices based on modern instrumentation shall influence the nature of selective analyses of lanthanides in the future.

The following chapters tend to provide the reader with ample, yet not very exhaustive, information on the principles of the application of each of the mentioned techniques. In doing so, a simplistic approach has been adopted, and the main purpose has been to provide the reader with as much information as might be needed to comprehend the strengths and limitations of each method, as well as the technical terminology used in the review sections.

A huge part of each chapter is further dedicated to an exhaustive review of the publications on the analysis of lanthanides using each technique. In case of techniques where instrumentation is the predominant element, the figures of merit (e.g., the detection limit, sensitivity, linearity ranges, etc.) have been included to provide the reader with an insight on the results acquired in each case. Some procedural detail has also been included where necessary.

On the other hand, in cases like optical sensors (optodes), potentiometric electrodes, or colorimetric methods, where ionophores, luminescence agents, or chemical reagents have been used, more information on the procedures has been included. This can certainly offer the reader a deep practical insight as to the design and construction of such analytical devices. The descriptive information on the mechanisms of action, where necessary, is of further help to those interested in the details of each area. Naturally, in these cases, the figures of merit have also been included to help the reader compare the methods quantitatively.

CHAPTER

5

Electrochemical Determination of Lanthanides Series

OUTLINE

Lanthanides Series Determination by Various Analytical Methods. http://dx.doi.org/10.1016/B978-0-12-804704-0.00005-0

Electrochemistry is the science and practice of studying chemical reactions at the interface of a conducting element and an ionic conductor. It includes methods that are based on the determination of analytes through measuring the electrochemical potentials and/or currents in a so-called electrochemical cell that contains the analyte. In general, the electrochemical methods can be categorized into interfacial and bulk methods. Fig. 5.1 shows classification of electroanalytical methods.

The most common electrochemical method used for the determination of lanthanides is potentiometry. Potentiometric methods, based on the design and application of selective sensors for techniques, shall be covered in the first part of this chapter, while an overview of the application of potentiostatic methods for the determination of the elements shall be given in the second part of the chapter.

Overall, the four elements that make up a basic electrochemical cell are as follows:

- **The electrode,** which is an electron-conducting element, on which the electrochemical reactions and charge transfer occurs. A minimum of two conductive electrodes are present in an electrochemical cell, namely, the anode, where the oxidation reaction takes place, and the cathode, on which the reduction reaction happens. The materials used for making electrodes must be sufficiently conductive, such as graphite, semiconductors, and even conductive polymers.

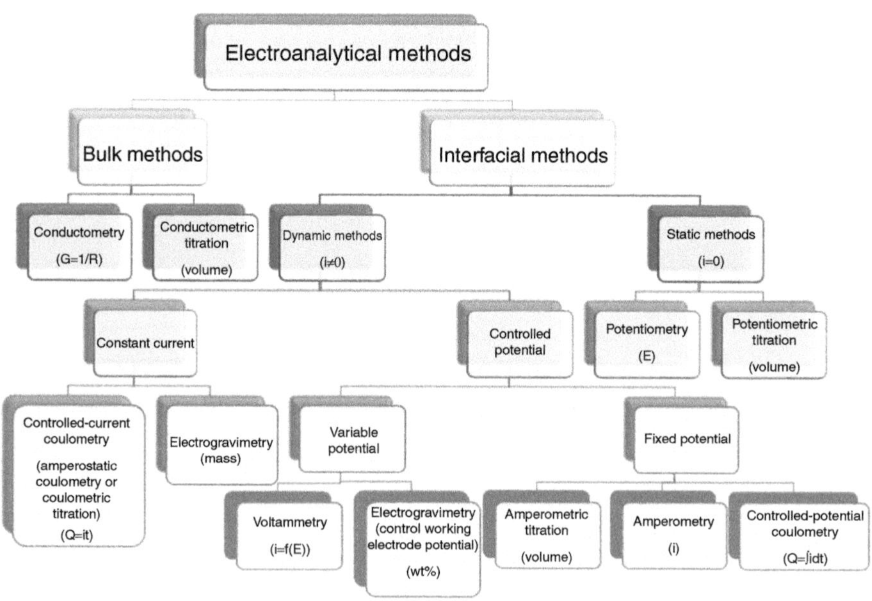

FIGURE 5.1 Electroanalytical methods classification.

- **The electrolyte**, which is an ionic conducting medium containing freely moving ions, can transport between the electrodes.
- **An external electronic circuit** connecting the electrodes, which should be able to detect or control the parameters under study (i.e., current, voltage, charge, or conductance).
- And, finally, **a housing compartment** for the electrodes and the electrolyte.

Electrochemical cells are divided into two classes: electrolytic and galvanic cells, the former requiring an externally applied voltage for its reactions to proceed (e.g., a typical electrolysis) and the latter spontaneously creates voltage based on a chemical reaction (as in a battery). Based on the definitions, the cathode and anode in a galvanic cell are positive and negative in charge, respectively, whereas in the electrolytic cells, the opposite is true. (The anode is positive and the cathode is negative.)

The cell potential, or in other terms its electromotive force (emf), is calculated based on the standard electrode potentials. These standard potentials are, by definition, the reduction potentials of the cathodic and anodic reactions as opposed to a standard hydrogen electrode (SHE) [1]. Because it is possible for reactions to be reversible (changing the role of an electrode depending on the relative oxidation/reduction potential of the other electrode), the oxidation potential for a particular electrode reaction can be achieved by simply making its reduction potential negative.

The standard cell potential is calculated using the standard electrode potentials for both electrodes, given the fact that the electrode with the smaller potential shall act as the anode whereas the opposite is true for the other electrode. The cell potential (emf) is given by

$$E^{\circ}_{\text{cell}} = E^{\circ}_{\text{red(cathode)}} - E^{\circ}_{\text{red(anode)}} = E^{\circ}_{\text{red(cathode)}} + E^{\circ}_{\text{ox(anode)}} \quad (5.1)$$

The electrical energy produced by a cell is equal to its emf times the electric charge transferred through the external circuit, that is,

$$\text{Electrical energy} = E_{\text{cell}} C_{\text{trans}} \quad (5.2)$$

E_{cell} being the emf in volts (V) and C_{trans} being summation or integral of the cell current over time expressed in coulombs (C). This latter value can be calculated by multiplying the total number of the transferred electrons (in moles) by the Faraday's constant (F).

The emf of a cell is maximum at zero current, and can hence be used to calculate the maximum electrical energy that the cell can potentially produce, through the equation below:

$$W_{\text{max}} = W_{\text{electrical}} = -nFE_{\text{cell}} \quad (5.3)$$

where work into the system is considered positive. This value also equals the free energy of the cell by definition:

$$\Delta G = -nFE_{\text{cell}} \tag{5.4}$$

If the cell potential is positive, the change in Gibbs free energy shall be negative (as in galvanic cells). A further relation between the equilibrium constant of the reaction, K, and the Gibbs free energy for an electrochemical cell is

$$\Delta G^\circ = -RT\ln K = -nFE^\circ_{cell} \tag{5.5}$$

Which gives

$$E^\circ_{cell} = \frac{RT}{nF}\ln K \tag{5.6}$$

or

$$E^\circ_{cell} = \frac{0.0591}{n}\log K \tag{5.7}$$

In case the concentrations of the reactants are different from the standard values, the cell potential will deviate from the standard potential.

$$\Delta G = \Delta G^\circ + RT\times\ln Q \tag{5.8}$$

ΔG being the change in the Gibbs free energy, ΔG° representing the standard cell potential, T being the temperature in Kelvin (K), R being the universal gas constant, and Q being the reaction quotient.

Based on these definitions, the German chemist Walther Nernst suggested a mathematical model for relating the effect of the reactant concentration on the electrochemical cell potential. Through including the contribution from the electric potential on charged species, he used the equation below, which is a combination of equations 5.5 and 5.8, to get:

$$nF\Delta E = nF\Delta E^\circ - RT\ln Q \tag{5.9}$$

and

$$\Delta E = \Delta E^\circ - (RT/nF)\ln Q \tag{5.10}$$

where n represents the number of electrons/moles of the product, F is the Faraday constant, R is the universal gas constant, T is the temperature in Kelvin, and ΔE and ΔE° are the formal and standard cell potentials. Considering the temperature to be 25°C and R to be 8.3145 J/(K mol) we have:

$$\Delta E_{cell} = \Delta E^\circ_{cell} - \frac{0.0591}{n}\log Q \tag{5.11}$$

which can be used for the electroanalytical measurements of concentrations based on the values of the electrochemical potentials at the working electrode. As a result, the design of the working electrodes of special characteristics has attracted a great deal of focus and has led to the development of an independent branch of electrochemical sensors and biosensors, which reportedly offer fast, rapid, continuous, and sensitive routes for the electrochemical determination of analytes.

POTENTIOMETRIC DETERMINATION METHODS

The basis of potentiometric methods is the measurement of the electrochemical potentials of the solutions between two electrodes, under zero or low current flows and relating it to the concentration of the species to be determined. Because of the absence of a current in the potentiometric systems, no redox reactions take place within the cell and hence the two electrodes used are not considered as cathode or anode. The electrodes are instead referred to as the indicator (working) and reference electrodes, and the signal recorded is actually the difference between potentials of the indicator and the reference electrodes, shown as follows:

$$E^{\circ}_{\text{cell}} = E^{\circ}_{\text{indicator}} - E^{\circ}_{\text{reference}} \tag{5.12}$$

A potentiometric cell is depicted in Fig. 5.2. The working electrode has a transducing element, which is conventionally an AgCl-coated Ag wire (called the internal reference electrode in addition to a sensing element), which is located at the interface of the electrode with the analyte solution. The other reference electrode is technically referred to as the external reference. The magnitude of the currents during these measurements ranges around 10^{-6} μA.

To conduct the measurements, the working (i.e., an ion-selective electrode) and the external reference electrodes are immersed into the analyte solution and the voltage difference between them is measured, using a DC voltmeter (e.g., a voltmeter with an input resistance of about 10 gigaohm, like the high-accuracy amplifier of a pH or ion meters). Conventional setups require at least 5–10 mL of the test solution, but the rather recent introduction of microelectrodes has made it viable to use smaller volumes.

One of the elements of potentiometric cells is a reference electrode (Fig. 5.3), which should be able to maintain a stable constant potential. The most conventional reference electrodes are based on silver wires coated with a layer of AgCl immersed in a saturated or 1 M potassium chloride solution. Although the change in the concentration of the electrolyte alters the electrode potential, its potential is constant for each concentration.

A second alternative is the mercury/mercurous chloride electrodes, which was very common in the recent past. The electrode has some

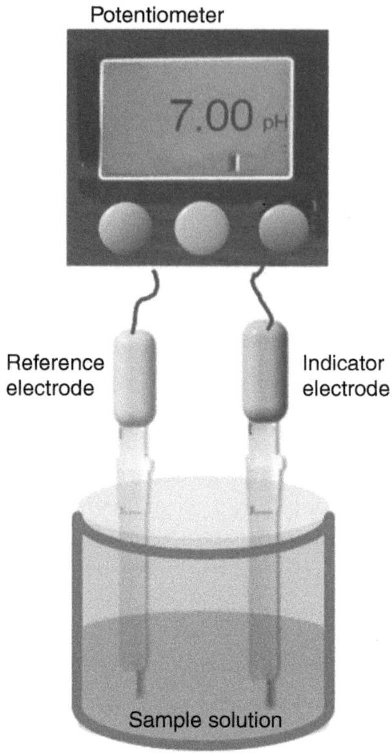

FIGURE 5.2 Picture and schematic illustration of a potentiometric cell.

advantages over the silver/silver chloride electrode in terms of better thermal hysteresis behavior, as well as being less prone to contamination due to the indirect contact between the mercury/mercurous chloride interface and the electrolyte. But it also has the disadvantages of being harder to construct and using the dangerous mercury compounds as opposed to the Ag/AgCl electrodes.

Some modifications on the Ag/AgCl electrodes have led to the construction of double-junction reference devices that enjoy the advantages of both electrodes.

The Working Electrodes

The use of commercially available working electrodes for the potential analyses of many ionic samples is rather common these days. The indicator electrodes designed for responding to specific ionic species are also referred to as ion-selective electrodes (ISEs) or sensors. In contrast with the reference electrodes, the potentials adopted by these during the

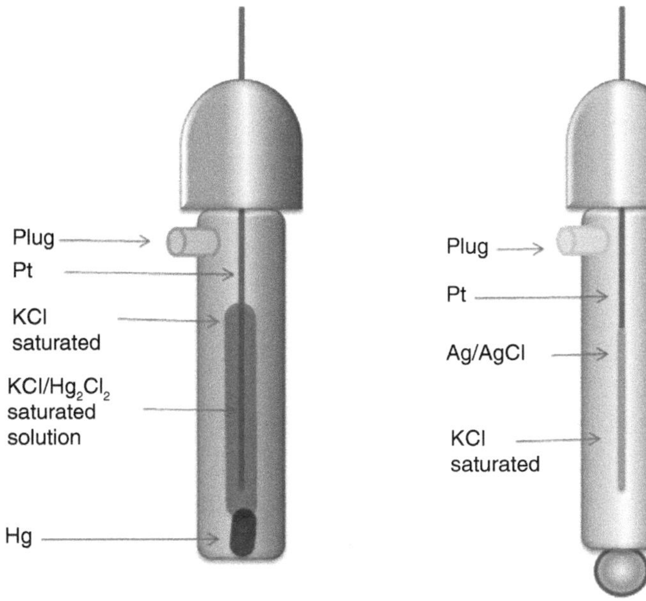

FIGURE 5.3 Scheme of reference electrodes.

potentiometric measurements is directly or indirectly determined by the concentration of the specific ionic species toward which they are sensitive. ISEs are divided into two groups, symmetrical and asymmetrical [1], the former referring to the classical ISEs where the sensing element acts as the interface between the internal and analyte solutions (which leads to larger electrode dimensions), whereas in the latter only the outer surface of the element is in contact with the analyte solution and the interface between the inner surface of the element and the inner reference electrode is in direct contact through a solid phase (as a result, these electrodes can be miniaturized). The two broad classes are subdivided into further classes as shown in Fig. 5.4.

The application of modified PVC membranes as the sensing element in the working electrodes led to the development of a wide range of sensors for different species (Fig. 5.5). The modification of the membranes is achieved through the incorporation of specific ion carriers (or ionophores) in the composition of the plasticized PVC membranes. These ion carriers or ionophores are capable of selective interactions with ionic species.

A plasticized PVC membrane is composed of powdered PVC, a suitable plasticizing solvent, a suitable amount of an ionophore and some ionic additive, which are dissolved in small amounts of tetrahydrofuran (THF) to give a homogenous mixture, which is coated on narrowed tips of Pyrex tubes to form symmetrical ISEs, or on Pt, Cu, or Ag wire or graphite rods; or the gate of ion-selective field effect transistors (ISFETs) to

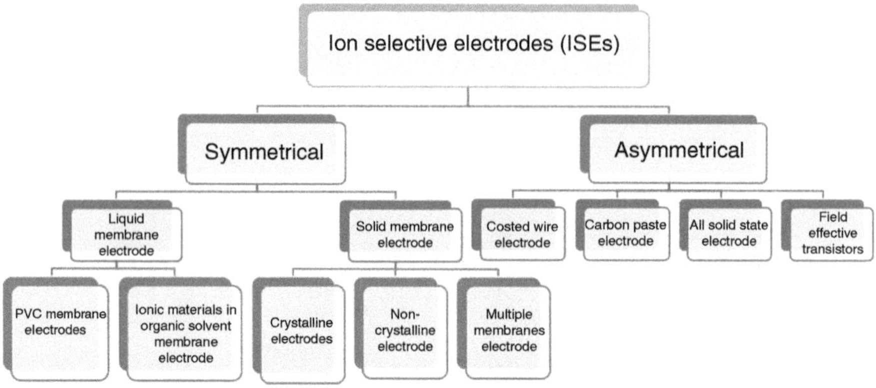

FIGURE 5.4 Classification of ISEs based on the structure of the electrode.

construct asymmetric ISEs. In the case of the symmetric ISEs, both sides of the membrane are next conditioned in contact with solutions of the target ions, whereas in asymmetric devices, only the external surface is subjected to conditioning. A silver/silver chloride electrode is used as the internal reference electrode in symmetric devices, whereas in the case of the asymmetric electrodes, a wire is used to this end [1,2].

In the case of carbon paste electrodes (CPEs), which form another group of ISEs, the electrode is composed of a mixture of graphite powder and the ionophore dispersed in a nonconductive mineral oil to form a paste-like material that is used for construction of the electrodes. Additives like nanostructured materials have also been used to modify the response of the CPEs. The paste is packed in a plastic or Teflon tube while strictly avoiding the formation of air gaps within its matrix, which can increase its electrical resistance to very high values, rendering the electrode useless. A copper wire is inserted into the paste to act as an electrical contact. The interfacial surface of the electrode is smoothed with soft abrasive paper before use.

All solid-state electrodes are also built using conducting polymers as the electrode matrix yielding durable ion sensors, in which the ionophore is covalently bound to the backbone of the conducting polymers [1,2].

ISFETs are another variation of asymmetric ISEs in which the ion-sensing membrane is directly coated on the gate of a field effect transistor (FET). The devices hence have high-input and low-output impedance, which can help monitor charge buildup on the ion-sensing membrane.

An important point is that these sensors show some degree of selectivity toward the target species. This can be due to the mechanism of interaction between the ionophore and the target species that almost leads to interferences with other ions present in the sample matrix, which could be a major source of error. (Enzyme-based sensors, however, are an exception.)

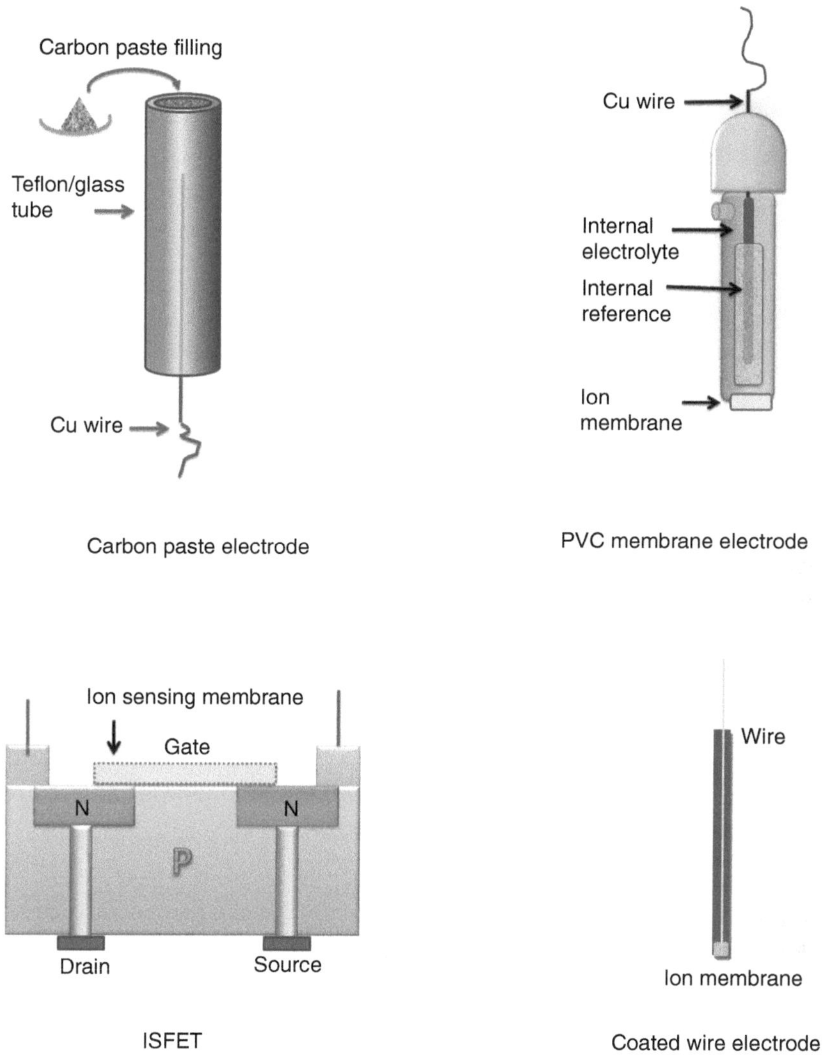

FIGURE 5.5 The structure of some conventional structures for ISEs.

This stems from the specific key-lock mechanism of their interaction with the analyte. This is also the case with gas-sensing electrode, which are actually a combination of conventional ISEs immersed in an electrolyte solution, which is in contact with the target gas through a gas-permeable membrane. The response mechanism includes the diffusion of the target gas through the membrane and its reaction with the internal electrolyte, which produces or consumes ionic species that can be detected with the ionic sensor.

Facts like the availability of various sensing materials that have been used in the construction of different electrodes, the ease of setting up the analytical system and using the devices, the good accuracy and precision of the results, and their robustness and durability has changed the devices into considerable methods for analytical purposes.

The Potential–Concentration Relations

As explained earlier, the Nernst equation relates the potential difference of an electrochemical cell with the chemical activity of the ions present in the solution. In case the electrode is an ISE, the ion-selective membrane is the most critical element in the setup. Once an ion is sensed by this element and the interaction between the layer and the ion reaches an electrochemical equilibrium, the phases at the two sides of the membrane will adopt two different potentials. The equilibrium state is a condition where rates of transfer of the target ions from the solution to the membrane and the opposite are equal, and hence there exists a near to zero current regime. The equilibrium potentials of the two phases are related to the concentrations of the ionic species on each side. Hence, if the ion-selective layer shows a good selectivity toward one ionic species in comparison with the other ones possibly present in the analyte solution, the equilibrium potential difference shall only be a function of the concentrations of the target ion on the two sides of the membrane.

Assuming a_1 and a_2 as the activities of the target ion in phases 1 and 2 respectively, the Nernst equation illustrating the relation between the potential differences (E) across the membrane can be written as

$$E = E^0 \pm \frac{RT}{n_x F} \ln\left(\frac{a_2}{a_1}\right) \tag{5.13}$$

where n_x is the charge of the target ion. Because the activity of the inner solution of the electrodes (a_1) is constant, a_2 can be simply measured through measuring the potential, so the equation can be rewritten as

$$E = E^0 \pm \frac{RT}{n_x F} \ln\left(\frac{a_x}{a_1}\right) \cong E = \text{const.} + S.\log(a_x) \tag{5.14}$$

where $S = 59.16/n_x$ (mV) at 298 K. If the internal and external reference electrodes (in phases 1 and 2 on the two sides of the membrane) are the same, the difference between the electrochemical potentials of the two phases is measured more easily. One could simply say that the difference is actually the difference between the electrochemical potential of the ion-selective and the external reference electrode present in the test solution.

Characteristics of a Potentiometric ISE

The measuring range, detection limit, response time, selectivity, lifetime, accuracy, and precision of the response of an ISE are among the most important characteristics considered when evaluating these devices [1,2].

The measuring or linearity range of an ISE indicates the concentration (activity) range, in which the potential–concentration logarithm relationship is linear. In the case of ISEs, the range is usually between 1 to 10^{-6} or even 10^{-7} M, but lower concentrations have also been reported. The upper and lower limits of detection of a sensor are calculated through extrapolating the linear parts of the calibration curve at both ends (Fig. 5.6). The two limits indicate the points where the responses of the ISE begin to considerably deviate from the Nernstian behavior.

The detection limits of the ISEs normally ranges around 10^{-5}–10^{-6} M, but even subnanomolar values have been reported using metal buffers, which eliminate the interferences in dilute solutions; cation-exchange resins in the inner filling solutions of ISEs, which help stabilize the activity of the target ion at a low level; lipophilic particles like silica gel 100 C18-reversed in the membrane compositions; sandwich membranes, and so on [3–6].

The deviation at the upper limit of detection of an ISE is due to the disturbed perm-selectivity of the ion-selective element, because of the increases in the concentration of the target ion, which increases the portion of the ionophores in complexed form. This leads to the coextraction of the counter ion together with the target ion and hence the response that is lower than that expected for Nernstian behavior.

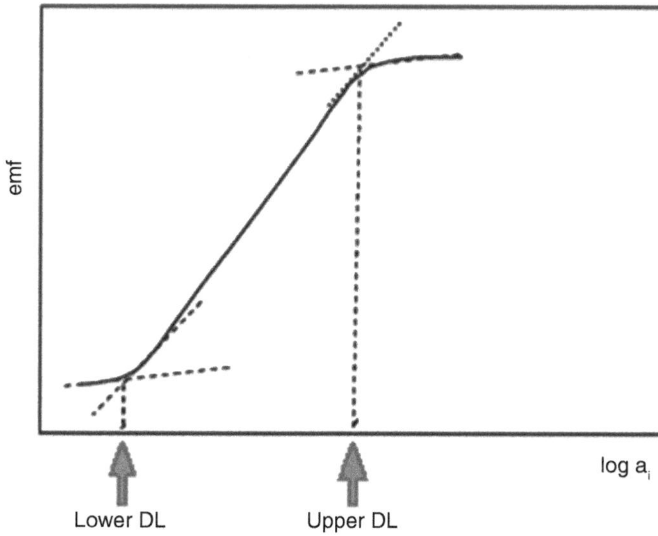

FIGURE 5.6 Typical calibration curve of a potentiometric electrode.

The response time is the indicator of the time required for achieving a stable response from the sensor after inserting it into the sample and is evaluated through measurements of the time needed for achieving potential readings in a range within ±0.1 mV of the steady-state potential, at different concentrations. The values usually range from 5 s to 2 min in conventional electrodes and are affected by stirring or flow rates, the concentration and composition of the analyte or the conditioning solutions, previous applications, operating temperature, and so on.

The selectivity of an ISE is a measure of the response of the electrode to the target ion as compared to the other ions, which might cause changes in the response of the device. The other ion is referred to as the "interfering ion." This selectivity can be regarded as the most prominent characteristic of an ISE. In the case of polymeric membrane sensors, the selectivity stems from the competition between the extraction of the target and interfering ions into the membrane. The extraction of the interfering ions is expected to be less, and hence the electrode response is more the result of the target ions extracted into the membrane. Selectivity is expressed as the logarithm of K_{xy} or the selectivity coefficient, the negative values of which indicate that the target ion is preferred over the interfering species, whereas positive values show that the interfering ion is preferred to the target ion. The most common methods used for the evaluation of the selectivity coefficients include the following [5]:

1. The separate solution method (SSM)
2. The mixed solution method (MSM)
   a. Fixed interference method
   b. Fixed primary method
3. The matched potential method (MPM)
4. The unbiased selectivity coefficients

For details on the procedures in each method, one can consult different references [1,2]. The average lifetimes of polymeric ISEs range from 4 to 10 weeks, during which the response of the electrode is reproducible. The response decreases and fluctuates more after this period, in addition to the increases in the lower detection limit. This is caused by loss of the ingredients of the polymeric film due to leaching. Lifetimes of asymmetric electrodes, however, exceed those of the symmetric ones.

As another factor used for evaluating the merits of the ISEs, the accuracy, precision, or reproducibility of the measurement conducted by them are very important and can depend on various factors. The errors in this method mostly arise from the slope of the calibration curve. As an example, in the case of an ISE for a monovalent ion, which should have a slope of 55 mV/decade of concentration, some ±1 mV error in measuring the potential can lead to errors of about ±4% in the resulting concentration. This error actually doubles to about 8% for bivalent ions. A point to mention is that because of the decrease in the sensitivity of ISEs, reflected by the drop in the slope at both ends of the calibration curve, the error is higher than average at these areas.

When working with an ISE, one should always bear in mind the effects of the interfering ions, the ionic strength of the solution on the relationship between the activity and concentration of the species present therein, as well as the changes in the electrode response on repeated use. The effects of the ionic strength are somehow compensated using ionic strength adjustment buffers (ISABs) in the standards used, for preparing the calibration data and the sample solutions. The overall idea here is to avoid potential drifts due to inhomogeneous ionic strengths. Another measure to this end is the preconditioning of the working and reference electrodes and the appropriate standard solutions before use.

Potentiometric Measurement

The methods used for the determination of different species through the use of ISEs can be divided into the subgroups of direct, incremental, and titration potentiometric techniques. The incremental methods are further subcategorized into standard addition, sample addition, and sample subtraction techniques. The techniques shall be covered in some detail below, but primarily one should be familiarized with the general practical considerations that should be considered when using these techniques.

It should be also noted that the methods can be used only for the evaluation of the concentration of water-soluble species, but working with organic ions also has the risk of contaminating the ISEs. Another issue of importance is that although the devices are expected to be selective toward one specific species, sample pretreatment is often very useful in getting good results. Also, the stirring of the analyte solution might help reach shorter response times and avoid formation of concentration gradients, but this might also lead to drawbacks, such as increased chance of contamination.

Direct Potentiometry

The simplest potentiometric procedure, direct potentiometry is based on measuring the potential response of the electrode in the analyte solution and using the readings for the calculation of the concentration by the calibration curve. The method is applicable to measuring of large numbers of samples over a wide concentration ranges.

Incremental Potentiometry

Standard Addition

The technique requires precise information on the volume of the sample solution. This is because after reading the initial potential of the sample (E_x), the next step includes the addition of a small but known volume of a standard solution to the sample and reading the potential of the new solution (E_{x+s}), which is given as

$$E_x = Con \pm \frac{RT}{Z_x F} \ln(a_x) \tag{5.15}$$

$$E_{x+s} = Con \pm \frac{RT}{Z_x F} \ln\left(\frac{a_x \cdot V_x + a_s \cdot V_s}{V_x + V_s} \right) \tag{5.16}$$

is used through the equation below to get to the concentration of the original sample:

$$E_{x+s} - E_x = \pm \frac{RT}{Z_x F} \ln\left(\frac{a_x \cdot \upsilon_x + a_s \cdot \upsilon_s}{\upsilon_x \cdot a_x + \upsilon_s a_x} \right) \tag{5.17}$$

where a_x and a_s represent the activities or concentrations of the sample and the standard solution respectively; υ_x and υ_s are the sample standard volumes, and R, F, and n_x are identical to those in the Nernst value. A plus sign is used for cation measurements and a minus sign is used for the anion samples.

The technique requires observing certain operational considerations like the fact that the concentration of the added standard solution should be enough to lead to noticeable changes in the voltage readings, whereas its volume should not be so large that it causes changes in the ionic strength of the sample.

Sample Addition

The technique is actually a complementary form of the standard addition technique, where the potential response of a relatively large but known volume of a standard solution (E_s) is measured before adding a small, but known, volume of the sample solution and measuring the potential of the new solution (E_{s+x}). The data can then be used for calculating the concentration of the sample using the equation below:

$$E_{s+x} - E_s = \pm \frac{RT}{Z_x F} \ln\left(\frac{a_s \cdot \upsilon_s + a_x \cdot \upsilon_x}{\upsilon_x \cdot a_s + \upsilon_s a_s} \right) \tag{5.18}$$

which is an analogue of the equation above (Eq. 5.17).

There is no need to mention that identical operating considerations as the standard addition technique are required in this case too. The method does not require repetitive measurements and also eliminates the need for the application of ISAB. Also, because of the fact that the electrodes remain in the solution during the measurement process, the changes in the liquid junction potential of the reference electrode and the consequent error are minimized. Using this method, even old electrodes that can of course still produce completely linear response can be used, even in the case of samples with low ionic strengths.

Sample Subtraction

Sample subtraction is actually based on the elimination of the analyte using a chemical reaction. In this method, a small but known volume of the sample is added to a standard solution of another that reacts with the target ion to form a complex or precipitate. Using ISEs that are selective for the second ion, one can indirectly measure the concentration of the target ion for which there might not exist a good ISE.

Potentiometric Titrations

Potentiometric measurements can also be used as very accurate means for indicating the end point of titrimetric measurements. This is due to the sharp changes that may occur at the end point of titrations because of the changes in the nature or physical form of the reactants. For instance, in the titration of Ce^{3+} solutions with EDTA, the concentration of free Ce^{3+} ions gradually decreases on the addition of the chelating agent, whereas that of the complexed form of the ion increases in parallel, so in this case, a Ce^{3+} selective electrode is used as the indicator electrode. The changes in the potential readings will be rather sharp at the end point, where the free Ce^{3+} concentration reaches to undetectable levels, gravely changing the signal. Such methods can also be used for the alternative measurement of the species for which the sensors are not available. For example, the same method can be used for the determination of the concentration of EDTA solution by titrating the solution with a Ce^{3+} solution and using a Ce^{3+} selective sensor.

SELECTOPHORES FOR LANTHANIDES

The development of lanthanide selective ISEs can be regarded as the main challenge in the area of potentiometric sensors, which mainly stem from the closeness of the physical properties of the elements in this group. Scientists first tested ligands with cavities (like crown ethers) as the ionophores for these elements, but their efforts were in vain.

Based on the ionic radii of the lanthanide ions, which vary from 1.032 Å in La^{3+} to 0.861 Å in Lu^{3+} (Table 5.1), there are some differences in properties like the charge density, size, and hydration energy of these elements (e.g., the hydration energies of La^{3+} to Lu^{3+} range from 3278 to 3722 kJ/mol) [6]. In the light of the extensive research by Ganjali et al., finally a solution was suggested to this problem in 2000. They suggested the use of flexible ionophores with semicavities and heteroatoms (mostly S and N) as the donor sites that can form selective complexes with specific lanthanides. This is due to the fine-tuning of the interactions of the ionophores with the

TABLE 5.1 The Data of the Ionic Radii and the Hydration Energy of the Lanthanide Ions

Ion	Radii (Å)	$\Delta H_{hydration}$ (kJ/mol)
La^{3+}	1.032	3278
Ce^{3+}	1.010	3326
Pr^{3+}	0.990	3373
Nd^{3+}	0.983	3403
Pm^{3+}	0.970	3427
Sm^{3+}	0.958	3449
Eu^{3+}	0.947	3501
Gd^{3+}	0.938	3517
Tb^{3+}	0.923	3559
Dy^{3+}	0.912	3567
Ho^{3+}	0.901	3623
Er^{3+}	0.890	3637
Tm^{3+}	0.880	3664
Yb^{3+}	0.868	3706
Lu^{3+}	0.861	3722

species so as to minimize the energy level of the resulting complexes. The complexation behavior of these ionophores is determined by the type, the number, and the location of the donor atoms, as well as the cavity size and flexibility of the ionophores. Based on this idea, rather selective ISEs were designed and constructed for 10 lanthanide elements.

Studies revealed that such ionophores can form stronger complexes with one of the lanthanide cations compared with the rest, leading to selective behaviors toward the specific element. So, finding a suitable ionophore with an optimal semicavity and flexibility for selecting one ionic species can be the main stage of constructing ISEs for lanthanides.

On the other hand, the tendency of the lanthanide ions to form complexes is influenced by several factors. Based on the electronic configuration of such elements, the 4f orbitals are actually inner orbitals compared to 5s and 5p, which means the two outer orbitals shield the inner 4f electrons, greatly influencing many of the properties of lanthanide ions. For instance, the trivalent lanthanide ions are essentially spherical, and their complexation behaviors are very similar to those of the alkali and alkaline earth ions. Further, the unavailability of the 4f orbitals in many chemical reactions leads to rather low crystal field stabilization energies for these elements, which fall in the order of 100 cm^{-1} compared to values of 30,000 cm^{-1} that are common for transition elements, where the d orbital is the outer shell.

Because of these properties, the lanthanide complexes have an ionic nature. Further, the ionic nature of the complexes is due to the ionic radii of ionic forms of these elements, which are rather high in comparison to the trivalent ions, leading to comparatively lower charge densities. The size of the ionic species also leads to higher coordination numbers (>6) in the lanthanide complexes [7].

POTENTIOMETRIC SENSORS FOR LANTHANIDE SERIES

Lanthanum ISEs

There exist some 25 reports on PVC membrane electrodes (PMEs), CPEs, coated graphite electrodes (CGEs), and coated wire electrodes (CWEs) for the potentiometric determination of lanthanum [8–32]. The first lanthanum ISE was developed by Shamsipur et al. in 2002 [8], which were plasticized polymeric membrane and membrane-coated graphite electrodes based on 1,3,5-trithiacyclohexane (**La-1**) as the ionophore. Fig. 5.7 illustrates the

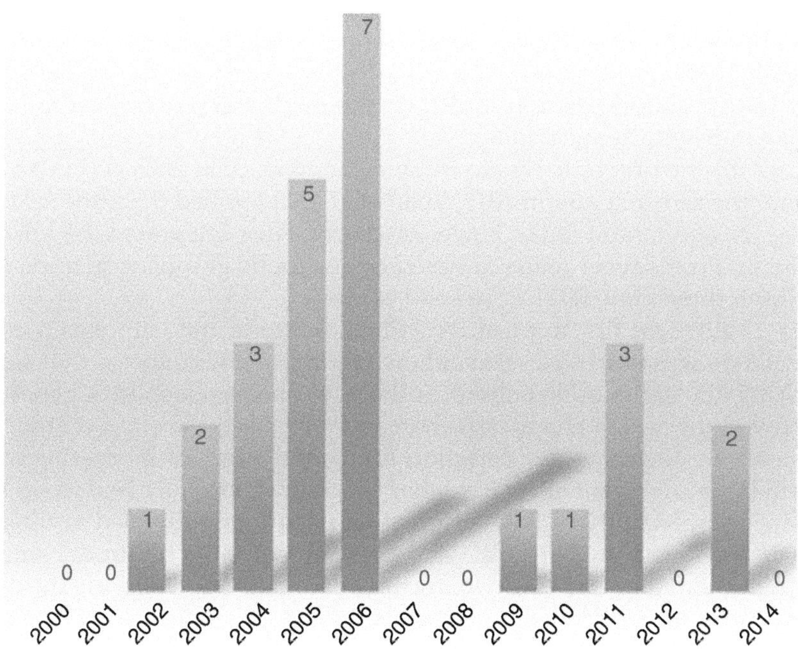

FIGURE 5.7 Historical distribution of the reports on the development of La^{3+} sensors.

historical distribution of the reports on the development of lanthanum sensors over time.

Table 5.2 further provides the details on the nature, composition, and characteristics of the lanthanum sensors reported so far. The linear range (LR), detection limit (DL), Nernstian slope, pH range, response time (RT), lifetime (LT), the most interfering ions (II) considering log $K_{sel} \geq -2.5$ are summarized.

Fig. 5.8 shows the chemical structure of the lanthanum selective ionophore. As previously discussed in this chapter, a good selectophore for lanthanide series should demonstrate a semicavity (not to be a rigid cycle) encompassing soft to intermediate heteroatoms (mostly S and N as donor atoms) and a high flexibility in its structure. Such structure can easily template itself to correspond to the exact size of the selected lanthanide and form a stronger complex, which is more stable. Also, because of high values of lanthanide solvation energy, the selectophores should be able to form a strong complex, which leads to the efficient extraction of the ion from the solution leading to the good selectivity of the sensors.

Among the 22 selectophores reported for lanthanide ions, the cyclic ones with a cavity, that is, La-1, La-2, La-3, La-6, La-17, and La-21, shows low selectivity. Similarly, La-7, La-10, La-15, La-20, and La-22, though having several S and N donating atoms, still lead to low selectivity because of their inability to form well-fitting templates around lanthanum (based on the results in Table 5.2). In the case of La-5, La-9, and La-11, regardless of the semicavity and heteroatoms in the structure of the ligands, the rigidity of the molecules originating from the steric hindrances still cause low selectivity.

La-4, La-8, La-12, La-13, La-14, La-16, La-18 and La-19, on the other hand, show good selectivity, based on the mechanism described for flexible ligands with semicavities.

La-7 for the first time was used by Akhond et al. in 2005 [14] in a PVC membrane sensor. Later in 2013, Yuan et al. [32] applied the same selectophore for lanthanum ion in a new assembly. They integrated the selectophore in a transducer material based on magnetic graphene hybrids and 2,2-dithiodipyridine (DTDP) in solid contact.

La-13 also was first applied for a PME electrode and later was used in CPE. La-19 was used as a selectophore in PME and CPE too.

Using the same selectophore in the asymmetric electrodes generally improved the sensor characterization and decreased the effect of interfering ions. By lowering the detection limits, the effect of interfering ions, which is an important effect in higher concentrations, may be decreased.

The nonselectivity of La-1, originating from its rigidity, led to the fact that it was used for both Ce^{3+} and La^{3+} ion selective electrode, simply through changing the polarity of the membrane through altering the plasticizer from acetophenone with a dielectric constant of 17.4 for inducing selectivity of La^{3+} to benzyl acetate (BA) with a lower dielectric constant (5.1) as the plasticizer for inducing Ce^{3+} selectivity. Although this shows

TABLE 5.2 The Characterization of the Reported Lanthanum Ion Membrane Sensors

Electrode type	Selectophore	Electrode composition	Characterizations of the electrode	Ref.
PME and CGE	La-1 1,3,5-Trithiacyclohexane (TTCH)	PVC: 30% Plasticizer: AP 53% Selectophore: 2% Additive: OA 15%	LR: 8.0×10^{-6} to 5.0×10^{-2} M for PME LR: 4.0×10^{-8} to 1.0×10^{-2} M for CGE Slope: 19.7 ± 0.2 mV/decade DL: 5.0×10^{-6} M for PME DL: 2.0×10^{-8} M for CGE pH 5.0–8.0 RT: 10 s LT: at least 6 months II: Ce^{3+}, Ag^{+}	[8]
PME	La-2 Monoaza-12-crown-4	PVC: 48% Plasticizer: DBP 378% Selectophore: 7% Additive: NaTPB 8%	LR: 3.16×10^{-5} to 1×10^{-1} M Slope: 20.5 ± 1.0 mV/decade DL: 2.0×10^{-8} M pH 3.0–7.0 RT: <15 s LT: >5 months II: K^{+}, Na^{+}, Rb^{+}, Cs^{+}, Cu^{2+}, Fe^{3+}, Cr^{3+}, Cd^{2+}, Zn^{2+}, Ca^{2+}, Pb^{2+}, Sr^{2+}	[9]
PME	La-3 5,14-*N*,*N*′-Hydroxyphenyl-4, 15-dioxo-1,5,14,18-tetraaza hexacosane (NHDTH)	PVC: 10% Plasticizer: DBP 7% Selectophore: 2% Additive: NaTPB 1%	LR: 1.77×10^{-6} to 0.1 M Slope: 20.0 mV/decade DL: 2.0×10^{-8} M pH 3.0–6.5 RT: 15 s LT: >4 months II: NH_4^{+}, Li^{+}, K^{+}, Na^{+}, Rb^{+}, Hg^{2+}, Ca^{2+}, Zn^{2+}, Pb^{2+}	[10]

(Continued)

TABLE 5.2 The Characterization of the Reported Lanthanum Ion Membrane Sensors (*cont.*)

Electrode type	Selectophore	Electrode composition	Characterizations of the electrode	Ref.
PME and CGE	La-4 Bis(2-mercaptoanil) diacetyl (BMDA)	PVC: 30% Plasticizer: AP 66% Selectophore: 2% Additive: NaTPB 2%	LR: 1.0×10^{-5} to 1.0×10^{-1} M for PBM LR: 1.0×10^{-6} to 1.0×10^{-1} M for CGM Slope: 19.7 mV/decade for PBM and CGM DL: 6.5×10^{-6} M for PBM DL: 5.0×10^{-7} M for CGM pH 4.0–8.0 RT: <15 s LT: at least 2 months II: Ce^{3+} for PBM II: None for CGM	[11]
PME	La-5 Bis(thiophenal)phenylen-1, 3-diamine (TPD)	PVC: 35% Plasticizer: BA 54% Selectophore: 8% Additive: NaTPB 3%	LR: 1.0×10^{-7} to 1.0×10^{-1} M Slope: 19.6 mV/decade DL: 2.0×10^{-8} M pH 3.5–9.5 RT: 10 s LT: at least 2 months II: Sm^{3+}, Ce^{3+}	[12]
PME	La-6 Dicyclohexano-18-crown-6 (DC18C6)	PVC: 33% Plasticizer: *o*-NPOE 61% Selectophore: 6%	LR: 1.0×10^{-6} to 1.0×10^{-1} M Slope: 19.0 mV/decade DL: 5.0×10^{-7} M pH 4.0–9.0 RT: <30 s LT: at least 5 months II: Ce^{3+}, Nd^{3+}, Sm^{3+}, Eu^{3+}, Pr^{3+}, Tb^{3+}, Dy^{3+}, Yb^{3+}	[13]

PME	La-7 2,2-Dithiodipyridine	PVC: 30.2% Plasticizer: *o*-NPOE 60.6% Selectophore: 2.4% Additive: OA 6.8%	LR: 7.6×10^{-6} to 2.2×10^{-2} M Slope: 20.0 ± 0.1 mV/decade DL: 3.1×10^{-6} M pH 3.3–8.0 RT: 7 s LT: at least 2 months II: Cu^{2+}, Ce^{3+}	[14]
PME	La-8 *N*-2,4-Dimethylphenyl-*N*′-ethylformamidine (amitraz)	PVC: 30% Plasticizer: BA 58% Selectophore: 8% Additive: NaTPB 4%	LR: 1.0×10^{-7} to 1.0×10^{-1} M Slope: 19.8 ± 0.2 mV/decade DL: 8.0×10^{-8} M pH 3.0–9.0 RT: <8 s LT: at least 2 months II: None	[15]
PME	La-9 Bis(2-methylbenzaldehyde) butane-2,3-dihydrazone (TDSB)	PVC: 30% Plasticizer: BA 60% Selectophore: 6% Additive: NaTPB 4%	LR: 1.0×10^{-5} to 1.0×10^{-1} M Slope: 19.8 mV/decade DL: 7.0×10^{-6} M pH 3.0–9.4 RT: 5 s LT: at least 2 months II: Cu^{2+}	[16]
PME	La-10 Rubeanic acid	PVC: 43.8% Plasticizer: *o*-NPOE 50.6% Selectophore: 1.2% Additive: OA 4.4%	LR: 3.2×10^{-8} to 1.0×10^{-2} M Slope: 19.4 mV/decade DL: 2.5×10^{-8} M pH 3.5–9.5 RT: <15 s LT: 5 months II: Ce^{3+}, Tb^{3+}	[17]

(*Continued*)

TABLE 5.2 The Characterization of the Reported Lanthanum Ion Membrane Sensors (*cont.*)

Electrode type	Selectophore	Electrode composition	Characterizations of the electrode	Ref.
PME	La-11 *N,N′*-Adipylbis(5-plenylazo salicylaldehyde hydrazone)	PVC: 27% Plasticizer: *o*-NPOE 54% Selectophore: 3% Additive: OA 16%	LR: 1.0×10^{-6} to 1.0×10^{-2} M Slope: 19.4 mV/decade DL: 4.0×10^{-7} M pH 4.0–8.0 RT: 10 s LT: at least 3 months II: Ce^{3+}	[18]
PME	La-12 Bis(5-nitro-2-furaldehyde) butane-2,3-dihydrazone (BNFB)	PVC: 30% Plasticizer: BA 56% Selectophore: 8% Additive: OA 6%	LR: 1.0×10^{-7} to 1.0×10^{-1} M Slope: 19.8 ± 0.2 mV/decade DL: 4.0×10^{-8} M pH 4.0–7.5 RT: <12 s LT: at least 2 months II: None	[19]
PME	La-13 8-Amino-*N*-(2-hydroxybenzylidene) naphthyl amine (ANA)	PVC: 30% Plasticizer: *o*-NPOE 61% Selectophore: 7% Additive: KTpClPB 2%	LR: 1.0×10^{-7} to 1.0×10^{-1} M Slope: 20.3 ± 0.3 mV/decade DL: 8.0×10^{-8} M pH 4.0–9.0 RT: <12 s LT: at least 7 weeks II: Pr^{3+}	[20]
PME	La-14 *N*-(2-Pyridyl)-*N′*-(4-methoxyphenyl)-thiourea (PMPT)	PVC: 35% Plasticizer: BA 54% Selectophore: 8% Additive: KTpClPB 3%	LR: 4.0×10^{-8} to 1.0×10^{-1} M Slope: 19.6 ± 0.2 mV/decade DL: 2.0×10^{-8} M pH 4.0–9.0 RT: 12 s II: None	[21]

PME	La-15 4-Methyl-2-hydrazinobenzothiazole (MHBT)	PVC: 30% Plasticizer: DBP 61% Selectophore: 6% Additive: NaTPB 3%	LR: 1.0×10^{-7} to 1.0×10^{-1} M Slope: 19.8 mV/decade DL: 2.5×10^{-8} M pH 3.0–10.0 RT: <6 s II: Co^{2+}, Cu^{+}, Ce^{3+}, Eu^{3+}	[22]
PME	La-16 3-Hydroxy-*N*′-(pyridin-2-ylmethylene)-2-naphthohydrazide (HPMN)	PVC: 30% Plasticizer: *o*-NPOE 62% Selectophore: 6% Additive: KTpClPB 2%	LR: 1.0×10^{-7} to 1.0×10^{-2} M Slope: 19.2 mV/decade DL: 7.0×10^{-8} M pH 3.0–9.0 RT: 15 s from 1.0×10^{-2} to 1.0×10^{-4} M and 30 s in range of 1.0×10^{-5} to 1.0×10^{-7} M LT: 7 weeks II: None	[23]
PME	La-17 Hexaaza macrocycle, 8,16-dimethyl-6,14-diphenyl-2,3,4:10,11,12-dipyridine-1,3,5,9,11,13-hexaazacyclohexadeca-3,5,8,11,13,16-hexaene	PVC: 30.4% Plasticizer: DBBP 65.8% Selectophore: 2.5% Additive: 1.3% NaTPB	LR: 7.94×10^{-7} to 1.0×10^{-1} M Slope: 19.8 ± 0.2 mV/decade DL: 4.5×10^{-10} M pH 2.5–10.0 RT: 12 s LT: 4 months II: K^{+}, Na^{+}, Ca^{2+}, Co^{2+}, Ni^{2+}, Zn^{2+}, Cu^{2+}, Ce^{3+}, Yb^{3+}	[24]
PME	La-18 *N*′-(1-Pyridin-2-ylmethylene)-2-furohydrazide (NPYFH)	PVC: 30% Plasticizer: NB 62% Selectophore: 5% Additive: NaTPB 3%	LR: 1.0×10^{-6} to 1.0×10^{-1} M Slope: 19.2 ± 0.6 mV/decade DL: 7.0×10^{-7} M pH 3.5–10.0 II: None	[25]

(Continued)

TABLE 5.2 The Characterization of the Reported Lanthanum Ion Membrane Sensors (*cont.*)

Electrode type	Selectophore	Electrode composition	Characterizations of the electrode	Ref.
CWE	La-19 1-[({2-[2-(2-Hydroxy-1-naphtyl)-3-(2-{[(E)-1-(2-hydroxy-1-naphtyl) methylidene] amino}ethyl)-1-imidazolidinyl]ethyl}imino) methyl]-2-naphthol (HHN)	PVC: 20% Plasticizer: *o*-NPOE 73% Selectophore: 4% Additive: KTpClPB 3%	LR: 1.0×10^{-9} to 4.0×10^{-4} M Slope: 19.5 ± 0.3 mV/decade DL: 4.5×10^{-10} M pH 3.5–8.5 LT: 2 months II: None	[26]
CPE	La-19	Graphite powder: 60% MWCNT: 10% RTIL: 20% Selectophore: 10%	LR: 1.0×10^{-9} to 1.0×10^{-2} M Slope: 19.7 ± 0.3 mV/decade DL: 8.0×10^{-9} M pH 3.0–9.8 RT: <25 s LT: about 3 months II: None	[27]
CPE	La-13	Graphite powder: 30% MWCNT: 5% RTIL: 15% Selectophore: 10%	LR: 1.0×10^{-6} to 1.0×10^{-2} M Slope: 19.8 ± 0.3 mV/decade DL: 8.0×10^{-7} M pH 4.0–9.0 RT: <20 s LT: at least 3 months II: None	[28]
PME	La-20 Ethyl-1,2,3,4-tetrahydro-6-methyl-4-phenyl-2-thioxopyrimidine-5-carboxylate (ETMPTC)	PVC: 30% Plasticizer: AP 60% Selectophore: 5% Additive: OA 5%	LR: 9.3×10^{-8} to 1.0×10^{-1} M Slope: 19.9 ± 0.3 mV/decade pH 3.0–10.0 DL: 1.7×10^{-8} M RT: <11 s LT: at least 3 months II: Cr^{3+}, Pr^{3+}	[29]

PVC on disk electrode	La-21 Benzo-15-crown-5 (B15C5)	PVC: 30% Plasticizer: *o*-NPOE 60.5% Selectophore: 2.5% Additive: OA 7%	LR: 5×10^{-7} to 1×10^{-2} M Slope: 19.5 mV/decade DL: 8.74×10^{-7} M LT: at least 5 months pH 3.0–6.0 RT: <15 s II: Cr^{3+}, Ca^{2+}	[30]
PME and CGE	La-22 3-Amino-2-mercapto-3H-quinazolin-4-one (AMQ)	PVC: 31% Plasticizer: *o*-NPOE 63% Selectophore: 5% Additive: KTpClPB 1%	LR: 3.0×10^{-7} to 1.0×10^{-1} M PME LR: 1.0×10^{-7} to 1.0×10^{-1}M CGE Slope: 20.1 ± 0.3 mV/decade PME Slope: 23.4 ± 0.4 mV/decade CGE DL: 2.0×10^{-7} M PME DL: 7.1×10^{-8} M CGE pH 3.0–9.0 RT: <10 s II: Ce^{3+}, Cr^{3+}, Sm^{3+}, Zn^{2+} for PME II: Zn^{2+} for CGE	[31]
CPE	La-7	Graphite powder: 66% Paraffin oil: 25% Selectophore: 9%	LR: 1.0×10^{-9} to 1.0×10^{-3} M Slope: 17.81 mV/decade DL: 2.75×10^{-10} M pH 3.0–8.0 RT: 10 s LT: 2 months II: None	[32]

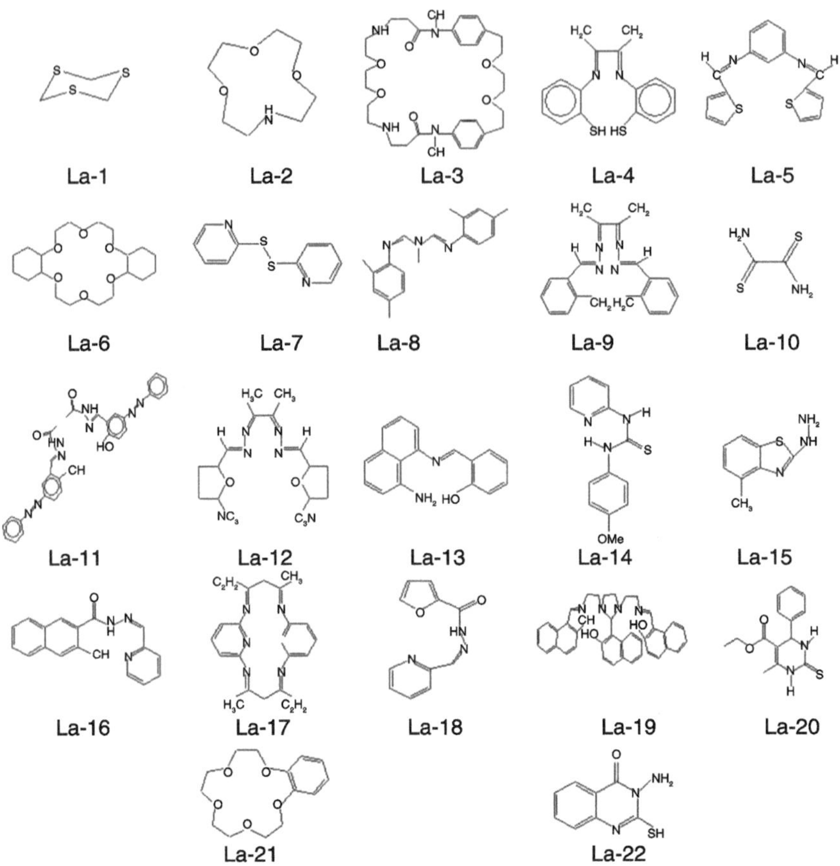

FIGURE 5.8 Chemical structures of the ionophores used for lanthanide ISEs.

that the rigidity of the ligands actually leads to nonselectvity between the two mentioned cations, the relative selectivity induced by altering the dielectric constant of the two membranes is attributed to the decrease in the solvation energy of La^{3+} as compared to Ce^{3+} by some 48 kJ mol^{-1}.

Cerium ISEs

The total number of reported potentiometric cerium ISEs are 17 [33–49]. The first report on cerium potentiometric ion selective sensor was the work of Shamsipur et al. in 2000 [33]. The sensors are PMEs, CPEs, CGEs, CWEs; a gold nanoparticle-screen-printed (GNPs-SPE) and modified screen-printed electrodes (MSPEs). Fig. 5.9 shows the historical distribution of the reports on the development of Ce^{3+}sensors.

Table 5.3 summarizes the reports on the construction of cerium ISEs together with the characteristics and properties of the devices, and

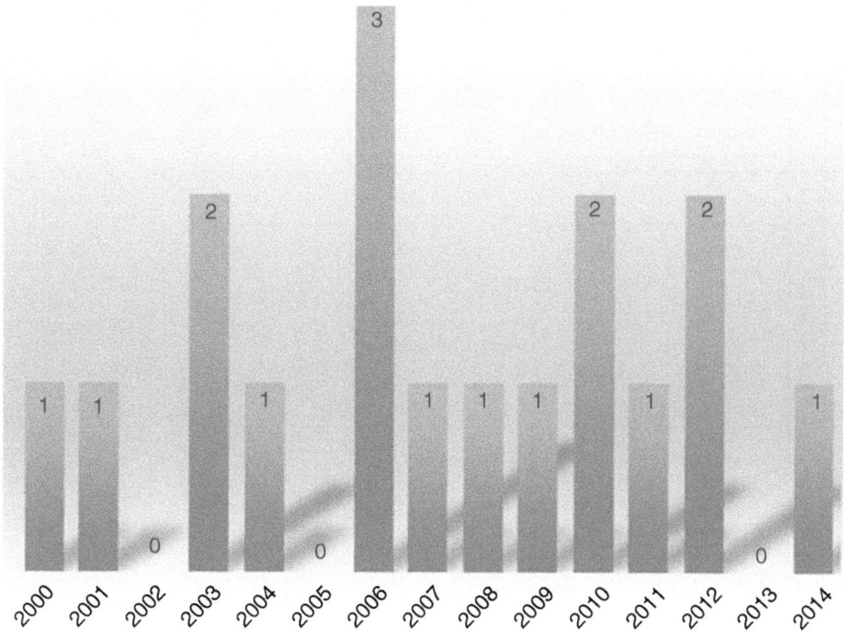

FIGURE 5.9 Historical distribution of the reports on the development of Ce^{3+} sensors.

Fig. 5.10 illustrates the chemical structure of the used ionophores for Ce^{3+}.

As discussed in the case of lanthanum, among the 15 ligands used in the reports for Ce^{3+} ions, the cyclic selectophores, that is, Ce-1, Ce-6, Ce-10, and Ce-11, show too low selectivity. The same selectivity failures are observed among Ce-2, Ce-3, Ce-4, Ce-7, Ce-9, and Ce-14, which possess several S and N donating heteroatoms but lack the flexibility required for forming wraparound complexes, which are most stable with cerium ions. Ce-5, Ce-8, Ce-12, and Ce-13, on the other hand, show a rather good selectivity, based on favorable structural features.

Ce-1 was the Ce^{3+} ionophore used in a PVC membrane and coated glassy electrode. Ce-8 was also used in both PME and CPE devices. In the case of Ce-15 (1,4-bis(8-mercapto octyloxy)-benzene), the ligand was used in a modified screen-printed electrode (SPE) in which the ligand was self-assembled to GNPs. The potentiometric GNPs-SPE device was used for the selective determination of Ce^{3+} and showed short response times of about 4 and 7 s. The device had a long lifetime of 7 and 5 months.

Praseodymium ISEs

The first report on potentiometric praseodymium ISE was the work of Ganjali et al. in 2006. The remaining 10 sensors are PMEs, CPEs, CGEs,

TABLE 5.3 The Characterization of the Reported Cerium Ion Membrane Sensors

Electrode type	Selectophore	Electrode composition	Characterizations of the electrode	Ref.
PME	Ce-1 1,3,5-Trithiane (TT)	PVC: 33% Plasticizer: BA 44% Selectophore: 5% Additive: OA 18%	LR: 5.0×10^{-5} to 1.0×10^{-1} M Slope: 19.4 ± 0.4 mV/decade DL: 3.0×10^{-5} M pH 5.0–8.0 RT: <15 s LT: at least 5 months II: Ag^{+}, Ca^{2+}, Ba^{2+}, Co^{2+}, Ni^{2+}, Cu^{2+}, Zn^{2+}, Cd^{2+}, Pb^{2+}, La^{3+}	[33]
CGE	Ce-1	PVC: 33% Plasticizer: BA 44% Selectophore: 5% Additive: OA 18%	LR: 2.5×10^{-8} to 4.7×10^{-4} M Slope: 19.4 ± 0.4 mV/decade DL: 2.0×10^{-8} M pH 4.5–8.0 RT: 10 s II: Cd^{2+}, Pb^{2+}, La^{3+}	[34]
PME	Ce-2 Azomethine of piperonylidine-4-[2.2] paracyclophanylamine	PVC: 33% Plasticizer: *o*-NPOE 66% Selectophore: 0.66% Additive: KTClPB 0.33%	LR: 2.5×10^{-5} to 1.0×10^{-1} M Slope: 19.3 ± 0.3 mV/decade DL: 1.2×10^{-5} M pH 4.5–8.0 RT: 30 s LT: at least 2 months II: Sm^{3+}, La^{3+}, K^{+}, Hg^{2+}, Fe^{3+}	[35]
PME	Ce-3 *N*-[(Z)-2-Chloro-2-(1-hydroxy-1,1,1-triphenyl phosphoranyl)-1-ethenyl]-4-ethyl-1-benzene sulfonamide	PVC: 40% Plasticizer: *o*-NPOE 45% Selectophore: 10% Additive: KT_pClPB 5%	LR: 6.6×10^{-7} to 6.2×10^{-2} M Slope: 19.5 ± 0.3 mV/decade DL: 2.3×10^{-7} M pH 4.5–8.5 RT: <10 s LT: at least 4 months II: Fe^{3+}, Al^{3+}, Cr^{3+}, $Ga^{3}+$, La^{3+}, Sm^{3+}, Gd^{3+}, Yb^{3+}	[36]

PME	Ce-4 2-Aminobenzothiazole	PVC: 24% Plasticizer: *o*-NPOE 59.8% Selectophore: 3.3% Additive: OA 13%	LR: 2.0×10^{-6} to 2.0×10^{-2} M Slope: 19.6 ± 1.0 mV/decade DL: 1.8×10^{-6} M pH 4.1–7.3 RT: ~13 s LT: at least 3 months II: Ag^{+}, Cu^{2+}, Pb^{2+}, Hg^{2+}, La^{3+}, Fe^{2+}, Al^{3+}	[37]
PME	Ce-5 *N*,*N*-Bis[2-(salicylideneamino) ethyl] ethane-1,2-diamine	PVC: 35.3% Plasticizer: *o*-NPOE 58.8% Selectophore: 1.2% Additive: OA 4.7%	LR: 1.41×10^{-7} to 1.0×10^{-2} M Slope: 27 mV/decade DL: 8.91×10^{-8} M pH 3.0–8.0 RT: <10 s LT: 4 months II: La^{3+}	[38]
PME and CWE	Ce-6 1,4,7-Trithiacvclononane (trithia-9-crown-3, TT9C3)	PVC: 30% Plasticizer: BA 49% Selectophore: 6% Additive: OA 15%	LR: 5.0×10^{-6} to 5.0×10^{-2} M for PME LR: 1.0×10^{-7} to 1.0×10^{-2} M for CPW Slope: 19.4 mV/decade for both DL: 3.5×10^{-6} M for PME DL: 8.0×10^{-8} M for CPW pH 5.0–8.0 for both RT: 15 s for both LT: at least 3 months for both II: La^{3+}, Ag^{+} for both	[39]
PME	Ce-7 4-(4′-Nitrobenzyl)-1-phenyl-3,5-pyrazolidinedion) (NBPP)	PVC: 33% Plasticizer: *o*-NPOE 66% Selectophore: 0.66% Additive: KTClPB 0.33%	LR: 2.5×10^{-6} to 1.0×10^{-1} M Slope: 19.5 ± 0.2 mV/decade DL: 1.6×10^{-6} M pH 4.0–8.0 RT: <10 s LT: 3 months II: K^{+}, Na^{+}, Li^{+}, Cs^{+}, Rb^{+}, NH_4^{+}, Ni^{2+}, Ca^{2+}, Sm^{3+}, Nd^{3+}	[40]

(*Continued*)

TABLE 5.3 The Characterization of the Reported Cerium Ion Membrane Sensors (*cont.*)

Electrode type	Selectophore	Electrode composition	Characterizations of the electrode	Ref.
PME	Ce-8 *N*-[(2-Hydroxyphenyl) methylidene]-2-furohydrazide (NHMF)	PVC: 30% Plasticizer: NB 62% Selectophore: 5% Additive: NaTPB 3%	LR: 1.0×10^{-5} to 1.0×10^{-1} M Slope: 19.4 ± 0.3 mV/decade DL: 7.6×10^{-6} M pH 3.5 – 10.0 RT: < 15 s II: None	[41]
PME	Ce-9 2,5-Dioxo-4-imidazolidinyl (DI)	PVC: 32% Plasticizer: *o*-NPOE 60.5% Selectophore: 5% Additive: NaTPB 2.5%	LR: 1.0×10^{-6} to 1.0×10^{-1} M Slope: 19.6 ± 0.3 mV/decade DL: 5.7×10^{-7} M pH 3.1–9.8 RT: <10 s LT: at least 10 weeks II: Nd^{3+}, La^{3+}	[42]
CPE	Ce-8	[bmim] BF_4: 25% Selectophore: 16% Carbon powder: 44% MWCNT: 15%	LR: 8.0×10^{-7} to 1.0×10^{-1} M Slope: 19.9 ± 0.2 mV/decade DL: 3.6×10^{-7} M pH 3.5–9.0 RT: 20 s LT: 4 months II: None	[43]
PME	Ce-10 Dibenzo-24-crown-8 (DB24C8)	PVC: 33% Plasticizer: *o*-NPOE 62.5% Selectophore: 4.5%	LR: 1.0×10^{-5} to 1.0×10^{-1} M Slope: 19 mV/decade DL: 3.0×10^{-5} M pH 3.5–8.0 RT: 20 s LT: 4 months II: La^{3+}, Dy^{3+}, Pr^{3+}, Tb^{3+}, Sm^{3+}, Yb^{3+}, Cu^{2+}, Nd^{3+}, Pb^{2+}, K^{+}, Ca^{2+}, Eu^{3+}, Ag^{+}, Mg^{2+}, Fe^{3+}, Hg^{2+}	[44]

PME and CGE	Ce-11 2,3,4:12,13,14-Dipyridine-1,3,5,8,11,13,15,18-octaazacycloicosa-2,12-diene (L(1)) 2,3,4:12,13,14-Dipyridine-1,5,8,11,15,18-hexamethylacrylate-1,3,5,8,11,13,15,18-octaazacycloicosa-2,12-diene (L(2))	PVC: 34% Plasticizer: *o*-NPOE 58% Selectophore: 6% Additive: KTpClPB 2%	LR: 1.4×10^{-7} to 1.0×10^{-1} M for PME LR: 1.0×10^{-8} to 1.0×10^{-1} M for CGE Slope: 19.5 ± 0.3 for PME mV/decade Slope: 19.7 ± 0.1 for CGE DL: 8.3×10^{-8} M for PME DL: 7.7×10^{-9} M for CGE pH 3.5–7.5 for PME pH 2.5–8.5 for CGE RT: 12 s for PME RT: 10 s for CGE LT: 3 months for PME LT: 5 months for CGE II: La^{3+}, Dy^{3+}, Sm^{3+}, Gd^{3+}, Nd^{3+}, Pb^{2+}, K^{+}, Sr^{2+}, Eu^{3+}, Na^{+}, Mg^{2+}, Ni^{2+} for PME II: Mg^{2+}, Cu^{2+}, Eu^{3+}, Pb^{2+}, Ni^{2+} for CGE	[45]
PME	Ce-12 2,9-Dihydroxy-1,10-diphenoxy-4,7-dithia decane	PVC: 30% Plasticizer: *o*-NPOE 61.5% Selectophore: 3% Additive: PA 5.5%	LR: 1.0×10^{-8} to 1.0×10^{-1} M Slope: 19.3 ± 1.0 mV/decade DL: 2.1×10^{-9} M pH 5.0–8.0 RT: 10 s II: Li^{+}	[46]
CPE	Ce-13 (*Z*)-2-((1*H*-1,2,4-Triazol-3-ylimino)methyl)phenol	$[BMP]Tf_2N$: 20% Selectophore: 16% Carbon powder: 51% MWCNT: 10% Nanosilica: 3%	LR: 2.0×10^{-8} to 1.0×10^{-1} M Slope: 19.79 ± 0.10 mV/decade DL: 6.45×10^{-9} M pH 3.0–7.5 RT: 5 s LT: at least 17 weeks II: None	[47]

(*Continued*)

TABLE 5.3 The Characterization of the Reported Cerium Ion Membrane Sensors (*cont.*)

Electrode type	Selectophore	Electrode composition	Characterizations of the electrode	Ref.
CPE	Ce-14 4-(4-Methylbenzylideneamino)-5-methyl-2*H*-1,2,4-triazole-3(4*H*)-thione	$[Bmim]PF_6$: 20% Selectophore: 17% Carbon powder: 50% MWCNT: 10% Nanosilica: 3%	LR: 2.5×10^{-8} to 1.0×10^{-1} M Slope: 19.32 ± 0.10 mV/decade DL: 7.0×10^{-9} M pH 3.0–8.0 RT: 5 s LT: at least 11 weeks II: Ni^{2+}, Mg^{2+}, La^{3+}	[48]
GNPs-SPE and MSPE	Ce-15 1,4-Bis-(8-mercaptooctyloxy)-benzene	PVC: 50% Paraffin oil: 13.75% Selectophore: 2.7% Carbon powder: 33.5% Nanoparticles: 200 (μL) (for GNPs-SPE) PVC: 50% Paraffin oil: 14.75% Selectophore: 3.7% Carbon powder: 31.5% (for MSPE)	LR: 3.25×10^{-10} to 1.0×10^{-1} M for GNPs-SPE LR: 1.0×10^{-7} to 1.0×10^{-2} M for MSPE Slope: 19.95 ± 0.97 mV/decade for GNPs-SPE Slope: 17.04 ± 1.02 mV/decade for MSPE DL: 3.2×10^{-10} M for GNPs-SPE DL: 9.5×10^{-8} M for MSPE pH 2.8–8.5 for GNPs-SPE pH 3.5–7.5 for MSPE RT: 4 s for GNPs-SPE RT: 7 s for MSPE LT: at least 7 months for GNPs-SPE LT: at least 5 months for MSPE II: Fe^{3+} and I^- for GNPs-SPE and MSPE, Cl^- for MSPE	[49]

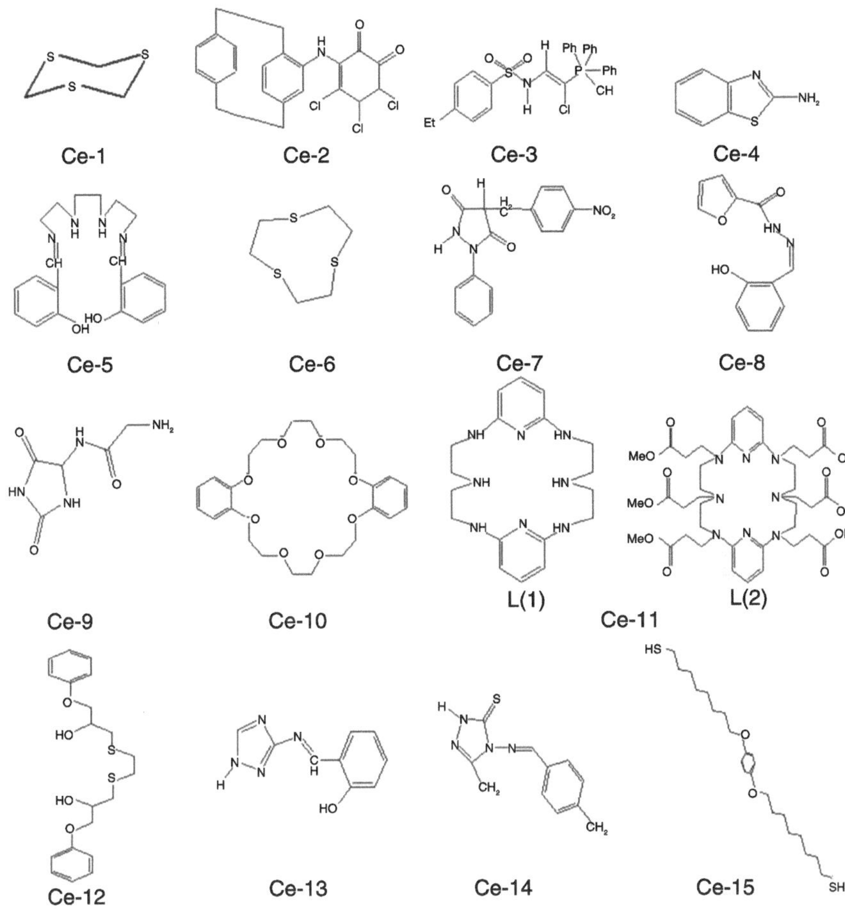

FIGURE 5.10 Chemical structures of the reported selectophores for cerium ions.

and CWE. Fig. 5.11 shows the historical trend of the reports on the development of Pr^{3+} sensors.

Table 5.4 summarizes the information on the selectophores used for the construction of praseodymium ion selective electrodes and the characteristics of the resulting sensors.

Fig. 5.12 illustrates the chemical structures of the selectophores used for the praseodium ions. It can be seen among the eight selectophores reported that the cyclic selectophores like Pr-5 shows very low selectivity. Likewise, the acyclic, yet rigid, ionophores like Pr-1, Pr-4, and Pr-8 do not show selective tendencies contrary to the presence of the donor S and N atoms in their structures (Table 5.4). Pr-3 and Pr-7, on the other hand, show good selectivity because of the flexibility of the semicavity in forming fine-tuned templates with Pr^{3+}. In the case of Pr-6, which is a

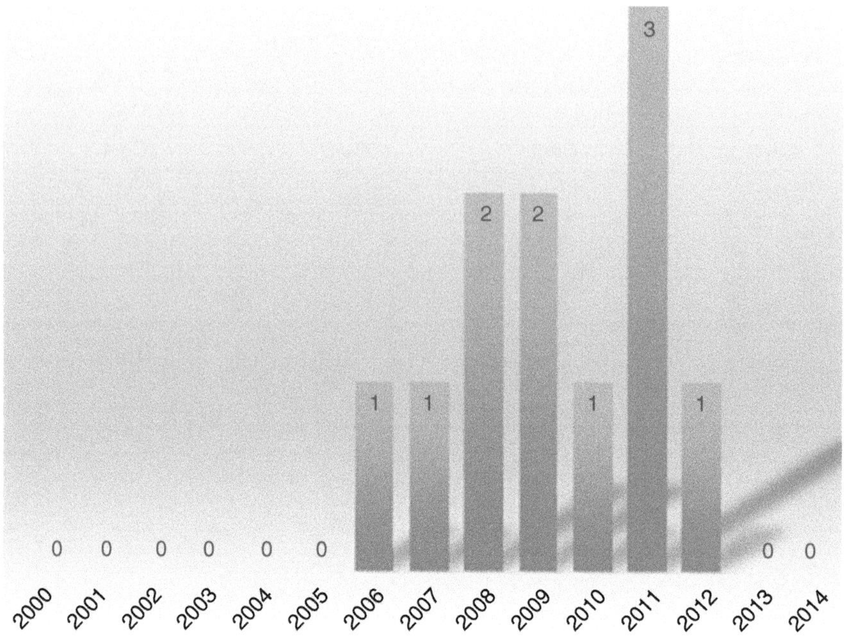

FIGURE 5.11 Historical trend of the reports on the development of Pr^{3+} sensors.

symmetric Schiff with a rigid structure, can form stable complexes with many metal ions and is hence nonselective. Pr-1 has been used in PME, CWE, and CPE forms, and Pr-3 has been used in PME and CPE. It should be noted that Pr-2, which is a zirconium(IV) antimoniotungstate crystal, cannot be compared with the PVC-based membrane with ionophores.

Neodymium ISEs

There are 17 reports on potentiometric neodymium selective sensors. The first report on neodymium potentiometric ion selective sensor dates back to the work of Gupta et al. [61], and another report Agarwal and Chandra [62], in 2004. The mechanism of response in these sensors is based on ion-exchanging. Fig. 5.13 shows the historical trend of the reports on the development of Nd^{3+} sensors.

Table 5.5 illustrates the selectophores used in the construction of neodymium ion selective electrodes and their characteristics.

Based on the structural information of the Nd^{3+} selective ionophores that are shown in Fig. 5.14 and as previously discussed in this chapter, a good ionophore should have a flexible semicavity with soft to intermediate heteroatoms (mostly S and N as donor atoms) to form finely optimized wraparound complexes that can lead to selectivity of one species

TABLE 5.4 The Characteristics of the Reported Praseodymium Ion Membrane Sensors

Electrode type	Selectophore	Electrode composition	Characterizations of the electrode	Ref.
PME	Pr-1 *N′*-(Pyridin-2-ylmethylene) benzohydrazide (PBA)	PVC: 30% Plasticizer: *o*-NPOE 59% Selectophore: 7% Additive: NaTPB 4%	LR: 1.0×10^{-6} to 1.0×10^{-2} M Slope: 21.1 mV/decade DL: 8.0×10^{-7} M pH 3.5–8.0 RT: 20 s LT: 6 weeks II: Er^{3+}, Sm^{3+}, Lu^{3+}, Eu^{3+}, Ce^{3+}, Gd^{3+}	[50]
CWE	Pr-1	PVC: 20% Plasticizer: *o*-NPOE 70% Selectophore: 7% Additive: KTpClPB 3%	LR: 1.0×10^{-8} to 1.0×10^{-3} M Slope: 21.1 ± 0.3 mV/decade DL: 7.0×10^{-9} M pH 3.0–8.5 RT: ~10 s II: None	[51]
PME	Pr-2 Zirconium(IV) antimoniotungstate	Full text not available	LR: 5.0×10^{-5} to 1.0×10^{-1} M Slope: 25.0 mV/decade pH 4.5–9.0 RT: 10–15 s II: Full text not available	[52]
PME	Pr-3 *N,N*-Bis(alpha-methylsalicylidene) diethylenetriamine (BMT)	PVC: 32% Plasticizer: *o*-NPOE 59% Selectophore: 6% Additive: NaTPB 3%	LR: 1.0×10^{-6} to 1.0×10^{-2} M Slope: 19.8 ± 0.2 mV/decade DL: 6.5×10^{-7} M pH 3.0–8.4 RT: <15 s II: None	[53]

(*Continued*)

TABLE 5.4 The Characteristics of the Reported Praseodymium Ion Membrane Sensors (*cont.*)

Electrode type	Selectophore	Electrode composition	Characterizations of the electrode	Ref.
CPE	Pr-1	[bmim]BF_4: 30% Selectophore: 17% Carbon powder: 38% MWCNT: 15%	LR: 2.5×10^{-7} to 1.0×10^{-1} M Slope: 19.9 ± 0.3 mV/decade DL: 9.0×10^{-8} M pH 3.5–7.5 RT: ~15 s LT: 8 weeks II: Pb^{2+}	[54]
PME	Pr-4 *N′*1,*N′*2-Bis(2-*oxo*-1,2-diphenylethylidene) ethanedihydrazide (BEH)	PVC: 30% Plasticizer: AP 66% Selectophore: 2% Additive: NaTPB 2%	LR: 1.0×10^{-6} to 1.0×10^{-2} M Slope: 19.6 ± 0.4 mV/decade DL: 4.3×10^{-7} M pH 2.8–8.7 RT: about 10 s LT: 2 months II: None	[55]
PME	Pr-5 1,6,7,12-Tetramine-2,5,8,11-tetraoxo-1(12),6(7)-di(biphenyl)dodecane (TATODBDD)	PVC: 35% Plasticizer: *o*-NPOE 25% Selectophore: 7% Additive: OA 6%	LR: 3.9×10^{-7} to 1.0×10^{-1} M Slope: 19.5 mV/decade DL: 5.0×10^{-8} M pH 2.9–9.5 RT: <10 s LT: 6 weeks II: Mg^{2+}, Sm^{3+}, La^{3+}, Dy^{3+} (interfering effect of 8 Ln were not studied)	[56]
CPE	Pr-3	IL: 25% Carbon powder: 45% Selectophore: 25% MWCNT: 5%	LR: 1.0×10^{-6} to 1.0×10^{-2} M Slope: 20.1 ± 0.4 mV/decade DL: 6.3×10^{-7} M pH 3.0 – 8.0 RT: < 10 s LT: at least 10 weeks II: None	[57]

PME	Pr-6 *N-N′-o*-Phenylene-bis(salicylideneimine)	PVC: 50% Selectophore: 50%	LR: 1.0×10^{-6} to 1.0×10^{-1} M for Epoxy resin and PVC Slope: 19.9 mV/decade for Epoxy resin Slope: 14.0 mV/decade for PVC pH 3.0–8.62 for Epoxy resin pH 2.97–8.90 for PVC RT: <10 s for Epoxy resin RT: 20 s for PVC LT: about 6 months for Epoxy resin LT: 4 months for PVC II: Nd^{3+}, Sm^{3+}, Eu^{3+}, Ce^{3+}, Fe^{3+}, La^{3+}, Al^{3+}, Na^{+} with Epoxy binder II: Nd^{3+}, Sm^{3+}, Eu^{3+}, Ce^{3+}, Fe^{3+}, Al^{3+} with PVC binder	[58]
PME	Pr-7 *N,N′*-Bis(4-hydroxysalicylidene)-1-3-phenylenediamine (HSPDA)	PVC: 30% Plasticizer: BA 65% Selectophore: 3% Additive: NaTPB 2%	LR: 1.0×10^{-6} to 1.0×10^{-2} M Slope: 19.8 ± 0.4 mV/decade DL: 5.7×10^{-7} M pH 3.1 – 9.8 RT: ~5 s LT: at least 2 months II: None	[59]
PME	Pr-8 3-Nitro-4-amino-5-naphtho[2,1-*b*]furan-2-yl-4H-1,2,4-triazole-3-thiol (NAFT)	PVC: 30% Plasticizer: *o*-NPOE 59% Selectophore: 8% Additive: NaTPB 3%	LR: 1.0×10^{-7} to 5.0×10^{-1} M Slope: 20.5 ± 0.2 mV/decade DL: 5.2×10^{-8} M pH 3.0–8.8 RT: ~8 s LT: at least 3 months II: Nd^{3+}, Ce^{3+}	[60]

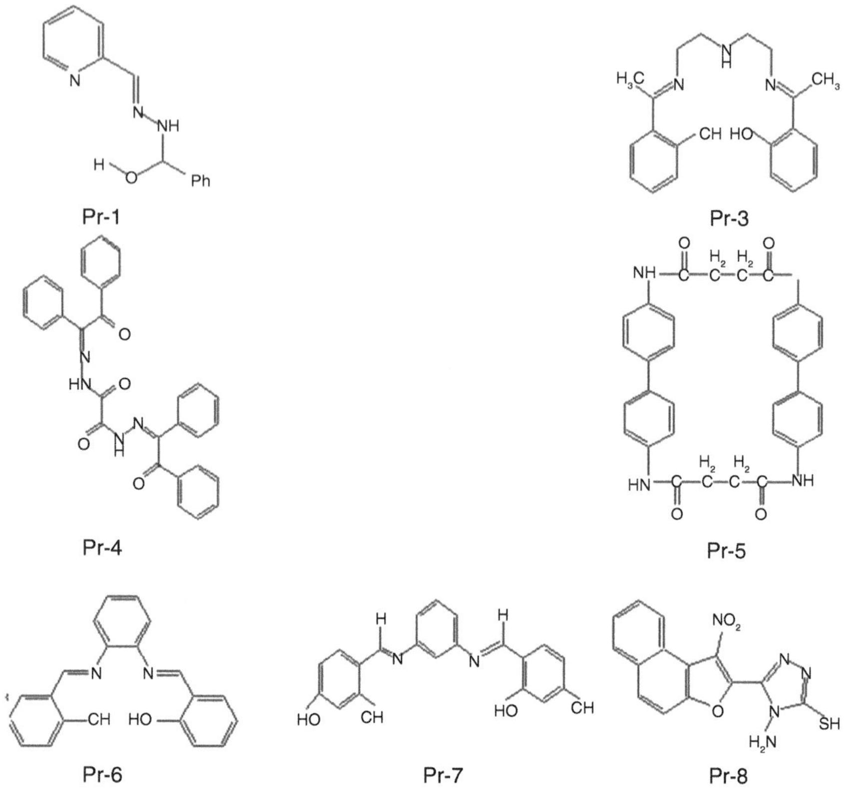

FIGURE 5.12 Chemical structures of the reported selectophores for praseodymium ions.

compared to the others. From about the 15 ionophores used to construct Nd^{3+} selective sensors, and following the pattern in the previous sections, the cyclic compounds, that is, Nd-2, Nd-6, and Nd-8, show too low selectivity. A similar behavior is observed among Nd-4 and Nd-12, which have bulky groups or small size avoiding the formation of wraparound complexes, or with Nd-5 and Nd-7, which are rigid Schiff bases. Nd-10, Nd-11, Nd-13, Nd-14, and Nd-15 on the other hand have good selectivity because of favorable structural properties.

Promethium ISEs

In the case of promethium, only two ionophores have been reported by Gupta et al. [78] in 2010 (Fig. 5.15). As can be seen in Fig. 5.15, which also contains the characteristics of the sensor, the ionophores are

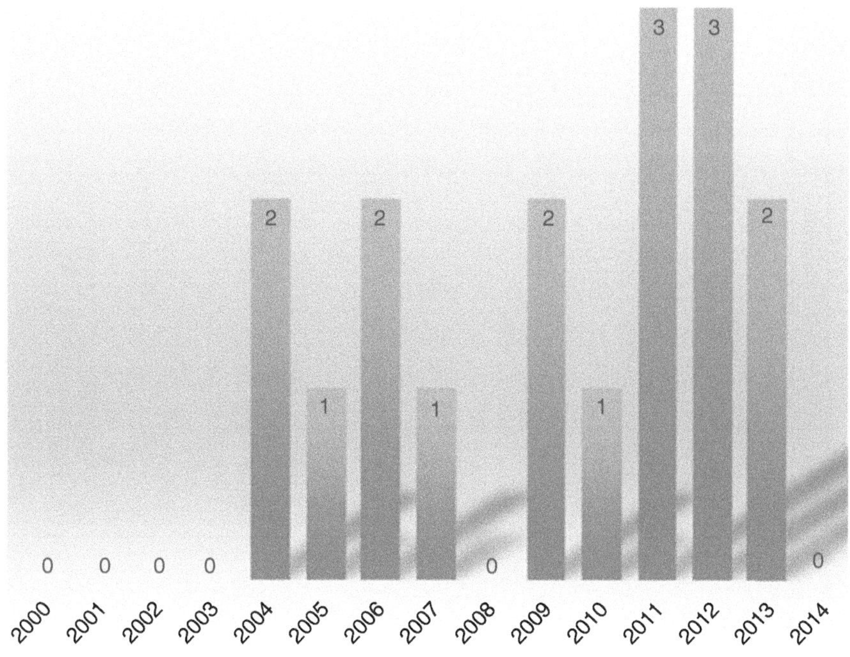

FIGURE 5.13 Historical trend of the reports on the development of Nd^{3+} sensors.

two Schiff base ligands. The researchers used 2-nitrophenyloctylether (*o*-NPOE), dibutyl phthalate, dioctylphthalate, tri-(2-ethylhexyl) phosphate, dibutyl butylphosphonate, chloronaphthalene (CN), and anionic additives like tetrakis(*p*-chlorophenyl)borate, sodiumtetraphenylborate, and oleic acid, and observed sensors results. This might seem contrary to the trends discussed so far, because the cyclic ligands show good selectivity. This exception can be attributed to the rather large ring size, which gives it some degree of flexibility and which helps the ligand to form stable wraparound complexes with Pm^{3+} ions, as compared to the rest of the ions tested.

Samarium ISEs

The first report on samarium potentiometric ion selective sensor was by Ganjali et al. using *N*-[2-[4-[[[(cyclohexyl amino)carbonyl]amino]sulfonyl]phenyl]ethyl]-5-methylpyrazine-2-carboxamide (glipizide) as the ionophore. This work was followed by that of Shamsipur et al. in 2003 [79]. The 17 ISEs developed for samaruim ISEs include PMEs, CPEs, CGE,

TABLE 5.5 The Characterization of the Reported Neodymium Ion Membrane Sensors

Electrode type	Selectophore	Electrode composition	Characterizations of the electrode	Ref.
PME	Nd-1 Tetracycline-sorbed zirconium(IV) tungstophosphate (TC-ZWP)	PVC: 75% Selectophore: 25%	LR: 5.0×10^{-5} to 1.0×10^{-1} M Slope: 21.7 mV/decade DL: 5.0×10^{-5} pH 3.0–7.0 RT: <30 s LT: 45 days II: La^{3+}, Ce^{3+}, Pr^{3+}, Sm^{3+}, Eu^{3+}, Gd^{3+}, Tb^{3+}, Dy^{3+}	[61]
PME	Nd-1	PVC: 75% Selectophore: 25%	LR: 5.0×10^{-5} to 1.0×10^{-1} M Slope: 21.7 mV/decade DL: 5.0×10^{-5} pH 3.0–7.0 RT: <25 s LT: 45 days II: La^{3+}, Ce^{3+}, Pr^{3+}, Sm^{3+}, Eu^{3+}, Gd^{3+}, Tb^{3+}, Dy^{3+}	[62]
PME	Nd-2 5-Pyridino-2,8-dithia[9] (2,9)-1,10-phenanthrolinephane (L)	PVC: 30% Plasticizer: *o*-NPOE 50% Selectophore: 4% Additive: OA 16%	LR: 1.0×10^{-6} to 1.0×10^{-2} M Slop: 19.6 ± 0.3 mV/decade DL: 7.9×10^{-7} pH 4.0–6.5 RT: <5 s LT: 9 weeks II: Li^{+}, Na^{+}, K^{+}, Ni^{2+}, Cu^{2+}, Zn^{2+}, Tl^{+}, Pb^{2+}, Ag^{+}, Hg^{2+}, Be^{2+}, La^{3+}, Mg^{2+}, Sm^{3+}, Ca^{2+}, Gd^{3+}, Sr^{2+}, Yb^{3+}, Ba^{2+}, Ce^{3+}, Co^{2+}	[63]

PME	Nd-3 2-{[(6-Aminopyridin-2-yl) imino] methyl}phenol (APIMP)	PVC: 30% Plasticizer: BA 62% Selectophore: 6% Additive: NaTPB 2%	LR: 1.0×10^{-5} to 1.0×10^{-2} M Slop: 19.6 ± 0.3 mV/decade DL: 2.0×10^{-6} pH 3.5–8.5 RT: <10 s LT: 6 weeks II: La^{3+}, Gd^{3+}, Sm^{3+}, Ce^{3+}	[64]
PME	Nd-4 *N*-(2-Furylmethylene) pyridine-2,6-diamine (FPD)	PVC: 33% Plasticizer: NB 59% Selectophore: 6% Additive: KTpClPB 2%	LR: 1.0×10^{-5} to 1.0×10^{-2} M Slop: 19.6 ± 0.3 mV/decade DL: 7.0×10^{-6} pH 4.0–8.0 RT: <15 s LT: 6 weeks II: La^{3+}, Gd^{3+}, Sm^{3+}, Yb^{3+}, Dy^{3+}, Cu^{2+}, Na^{+}, K^{+}, Mg^{2+}, Ca^{2+}	[65]
PME	Nd-5 Benzyl bisthiosemicarbazone (BTC)	PVC: 30% Plasticizer: *o*-NPOE 61% Selectophore: 6% Additive: NaTPB 3%	LR: 1.0×10^{-6} to 1.0×10^{-2} M Slop: 19.7 ± 0.4 mV/decade DL: 6.2×10^{-7} pH 3.7–8.3 RT: <10 s LT: 7 weeks II: La^{3+}, Gd^{3+}, Tb^{3+}, Dy^{3+}, Pb^{2+}, Hg^{2+}, Ag^{+}, Ni^{2+}, Cd^{2+}	[66]
PME	Nd-6 11,13-Diaza-4,7,12-trioxo-2(3),8(9)-dibenzoyl-cyclotetridecane-1,11-diene (DATODBCT)	PVC: 29% Plasticizer: DBP 59% Selectophore: 9% Additive: NaTpClPB 3%	LR: 1.0×10^{-8} to 1.0×10^{-1} M Slope: 19.4 ± 0.3 mV/decade DL: 8.0×10^{-7} pH 3.0–7.0 RT: <15 s LT: 49 days II: La^{3+}, Gd^{3+}, Sm^{3+}, Dy^{3+}, Yb^{3+}, Pr^{3+}	[67]

(*Continued*)

TABLE 5.5 The Characterization of the Reported Neodymium Ion Membrane Sensors (*cont.*)

Electrode type	Selectophore	Electrode composition	Characterizations of the electrode	Ref.
PME	Nd-7 *N,N'*-Bis((1*H*-pyrrol-2-yl)methylene) cyclohexane-1,2-diamine (L_1) 3,3'-(cyclohexane-1,2-diylbis(azan-1-yl-1-ylidene)bis(methan-1-yl-1-ylidene)bis(5-hydroxymethyl) pyridine-2-ol) (L_2)	PVC: 32.6% Plasticizer: *o*-NPOE 65.2% Selectophore: 1.1% Additive: 1.1% NaTPB for L_1 and L_2	LR: 5.0×10^{-7} to 1.0×10^{-2} M for L_1 LR: 1.0×10^{-5} to 1.0×10^{-2} M for L_2 Slope: 19.8 ± 0.3 for L_1 mV/decade Slope: 19.2 ± 0.2 for L_2 mV/decade DL: 1.0×10^{-7} for L_1 DL: 5.0×10^{-6} for L_2 pH 4.0–8.0 for L_1 and L_2 RT: 10 s for L_1 RT: 18 s for L_2 LT: 30 months for L_1 LT: 1.2 months for L_2 II: Co^{2+}, Hg^{2+}, Ni^{2+}, Zn^{2+}, Ca^{2+}, Fe^{3+}, Al^{3+}, K^+ for L_1 II: Co^{2+}, Yb^{3+}, Cu^{2+}, Hg^{2+}, Ni^{2+}, Zn^{2+}, Ca^{2+}, Fe^{3+}, Al^{3+}, Na^+, K^+ for L_2	[68]
CGE	Nd-8 1,5-Di(cyanoethane)-2,4:7,8:13,14-tribenzo-1,5-diaza-9,12-dioxacyclopentadeca-2,7,13-triene (L_1) and 1,5-di(cyanoethane)-2,3,4-pyridine-7,8:13,14-dibenzo-1,3,5-triaza-9,12-dioxa cyclopentadeca-2,7,13-triene (L_2)	PVC: 35% Plasticizer: *o*-NPOE 57% Selectophore: 5% Additive: NaTPB 3% for L_1 PVC: 39% Plasticizer: *o*-NPOE 53% Selectophore: 5% Additive: NaTPB 3% for L_2	LR: 8.4×10^{-8} to 1.0×10^{-2} M for L_1 LR: 4.6×10^{-8} to 5.0×10^{-2} M for L_2 DL: 3.8×10^{-8} for L_1 DL: 1.6×10^{-8} for L_2 Slope: 19.8 ± 0.4 for L_1 mV/decade Slope: 19.7 ± 0.5 for L_2 mV/decade RT: 14 for L_1 RT: 10 for L_2 LT: 3 months pH 3.5–7.6 II: La^{3+}, Yb^{3+}, Tb^{3+}, Dy^{3+}, Gd^{3+}, Co^{2+} for L_1 and L_2	[69]

CPE	Nd-3	RTIL: 20% Carbon powder: 53% Selectophore: 20% NS: 2% MWCNT-NH2: 5%	LR: 1.0×10^{-6} to 1.0×10^{-2} M Slope: 19.8 ± 0.12 mV/decade DL: 8.0×10^{-7} pH 4.0–8.0 II: La^{3+}, Sm^{3+}, Gd^{3+}, Ce^{3+}	[70]
PME	Nd-9 Tin (IV) antimoniotungstate (SnSbW) and zirconium (IV) antimoniotungstate (ZrSbW)	Antimoniotungstate: 50% Epoxy resin: 50% for SnSbW Antimoniotungstate: 40% Epoxy resin: 60% for ZrSbW	LR: 1.0×10^{-4} to 1.0×10^{-1} M for SnSbW LR: 1.0×10^{-5} to 1.0×10^{-1} M for ZrSbW Slope: 40.0 for SnSbW mV/decade Slope: 25.0 for ZrSbW mV/decade RT: 20 s for SnSbW RT: 10 s for ZrSbW pH 2.9–9.9 for SnSbW pH 3.0–9.0 for ZrSbW LT: 6 months II: La^{3+}, Sm^{3+}, Gd^{3+}, Eu^{3+}, Pr^{3+}, Ce^{3+}, Dy^{3+}, Y^{3+}, Fe^{3+} for SnSbW II: La^{3+}, Sm^{3+}, Eu^{3+}, Pr^{3+}, Ce^{3+}, Tb^{3+}, Al^{3+} for ZrSbW	[71]
PME	Nd-10 *N,N'*-Bis(quinoline-2-carboxamido)-4,5-dimethylbenzene (H_2L_4)	PVC: 30% Plasticizer: DBP 65% Selectophore: 3% Additive: NaTPB 2%	LR: 5.0×10^{-6} to 1.0×10^{-2} M Slope: 19.5 ± 0.4 mV/decade DL: 4.8×10^{-6} pH 2.9–9.2 RT: <10 s II: None	[72]

(*Continued*)

TABLE 5.5 The Characterization of the Reported Neodymium Ion Membrane Sensors (*cont.*)

Electrode type	Selectophore	Electrode composition	Characterizations of the electrode	Ref.
PME	Nd-11 *N*-(Furfuralidene)-*N*-isonicotinoylhydrazine (FIH)	PVC: 30% Plasticizer: *o*-NPOE 63% Selectophore: 4% Additive: NaTPB 3%	LR: 1.0×10^{-7} to 5.0×10^{-1} M Slope: 20.2 ± 0.2 mV/decade DL: 8.6×10^{-8} pH 3.0–8.4 RT: 8 s LT: 11 weeks II: None	[73]
CWE	Nd-12 5-(Methylsulfanyl)-3-phenyl-1*H*-1,2,4-triazole (MPT)	PVC: 30% Plasticizer: NB 66% Selectophore: 2% Additive: NaTPB 2%	LR: 1.0×10^{-6} to 1.0×10^{-2} M Slope: 20.2 ± 0.2 mV/decade DL: 3.7×10^{-7} pH 2.7–8.1 RT: 5 s II: Sm^{3+}, Co^{3+}	[74]
PME	Nd-13 *N*-Benzoylethylidene-2-aminobenzylamine (BEA)	PVC: 30% Plasticizer: NB 58% Selectophore: 2% Additive: OA 10%	LR: 1.0×10^{-6} to 1.0×10^{-2} M Slope: 21.2 ± 0.2 mV/decade DL: 6.3×10^{-7} pH 2.4–8.5 RT: 7 s LT: 8 weeks II: None	[75]

PME and CGE	Nd-14 Multidentate Schiff bases 5-((1-(3-(3-((5-mercapto-1,3,4-thiadiazol-2-ylimino)methyl)-1*H*-indol-1-yl)propyl)-1*H*-indol-3-yl)methyleneamino)-1,3,4-thiadiazole-2-thiol) (S-1) and *N*-((1-(3-(3-((thiazol-2-ylimino)methyl)-1*H*-indol-1-yl)propyl)-1*H*-indol-3-yl)methylene)thiazol-2-amine (S-2)	PVC: 36% Plasticizer: *o*-NPOE 58% Selectophore: 4% Additive: KTpClPB 2% for S-1 PVC: 37% Plasticizer: *o*-NPOE 55% Selectophore: 6% Additive: KTpClPB 2% for S-2	LR: 2.6×10^{-7} to 1.0×10^{-1} M for PME S_1 LR: 3.1×10^{-6} to 1.0×10^{-1} M for PME S_2 LR: 3.1×10^{-8} to 1.0×10^{-1} M for CGE S_1 Slope: 19.3 ± 0.7 mV/decade for PME S_1 Slope: 19.5 ± 0.4 mV/decade for PME S_2 Slope: 19.8 ± 0.3 mV/decade for CGE S_1 DL: 8.5×10^{-8} PME S_1 DL: 7.4×10^{-7} PME S_2 DL: 1.1×10^{-8} CGE S_1 pH 3.5–8 for both PME and CGE based on S_1 RT: 10 s for CGE RT: 14 s for PME LT: 105 days for CGE LT: 90 days for PME II: None	[76]
PME	Nd-15 1,2-Dimaleimidobenzene (DMB)	PVC: 30% Plasticizer: DBP 61% Selectophore: 2% Additive: NaTPB 2% and 5% OA	LR: 1.0×10^{-6} to 1.0×10^{-2} M Slope: 20.7 ± 0.3 mV/decade DL: 5.8×10^{-7} pH 2.3–9.1 RT: 5 s II: None	[77]

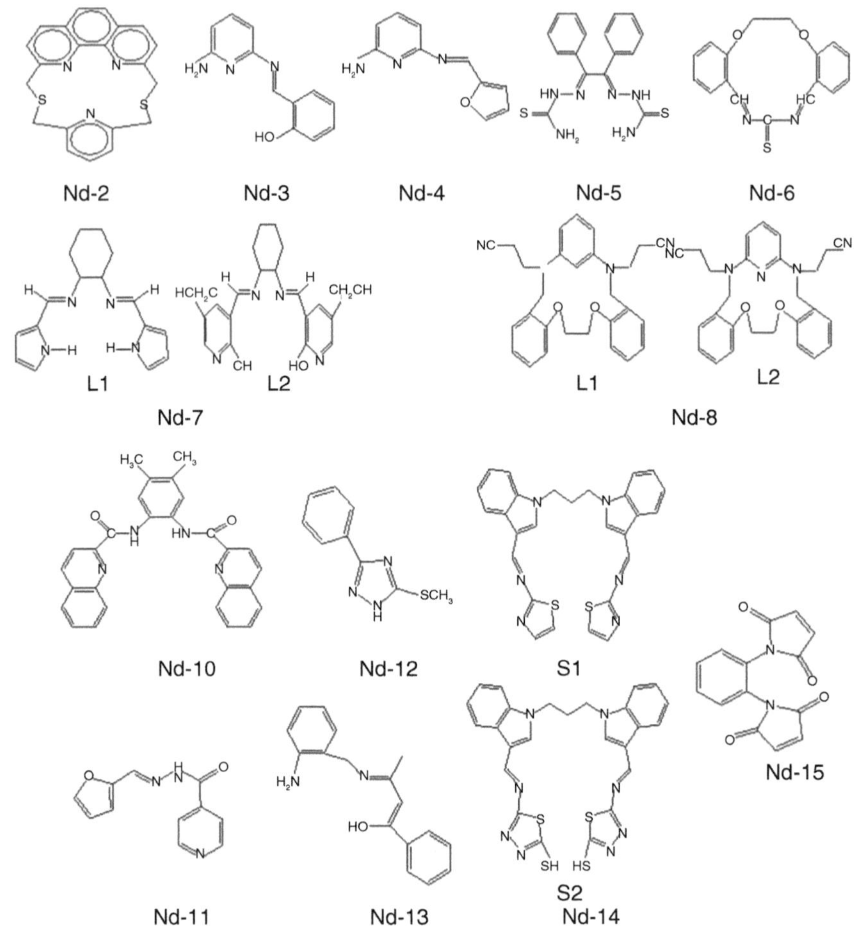

FIGURE 5.14 Chemical structures of the reported selectophores for neodymium ions.

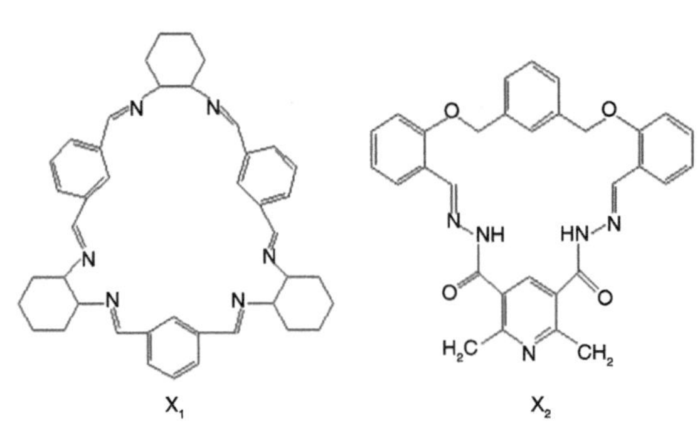

LR: 4.5×10^{-7}–1.0×10^{-2} M for X_1
LR: 3.5×10^{-6}–1.0×10^{-2} M for X_2
DL: 3.2×10^{-7} M for X_1
DL: 2.3×10^{-6} M for X_2
Slope: 20.0 ± 0.5 mV/decade for X_1
Slope:19.5 ± 0.5 mV/decade for X_2
pH: 4–7.5 forX_1
pH: 4–6.5 for X_2
RT: 10s for X_2
RT: 14s for X_2

FIGURE 5.15 Chemical structures of the reported selectophores for promethium ions.

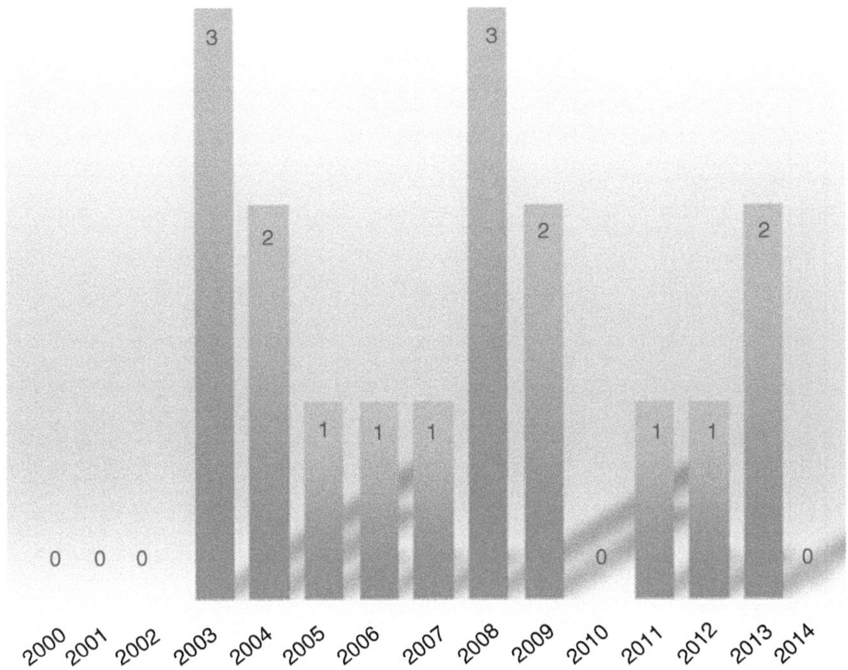

FIGURE 5.16 Historical trend of the reports on the development of Sm^{3+} ISEs.

and CWE. Fig. 5.16 illustrates the historical trend of the reports on the development of Sm^{3+} ISEs in the literature.

Table 5.6 shows the properties of the sensors developed based on the different selectophores, the structures of which are illustrated in Fig. 5.17.

As described earlier, ionophores with donating heteroatoms but lacking semicavities in their structures, such as Sm-2, Sm-3, Sm-6, Sm-10, Sm-11, and Sm-13, cannot form optimized wraparound complexes with one species (here Sm^{3+}) and hence lead to sensors with low selectivity (Table 5.6). The structures of Sm-7, Sm-8, Sm-9, and Sm-13, however, allow for such complexes with samarium cation, and hence the sensors based on these ionophores show relatively good selectivity. Sm-1 and Sm-8 have been used in PMEs as well as a GCE and CPE, respectively. In the case of Sm-12 (i.e., 7,16-dibenzyl-1,4,10,13-tetraoxa-7,16-diazacyclooctadecane [DBDA18C6]), the ionophore was used in a composite multiwalled carbon nanotube polyvinyl chloride matrix. The optimum electrode membrane was composed of a mixture of DBDA18C6 (2.5 wt%), MWNT (2.0 wt%), NaTPB (2.0 wt%), *o*-NPOE (60.5 wt%), and PVC (33 wt%). The sensors corresponding to Sm-4 and Sm-5 (which are

TABLE 5.6 The Characterization of the Reported Samarium Ion Membrane Sensors

Electrode type	Selectophore	Electrode composition	Characterizations of the electrode	Ref.
PME	Sm-1 *N*-[2-[4-[[[(Cyclohexyl amino) carbonyl] an-lino] sulfonyllphenyll ethyl] -5 -methyl pyrazine carboxamide (glipizid)	PVC: 30% Plasticizer: BA 53% Selectophore: 11% Additive: NaTPB 6%	LR: 1×10^{-6} to 1×10^{-1} M Slope: 19.8 mV/decade DL: 6.0×10^{-7} M pH 4.0–8.0 RT: <10 s LT: 2 months II: None	[79]
PME and CGE	Sm-2 Isopropyl 2-[isopropoxycarbothioyl) disulfanyl]ethanethioate	PVC: 30% Plasticizer: BA 57% Selectophore: 3% Additive: OA 10%	LR: 1×10^{-5} to 1×10^{-1} M for PME LR: 1×10^{-6} to 1×10^{-1} M for CGE Slope: 19.5 ± 0.4 mV/decade for PME Slope: 19.3 ± 0.3 mV/decade for CGE DL: 3.1×10^{-6} M for PME DL: 5.0×10^{-7} M for CGE pH 4.0–7.0 RT: 20 s for PME RT: 18 s for CGE II: Ce^{3+}, La^{3+}, Gd^{3+}, Cu^{2+}, Pb^{2+}, Zn^{2+}, Cd^{2+}, Ni^{2+}, Co^{2+}, Mg^{2+}, Ag^{+} for PME	[80]
PME and CGE	Sm-3 4,5,6,7-Tetrathiocino[1,2-b:3,4-b′] diimidazolyl-1,3,8,10-tetraethyl-2,9-dithione	PVC: 30% Plasticizer: BA 57% Selectophore: 3% Additive: OA 10%	LR: 1×10^{-5} to 1×10^{-1} M for PME LR: 1×10^{-7} to 1×10^{-1} M for CGE Slope: 19.6 ± 0.5 mV/decade for PME Slope: 19.3 ± 0.3 mV/decade for CGE DL: 8.0×10^{-6} M for PME DL: 1.6×10^{-8} M for CGE pH 4.0–6.5 RT: 20 s for PME RT: 18 s for CGE II: Ce^{3+}, La^{3+}, Gd^{3+}, Pb^{2+}, Hg^{2+} for PME	[81]

CGE	Sm-1	PVC: 30% Plasticizer: BA 53% Selectophore: 11% Additive: NaTPB 6%	LR: 1×10^{-10} to 1×10^{-5} M Slope: 19.3 ± 0.4 mV/decade DL: 8.0×10^{-11} M pH 3.0 – 10.5 RT: < 15 s LT: 10 weeks	[82]
PME	Sm-4 Tin (IV) boratophosphate	Selectophore: 40% Epoxy resin: 60%	LR: 1×10^{-5} to 1×10^{-1} M Slope: 40.0 mV/decade DL: 1.0×10^{-5} M pH 4.0–10.0 RT: 10 s LT: 6 months II: Pr^{3+}, Ca^{2+}, Na^{+}, Eu^{3+}, Nd^{3+}	[83]
PME	Sm-5 Zirconium (IV) boratophosphate	Selectophore: 10% PVC: 90%	LR: 1×10^{-5} to 1×10^{-1} M Slope: 20.2 mV/decade DL: 1.0×10^{-5} M pH 6.0–10.0 RT: 15 s II: Pr^{3+}, Na^{+}, Eu^{3+}, Ca^{2+}, Nd^{3+}	[84]
PME	Sm-6 1-Phenyl-3′(2-nitrophenyl) spiro[oxirane-2.4-pyrazoline]-3,5-dione (PNSOP)	PVC: 60% Plasticizer: *o*-NPOE 120% Selectophore: 1% Additive: KTClPB 0.5%	LR: 1×10^{-6} to 1×10^{-1} M Slope: 19.3 ± 0.2 mV/decade DL: 6.1×10^{-7} M pH 4.0–8.2 RT: ~5 s LT: 4 months II: Ce^{3+}, Fe^{3+}, Nd^{3+}	[85]

(*Continued*)

TABLE 5.6 The Characterization of the Reported Samarium Ion Membrane Sensors (*cont.*)

Electrode type	Selectophore	Electrode composition	Characterizations of the electrode	Ref.
PME	Sm-7 3-{[2-*oxo*-1(2*H*)-Acenaphthylenyliden]amino}-2-thioxo- 1,3-thiazolidin-4-one (ATTO)	PVC: 30% Plasticizer: *o*-NPOE 63.5% Selectophore: 5% Additive: NaTPB 1.5%	LR: 1×10^{-6} to 1×10^{-1} M Slope: 19.3 ± 0.6 mV/decade DL: $(5.5 \pm 0.3) \times 10^{-7}$ M pH 3.5–7.5 RT: ~10 s LT: 10 weeks II: None	[86]
PME	Sm-8 2-((2-Thioxothiazolidin-4-one) methyl)phenol (TMP)	PVC: 30% Plasticizer: DBP 63% Selectophore: 5% Additive: NaTPB 2%	LR: 1×10^{-7} to 1×10^{-1} M Slope: 19.35 ± 0.2 mV/decade DL: 1.0×10^{-7} M pH 3.0–8.0 RT: 10–15 s LT: 7 weeks II: Gd^{3+}	[87]
CWE	Sm-8	PVC: 20% Plasticizer: DBP 73% Selectophore: 5% Additive: NaTPB 2%	LR: 1×10^{-9} to 1×10^{-4} M Slope: 18.27 ± 0.3 mV/decade DL: 8.0×10^{-10} M pH 3.0–8.0 RT: 20–25 s LT: 6 weeks II: None	[88]
PME	Sm-9 2-[{{4-[{3-{[(*z*)-1-(2-Hydroxy phenyl) methylidene] amino}phenyl)oxy] phenyl}imino)methyl]-1-benzenol (APOPIB)	PVC: 30% Plasticizer: NB 62% Selectophore: 5% Additive: NaTPB 3%	LR: 1×10^{-6} to 1×10^{-1} M Slope: 19.3 ± 0.2 mV/decade DL: 6.1×10^{-7} M pH 5.5–9.5 RT: 10–15 s LT: 7 weeks II: Ca^{2+}, Gd^{3+}, Tm^{3+}	[89]

PME	Sm-10 2-Ethoxy-1-ethoxycarbonyl-1,2-dihydroquinoline (EED)	PVC: 30% Plasticizer: DBP 66% Selectophore: 2% Additive: NaTPB 2%	LR: 1×10^{-6} to 1×10^{-2} M Slope: 19.8 ± 0.4 mV/decade DL: 5.0×10^{-7} M pH 3.7–8.2 RT: 10 s II: Co^{2+}, Pb^{2+}, La^{3+}, Yb^{3+}, Tb^{3+}, Lu^{3+}, Eu^{3+}, Dy^{3+}, Ce^{3+}, Nd^{3+}, Pr^{3+}	[90]
PME	Sm-11 2-[(*E*)-1-(1*H*-pyrrol-2-yl)methylidene]-1-hydrazinecarbothioamide (PMH)	PVC: 30% Plasticizer: NB 66% Selectophore: 2% Additive: NaTPB 2%	LR: 1×10^{-6} to 1×10^{-2} M Slope: 19.8 ± 0.3 mV/decade DL: 5.2×10^{-7} M pH 4.2–8.5 RT: 10 s LT: 10 weeks II: La^{3+}, Yb^{3+}, Gd^{3+}, Tb^{3+}, Ce^{3+}, Er^{3+}, Eu^{3+}, Ho^{3+}, Dy^{3+}, Ni^{2+}, Cu^{2+}, Hg^{2+}	[91]
GCE	Sm-12 7,16-Dibenzyl-1,4,10,13-tetraoxa-7,16-diazacyclooctadecane (DBDA18C6)	PVC: 33% Plasticizer: *o*-NPOE 60.5% Selectophore: 2.5% Additive: NaTPB 2% MWNT: 2.0%	LR: 1×10^{-8} to 1×10^{-2} M Slope: 20.2 ± 0.48 mV/decade DL: $6.3 \pm 0.36 \times 10^{-9}$ M pH 2.5–8.5 RT: 8 s LT: 5 months II: Not reported.	[92]

(*Continued*)

TABLE 5.6 The Characterization of the Reported Samarium Ion Membrane Sensors (*cont.*)

Electrode type	Selectophore	Electrode composition	Characterizations of the electrode	Ref.
PME and CGE	Sm-13 *N*2-((3*H*-Indol-3-yl)methylene)-6-phenyl-1,3,5-triazine-2,4-diamine (L_1) and *N*2,*N*4-bis((3*H*-indol-3-yl) methylene)-6-phenyl-1,3,5-triazine-2,4-diamine (L_2)	PVC: 37% Plasticizer: *o*-NPOE 57% Selectophore: 4% Additive: NaTPB 2% for PME L_1 PVC: 40% Plasticizer: *o*-NPOE 54% Selectophore: 4% Additive: NaTPB 2% for PME L_2	LR: 1.5×10^{-7} to 1×10^{-1} M for L_1 PME LR: 8.1×10^{-8} to 1×10^{-1} M for L_2 PME LR: 4.1×10^{-8} to 1×10^{-1} M for L_2 CGE Slope: 19.6 ± 0.4 mV/decade for L_2 CGE Slope: 19.8 ± 0.5 mV/decade for L_1 PME Slope: 19.5 ± 0.2 mV/decade for L_2 PME DL: 1.0×10^{-7} M for L_1 PME DL: 5.3×10^{-8} M for L_2 PME DL: 1.8×10^{-8} M for L_2 CGE pH 3.0–9.0 for CGE L_2 RT: 11 s for CGE RT: 14 s for PME LT: 9 weeks II: Eu^{3+}, Cu^{2+}, Er^{3+}, Cr^{3+}, Co^{2+}, Ca^{2+}, Mg^{2+}, Ag^{+} for L_1 II: Ca^{2+}, Mg^{2+}, Ag^{+} for L_2	[93]
PME	Sm-14 Hexadentate macrocyclic ligand (3,5,13,15-tetramethyl 2,6,12,16,21-22-hexaazatricyclo [15.3.1.(1-17) 1(7-11)]cosa-1(21), 2,5,7,9,11(22),12,15,17,19-decane)	PVC: 30% Plasticizer: DBP 65% Selectophore: 3% Additive: NaTPB 2%,	LR: 1×10^{-8} to 1×10^{-1} M Slope: 20.7 mV/decade DL: 1.58×10^{-7} M RT: 10 s LT: 9 weeks II: full text not available	[94]
PME	Sm-15 *N*,*N*′,*N*″-Tris(4-pyridyl) trimesic amide (TPTA)	PVC: 30% Plasticizer: NB 55% Selectophore: 3% Additive: NaTPB 2%, OA 10%	LR: 1×10^{-6} to 1×10^{-2} M Slope: 19.8 ± 0.5 mV/decade DL: 4.7×10^{-7} M pH 3.6–9.2 RT: 10 s LT: 10 weeks	[95]

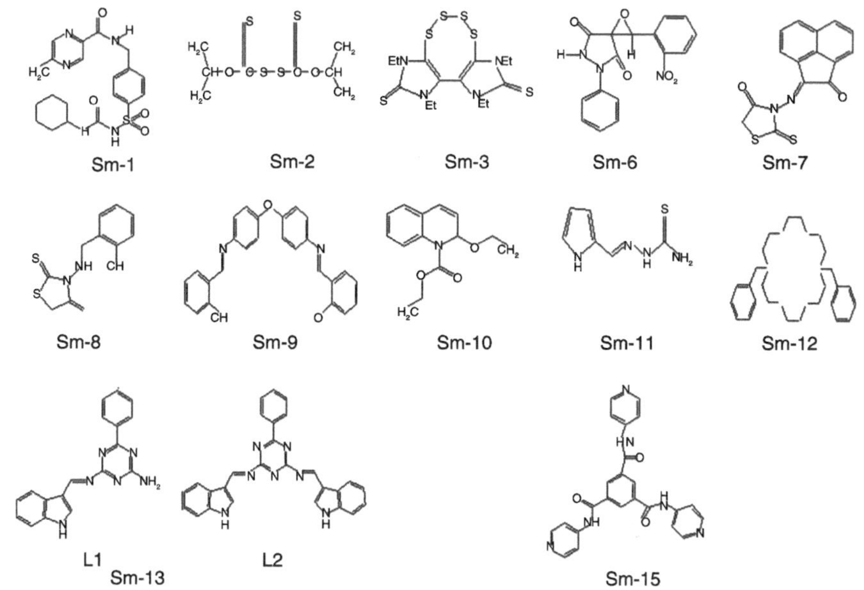

FIGURE 5.17 Chemical structures of the reported selectophores for samarium ions.

tin(IV) boratophosphate and zirconium boratophosphate crystals) work based on in-exchange mechanisms.

Europium ISEs

The first report on europium potentiometric ISE was published by Ganjali et al. in 2004 [96], and the remaining eight sensors range from PMEs and CPEs to CGEs and CWEs. Fig. 5.18 illustrates the history of reports on europium sensors.

Table 5.7 summarizes the properties of the sensors based on different ionophores, which are also illustrated in Fig. 5.19.

Eu-1, with several S and N atoms in its structure, suffers low selectivity due to the lack of a semi-cavity in its structure and consequently leads unselective responses according to the data in Table 5.7. The same is true with Eu-3, Eu-5, and Eu-6, which cannot form a wraparound complex with only europium ions, although they have S and N donor atoms and a semicavity. This is due to the lack of enough flexibility as a result of the structural limitations of the ligands. Eu-4 and Eu-8 (L2), on the other hand, show a rather good selectivity because of the structural characteristics described for a good ionophore earlier. In the case of Eu-8, the more flexible L2 had a better selectivity than L1, because of the replacement

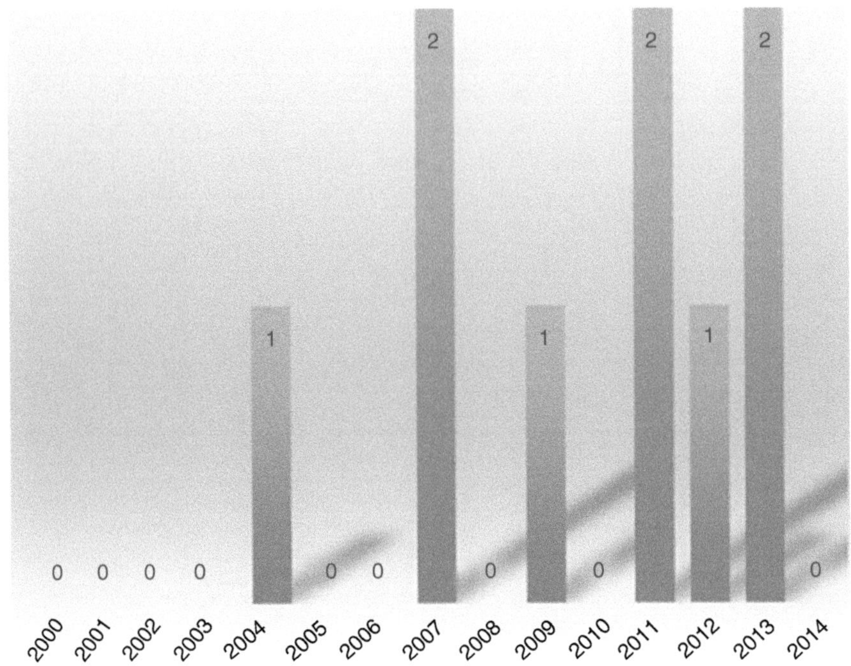

FIGURE 5.18 The history of reports on Eu^{3+} sensors.

of pyridine in L2 with a phenol group, leading to better selectivity behaviors.

Eu-2 has been used in two types of devices, a PME and a CPE. The mechanism is different with the inorganic ion exchanger titanium(IV) tungstophosphate (TiWP) (Eu-7), which works based on the ion exchange properties.

Gadolinium ISEs

The first report on a potentiometric gadolinium ISE was made by Ganjali et al. in 2003 [105], and the remaining 14 have been in the forms of PMEs, CPEs, CGEs, and CWEs. Fig. 5.20 illustrates the historical trend of the reports on gadolinium sensors.

Like the case of the previously reported elements, Table 5.8 summarizes the properties of the reported sensors. The structures of the ionophores are also shown in Fig. 5.21.

Fig. 5.21 shows the chemical structure of the used selectophores for the gadolinium ions. As anticipated, Gd-2, Gd-5, Gd-9, Gd-10, Gd-12, and Gd-13 lack selective interactions because of the lack of a semicavity and their

TABLE 5.7 The Characterization of the Reported Europium Ion Membrane Sensor

Electrode type	The selectophore	Electrode composition	Characterizations of the electrode	Ref.
PME	Eu-1 *N*,*N*-Diethyl-*N*-(4-hydroxy-6methylpyfidin-2-yl)guanidine (GD)	PVC: 30% Plasticizer: BA 55% Selectophore: 5% Additive: OA 10%	LR: 7.0×10^{-5} to 1.0×10^{-1} M Slope: 19.8 ± 0.3 mV/decade DL: 5.0×10^{-5} M RT: <20 s pH 3.0–7.0 II: Gd^{3+}, Ce^{3+}, Tb^{3+}, Co^{2+}, Cu^{2+}, Sm^{3+}, La^{3+}, Pr^{3+}, Zn^{3+}, Ca^{2+}, Yb^{3+}	[96]
PME	Eu-2 S-N hexadentates Schiff's base, bis(thiophenol)butane-2,3-dihydrazone (SNSB)	PVC: 30% Plasticizer: *o*-NPOE 63% Selectophore: 5% Additive: KTpClPB 5%	LR: 1.0×10^{-5} to 1.0×10^{-2} M Slope: 19.8 ± 0.3 mV/decade DL: 5.0×10^{-6} M RT: <5 s pH 3.0–8.5 LT: 8 weeks II: Gd^{3+}, Sm^{3+}	[97]
PME	Eu-3 4-(2-Hydroxybenzylideneamino)-6-methyl-3-thioxo-3,4-dihydro-1,2,4-triazin-5(2*H*)-one (HMTDT)	PVC: 30% Plasticizer: o-NPOE 64% Selectophore: 4% Additive: NaTPB 2%	LR: 1.0×10^{-6} to 1.0×10^{-1} M Slope: 19.7 ± 0.4 mV/decade DL: 7.8×10^{-7} M RT: <15 s pH 3.5–7.8 LT: 8 weeks II: Tb^{3+}, Yb^{3+}, Cr^{3+}	[98]

(*Continued*)

TABLE 5.7 The Characterization of the Reported Europium Ion Membrane Sensor (*cont.*)

Electrode type	The selectophore	Electrode composition	Characterizations of the electrode	Ref.
PME	Eu-4 4*E*-4-(2-Phenylviazenyl)-2-((*E*)-(2-aminoethylimino)methyl) phenol (PMP)	PVC: 33% Plasticizer: o-NPOE 60% Selectophore: 4% Additive: KTpClPB 3%	LR: 4.0×10^{-7} to 1.0×10^{-2} M Slope: 18.8 ± 0.2 mV/decade DL: 1.5×10^{-7} M RT: 25 s pH 3.5–8.0 LT: 8 weeks II: None	[99]
PME	Eu-5 1,2-Diaminopropane-*N*,*N*,*N*′, *N*′-tetraacetic acid (DAPTA)	PVC: 30% Plasticizer: NB 65% Selectophore: 3% Additive: NaTPB 2%	LR: 1.0×10^{-6} to 1.0×10^{-2} M Slope: 19.7 ± 0.5 mV/decade DL: 7.2×10^{-7} M RT: <10 s pH 2.5–9.1 II: Sm^{3+}, Gd^{3+}	[100]
CPE	Eu-2	RTIL: 15% Carbon powder: 63% Selectophore: 15% MWCNT: 5% Nanosilica: 2%	LR: 1.0×10^{-6} to 1.0×10^{-2} M Slope: 19.9 ± 0.2 mV/decade DL: 4.0×10^{-7} M RT: <10 s pH 3.5–9.0 LT: 13 weeks II: Gd^{3+}, Tb^{3+}, La^{3+}, Sm^{3+}, Ce^{3+}	[101]
PME	Eu-6 *N*-Pyridine-2-carboxamido-8-aminoquinoline (PCQ)	PVC: 30% Plasticizer: NB 56% Selectophore: 2% Additive: NaTPB 2%, OA 10%	LR: 1.0×10^{-6} to 1.0×10^{-2} M Slope: 19.8 ± 0.3 mV/decade DL: 6.4×10^{-7} M RT: <10 s pH 2.4–7.4 LT: 2 months II: Gd^{3+}, Tb^{3+}, La^{3+}, Lu^{3+}, Sm^{3+}, Fe^{3+}	[102]

PME	Eu-7 Titanium [IV] tungstophosphate (TiWP)	Titanium tungstophosphate: 40% Epoxy resin: 60%	LR: 1.0×10^{-5} to 1.0×10^{-1} M Slope: 21.0 mV/decade DL: 1.0×10^{-6} M RT: 10 s pH 2.9–8.9 II: full text not available	[103]
CGE	Eu-8 (2((2-Phenyl-2-(pyridin-2-yl) hydrazono) methyl)pyridine) [L_1], (2-((2-phenyl-2-(pyridin-2-yl)hydazono) methyl)phenol) [L_2]	PVC: 44% Plasticizer: 1-CN 53% Selectophore: 3% Additive: NaTPB 2% for L_1 PVC: 40% Plasticizer: 1-CN 52.5% Selectophore: 5% Additive: NaTPB 2.5% for L_2	LR: 1.3×10^{-6} to 1.0×10^{-1} for L_1 LR: 5.7×10^{-8} to 1.0×10^{-1} for L_2 DL: 6.1×10^{-7} for L_1 DL: 3.5×10^{-8} for L_2 Slope: 19.8 ± 0.4 for L_1 Slope: 19.5 ± 0.4 for L_2 RT: 14 s for L_1 RT: 10 s for L_2 pH 3.5–8.5 for L_1 pH 2.7–9.0 for L_2 LT: 4 weeks for L_1 LT: 4 weeks for L_2 II: Fe^{3+}, Cr^{3+}, K^{+}, Sm^{3+}, La^{3+} for L_1 II: Sm^{3+} for L_2	[104]

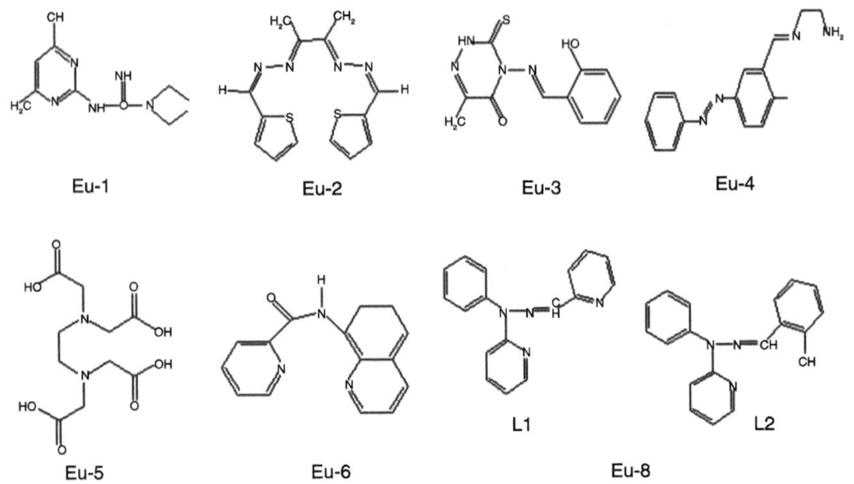

FIGURE 5.19 Chemical structures of the selectophores reported for europium ions.

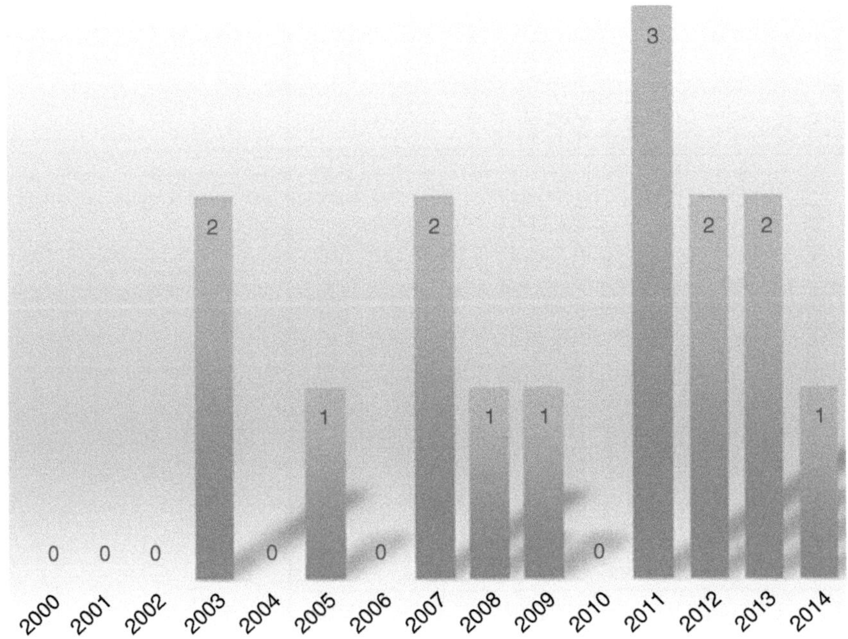

FIGURE 5.20 The historical trend of the reports on Gd^{3+} ISEs.

TABLE 5.8 The Characterization of the Reported Gadolinium Ion Membrane Sensors

Electrode type	The selectophore	Electrode composition	Characterizations of the electrode	Ref.
PME	Gd-1 S-N Schiff s base (2-[{3-[(2-sulfanylphenyl) imino)-1-methylbutylidene}amino]phenyl hydrosulfide (SMPH)	PVC: 33% Plasticizer: BA 61% Selectophore: 5% Additive: NaTPB 2%	LR: 1.0×10^{-5} to 1.0×10^{-1} M Slope: 19.3 ± 0.3 mV/decade DL: 3.0×10^{-6} M pH 4.0–8.0 RT: >10 s LT: 9 weeks II: Tb^{3+}, Dy^{3+}, Eu^{3+}	[105]
PME	Gd-2 Omeprazole	PVC: 35% Plasticizer: DBP 54% Selectophore: 5% Additive: TPB 6%	LR: 1.0×10^{-5} to 1.0×10^{-1} M Slope: 19.3 ± 0.3 mV/decade DL: 5.0×10^{-6} M pH 4.5–10.0 RT: <3 s LT: 2 months II: Ce^{3+}, UO_2^{2+}, Cd^{2+}, Cu^{2+}	[106]
PME	Gd-3 Bis(thiophenal) pyridine-2,6-diamine (BPD)	PVC: 30% Plasticizer: BA 62% Selectophore: 5% Additive: KTKpClPB 3%	LR: 1.0×10^{-6} to 1.0×10^{-1} M Slope: 19.4 ± 0.4 mV/decade DL: 7.0×10^{-7} M pH 3.5–8.0 RT: <20 s LT: 8 weeks II: None	[107]
PME	Gd-4 *N*-(2-Pyridyl)-*N*′-(4-nitrophenyl)thiourea ($PyTu4NO_2$)	PVC: 30% Plasticizer: BA 63% Selectophore: 5% Additive: KTpClPB 2%	LR: 3.0×10^{-7} to 1.0×10^{-1} M Slope: 19.95 ± 0.3 mV/decade DL: 3.0×10^{-7} M pH 4.0–9.0 RT: 10 s LT: 8 weeks II: None	[108]

(Continued)

TABLE 5.8 The Characterization of the Reported Gadolinium Ion Membrane Sensors (*cont.*)

Electrode type	The selectophore	Electrode composition	Characterizations of the electrode	Ref.
PME	Gd-5 6-Methyl-4-{[1-(2-thienyl)methylidene] amino}3-thioxo-3,4-dihydro-1,2,4-triazin-5-(2H)-one (MATDTO)	PVC: 32% Plasticizer: *o*-NPOE 63% Selectophore: 3.5% Additive: NaTPB 1.5%	LR: 1.0×10^{-6} to 1.0×10^{-1} M Slope: 19.8 ± 0.2 mV/decade DL: 5.8×10^{-7} M pH 3.2–8.7 RT: <10 s LT: 2 months II: Hg^{2+}	[109]
PME	Gd-4	PVC: 20% Plasticizer: BA 72% Selectophore: 5% Additive: KTpClPB 3%	LR: 1.0×10^{-8} to 1.0×10^{-3} M Slope: 17.46 ± 0.3 mV/decade DL: 3.0×10^{-9} M pH 4.0–9.0 RT: 10–15 s LT: 6 weeks II: Sm^{3+}, Eu^{3+}, Dy^{3+}, Er^{3+}	[110]
PME	Gd-6 Zirconium (IV) tungstophosphate [ZrWP]	Zirconium (IV) tungstophosphate: 40% Epoxy resin: 60%	LR: 1.0×10^{-5} to 1.0×10^{-1} M Slope: 30 mV/decade pH 4.0–10.0 for 10^{-1} M and 3.0–7.0 for 10^{-2} M RT: 15 s II: Nd^{3+}, Dy^{3+}, Pb^{2+}, Fe^{3+}	[111]
CGE	Gd-7 2,6-Bis-[1-{*N*-cyanopropyl,*N*-(2-methylpridyl)}aminoethyl]pyridine [P]	PVC: 30% Plasticizer: *o*-NPOE 58% Selectophore: 8% Additive: NaTPB 4%	LR: 2.8×10^{-7} to 5.0×10^{-2} M Slope: 19.6 ± 0.1 mV/decade DL: $(6.3 \pm 0.1) \times 10^{-8}$ M pH 2.0–8.0 RT: 12 s LT: 2.5 months II: None	[112]

PME	Gd-8 Tin (II) tungstophosphate [SnWP]	Tin (II) tungstophosphate: 40% Epoxy resin: 60%	LR: 1×10^{-6} to 1×10^{-2} M Slope: 28 mV/decade pH 3.4 – 10.2 for 10^{-2} M and 3.3 – 8.0 for 10^{-3} M RT: <5 s II: not found PDF article	[113]
PME	Gd-9 4-Methoxyphenylcyanamide (4-MeOpcyd)	PVC: 30% Plasticizer: TBP 60% Selectophore: 6% Additive: NaTPB 4%	LR: 1×10^{-6} to 1×10^{-2} M Slope: 19.5 mV/decade DL: 6.2×10^{-7} M pH 4.1–8.3 RT: 7 s LT: 6 weeks II: Mg^{2+}, Ca^{2+}, Eu^{3+}, Y^{3+}	[114]
PME	Gd-10 3-Methyl-1*H*-1,2,4-triazole-5-thiol (MTH)	PVC: 30% Plasticizer: NB 66% Selectophore: 2% Additive: NaTPB 2%	LR: 1×10^{-7} to 1×10^{-2} M Slope: 19.8 ± 0.4 mV/decade DL: 7.3×10^{-8} M pH 2.9–8.4 RT: <10 s II: Not reported.	[115]
PME	Gd-11 *N,N′*-Bis(methylsalicylidene)-2-aminobenzylamine (MSAB)	PVC: 30% Plasticizer: AP 66% Selectophore: 2% Additive: NaTPB 2%	LR: 1×10^{-6} to 1×10^{-2} M Slope: 19.7 ± 0.3 mV/decade DL: 5.0×10^{-7} M pH 3.5–10.1 RT: <10 s LT: 10 weeks II: None	[116]

(*Continued*)

TABLE 5.8 The Characterization of the Reported Gadolinium Ion Membrane Sensors (*cont.*)

Electrode type	The selectophore	Electrode composition	Characterizations of the electrode	Ref.
PME	Gd-12 2-(2-Aminothiazol-4-yl)phenol (L_1) 2-(4-phenyl-1,3-thiazol-2-yliminomethyl) phenol (L_2)	PVC: 30% Plasticizer: AP 62% Selectophore: 5% Additive: NaTPB 3% for L_1 PVC: 33% Plasticizer: AP 60% Selectophore: 5% Additive: NaTPB 2% for L_2	LR: 4.72×10^{-6} to 1×10^{-2} M for L_1 LR: 9.38×10^{-7} to 1×10^{-2} M for L_2 Slope: 19.0 mV/decade for L_1 Slope: 19.5 ± 0.5 mV/decade for L_2 DL: 5.0×10^{-7} M for L_2 pH 3.5–8.5 RT: 10 s for L_1 RT: 8 s for L_2 LT: 3 months II: Nd^{3+}, Ce^{3+}, La^{3+}	[117]
PME	Gd-13 1-(4-Nitrophenyl)-3-(4-phenylazophenyl) triazene (PPAT)	PVC: 30% Plasticizer: NB 60% Selectophore: 6% Additive: NaTPB 4%	LR: 1×10^{-6} to 1×10^{-2} M Slope: 19.9 ± 0.2 mV/decade DL: 4.7×10^{-7} M pH 2.6–8.7 RT: ~10 s LT: 10 weeks II: Yb^{3+}, Ca^{2+}	[118]
PME	Gd-14 *N′*-(2-*oxo*-1,2-di(pyridin-2-yl)ethylidene) furan-2-carbohydrazide (L)	PVC: 30% Plasticizer: *o*-NPOE 51% Selectophore: 6% Additive: NaTPB 3%, OA 10%	LR: 1×10^{-6} to 1×10^{-2} M Slope: 19.9 ± 0.6 mV/decade DL: 4.2×10^{-7} M pH 4.2–8.0 RT: ~10 s LT: 10 weeks II: None	[119]

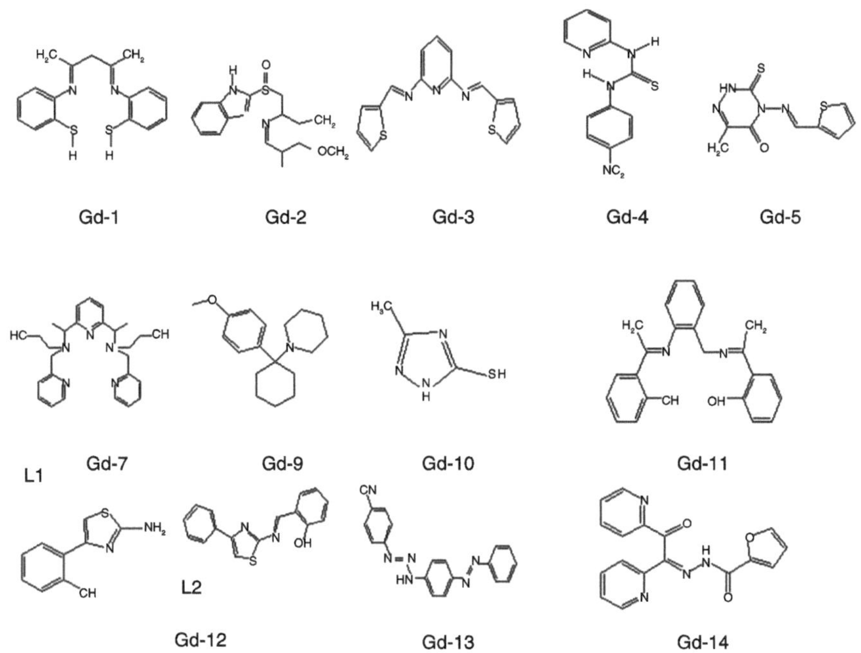

FIGURE 5.21 Chemical structures of the reported selectophores for gadolinium ions.

inability to form wraparound complexes with Gd^{3+}. The same behavior is observed with Gd-1, which is a symmetric Schiff base with high rigidity and hence low selectivity, or with Gd-6 and Gd-8. Gd-3, Gd-4, Gd-7, Gd-11, and Gd-14, however, show good selectivity because of the same structural reasons as mentioned in their cases. Gd-4 was used both in a PME and a CWE.

Terbium ISEs

The first report on a terbium potentiometric ISE was also made by Ganjali et al., in 2005 [120], and a further 11 PMEs, CPEs, CGEs, and CWEs followed in the next years (Fig. 5.22), the details on the properties of which are given in Table 5.9.

Fig. 5.23 shows the chemical structure of the selectophores used for Tb(III) ions. As we can anticipate based on the previous discussions, a cyclic ionophore such as Tb-6 shows a very low selectivity, and a similar behavior is observed with Tb-1, Tb-2, Tb-4, Tb-5, Tb-7, Tb-8 and Tb-11, which lack enough flexibility to form wraparound complexes. Tb-9, Tb-10, and Tb-12, however, show good selectivity because of their structural fitness. The ion-exchange zirconium(IV) boratophosphate (Tb-3)

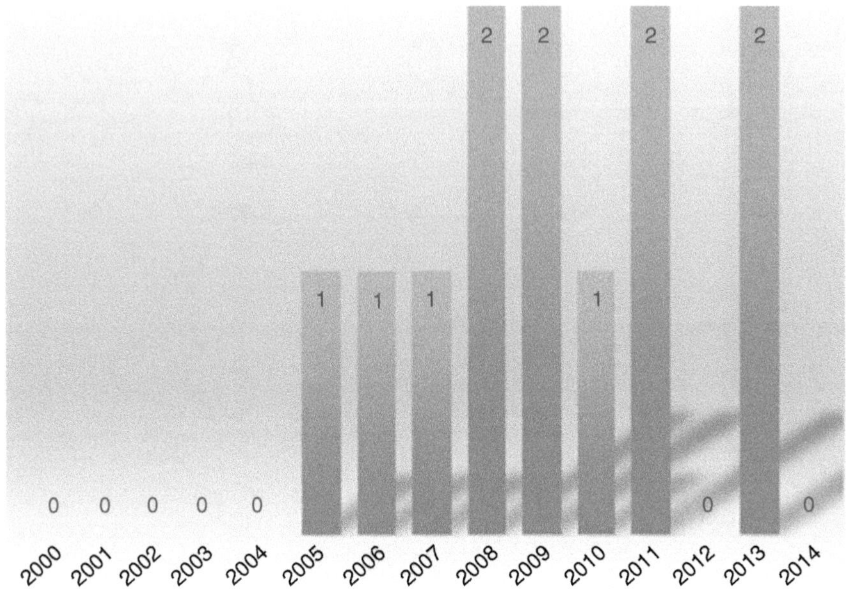

FIGURE 5.22 Historical distribution of reports on the potentiometric sensors for Tb^{3+}.

has a different structure and mechanism of action as compared to membrane electrodes.

DYSPROSIUM ISEs

The first report on a dysprosium ISE was made by Gupta et al. in 2003 [132] and the remaining 10 PMEs, CPEs, CGEs, and CWEs followed afterwards. Fig. 5.24 shows the historical trend of the development of Dy^{3+} sensors.

Table 5.10 includes the structures of all ionophores used for the construction of Dy^{3+} sensors.

Fig. 5.25 shows the chemical structure of the selectophores used for the construction of Dy^{3+} ISEs. About 11 selectophores have been reported for Dy^{3+} ions, so far. Among them, Dy-3, Dy-5, and Dy-11 show rather good selectivity because of preferred size and structural properties as well as flexibility.

One report is on the dispersion of Dy^{3+} imprinted polymer (Dy-4) particles in 2-nitrophenyloctyl ether plasticizer and then embedding the resulting mixture in a polyvinyl chloride matrix, which led to good selectivity.

TABLE 5.9 The Characterization of the Reported Terbium Ion Membrane Sensors

Electrode type	The selectophore	Electrode composition	Characterizations of the electrode	Ref.
PME	Tb-1 *N,N,*-Bis(pyrrolidene) benzene-1,2-diamine	PVC: 30% Plasticizer: BA 63% Selectophore: 5% Additive: TPB 2%	LR: 1.0×10^{-5} to 1.0×10^{-1} M Slope: 19.8 mV/decade DL: 7.0×10^{-6} M pH 2.0–8.0 RT: <20 s LT: 2 months II: La^{3+}, Dy^{3+}, Yb^{3+}, Ce^{3+}, Sm^{3+}	[120]
PME	Tb-2 4-Amino-3-{2-[4-amino-6-methyl-5-*oxo*-4,5-dihydro-1,2,4-triazin-3(2*H*)-yliden] hydrazono}-6-methyl-3,4-dihydro-1,2,4-triazin-5(2*H*)-one (ATO)	PVC: 30% Plasticizer: DBP 61% Selectophore: 5.5% Additive: NaTPB 3.5%	LR: 1.0×10^{-6} to 1.0×10^{-1} M Slope: 19.4 ± 0.5 mV/decade DL: 8.6×10^{-7} M pH 2.0–8.0 RT: ~15 s LT: 2 months II: Gd^{3+}	[121]
PME	Tb-3 Zirconium (IV) boratophosphate	ZrBP: 60% Epoxy resin: 40%	LR: 1.0×10^{-4} to 1.0×10^{-1} M Slope: 45 mV/decade pH 2.0–10.0 RT: 15 s II: full text not available	[122]
PME	Tb-4 *S*-2-Benzothiazolyl-2-amino-alpha-(methoxyimino)-4-thiazolethiol acetate	PVC: 31.7% Plasticizer: BA 63.4% Selectophore: 3.17% Additive: KTpClPB 1.58%	LR: 1.5×10^{-7} to 1.0×10^{-2} M Slope: 19.5 mV/decade DL: 9.3×10^{-8} M pH 2.0–8.0 RT: ~10 s LT: 3 months II: Yb^{3+}	[123]

(*Continued*)

TABLE 5.9 The Characterization of the Reported Terbium Ion Membrane Sensors (*cont.*)

Electrode type	The selectophore	Electrode composition	Characterizations of the electrode	Ref.
PME	Tb-5 *N*,*N*-Dimethyl-*N*′,*N*″-bis(4-methoxyphenyl)phospho ramidate (DMP)	PVC: 32% Plasticizer: o-NPOE 59% Selectophore: 6% Additive: NaTPB 3%	LR: 1.0×10^{-6} to 1.0×10^{-1} M Slope: 19.7 ± 0.4 mV/decade DL: 8.0×10^{-7} M pH 3.5–8.0 RT: ~10s LT: 2 months II: Gd^{3+}	[124]
PME	Tb-6 Hematoporphyrin (HP)	PVC: 30% Plasticizer: AP 66% Selectophore: 2% Additive: NaTPB 2%	LR: 1.0×10^{-6} to 1.0×10^{-2} M Slope: 19.8 ± 0.3 mV/decade DL: 7.4×10^{-7} M pH 3.3–8.7 RT: ~10 s II: Fe^{3+}, Eu^{3+}, Tm^{3+}, Pr^{3+}, Nd^{3+}, Co^{2+}	[125]
PME	Tb-7 2-{2-(2-Anilinocarbothioyl) hydrazino)-2-oxoacetyl}-*N*-phenyl-1-hydrazinecarbothioamide (AHC)	PVC: 30% Plasticizer: AP 66% Selectophore: 2% Additive: NaTPB 2%	LR: 1.0×10^{-6} to 1.0×10^{-2} M Slope: 19.6 ± 0.3 mV/decade DL: 6.8×10^{-7} M pH 3.5–8.5 RT: 10 s LT: 10 weeks II: Pr^{3+}	[126]
CPE	Tb-8 *N*′-(2-Naphthoyl)-8-(dimethylamino) naphthalene-2-sulfonohydrazide (L)	[bmim]BF_4: 15% Selectophore: 15% Carbon powder: 60% MWCNT: 10%	LR: 7.5×10^{-8} to 1.0×10^{-2} M Slope: 19.9 ± 0.2 mV/decade DL: 6.0×10^{-8} M pH 3.0–8.0 RT: <16 s LT: 3 months II: Er^{3+}, Tm^{3+}, Ho^{3+}, Cu^{2+}, Nd^{3+}, Yb^{3+}, Lu^{3+}, Gd^{3+}, Pr^{3+}, Eu^{3+}, Cr^{3+}	[127]

PME	Tb-9 *N,N′*-Bis(pyrrolylmethylene)-2-aminobenzylamine (PMA)	PVC: 30% Plasticizer: AP 58% Selectophore: 2% Additive: OA 10%	LR: 1.0×10^{-5} to 1.0×10^{-2} M Slope: 19.7 ± 0.4 mV/decade DL: 4.6×10^{-6} M pH 2.9–8.1 RT: ~5 s II: None	[128]
PME	Tb-10 *N,N′*-Bis(5-nitrosalicylidene)-2-aminobenzylamin (L)	PVC: 30% Plasticizer: NB 65% Selectophore: 3% Additive: NaTPB 2%	LR: 1.0×10^{-6} to 1.0×10^{-2} M Slope: 20.1 mV/decade DL: 6.3×10^{-7} M pH 2.6–9.4 RT: ~10 s LT: 2 months II: None	[129]
PME	Tb-11 Di(*N*-succinimidyl) carbonate (NSC)	PVC: 30% Plasticizer: NB 58% Selectophore: 2% Additive: OA 10%	LR: 1.0×10^{-6} to 1.0×10^{-2} M Slope: 19.8 ± 0.3 mV/decade DL: 5.8×10^{-7} M pH 4.0–9.0 RT: ~5 s II: Fe^{3+}, Yb^{3+}, Gd^{3+}, Pr^{3+}	[130]
PME	Tb-12 3-Hydroxybenzaldehyde azine (HBH)	PVC: 30% Plasticizer: NB 66% Selectophore: 2% Additive: NaTPB 2%	LR: 1.0×10^{-6} to 1.0×10^{-2} M Slope: 20.2 ± 0.2 mV/decade DL: 8.3×10^{-7} M pH 2.8 – 9.2 RT: ~5 s II: None	[131]

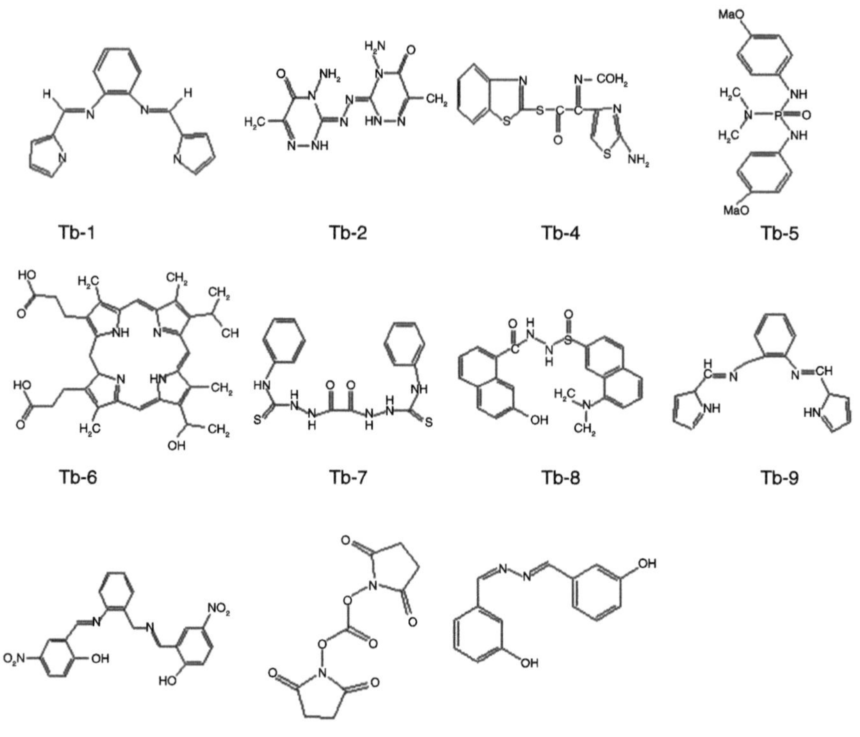

FIGURE 5.23 Chemical structures of the reported selectophores for terbium ions.

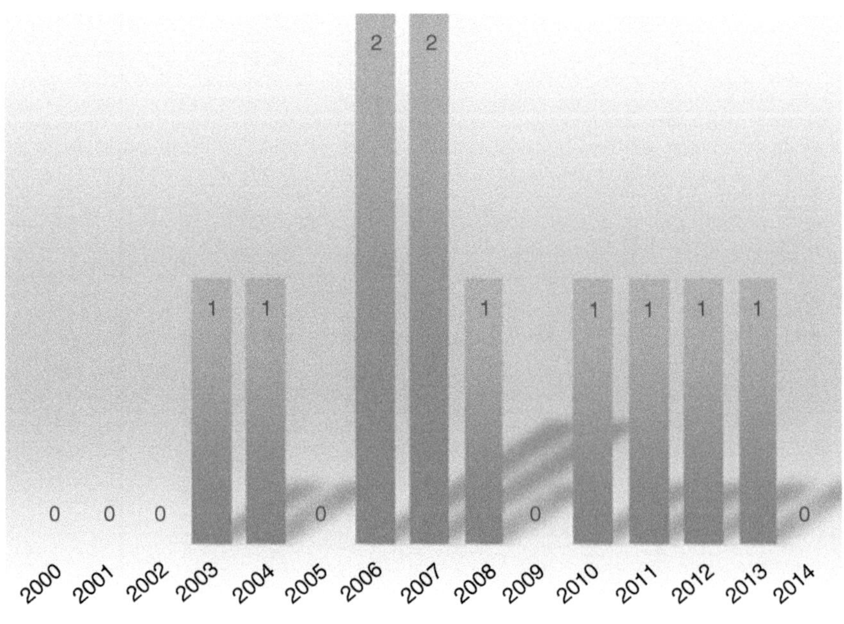

FIGURE 5.24 Historical trend of Dy^{3+} potentiometric sensors.

TABLE 5.10 The Properties of the Reported Dysprosium ISEs

Electrode type	Selectophore	Electrode composition	Characterizations of the electrode	Ref.
PME	Dy-1 Zirconium (IV) selenomolybdate	PVC: 75% Selectophore: 25%	LR: 5.0×10^{-5} to 1.0×10^{-1} M Slope: 18.3 mV/per decade DL: 3.8×10^{-5} pH 2.0–9.0 RT: 15 s II: almost all lanthanide	[132]
PME	Dy-2 *N,N*-Bis(pyrrolidene) benzne-1,2-diamine	PVC: 30% Plasticizer: BA 63% Selectophore: 5% Additive: TPB 2%	LR: 1.0×10^{-5} to 1.0×10^{-1} M Slope: 20.6 ± 0.2 mV/per decade DL: 6.0×10^{-6} pH 3.5–8.0 RT: <20 s LT: 2 months II: La^{3+}, Ce^{3+}, Yb^{3+}, Sm^{3+}, Gd^{3+}	[133]
PME	Dy-3 (*E*)-*N*-(2-Hydroxybenzylidene) benzohydraide (BBH)	PVC: 30% Plasticizer: NB 61% Selectophore: 5% Additive: NaTPB 4%	LR: 1.0×10^{-6} to 1.0×10^{-2} M Slope: 20.1 ± 0.8 mV/per decade DL: 8.0×10^{-7} pH 3.0–8.0 RT: <20 s LT: 6 weeks II: Yb^{3+}	[134]
PME	Dy-4 Dysprosium (III) ion imprinted polymer	PVC: 22% Plasticizer: *o*-NPOE 50% Selectophore: 22% Additive: NaTPB 6%	LR: 8.0×10^{-6} to 1.0×10^{-1} M Slope: 21.7 mV/per decade DL: 2.0×10^{-6} pH 6.5 – 8.5 RT: ~10 s II: None	[135]

(Continued)

TABLE 5.10 The Properties of the Reported Dysprosium ISEs (*cont.*)

Electrode type	Selectophore	Electrode composition	Characterizations of the electrode	Ref.
PME	Dy-5 Procaine	PVC: 30% Plasticizer: DBP 63% Selectophore: 5% Additive: NaTPB 2%	LR: 1.0×10^{-6} to 1.0×10^{-1} M Slope: 19.6 ± 0.3 mV/per decade DL: 7.9×10^{-7} pH 4.0–8.0 RT: <10 s LT: 8 weeks II: None	[136]
PME	Dy-6 6-Hydrazino-1,5-diphenyl-6,7-dihydropyrazolo-[3,4-d] pyrimidine-4(5*H*)-imine (HDDPI)	PVC: 30% Plasticizer: *o*-NPOE 63% Selectophore: 5% Additive: NaTPB 2%	LR: 8.0×10^{-7} to 1.0×10^{-1} M Slope: 19.6 ± 0.3 mV/per decade DL: 4.2×10^{-7} pH 3.5–8.3 RT: <10 s LT: 10 weeks II: Full text not available	[137]
PME	Dy-7 Benzoxazoleguanidine (BG)	PVC: 30% Plasticizer: o-NPOE 60.5% Selectophore: 6.5% Additive: NaTPB 3%	LR: 1.0×10^{-6} to 1.0×10^{-1} M Slope: 19.5 ± 0.4 mV/per decade DL: 4.7×10^{-7} pH 3.3 – 8.4 RT: <10 s LT: 2 months II: None	[138]
PME	Dy-8 Zirconium (IV) antimonomolybdate	ZrSbMo: 60% Epoxy resin: 40%	LR: 1.0×10^{-4} to 1.0×10^{-1} M Slope: 44 mV/per decade DL: 3.9×10^{-8} pH 2.10–9.8 RT: <10 s II: not recognized	[139]

PME	Dy-9 *N*, *N′*-Ethylene bis(salicylideneimine)	Not found PDF article 60% Schiff base	LR: 1.0×10^{-6} to 1.0×10^{-1} M Slope: 20 mV/per decade DL: 3.9×10^{-8} pH 2.3–10.8 RT: 10 s II: Not found, PDF article	[140]
PME	Dy-10 Thiourea (DPDP)	PVC: 30% Plasticizer: BA 63% Selectophore: 5% Additive: NaTPB 2%	LR: 1.0×10^{-7} to 5.0×10^{-1} M Slope: 20.5 ± 0.3 mV/per decade DL: 3.9×10^{-8} pH 3.4–8.2 RT: 6–9 s LT: 11 weeks II: Tb^{3+}	[141]
PME	Dy-11 3,4-Diamino-*N′*-((pyridin-2-yl) methylene)benzohydrazide (L)	PVC: 30% Plasticizer: *o*-NPOE 54% Selectophore: 1% Additive: OA 15%	LR: 1.0×10^{-6} to 1.0×10^{-2} M Slope: 19.8 ± 0.6 mV/per decade DL: 5.5×10^{-7} pH 2.4–8.9 RT: <10 s LT: 8 weeks II: None	[142]

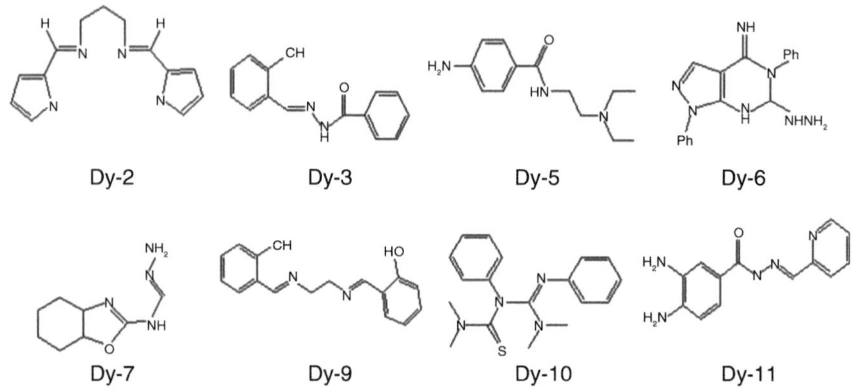

FIGURE 5.25 Chemical structures of the ionophores reported selectophores for Dy^{3+} ions.

HOLMIUM ISEs

The first report on a potentiometric holmium ISE was made by Ganjali et al. in 2006 [143], and about 13 PMEs, CPEs, CGEs, and CWEs were devised (Fig. 5.26).

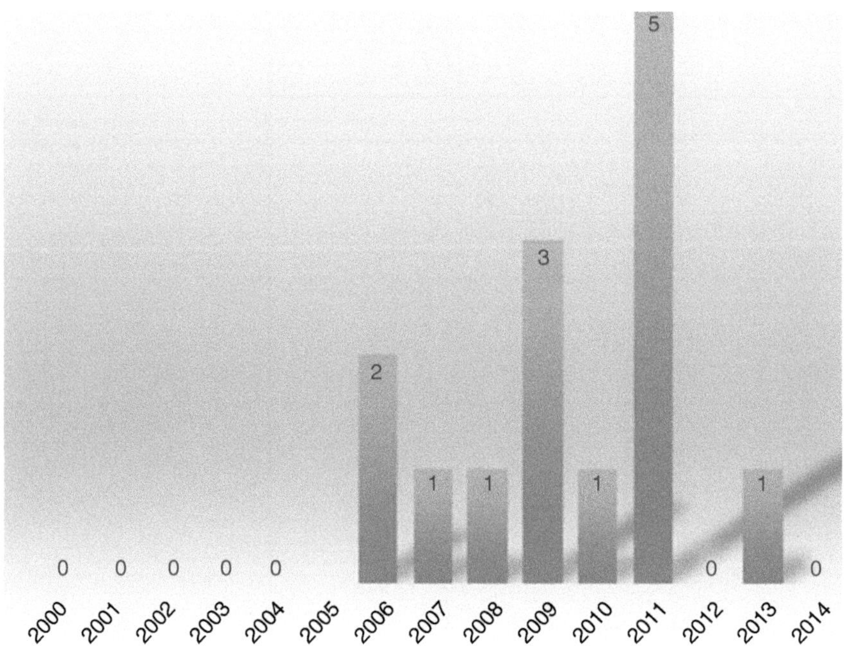

FIGURE 5.26 The trend of reports potentiometric sensor for Ho^{3+} ion over time.

TABLE 5.11 The Properties of the Reported Holmium Ion Membrane Sensors

Electrode type	The selectophore	Electrode composition	Characterizations of the electrode	Ref.
PME	Ho-1 *N*-(1-Thien-2-ylmethylene)-1,3-benzothiazol-2-amine (TBA)	PVC: 31% Plasticizer: BA 61% Selectophore: 6% Additive: NaTPB 2%	LR: 1.0×10^{-5} to 1.0×1^{-2} M Slope: 19.7 ± 0.2 mV/decade DL: 7.0×10^{-6} pH 4.0–9.5 RT: <15 s LT: 3 months II: Na^{+}, Mg^{2+}, Cu^{2+}, Pb^{2+}, La^{3+}, Sm^{3+}, Dy^{3+}, Gd^{3+}, Pr^{3+}, Lu^{3+}	[143]
PME	Ho-2 *N,N′*-Bis(2-pyridinecarboxamide)-1,2-benzene (H_2BPB)	PVC: 30% Plasticizer: BA 60% Selectophore: 6% Additive: KTpC1PB 4%	LR: 1.0×10^{-5} to 1.0×1^{-2} M Slope: 19.6 ± 0.2 mV/decade DL: 8.0×10^{-6} pH 3.0–11.5 RT: 5 s LT: 3 months II: Dy^{3+}, Sm^{3+}, Er^{3+}	[144]
PME	Ho-3 4-Chloro-1,2-bis(2-pyridinecarboxamido)benzene (CBPB)	PVC: 32% Plasticizer: *o*-NPOE 60% Selectophore: 5.5% Additive: NaTPB 2.5%	LR: 1.0×10^{-6} to 1.0×1^{-2} M Slope: 19.7 ± 0.3 mV/decade DL: 8.5×10^{-7} pH 2.7–9.8 RT: <15 s LT: 10 weeks II: None	[145]

(Continued)

TABLE 5.11 The Properties of the Reported Holmium Ion Membrane Sensors (*cont.*)

Electrode type	The selectophore	Electrode composition	Characterizations of the electrode	Ref.
PME	Ho-4 Diazatetrathia-18-crown-6 containing the 4,5-dithio-1,3-Dithiole-2-thione unit (ATCD)	PVC: 20% Plasticizer: BA 63% Selectophore: 5% Additive: NaTPB 2%	LR: 1.0×10^{-6} to 1.0×1^{-2} M Slope: 19.97 ± 0.3 mV/decade DL: 6.3×10^{-7} pH 4.5–9.0 RT: ~10 s LT: 7 weeks II: La^{3+}	[146]
CWE	Ho-1	PVC: 20% Plasticizer: BA 72% Selectophore: 6% Additive: NaTPB 2%	LR: 1.0×10^{-8} to 1.0×1^{-3} M Slope: 19.4 ± 0.2 mV/decade DL: 1.0×10^{-8} pH 3.0–8.0 RT: 10 s LT: 9 weeks II: Dy^{3+}, Er^{3+}, Tm^{3+}, Yb^{3+}, Lu^{3+}	[147]
CPE	Ho-5 *N′*-(2-Hydroxybenzylidene) furan-2-carbohydrazide (NFC)	Paraffin oil: 20% Carbon powder: 52% Selectophore: 20% MWCNTs: 5% Nanosilica: 3%	LR: 1.0×10^{-7} to 1.0×1^{-2} M Slope: 19.9 ± 0.3 mV/decade DL: 3.1×10^{-8} pH 3.0–8.0 RT: 10 s LT: 2 months II: Dy^{3+}	[148]
PME	Ho-6 *N′*-(1-Pyridin-2-ylmethylene)-2-furohydrazide (NFH)	PVC: 30% Plasticizer: DBP 60% Selectophore: 6% Additive: NaTPB 4%	LR: 1.0×10^{-6} to 1.0×1^{-1} M Slope: 19.97 ± 0.3 mV/decade DL: 7.5×10^{-7} pH 3.8–8.0 RT: 15 s LT: 7 weeks II: Er^{3+}, Tm^{3+}, Yb^{3+}, Lu^{3+}	[149]

CPE	Ho-1	Paraffin oil: 20% Carbon powder: 60% Selectophore: 15% MWCNTs: 5%	LR: 1.0×10^{-8} to 1.0×1^{-2} M Slope: 19.3 ± 0.3 mV/decade DL: 7.0×10^{-9} pH 3.8–7.5 RT: 13 s LT: 2 months II: None	[150]
PME	Ho-7 *N*-[(Z)-1-(2-Thienyl)-methylidene]-*N*-[4-(4-{[(Z)-1-(2-thienyl)methylidene]amino}phenoxy) phenyl] amine (TPA)	PVC: 30% Plasticizer: AP 64% Selectophore: 4% Additive: NaTPB 2%	LR: 1.0×10^{-6} to 1.0×1^{-2} M Slope: 19.5 ± 0.3 mV/decade DL: 4.6×10^{-7} pH 3.2–9.8 RT: ~5 s LT: 8 weeks II: None	[151]
PME	Ho-8 2-(4-Oxopentan-2-ylideneamino) isoindoline-1,3-dione (OID)	PVC: 30% Plasticizer: AP 64% Selectophore: 3% Additive: NaTPB 3%	LR: 1.0×10^{-6} to 1.0×1^{-2} M Slope: 19.6 ± 0.5 mV/decade DL: 5.8×10^{-7} pH 3.5–8.8 RT: ~5 s II: Nd^{3+}, Gd^{3+}, Sm^{3+}, Dy^{3+}, Yb^{3+}, Lu^{3+}, Cr^{3+}, Pb^{2+}	[152]
PME	Ho-9 *N*-Phenyl-2-(thiophen-2-ylmethylene) hydrazinecarbothioamide (PHC)	PVC: 30% Plasticizer: NB 66% Selectophore: 2% Additive: NaTPB 2%	LR: 1.0×10^{-6} to 1.0×1^{-2} M Slope: 20.4 ± 0.3 mV/decade DL: 6.2×10^{-7} pH 3.3–10.9 RT: ~5 s LT: 10 weeks II: Na^{+}, Pb^{2+}, La^{3+}, Sm^{3+}, Pr^{3+}, Gd^{3+}, Dy^{3+}	[153]

(*Continued*)

TABLE 5.11 The Properties of the Reported Holmium Ion Membrane Sensors (*cont.*)

Electrode type	The selectophore	Electrode composition	Characterizations of the electrode	Ref.
PME	Ho-10 *N*-1,*N*-2-Bis[1-(2-hydroxyphenyl)methylidene] ethanedihydrazide (MEH)	PVC: 30% Plasticizer: NB 66% Selectophore: 2% Additive: NaTPB 2%	LR: 1.0×10^{-6} to 1.0×1^{-2} M Slope: 19.8 ± 0.3 mV/decade DL: 5.8×10^{-7} pH 2.5 – 9.8 RT: <10 s II: Na^{+}, Pb^{2+}, La^{3+}	[154]
PME	Ho-11 *N*-Quinoline-2-carboxamido-8-aminoquinoline (QCA)	PVC: 30% Plasticizer: NB 56% Selectophore: 2% Additive: NaTPB 2%, OA 10%	LR: 1.0×10^{-6} to 1.0×1^{-2} M Slope: 20.4 ± 0.3 mV/decade DL: 4.2×10^{-7} pH 2.4–7.4 RT: ~5 s II: None	[155]
PME	Ho-12 *N*-(4,5-Dimethyl-2-(picolinamido) phenyl) picolinamide (H_2Me_2bpb)	PVC: 30% Plasticizer: *o*-NPOE 57% Selectophore: 3% Additive: OA 10%	LR: 1.0×10^{-6} to 1.0×1^{-2} M Slope: 20.1 ± 0.2 mV/decade DL: 5.0×10^{-7} pH 3.5–9.4 RT: <10 s LT: 8 weeks II: None	[156]

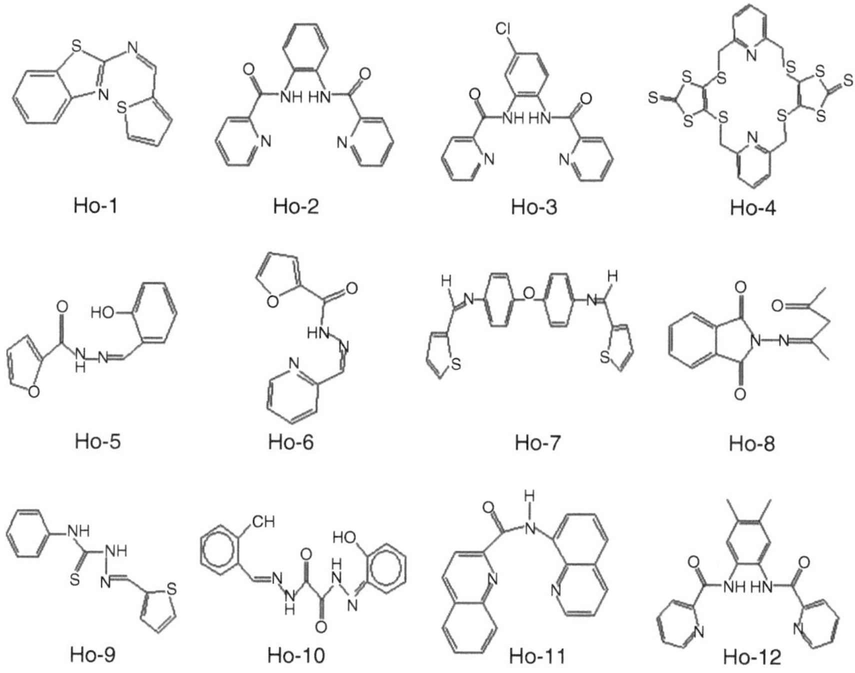

FIGURE 5.27 Chemical structures of the reported selectophores for holmium ions.

A summary of the reports published on Ho^{3+} sensors is given in Table 5.11.

Based on the information in Table 5.11 and the structure of the ionophore used for the sensors (Fig. 5.27), one can conclude that as anticipated, Ho-2, Ho-3, Ho-7, Ho-11, and Ho-12 show good selectivity, whereas Ho-1, Ho-5, Ho-8, Ho-9, and Ho-10 are not very selective because of their inability to form wraparound complexes with the target ion, although they have several S and N atoms in their structure. Ho-1 has been used in both a PME and a CPE.

Erbium ISEs

Like the case of many of the previous lanthanide ions, the first Er^{3+} sensor was developed by Ganjali et al., in 2006 [157], which was the basis for the development of a total of 11 PMEs, CPEs, CGEs, and CWEs (Fig. 5.28).

Table 5.12 shows the selectophores reported for the construction of erbium ion selective electrodes. Based on this information and the structures of the ionophores for Er^{3+} ions expressed in Fig. 5.29, the previous discussions on the structural requirements for establishing selectivity are

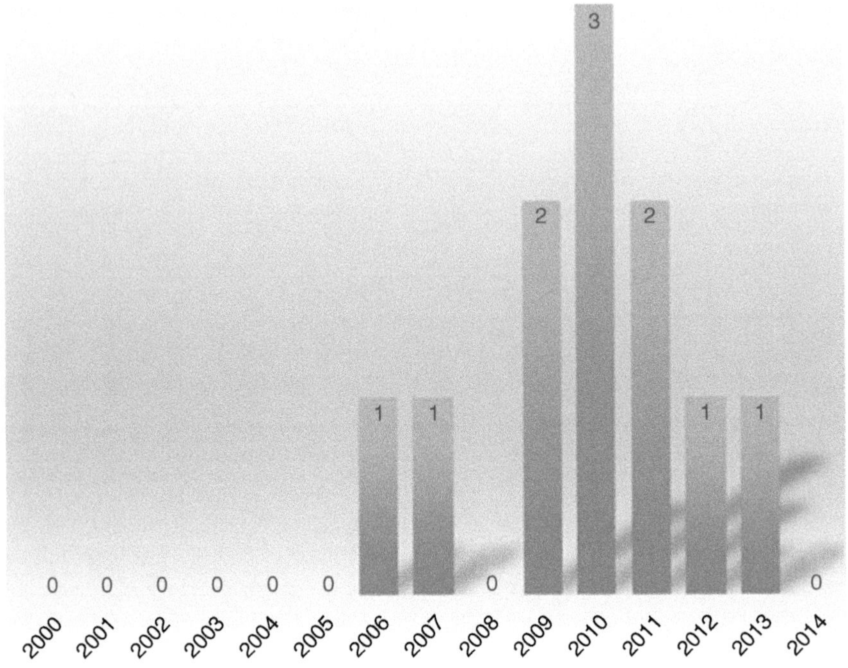

FIGURE 5.28 Number of reported potentiometric sensors for Er^{3+} ions.

confirmed. Among all of the ionophores used, Er-1 was the most selective selectophore and was used in a CWE and a CPE.

Thulium ISEs

The first report on a thulium potentiometric ion selective sensor was by of Ganjali et al. in 2005 [168]. Next, 10 other PMEs, CPEs, CGEs, and CWEs were reported (Fig. 5.30). The properties of the sensors and the structure of the ionophores used in each case are summarized and illustrated in Table 5.13 and Fig. 5.31, respectively. These data confirm the previous discussions on the structural roots of selectivity for lanthanides.

Ytterbium ISEs

Ganjali et al. reported the first of the 11 Yb^{3+} sensors developed so far (Fig. 5.32) [179], the data on the properties of which, as well as the structural information on the ligands are given in Table 5.14 and Fig. 5.33. PMEs, CPEs, CGE, and CWE were reported.

TABLE 5.12 The Properties of the Reported Er^{3+} ISEs

Electrode type	The selectophore	Electrode composition	Characterizations of the electrode	Ref.
PME	Er-1 *N′*-(2-Hydroxy-1,2-diphenylethylidene) benzohydrazide (HDEBH)	PVC: 30% Plasticizer: DBP 62% Selectophore: 5% Additive: KTpC1PB 3%	LR: 1.0×10^{-7} to 1.0×10^{-2} M Slope: 21.5 ± 0.5 mV/decade DL: 7.0×10^{-8} M pH 3.2–8.9 RT: <20 s LT: 49 days II: None	[157]
CWE	Er-1	PVC: 20% Plasticizer: DBP 72% Selectophore: 5% Additive: KTpC1PB 3%	LR: 1.0×10^{-10} to 1.0×10^{-3} M Slope: 17.5 ± 0.5 mV/decade DL: 2.0×10^{-10} M pH 3.0–9.0 RT: ~10 s LT: 10 weeks II: None	[158]
CPE	Er-2 5-(Dimethylamino) naphthalene-1-sulfonyl 4-phenylsemicarbazide (NSP)	[bmim]BF_4: 20% Carbon powder: 45% Selectophore: 20% MWCNT: 15%	LR: 1.0×10^{-7} to 1.0×10^{-1} M Slope: 19.8 ± 0.3 mV/decade DL: 5.0×10^{-8} M pH 3.5–9.0 RT: <16 s LT: 12 weeks II: None	[159]

(Continued)

TABLE 5.12 The Properties of the Reported Er^{3+} ISEs (*cont.*)

Electrode type	The selectophore	Electrode composition	Characterizations of the electrode	Ref.
PME	Er-3 4-(2-Thiazolylazo)resorcinol (TR)	PVC: 30% Plasticizer: AP 66% Selectophore: 2% Additive: NaTPB 2%	LR: 1.0×10^{-6} to 1.0×10^{-2} M Slope: 19.6 ± 0.6 mV/decade DL: 6.6×10^{-7} M pH 2.8–9.3 RT: 10 s LT: 12 weeks II: Nd^{3+}, Dy^{3+}, Ho^{3+}, Sm^{3+}, Tb^{3+}, Tm^{3+}, La^{3+}, Zn^{2+}	[160]
CPE	Er-1	[bmim]BF_4: 20% Carbon powder: 47% Selectophore: 20% MWCNT: 10% Nanosilica: 3%	LR: 7.5×10^{-8} to 1.0×10^{-2} M Slope: 19.6 ± 0.3 mV/decade DL: 6.0×10^{-8} M pH 3.0–8.0 RT: <16 s LT: 12 weeks II: None	[161]
PME	Er-4 Oxalic acid bis[2-[(phenyl amino)carbonyl]hydrazide] (AOPH)	PVC: 30% Plasticizer: NB 64% Selectophore: 4% Additive: NaTPB 2%	LR: 1.0×10^{-6} to 1.0×10^{-2} M Slope: 19.7 ± 0.4 mV/decade DL: 5.8×10^{-7} M pH 2.7–10.2 RT: <10 s LT: 2 months II: Fe^{3+}, Sm^{3+}, Lu^{3+}, Eu^{3+}, Tm^{3+}, Tb^{3+}, Gd^{3+}, Dy^{3+}, Nd^{3+}, La^{3+}, Ho^{3+}	[162]
PME	Er-5 Ethylene glycol(2-aminoethyl)*N*,*N*,*N*′,*N*′-tetraacetic acid (EGTA)	PVC: 30% Plasticizer: DBP 66% Selectophore: 2% Additive: NaTPB 2%	LR: 1.0×10^{-6} to 1.0×10^{-2} M Slope: 21.3 ± 0.5 mV/decade DL: 5.8×10^{-7} M pH 3.0–8.5 RT: 10 s LT: 2 months II: Lu^{3+}	[163]

PME	Er-6 *N*-(Benzyloxycarbonyloxy) succinimide (BCS)	PVC: 30% Plasticizer: BA 66% Selectophore: 2% Additive: NaTPB 2%	LR: 1.0×10^{-6} to 1.0×10^{-2} M Slope: 19.5 ± 0.2 mV/decade DL: 6.3×10^{-7} M pH 2.5–10.6 RT: <10 s II: None	[164]
PME	Er-7 *N*,*N*′-Bis(salicylidene)-2-aminobenzylamine (H_2L)	PVC: 30% Plasticizer: AP 66% Selectophore: 2% Additive: NaTPB 2%	LR: 1.0×10^{-6} to 1.0×10^{-2} M Slope: 19.5 ± 0.2 mV/decade DL: 5.0×10^{-7} M pH 2.6–9.5 RT: <10 s LT: 10 weeks II: Nd^{3+}, Gd^{3+}, Lu^{3+}, Tm^{3+}, Dy^{3+}, Ho^{3+}, Fe^{3+}	[165]
PME	Er-8 5,17-Bis(diphenylphosphinoyl)-25,26,27,28-tetrapropoxycalix[4] arene (Calix)	PVC: 30% Plasticizer: DOP 65% Selectophore: 3% Additive: NaTPB 2%	LR: 1.0×10^{-7} to 1.0×10^{-3} M Slope: 20.0 ± 0.5 mV/decade DL: 3.8×10^{-8} M pH 4.5–6.5 RT: ~7 s LT: 3 months II: Y^{3+}, Eu^{3+}, Gd^{3+}, Dy^{3+}, Ho^{3+}, Yb^{3+}	[166]
PME	Er-9 3-Hydroxy-*N*′-((pyridin-2-yl) methylene)naphthalene-2-carbohydrazide (L)	PVC: 30% Plasticizer: DBP 67% Selectophore: 2% Additive: NaTPB 1%	LR: 1.0×10^{-6} to 1.0×10^{-2} M Slope: 20.0 ± 0.5 mV/decade DL: 5.3×10^{-7} M pH 2.8–9.5 RT: ~5 s LT: 3 months II: full text not available	[167]

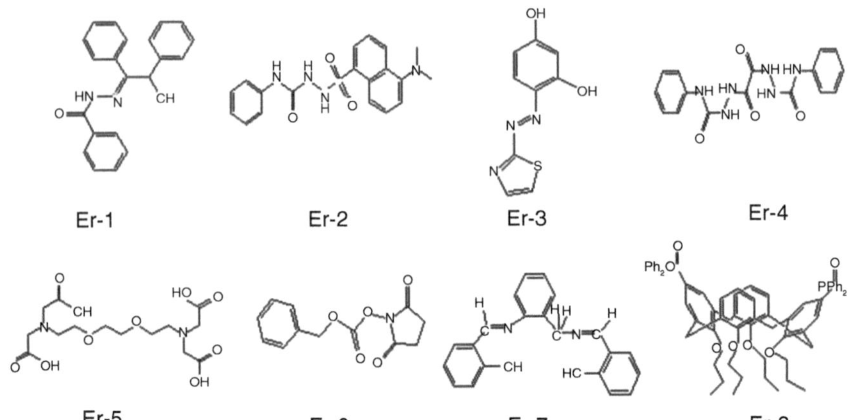

FIGURE 5.29 Chemical structures of the reported selectophores for erbium ions.

The information in Table 5.14 and Fig. 5.33 confirm the discussions on the previous cases. For example, in the case of Yb-7, which is a cyclic molecule, the selectivity is very low because many ions can enter the cavity, which cannot alter itself to choose one element over the others, and hence the selectivity is low.

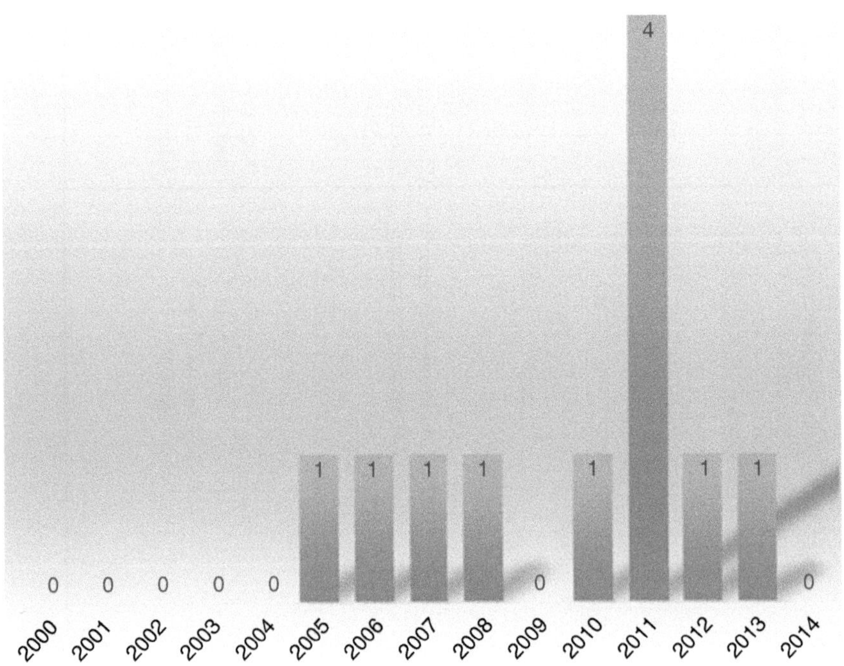

FIGURE 5.30 The historical record of the records on Tm^{3+} sensors.

TABLE 5.13 The Characterization of the Reported Thulium Ion Membrane Sensors

Electrode type	The selectophore	Electrode composition	Characterizations of the electrode	Ref.
PME	Tm-1 Thiophene-2-carbaldehyde-(7-methyl-1,3-benzothiazol-2-yl)hydrazone (TCMH)	PVC: 30% Plasticizer: NB 58% Selectophore: 7% Additive: NaTPB 5%	LR: 1.0×10^{-5} to 1.0×10^{-2} M Slope: 19.5 ± 0.3 mV/decade DL: 8.0×10^{-6} pH 4.8–8.5 RT: <7 s LT: 7 weeks II: Nd^{3+}, Ho^{3+}, Er^{3+}, Mg^{2+}	[168]
PME	Tm-2 2,2′-Dianiline disulfide (DADS)	PVC: 32% Plasticizer: *o*-NPOE 59.5% Selectophore: 5.5% Additive: NaTPB 3%	LR: 1.0×10^{-6} to 1.0×10^{-2} M Slope: 19.5 ± 0.3 mV/decade DL: 4.0×10^{-7} pH 4.8–8.5 RT: <15 s LT: 4 weeks II: Pb^{3+}, Yb^{3+}, Pr^{3+}, Lu^{3+}, Er^{3+}	[169]
CWE	Tm-1	PVC: 25% Plasticizer: NB 63% Selectophore: 7% Additive: KTpClPB 5%	LR: 1.0×10^{-11} to 4.0×10^{-6} M Slope: 19.2 ± 0.3 mV/decade DL: 1.0×10^{-11} pH 4.0–11.0 RT: <5 s II: None	[170]

(*Continued*)

TABLE 5.13 The Characterization of the Reported Thulium Ion Membrane Sensors (*cont.*)

Electrode type	The selectophore	Electrode composition	Characterizations of the electrode	Ref.
PME	Tm-3 4-Methyl-1,2-bis(2-pyridinecarboxamido) benzene (MPB)	PVC: 30% Plasticizer: NB 60% Selectophore: 6% Additive: NaTPB 4%	LR: 1.0×10^{-7} to 1.0×10^{-1} M Slope: 21.2 ± 0.3 mV/decade DL: 9.0×10^{-8} pH 3.6–9.0 RT: 15 s LT: 8 weeks II: None	[171]
PME	Tm-4 1-(2-Thiazolylazo)-2-naphthol (TN)	PVC: 30% Plasticizer: BA 65% Selectophore: 3% Additive: NaTPB 2%	LR: 1.0×10^{-6} to 1.0×10^{-2} M Slope: 19.5 ± 0.2 mV/decade DL: 8.7×10^{-7} pH 3.2–9.5 RT: ~10 s LT: 2 months II: Yb^{3+}, Lu^{3+}	[172]
CPE	Tm-3	RTIL: 25% Carbon powered: 50% Selectophore: 20% NG: 5%	LR: 1.0×10^{-6} to 1.0×10^{-2} M Slope: 19.9 ± 0.3 mV/decade DL: 8.3×10^{-7} pH 3.5–9.0 RT: 15 s LT: 12 weeks II: Ho^{3+}, Lu^{3+}, Gd^{3+}	[173]

CPE	Tm-2	RTIL: 20% Carbon powder: 57% Selectophore: 15% NG: 6% NS: 2%	LR: 1.0×10^{-6} to 1.0×10^{-2} M Slope: 19.7 ± 0.2 mV/decade DL: 2.0×10^{-7} pH 4.0–8.5 RT: 25 s LT: 12 weeks II: Yb^{3+}, Pr^{3+}, Lu^{3+}, Er^{3+}	[174]
PME	Tm-5 1-Ethyl-3-(2,5-dihydro-4-(3,5-dimethyl-1H-pyrazol-4-yl)5-*oxo*-1H-pyrazol-3-yl)thiourea (ET)	PVC: 30% Plasticizer: DBP 65% Selectophore: 3% Additive: NaTPB 2%	LR: 1.0×10^{-6} to 1.0×10^{-2} M Slope: 19.6 ± 0.4 mV/decade DL: 7.2×10^{-7} pH 4.3–10.4 RT: <10 s LT: 2 months II: None	[175]
PME	Tm-6 1-Butyl-3-(2,5-dihydro-4-(3,5-dimethyl-1H-pyrazol-4-yl)-5-*oxo*-1H-pyrazol-3-yl)thiourea (BT)	PVC: 30% Plasticizer: DBP 63% Selectophore: 5% Additive: NaTPB 2%	LR: 1.0×10^{-6} to 1.0×10^{-2} M Slope: 19.6 ± 0.5 mV/decade DL: 6.8×10^{-7} pH 3.8–9.2 RT: ~10 s LT: 2 months II: Lu^{3+}	[176]

(Continued)

TABLE 5.13 The Characterization of the Reported Thulium Ion Membrane Sensors (*cont.*)

Electrode type	The selectophore	Electrode composition	Characterizations of the electrode	Ref.
PME	Tm-7 *N*-(4-Acetyl-3-cyano-5-methlfuran-2-yl) acetamide (ACMA)	PVC: 30% Plasticizer: DBP 65% Selectophore: 3% Additive: NaTPB 2%	LR: 1.0×10^{-6} to 1.0×10^{-2} M Slope: mV/decade DL: 6.5×10^{-7} pH 3.1–9.5 RT: <5 s LT: 10 weeks II: Cr^{3+}	[177]
PME	Tm-8 4-(3-Nitrophenyl)-2,6-di-2-thienylpyridine (PTP)	PVC: 30% Plasticizer: NB 64% Selectophore: 4% Additive: NaTPB 2%	LR: 1.0×10^{-7} to 1.0×10^{-2} M Slope: 20.8 ± 0.4 mV/decade DL: 8.0×10^{-8} pH 2.4–9.4 RT: <10 s LT: 2 months II:Ni^{2+}, Lu^{3+}	[178]

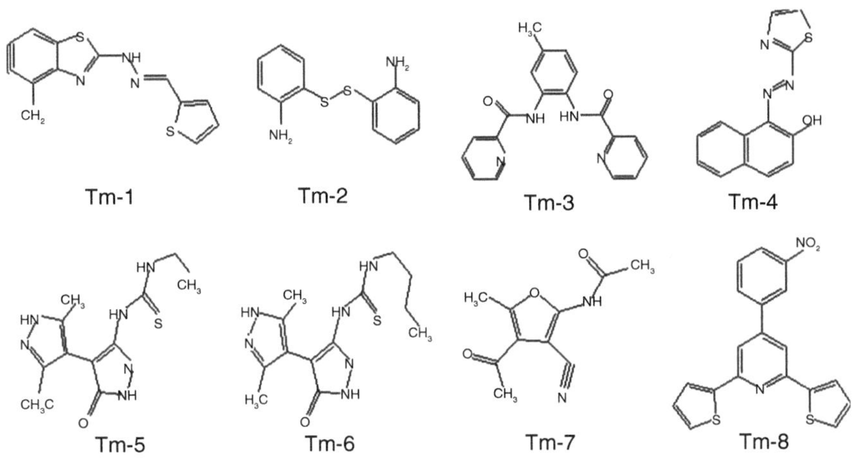

FIGURE 5.31 Chemical structures of the reported selectophores for thulium ions.

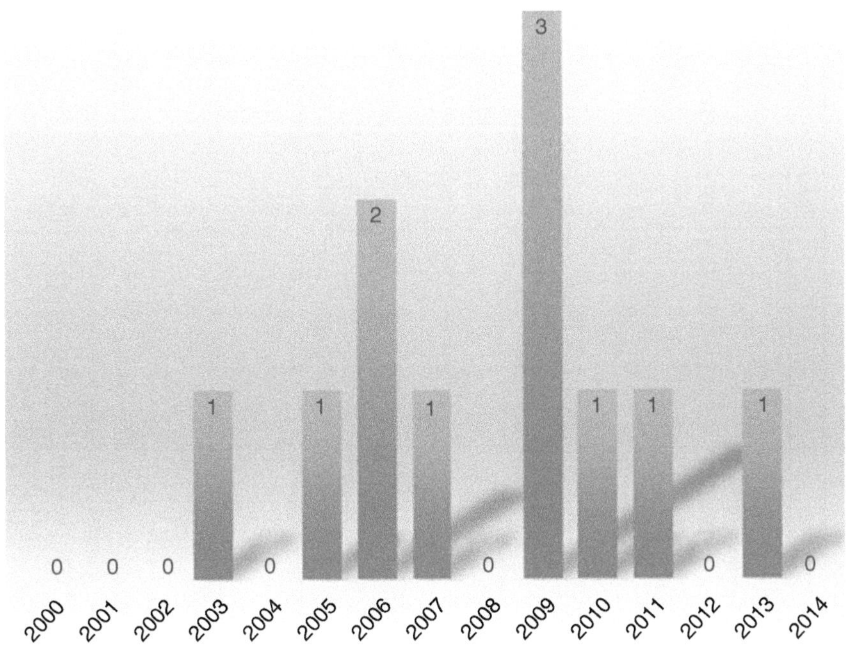

FIGURE 5.32 Historical record of the reported potentiometric sensors for Yb^{3+}.

TABLE 5.14 The Characteristics of the Reported Ytterbium Ion Membrane Sensors

Electrode type	Selectophore	Electrode composition	Characterizations of the electrode	Ref.
PME	Yb-1 Cefixime (CFI)	PVC: 33% Plasticizer: BA 50% Selectophore: 11% Additive: NaTPB 6%	LR: 1.0×10^{-6} to 1.0×10^{-2} M Slope: 19.5 mV/decade DL: 7.0×10^{-7} M pH 3.0–8.0 RT: 15 s LT: 8 weeks II: Ce^{3+}	[179]
PME	Yb-2 *N*-(2-Pyridyl)-*N* ′-(2methoxyphenyl)-thiourea (PMT)	PVC: 30% Plasticizer: *o*-NPOE 59% Selectophore: 7% Additive: NaTPB 4%	LR: 1.0×10^{-6} to 1.0×10^{-2} M Slope: 19.0 ± 0.2 mV/decade DL: 5.0×10^{-7} M pH 3.5–9.0 RT: 8–10 s LT: 10 weeks II: Dy^{3+}	[180]
PME	Yb-3 *N*-(6-Picolyl)*N*′-(4-metoxyphenyl)-thiourea (PCMPT)	PVC: 32% Plasticizer: BA 60% Selectophore: 4% Additive: NaTPB 4%	LR: 1.0×10^{-6} to 1.0×10^{-2} M Slope: 19.0 ± 0.2 mV/decade DL: 6.0×10^{-7} M pH 3.3–10.6 RT: ~15 s LT: 2 months II: Dy^{3+}, Gd^{3+}	[181]
PME	Yb-4 3-Hydroxy-*N*′-[(2-hydroxyphenyl) methylene]-2-naphthohydrazide (HPMN)	PVC: 33% Plasticizer: DBP 57% Selectophore: 4% Additive: NaTPB 4%	LR: 1.0×10^{-7} to 1.0×10^{-2} M Slope: 19.2 ± 0.2 mV/decade DL: 4.0×10^{-8} M pH 3.0–8.5 RT: <6 s LT: 2 weeks II: Pb^{2+}, Nd^{3+}, Gd^{3+}	[182]

PME	Yb-5 6-Methy-4-{[1-(1*H*-pyrrol-2-yl) methylidene]amino}-3-thioxo-3,4-dihydro-1,2,4-triazin-5(2*H*)-one (PMTO)	PVC: 30% Plasticizer: NB 63% Selectophore: 5% Additive: NaTPB 2%	LR: 1.0×10^{-6} to 1.0×10^{-1} M Slope: 19.5 ± 0.3 mV/decade DL: 4.6×10^{-7} M pH 3.3–8.0 RT: 10 s LT: 2 months II: Sm^{3+}, Cd^{2+}	[183]
PME	Yb-6 2,5-Bis(5-*tert*-butyl-benzoxazol-2-yl) thiophene (BBT)	PVC: 30% Plasticizer: *o*-NPOE 61.5% Selectophore: 5% Additive: NaTPB 3.5%	LR: 1.0×10^{-6} to 1.0×10^{-2} M Slope: 19.7 ± 0.5 mV/decade DL: 4.4×10^{-7} M pH 3.5–8.4 RT: 10 s LT: 2 months II: Sm^{3+}, Cr^{3+}	[184]
PME and CGE	Yb-7 2,12-(2-Methoxyaniline)(2)-4,14-Me(2)-[20]-1,4,11,14-tetraene-1,5,8,11,15,18-*N*(G) (L_1) 2,12-(2-Methoxyaniline)(2)-4,14-Me(2)-8,18-dimethylacrylate-[20]-1,4,11,14-tetraene-1,5,8,11,15,18-*N*(G) (L_2)	PVC: 30% Plasticizer: DBP 60% Selectophore: 6% Additive: KTpClPB 4% for L_1 PME PVC: 40% Plasticizer: BA 52% Selectophore: 5% Additive: NaTPB 3% for L_2 PME and CGE	LR: 1.6×10^{-7} to 1.0×10^{-1} M for L_1 PME LR: 8.0×10^{-8} to 1.0×10^{-1} M for L_2 PME LR: 9.4×10^{-9} to 1.0×10^{-1} M for L_2 CGE Slope: 19.4 mV/decade for L_1 PME Slope: 19.8 ± 0.2 mV/decade for L_2 PME Slope: 19.7 ± 0.1 mV/decade for L_2 CGE DL: 1.1×10^{-7} M for L_1 PME DL: 4.3×10^{-8} M for L_2 PME DL: 5.8×10^{-9} M for L_2 CGE pH 3.0–8.0 for L_2 PME pH 2.5–8.5 for L_2 CGE RT: 10 s for L_2 PME RT: 8 s for L_2 CGE LT: 4 months for L_2 PME LT: 5 months for L_2 CGE II: Sm^{3+}, Tb^{3+}, Gd^{3+}, Mg^{2+}, Ni^{2+}, Al^{3+}, Sr^{2+}, K^+	[185]

(Continued)

TABLE 5.14 The Characteristics of the Reported Ytterbium Ion Membrane Sensors (*cont.*)

Electrode type	Selectophore	Electrode composition	Characterizations of the electrode	Ref.
PME	Yb-8 *N*1,*N*2-Bis-[1-(2-hydroxy-1,2-diphenyl) ethylidene]ethanedihydrazide (BHDEH)	PVC: 30% Plasticizer: DBP 66% Selectophore: 2% Additive: NaTPB 2%	LR: 1.0×10^{-6} to 1.0×10^{-2} M Slope: 19.6 ± 0.3 mV/decade DL: 4.2×10^{-7} M pH 3.2–8.3 RT: <10 s LT: 2 months II: Pr^{3+}, Fe^{3+}	[186]
CPE	Yb-9 *N*′-(1-oxoacenaphthylen-2(1*H*)-ylidene)furan-2-carbohydrazide (L)	[bmim]BF_4: 25% Selectophore: 25% Carbon powder: 45%, MWCNT: 5%	LR: 1.0×10^{-8} to 1.0×10^{-2} M Slope: 19.6 ± 0.3 mV/decade DL: 1.0×10^{-8} M pH 3.5–8.5 RT: <20 s LT: 8 weeks II: None	[187]
PME	Yb-10 1,10-Phenanthroline-5,6-dione (PDO)	PVC: 31% Plasticizer: TPB 61% Selectophore: 3% Additive: NaTPB 5%	LR: 5.0×10^{-6} to 1.0×10^{-2} M Slope: 19.7 ± 0.2 mV/decade DL: 5.7×10^{-7} M pH 4.0–12.0 RT: <5 s LT: 6 weeks II: Ag^{+}, Ni^{2+}, Mg^{2+}, Sr^{2+}, Fe^{3+}	[188]
PME	Yb-11 4-Methoxybenzyl carbonate (MBC)	PVC: 30% Plasticizer: NB 66% Selectophore: 2% Additive: NaTPB 2%	LR: 1.0×10^{-7} to 1.0×10^{-2} M Slope: 19.8 ± 0.4 mV/decade DL: 8.3×10^{-8} M pH 2.2–9.7 RT: ~5 s LT: 10 weeks II: Tm^{3+}, Tb^{3+}	[189]

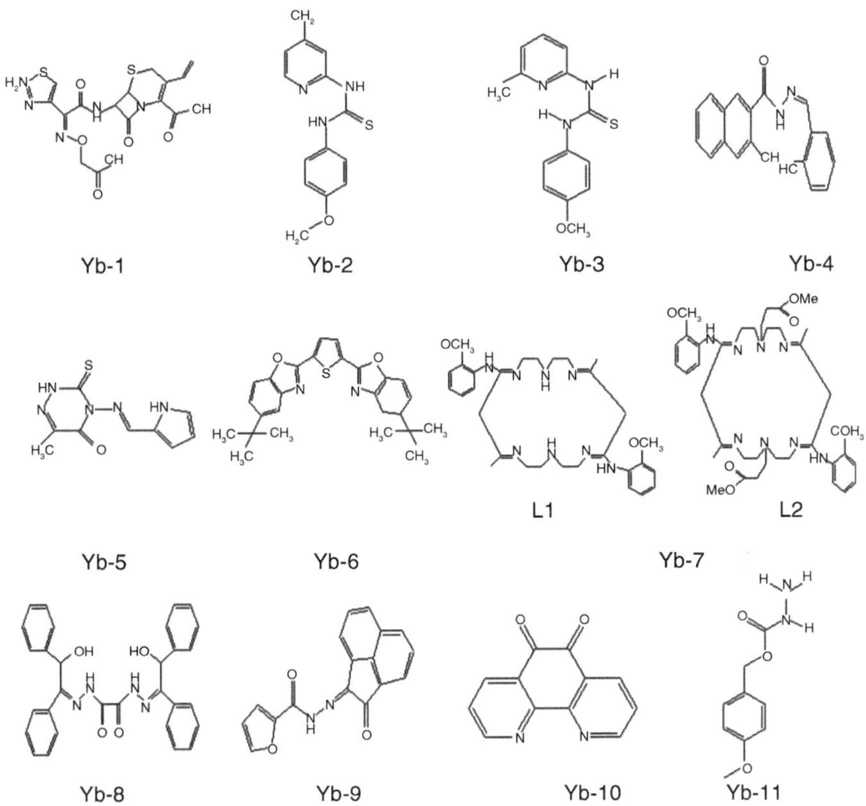

FIGURE 5.33 Chemical structures of the reported selectophores for ytterbium ions.

Lutetium ISEs

The first report on a lutetium potentiometric ion selective sensor was by Ganjali et al., in 2006 [190], and 11 PMEs, CPEs, CGEs, and CWE were developed (Fig. 5.34) afterwards.

The information on the properties of the sensors (Table 5.15) and the structure of the ionophores (Fig. 5.35) confirms the previous discussions (e.g., like the previous example, the cyclic Lu-9 is not very selective).

In conclusion, Fig. 5.36 shows the total trend of the 194 reports on potentiometric sensors for lanthanides developed so far.

POTENTIOSTATIC DETERMINATION METHODS

Voltammetry is an umbrella term for a family of electroanalytical techniques, where the signal is acquired through measuring the electrical current at fixed or varying potentials. In order for this to be possible, and

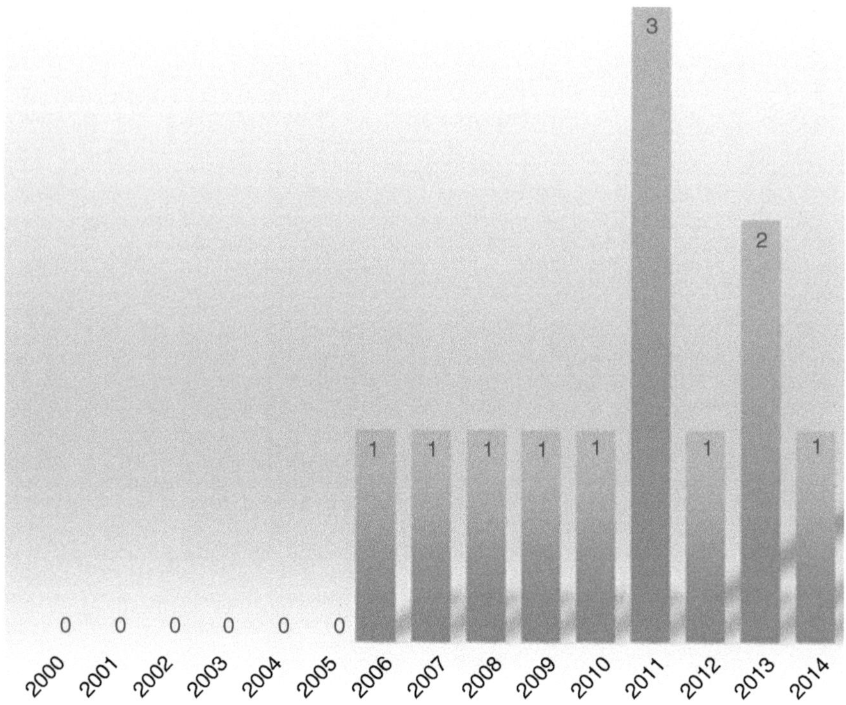

FIGURE 5.34 The historical trend of the reports on Lu^{3+} ISE.

opposite to potentiometric techniques, the analyte should be oxidizable or reducible in a specially applicable potential range, which in electroanalytical terminology is referred to as being electroactive. Yet again, in case a nonelectroactive compound can have a stoichiometric interaction with an electroactive one, the former can be measured through indirect voltammetric techniques. Voltammetric techniques are seen as nondestructive methods because they need only a small portion of the sample.

As mentioned, voltammetry is based on monitoring the current potential behavior of the sample, and the resulting current-potential ($i = f(E)$) curves are referred to as voltammograms. By varying the potential through stepwise or continuous regimes, the current is measured with at least two electrodes. The electrode whose potential is changed is called the working electrode (WE), whereas the other possesses a known potential to act as a reference against the potential of the WE and is adjusted with respect to the reference electrode (RE). This electrode also compensates for the changes in the charge of the solution, which is altered due to the reaction at the WE, forming the other half cell.

One important point is that although in theory this should form a complete cell, because the RE cannot practically cope with the tasks of both

TABLE 5.15 The Properties of the Reported Lutetium Ion Membrane Sensors

Electrode type	Selectophore	Electrode composition	Characterizations of the electrode	Ref.
PME	Lu-1 *N*-(Thien-2-ylmethylene)pyridine-2, 6-diamine (TPD)	PVC: 32% Plasticizer: *o*-NPOE 58% Selectophore: 6% Additive: NaTPB 4%	LR: 1×10^{-6} to 1×10^{-2} M Slope: 20.5 ± 0.4 mV/decade DL: 8.0×10^{-7} pH 4.5–8.0 RT: <10 s LT: 8 weeks II: Nd^{3+}, Dy^{3+}	[190]
CWE	Lu-2 *N*-(2-Furylmethylene)-pyridine-2, 6-diamine (FPD)	PVC: 32% Plasticizer: *o*-NPOE 60% Selectophore: 4% Additive: KTpClPB 4%	LR: 1×10^{-11} to 1×10^{-6} M Slope: 20.5 ± 0.2 mV/decade DL: 3.0×10^{-11} pH 4.0–10.5 RT: <10 s LT: 8 weeks II: None	[191]
PME	Lu-3 *N*,*N*′-Bis(2-pyridinecarboxamide)-1, 3-benzene (BPMB)	PVC: 30% Plasticizer: BA 63% Selectophore: 5% Additive: NaTPB 2%	LR: 1×10^{-6} to 1×10^{-1} M Slope: 20.1 ± 0.2 mV/decade DL: 6.3×10^{-7} pH 3.0–7.5 RT: ~5s LT: 8 weeks II: None	[192]

(*Continued*)

TABLE 5.15 The Properties of the Reported Lutetium Ion Membrane Sensors (*cont.*)

Electrode type	Selectophore	Electrode composition	Characterizations of the electrode	Ref.
PME	Lu-4 *N,N′*-Bis(2-pyridinecarboxamide)-1,2-ethane (PCAE)	PVC: 30% Plasticizer: o-NPOE 63% Selectophore: 5% Additive: NaTPB 2%	LR: 1×10^{-6} to 1×10^{-1} M Slope: mV/decade DL: 6.0×10^{-7} pH 4.0–9.0 RT: 5 s LT: 8 weeks II: Ho^{3+}, Ce^{3+}	[193]
PME	Lu-5 *N*-[(Z)-1-(2-Thienyl)methylidene]-*N*-[4-(4-{[(Z)-1-(2-thienyl) methylidene]amino} benzyl)phenyl] amine (TBPA)	PVC: 30% Plasticizer: AP 65% Selectophore: 3% Additive: NaTPB 2%	LR: 1×10^{-6} to 1×10^{-2} M Slope: 19.8 ± 0.5 mV/decade DL: 7.2×10^{-7} pH 2.7–10.6 RT: <10 s LT: 2 months II: Ho^{3+}, Er^{3+}, Tm^{3+}	[194]
PME	Lu-6 2,2′-Dithiobis(4-methylthiazole) (TMT)	PVC: 30% Plasticizer: AP 66% Selectophore: 2% Additive: NaTPB 2%	LR: 1×10^{-6} to 1×10^{-2} M Slope: 19.6 ± 0.4 mV/decade DL: 6.8×10^{-7} pH 2.7–9.6 RT: ~5 s II: Ho^{3+}, Gd^{3+}	[195]
CPE	Lu-5	RTIL: 20% Carbon powder: 49% Selectophore: 25% MWCNT-NH(2): 5% NS: 1%	LR: 1×10^{-6} to 1×10^{-2} M Slope: 19.8 ± 0.2 mV/decade DL: 9.5×10^{-7} pH 3.5–9.0 RT: 20 s LT: 11 weeks II: Ho^{3+}, Er^{3+}, Tm^{3+}	[196]

PME	Lu-7 (2-*oxo*-1,2-Diphenylethylidene)-*N*-phenylhydrazinecarbothioamide (PHCT)	PVC: 30% Plasticizer: DBP 65% Selectophore: 3% Additive: NaTPB 2%	LR: 1×10^{-6} to 1×10^{-2} M Slope: 19.8 ± 0.3 mV/decade DL: 6.8×10^{-7} pH 2.5–8.7 RT: 5 s LT: 2 months II: Ni^{2+}, Ca^{2+}	[197]
PME	Lu-8 2-Amino-4-(4-aminophenyl)thiazole (AAT)	PVC: 30% Plasticizer: BA 63% Selectophore: 5% Additive: NaTPB 2%	LR: 1×10^{-6} to 1×10^{-1} M Slope: 19.8 ± 0.4 mV/decade DL: 5.7×10^{-7} pH 3.2–8.8 RT: ~6 s II: Ni^{2+}	[198]
PME	Lu-9 4′-Carboxybenzo-18-crown-6 (4′C18C6)	PVC: 30% Plasticizer: DBP 63% Selectophore: 2% Additive: OA 5%	LR: 1×10^{-6} to 1×10^{-2} M Slope: 19.8 ± 0.4 mV/decade DL: 6.5×10^{-7} pH 2.8–8.7 RT: ~5 s II: Na^{+}, Eu^{3+}	[199]
PME	Lu-10 Di(*N*-succinimidyl) oxalate (DSO)	PVC: 30% Plasticizer: NB 60% Selectophore: 8% Additive: NaTPB 2%	LR: 1×10^{-7} to 1×10^{-2} M Slope: 20.5 ± 0.3 mV/decade DL: 7.5×10^{-8} pH 2.8–9.1 RT: ~5 s II: None	[200]
CPE	Lu-11 3-[(Pyridine-2-ylmethylene)-amino]-2-thioxo-4-one (PAT)	Paraffin oil: 20%, Carbon powder: 56% Selectophore: 18% f-MWCNTs: 6%	LR: 1×10^{-6} to 1×10^{-1} M Slope: 21.1 ± 0.3 mV/decade DL: 8×10^{-7} pH 3.5–8.0 RT: <15 s LT: 1 month II: Yb^{3+}, Fe^{3+}, Cr^{3+}, Mg^{2+}, Ni^{2+}	[201]

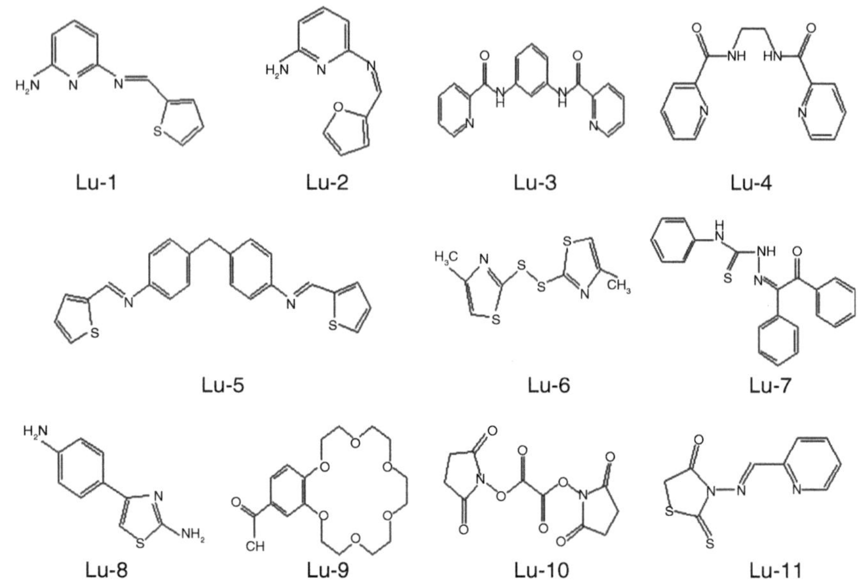

FIGURE 5.35 Chemical structures of the reported selectophores for lutetium ions.

keeping the potential constant and adjusting the electroneutrality of the solution. Consequently, the two tasks are separated from one another and assigned to two independent so-called "reference" and "auxiliary" electrodes, the former forming the other half cell with WE only used for monitoring and adjusting the potential of WE at zero current. The auxiliary

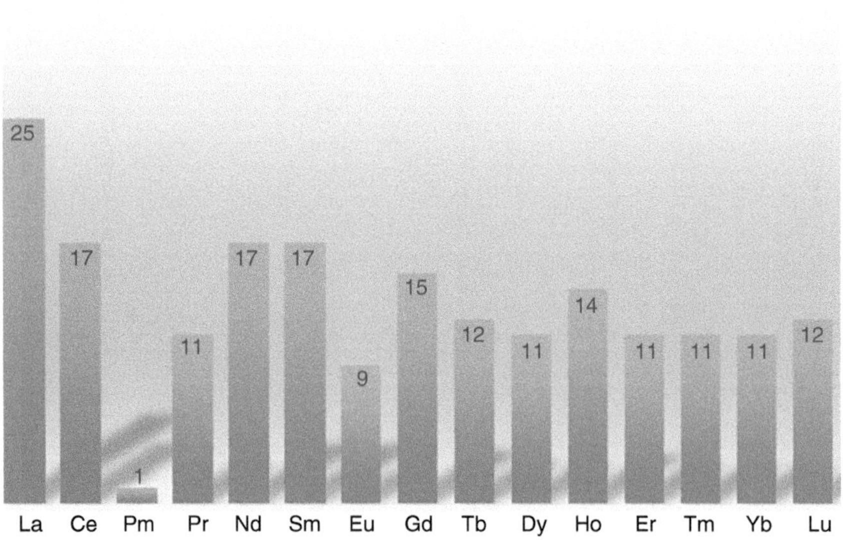

FIGURE 5.36 Total number of reported Ln ISEs.

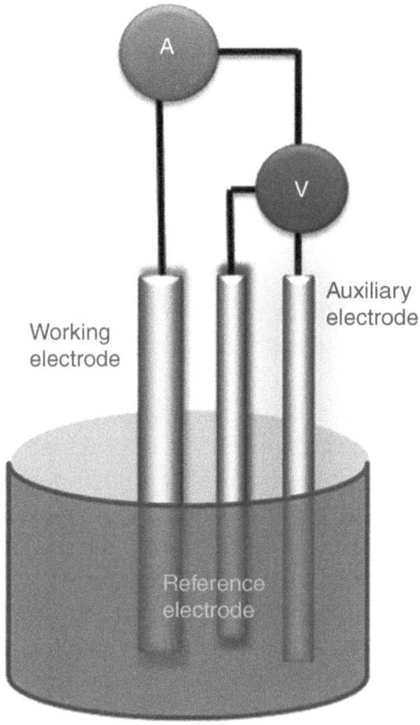

FIGURE 5.37 A schematic presentation of a voltammetric setup.

electrode (AE), on the other hand, takes charge of compensating for the electroneutrality of the solution, and the current needed at this end passes through that. This naturally leads to changes in the potential of the AE within the window of solvent, which illustrates the two potentials at which only the solvent molecules or the species of the supporting electrolyte are oxidized or reduced. Hence, voltammetric systems are actually three-electrode setups (Fig. 5.37) [202–204].

Upon the oxidation or reduction of the analyte on the surface of WE, an electron flux forms "from or to" the WE, along the external circuit and the potentiostat, "to or from" the AU. This leads to reduction or oxidation reactions at the surface of the AE, respectively. The reduced or oxidized compound is usually another electroactive species or the solvent itself. This electron flux that leads to a chemical reaction is called a Faradic current. That is further specified as anodic or cathodic based on the nature of the reaction at WE. If the analyte is reduced at WE, the current is referred to as a cathodic Faradic current, whereas in the opposite case it is an anodic one.

The analyte should, on the other hand, reach the surface of the WE from the bulk of the solution. Because of the intrinsic requirements of voltammetric techniques, measures are taken so that the current is only of diffusion in nature. This means upon applying a potential to the WE, and the onset of the reaction of the analytes already present on the surface of the WE, there is a concentration gradient between those at the surface of the WE and the solution bulk. Under a diffusion-only regime, only this concentration gradient, and no other mass transfer mechanisms, serves as the driving force for the diffusion of the species to and from the WE from and to the bulk. If the reactant species do not reach—and the products do not leave—the surface of the WE, the Faradic current shall go down once the concentration of the species at the surface reaches zero. This diffusion-based flux of electroactive species within the sample exists until the concentration gradients between the WE and the bulk exist in the solution.

Voltammetric techniques require the mass transfer to be only through diffusion, since they are based on Fick's law, which shall be elaborated later. To this end, the two other possible mechanisms, namely, convection and migration, should be eliminated. Contrary to diffusion, which is the result of concentration gradients that form in the solution upon the electrochemical reactions at the WE and the AE, convection is the result of unbalanced mechanical forces like stirring or even random movements as well as thermal gradients. Although stirring magnets or rotating electrodes, or flow-injection systems where concentration is a dominant phenomenon, are sometimes used with voltammetric techniques, these methods basically require that the phenomenon be eliminated or controlled, through keeping the systems as steady as possible or increasing its viscosity.

Migration, on the other hand, is caused by the electrostatic interactions between the charged species and the electrodes. In the case of positively charged electrodes, for example, anions tend to move toward the more positive electrode, whereas the cations naturally tend to take the opposite direction. To decrease the share of the analytes reaching the WE through migration, a so-called supporting electrolyte is used at much higher concentration than the analyte (e.g., >100 times), so that the higher number of nonelectroactive species competing to reach the surface of the WE decreases the chances of the electroactive ones doing so.

Based on the above discussion, one can easily conclude that two major processes control the current in voltammetric setups, one being the rate of mass transport in the electrolytic part of the circuit and the other being the movement of electrons through the electronic half of the circuit. It should be noted that the rate of electron transfer is actually determined by the kinetics of electron transfer at the electrodes, which is a decisive factor that determines the applicability of voltammetric techniques for desired species. This means if the kinetics of the reaction is fast enough, it can readily reach equilibrium, which can be considered electrochemically reversible. Under such conditions, the Nernst equation governs the process, whereas in the case of electron transfer kinetics that are sufficiently slow, the system changes into a so-called irreversible one. One, the equation cannot be used to predict the concentrations of the reactant and product at the surface of the electrode. As a result, it cannot predict the magnitude of the Faradic currents.

Another phenomenon observed in voltammetric systems is some unwanted Faradic currents due to the reactions of electroactive contaminants present in the solution. In addition, there is some nonfaradic current, which naturally causes errors. The nonfaradic current is attributed to the arrangement of the cations and anions with respect to the electrode. This leads to the formation of a so-called electrical double layer (EDL). After the establishment of equilibrium, a small nonfaradic current, which is referred to as the charging current, forms for a rather short period of time. This phenomenon happens upon every change in the electrode potential, and because of its nonfaradic nature, it is independent of the presence of the analyte. The dominance of each of the two possible regimes (i.e., mass-transfer control and charge transfer control) as well as the thermodynamics of the reactions influences the data analyses in voltammetry. The influence of time on the behavior of the system, which is ignored by the Nernst equation, is considered by the Tafel and Butler-Volmer models [204,205].

According to the models describing the voltammetric systems, the concentrations of the redox species present at the surface of the WE (c_O^0 and c_R^0), and the rate of the reaction (k^0) are greatly influenced by the applied potential. As mentioned before, in case the flux of the analyte to the surface of the WE is only through diffusion, the relationship between the Faradic current and the flux is given by Fick's law. For a reversible electrochemical reaction,

$$\mathrm{O} + \mathrm{ne}^- \leftrightarrow \mathrm{R}$$

The Nernst equation can be written as

$$E = E^0 - \frac{RT}{nF} \ln \frac{c_R^0}{c_O^0} \tag{5.19}$$

R is the molar gas constant (8.3144 J mol^{-1}K^{-1}), T is the temperature in Kelvin (K), in the number of electrons transferred during the reaction, F the Faraday constant (96485 C/mol), E^0 the standard reduction potential of the redox couple, E the applied potential, and c_O^0 and c_R^0 representing the concentrations of O and R at the WE. The equation clearly shows the interdependence between the applied potential and the ratio of R and O concentrations (c_R^0/ c_O^0).

On the other hand, in techniques where the relationship among current, potential, and concentration is required, the Butler–Volmer equation

$$\frac{i}{nFA} = k^0 \exp[-\alpha\theta] - C_R^0 \exp[(1-\alpha)\theta] \tag{5.20}$$

is used. In this equation, k^0 is the heterogeneous rate constant, $\theta = nF(E - E^0)/RT$, α is the transfer coefficient, and A represents the area of the surface of the electrode.

Species O or R, produced on the surface of the electrode, diffuses toward the bulk of the solution as a result of the conditions in the solution. If the species are consumed on the surface of the electrode, however, the flux will be in the opposite direction from the bulk with the higher concentration to the electrode with the lower concentration. The relationship between the flux (Φ) and the concentration gradients in a system is given by Fick's law for species O:

$$\Phi = -AD_O(\partial c_O/\partial x) \tag{5.21}$$

In which A represents the electrode surface, x is the distance from the electrode surface, and D_O is the diffusion coefficient for the species O.

Instruments and Reagents

Electrodes

Regarding the above models, and the role of the electrode dimensions in each, having WEs of known dimensions is very important. The same considerations exist about the surface properties of WEs, which might be altered by previous reactions in the case of solid-state WEs. Hence, cleaning and polishing the WE to renew its surface is a common practice. In a group of techniques referred to as polarography, a mercury drop is used as a WE. This creates advantages like a renewable surface. However, Pt, Au, and Ag and carbon paste WEs have also been used. It is noteworthy that based on its nature, each electrode has its own applicable range of potentials. Dropping mercury WEs, for instance, have a wide cathodic potential range. Solid and liquid mercury electrodes can be alternatively used, but in general solid WE are preferred in analyses requiring negative potentials (which is very limited in the case of the mercury electrodes), or at positive ones where the application of the mercury electrode is not possible.

CPEs are made by mixing carbon powder or nanostructures with a viscose solvent and filling a cavity with the paste, finally establishing electrical contact by inserting a wire into the paste. The advent of these chemically modified electrodes based on CPEs has opened new doors for a wide range of experiments. Another advance in the area of WEs has been the advent of micro- or nanoelectrodes, which can produce currents in the range of pico- to even nanoamperes.

The choice of the RE and the AE, on the other hand, is often more limited. Common REs are saturated calomel electrode (SCE) and Ag–AgCl, whereas any conducting metal that is inert in the sample solution can be chosen as an AE.

Electrolyte

As mentioned earlier, for purposes of elimination, the migration flux as well as increasing the conductivity of the solution, a bulk or supporting electrolyte is added to the sample solution in voltammetric experiments. Tests in the absence of this electrolyte are inaccurate. In cases where the solvent itself is an ionic liquid, the need for an electrolyte can be compromised. The nature of the electrolyte and the changes it introduces to the test solutions can influence the voltammetric results. As an example, the electrolytes can gravely influence the potential window of WEs, e.g., for a Platinum WE in acidic and basic pH, the applicable potential window ranges from about +1.2 V to –0.2 V and from +0.7 V to –1 V versus SCE, respectively.

Potentiostat

A potentiostat is an electrical device used for controlling the potential of the WE with respect to the RE, during the voltammetric analyses. The WE and the RE are connected to one another through a high-impedance potentiometer, and the control of the potential is achieved through adjustments in a slide wire resistor connected to the AE. In a constant potential regime, upon changes in the potential of the WE, the adjustments in the slide wire resistor returns the potential to its initially set value. The potentiostat is also equipped with an amperometer to measure the current between the WE and AE, and newer models of the devices are further equipped with electrometer circuits, microprocessors, as well as analog-to-digital (A/D) and digital-to-analog (D/A) converters, to help use analog data in digital environment and vice versa.

Apart from fixed potential methods, many dynamic regimes, which require altering the potential according to desired waveforms, are also used in voltammetric work. Consequently, accurate and robust control over the applied potential is very important. These methods have been so successful that in conventional devices, a "staircase" potential, in which small changes in the potential is applied at certain intervals, has replaced the conventional linear potential sweep in analog devices. The versatility of the digital devices in controlled changing of the potential has created the basis for new techniques like pulsed voltammetry with higher sensitivity. As a result of being equipped with waveform generators, which can apply different potential–time profiles (e.g., fixed or incremental) pulses to the WE, the capabilities of voltammetric techniques have been increased.

The advent of miniaturized electrodes has increased the need for potentiostats that can work under conditions in which negligible currents are created, as well as devices that can be used in fast scanning techniques such as square-wave voltammetry (SWV).

Voltammetric Techniques used for the Determination of Lanthanides

Depending on the type of electrode, the electrolyte, the potential–time profile, and the technique of measuring the current, a wide range of voltammetric methods can be developed. Among the voltammetric methods, stripping analysis can be regarded as the most sensitive technique (since it includes a preconcentration step), with detection limits in the order of parts per billion (ppb) or even sub–parts per billion.

As mentioned, the most common oxidation state of lanthanides is 3 +, although examples of stable lanthanide cations in a 2+ state exist and include $LnCl_2$ (Ln = Nd, Sm, Eu, Dy, Tm, and Yb), $LnBr_2$ (Ln = Sm, Yb, and Eu), LnI_2 (Ln = Nd, Sm, Eu, Dy, Tm, Yb, La, Ce, Pr, and Gd), LnF_2 (Ln = Sm, Eu, and Yb), LnS, LnSe, and LnTe (Ln = Sm, Eu, Tm, and Yb), $LnSO_4$ (Ln = Sm and Eu), $LnCO_2$ (Ln = Sm, Eu, and Yb), EuC_2O_4, LnD_2, and LnH_2 (Ln = Eu and Yb). Also, some rather stable tetravalent cationic species of Ce, Pr, Nd, Tb, and Dy, as in LnF4 (Ln = Ce, Pr, and Tb) or LnO_2 (Ln = Ce, Pr, and Tb) have been reported [7].

However, among the 15 members of lanthanide family, only six have two stable oxidation states (i.e. Ce^{4+}, Ce^{3+}; Pr^{4+}, Pr^{3+}; Sm^{2+}, Sm^{3+}; Eu^{2+}, Eu^{3+}; Tb^{4+}, Tb^{3+}; Yb^{2+}, Yb^{3+}) [7]. Among the divalent lanthanide ions, Eu^{2+} is the most stable in aqueous solutions whereas Ce^{4+}, Pr^{4+}, Tb^{4+}, and Ce^{4+} are all stable. The most electroactive member of the lanthanide series is cerium, which can be determined directly through voltammetric techniques. However, the electrochemical behavior of lanthanides makes their voltammetric determination very difficult. Lanthanides, especially the ones that are not electroactive, can be determined indirectly using the suitable electrochemical probes.

A review of the reports on the voltammetric determination of various lanthanides shall be covered in the lines below.

Brainina in 1974 [206] used cathodic stripping voltammetry at a graphite electrode that yielded a detection limit of 6×10^{-7} M for Cerium. In another report, Iikova et al. discussed the determination of cerium in oxides of REEs through inversion voltammetry on a graphite electrode in 1983 [207]. Also, Wang et al. [208] reported the application of adsorptive stripping for the determination of lanthanum, cerium, and praseodymium ions, through the controlled adsorptive accumulation of the complex of lanthanide ions and *o*-cresolphthalexon on a static mercury drop electrode for 20 min. Next a differential pulse potential was applied leading to detection limits down to 1×10^{-10} M. They reported that 0.5 to 4 min was enough for the accumulation of samples ranging from 2.5×10^{-8} to 2.5×10^{-9} M.

In another report on the determination of lanthanides through voltammetric techniques, Eu was determined through adsorptive cathodic stripping voltammetry (ACSV) after complexation with *N*-nitroso-*N*-phenylhydroxylamine (cupferron) [209]. The complex was accumulated into the hanging mercury drop electrode (HMDE) at 0.0 V for 60 s, after degassing the sample solution for 5 min. During the accumulation period, the sample solution was stirred. After the acquisition of the signal of the blank solution, known amounts of Eu solutions were added and the solution was purged again for 10 s, before another cycle of preconcentration scan. The results showed a reduction peak corresponding to Eu at –0.88 V, and a detection limit of 0.06 nmol dm^{-3} was reported. The authors suggested the same routine for the analysis of ytterbium and of the other rare earth elements after a separation step.

The report by Khoo and Zhu [210] on the oxidative electropolymerization of resorcinol, catechol, and pyrogallol at a glassy carbon electrode in different media concluded that in the case of catechol, at the GC electrode in 0.10 M NaOH, a very sensitive and selective film for Ce^{3+} was possible. This was claimed to be applicable to the selective preconcentration of Ce^{3+} for further determination through differential pulse anodic stripping voltammetry in batch and flow systems. Under optimal conditions, the response was found to be linear in the ranges of 2.00×10^{-9} M to 1.00×10^{-8} M and 2.00×10^{-8} to 1.00×10^{-7} M of Ce^{3+} in batch mode. For accumulation times of 10 min, the detection limit was 2.0×10^{-10} M (0.027 ppb) (S/N = 3) with relative standard deviations of 3.36%, 1.76%, and 4.08%, respectively.

In the continuous-flow analysis mode, the interference of different ions was also tested. The system was further used for the determination of Ce^{3+} in human urine, in batch and continuous systems. During each preparation, the substrate GCE was treated through polishing with 0.3-mm alumina–water slurry on a polishing cloth and subsequently in using with Millipore water, before sonication for 5 min in a water bath. Next, it was immersed in a catechol solution that was typically a 12.00 mM solution in 0.10 M NaOH. Subsequently, the potential of WE was cycled several times from 0.40 V to +1.00 V at 20 mV s^{-1}, in quiescent solution, and next extracted, washed with water, and immersed in a 0.10 M sodium hydroxide solution. Finally, the potential was cycled several times between the same potential, at 100 mV s^{-1}, to eliminate monomers. The prepared electrode was stored in air. The measurements included the accumulation step, in which the PC-GCE as WE was immersed in the analyte solution under stirring for typically 3 min at open circuit. This was followed by rinsing and drying of the electrode. Next, a stripping step was performed by introducing the WE into the electrochemical cell containing a degassed stripping solution, and the potential was scanned, from +0.40 V to +1.10V, in a differential pulse fashion with a pulse amplitude of 50 mV, at 10 mV s^{-1}. The results showed an anodic peak attributed to the oxidation of the accumulated Ce^{+3} to Ce^{+4}.

Finally, the removal step was performed by immersing the electrode in a 0.10 M HCl solution for 1 min (open circuit) while the solution was stirred to facilitate the removal of the accumulated cerium.

In another work in 2000, Araiza et al. [211] reported the determination of La^{3+} through DPV after the adsorptive accumulation of the complex of La^{3+} with carminic acid complex on an HMDE in 0.1 mol L^{-1} acetate buffer, at pH 5.8. The first step included the evaluation of the redox and adsorptive behavior of the species through cyclic voltammetry and electrocapillary curves. The study revealed the optimum conditions for the measurement to be a 3.2×10^{-7} M carminic acid, an accumulation potential of −0.70 V versus ESC, and a 2-min accumulation time. The method showed a linear response behavior down to 1.4×10^{-8} M and reported a detection limit of 1.7×10^{-9} M, with a relative standard deviation of 4.8%.

In another report, Moretto et al. [212] used a multiple square-wave voltammetry (MSWV) technique for the determination of nano- and sub-nanomolar concentrations of Eu^{3+} using polymer-coated WEs. MSWV can be defined as a multipulse technique using a combination of square-wave signals with a staircase waveform, which scans the entire range of the redox system. A measure was made to reduce the signal-to-noise ratio during the studies, to help lower the detection limit by sampling the charge during the pulse sequence. The study also included a comparison between the ion-exchange cyclic voltammetry using GC electrodes, modified with the poly(estersulfonate) Kodak AQ55 and Nafion. The results proved that the latter leads to better results. The study also indicated that the preconcentration has a more pronounced effect on the voltammetric response of the Nafion-coated electrode than the AQ-GCE.

One year later, the same group used MSWV double-differential (DD) ion-exchange voltammetry for the determination of the same analyte (Eu^{+3}), using Nafion-coated ion-exchange glassy carbon electrodes [213]. The DD mode of MSWV (Fig. 5.38) gives the background signals as straight lines, whereas that of the analyte is illustrated as a twin opposite peak system, approximately corresponding to the instrumental DD of the classic voltammetric wave. This way, the detection limit and sensitivity of the MSWV-DD were reported to be superior to those achievable by conventional pulsed techniques, and the method was reportedly used for the determination of Eu^{3+}, in the range of 10^{-11} to 10^{-10} M.

Fig. 5.38 shows a schematic illustration of the waveform used in the method. In the case of a reversible system, the twin peaks of opposite signs appear at $E_{1/2} - \Delta E_{amp/2}$ and $E_{1/2} + \Delta E_{amp/2}$. Equation (5.22) was used:

$$\Delta^2 I = 1/N\,\Sigma(Q/T_{int}) \tag{5.22}$$

in which T_{int} represents the integration time for each transient, and $\Delta^2 I$ represents the signal corresponding to the double derivative of I, to

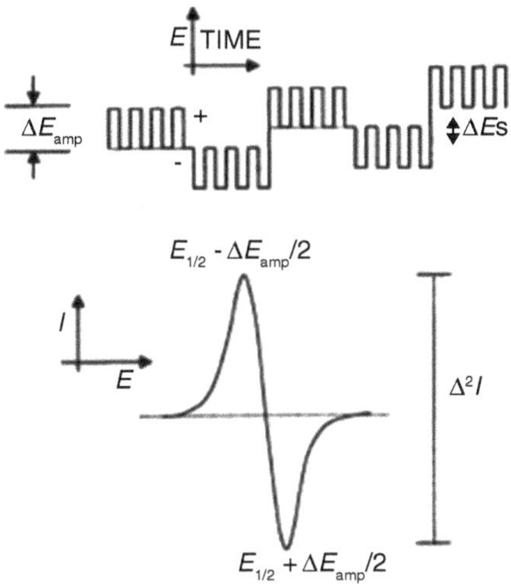

FIGURE 5.38 The schematic illustration of the sequence of pulses and the response for MSW-DD for a reversible one-electron process. The sequence is reduced to $N = 4$ [213].

normalize the recorded charges with respect to the sum of integration times.

This helps eliminate direct numerical effects of N, and hence $\Delta^2 I$ becomes directly proportional to the concentration of the analyte. The technique also allowed for the subtraction of the background component, further improving the sensitivity of the measurements.

In 2003, Mlakar [214] reported the determination of trace amounts of Yb^{3+} through cathodic stripping voltammetry (CSV). The analyte was complexed with 2-thenoyltrifluoroacetone and polyethyleneglycol in an ammonium chloride solution, and an HMDE with a potential of –1.0 V was used. Linear scan and SW voltammetries were used. It was found that a mixture of the 2-thenoyltrifluoroacetone and polyethyleneglycol led to successful determination of Yb^{3+}, because the mixed ligand complex showed a strong tendency to attach to the electrode surface under the optimal conditions, leading to the effective preconcentration of the species by the electrode. They also observed that NH_4Cl improved the sensitivity of the method because of its catalytic characteristics.

In 2005, Liu et al. [215] used adsorptive stripping voltammetric method for the trace determination of cerium based on the oxidation of the absorbed cerium (III)–alizarin complex at a CPE modified with cetyltrimethylammonium bromide. The detection limit after a 120 s preconcentration was obtained as 6.0×10^{-10} M (S/N = 3).

Rather recently, Javanbakht et al. [216] used an adsorptive stripping voltammetric method for the determination of Ce^{3+} using a CPE, which was chemically modified with dipyridyl-functionalized nanoporous

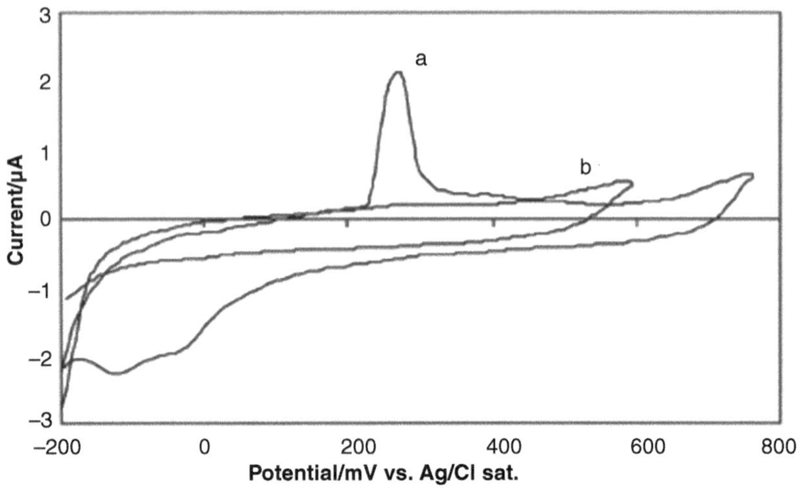

FIGURE 5.39 The cyclic voltammograms of Ce^{3+} at (a) DPNSG and (b) unmodified CPE. (0.05 M phosphate buffer solution pH 7.0; 0.02 M KNO_3 as the supporting electrolyte; Ce^{3+} concentration = 5 ng mL^{-1}; accumulation time = 10 min; scan rate from −100 to +600 mV, 100 mV s^{-1}). *Reprinted with permission from Ref. [217].*

silica gel (DPNSG). The method included two steps of chemical accumulation of Ce^{+3} at −200 mV, and the detection of its dipyridyl complex through anodic stripping voltammetry (ASV). The voltammograms obtained using the DPNSG-CPE working electrode (Fig. 5.39) revealed the superiority of the modified electrode in determination of Ce^{3+}. The anodic peak at 270 mV (vs. Ag/AgCl sat.) was attributed to the oxidation of the Ce^{3+} complex to the tetravalent cation complex. The calibration curve (Fig. 5.40) indicates the ASV at different Ce^{3+} concentrations.

Under the optimized conditions (including the amount of the modifier in the carbon paste, the electrolyte concentrations, the solution pH, and the accumulation potential or time), a linear response was observed in the concentration range of 1.0–28 ng mL^{-1}, with a relative standard deviation in the order of 3.2% or 2.5%.

The last report on the indirect determination of lanthanides through voltammetric systems, so far, is the work of Kumric et al. [217]. They determined lutetium through differential pulse anodic stripping voltammetry in an acidic solution also containing Zn-EDTA. The analyte cations freed the zinc cations, which was preconcentrated on an HMDE and measured through anodic stripping. This way, detection limits of less than 2.1×10^{-9} M was achieved after 2 min of preconcentration. The indirect method was chosen because Lu^{3+} is not electroactive at an HMDE. The results indicated that the method using 0.2 M $NaClO_4$ as the supporting

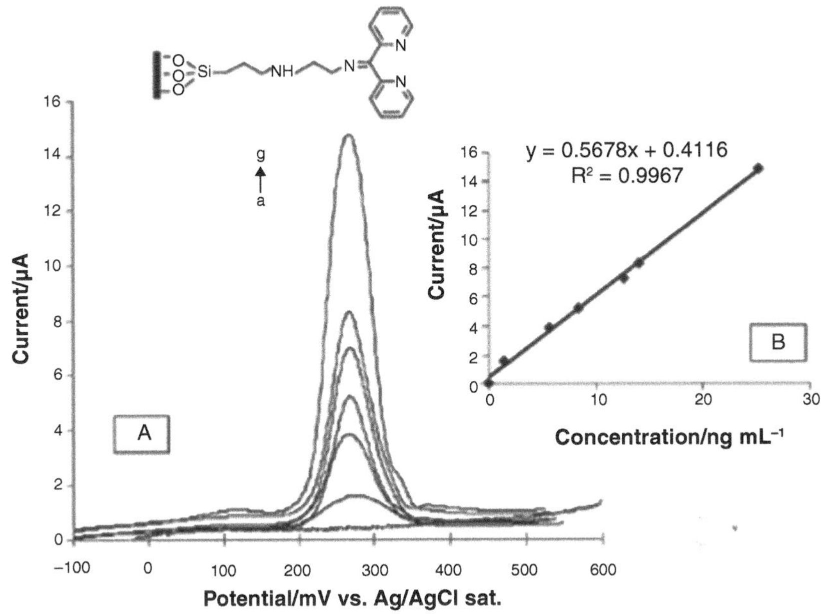

FIGURE 5.40 (A) The anodic stripping voltammograms, after preconcentration of different Ce^{3+} concentrations (accumulation time = 10 min; accumulation potential = −200 mV Ag/$AgCl_{sat}$.; scan rate: 10 mV s^{-1}; (a) 0.0; (b) 1.4 ng mL^{-1}; (c) 5.6 ng mL^{-1}; (d) 8.4 ng mL^{-1}; (e) 12.6 ng mL^{-1}; (f) 14 ng mL^{-1}; (g) 28 ng mL^{-1}. (B) Stripping peak current versus the Ce^{3+} concentration. *Reprinted with permission from Ref. [215].*

electrolyte in 0.05 M, pH 5.4 sodium-acetate buffer produces dependable results.

References

[1] Ganjali MR, Norouzi P, Rezapour M. Encyclopedia of sensors: potentiometric ion sensors, potentiometric ion sensors. Am. Sci. Publisher 2006;8:197–288.

[2] Ganjali MR, Norouzi P, Faridbod F. Research signpost transworld research network. Electrochemical Sensors, Including 11 Chapters, 2010.

[3] Püntener M, Vigassy T, Baier E, Ceresa A, Pretsch E. Improving the lower detection limit of potentiometric sensors by covalently binding the ionophore to a polymer backbone. Anal Chim Acta 2004;503:187–94.

[4] Qin W, Zwickl T, Pretsch E. Improved detection limits and unbiased selectivity coefficients obtained by using ion-exchange resins in the inner reference solution of ion-selective polymeric membrane electrodes. Anal Chem 2000;72:3236–40.

[5] Peper S, Ceresa A, Bakker E. Improved detection limits and sensitivities of potentiometric titrations. Anal Chem 2001;73:3768–75.

[6] Cotton SA. Lanthanide and actinide chemistry. Chichester: John Wiley & Sons Ltd, The Atrium, Southern Gate; 2006.

[7] Sastri VS, Bunzli JCG, Rao VR, Rayudu GVS, Perumareddi JR. Modern aspects of rare earths and their complexes. Elsevier Publication; 2003.

[8] Shamsipur M, Yousefi M, Hosseini M, Ganjali MR. Lanthanum(III) PVC membrane electrodes based on 1,3,5-trithiacyclohexane. Anal Chem 2002;74:5538–43.
[9] Gupta VK, Jain S, Chandra S. Chemical sensor for lanthanum(III) determination using aza-crown as ionophore in poly(vinyl chloride) matrix. Anal Chim Acta 2003;486:199–207.
[10] Khalil S. Ion-selective electrode for lanthanum determination in standard samples. Anal Lett 2003;36:1335–49.
[11] Ganjali MR, Kiani-Anbouhi R, Shamsipur M, Poursaberi T, Salavati-Niasari M, Talebpour Z, Emami M. Novel potentiometric PVC-membrane and coated graphite sensors for lanthanum(III). Electroanalysis 2004;16:1002–8.
[12] Ganjali MR, Qomi M, Daftari A, Norouzi P, Salavati-Niasari M, Rabbani M. Novel lanthanum(III) membrane sensor based on a new N-S Schiff's base. Sens Actuators B 2004;98:92–6.
[13] Mittal SK, Kumar SKA, Sharma HK. PVC-based dicyclohexano-18-crown-6 sensor for La(III) ions. Talanta 2004;62:801–5.
[14] Akhond M, Najafi MB, Tashkhourian J. Lanthanum-selective membrane electrode based on 2,2 ′-dithiodipyridine. Anal Chim Acta 2005;531:179–84.
[15] Ganjali MR, Akbar V, Ghorbani M, Norouzi P, Ahmadi A. Fluoride determination in some mouth wash preparations by a novel La(III) graphite coated membrane sensor based on amitraz. Anal Chim Acta 2005;531:185–91.
[16] Ganjali MR, Matloobi P, Ghorbani M, Norouzi P, Salavati-Niasari M. La(III) selective membrane sensor based on a new N-N Schiff's base. Bull Korean Chem Soc 2005;26:38–42.
[17] Jain AK, Singh AK, Mehtab S, Saxena P. Rubeanic acid as novel carrier in construction of selective membrane sensor for La(III). Anal Chim Acta 2005;551:45–50.
[18] Shamsipur M, Ershad S, Samadi N, Esmaeilbeig AR, Kia R, Abdolmaleki A. Polymeric membrane lanthanum(III)-selective electrode based on *N,N′*-adipylbis(5-plenylazo salicylaldehyde hydrazone). Electroanalysis 2005;17:1828–34.
[19] Ganjali MR, Gholivand MB, Rahimi-Nasrabadi M, Maddah B, Salavati-Niasari M, Ahmadi F. Synthesis of a new octadentates Schiff's base and its application in construction of a highly selective and sensitive lanthanum (III) membrane sensor. Sensor Lett 2006;4:356–63.
[20] Ganjali MR, Norouzi P, Alizadeh T, Adib M. Application of 8-amino-*N*-(2-hydroxybenzylidene)naphthyl amine as a neutral ionophore in the construction of a lanthanum ion-selective sensor. Anal Chim Acta 2006;576:275–82.
[21] Ganjali MR, Norouzi P, Alizadeh T, Tadjarodi A, Hanifehpour Y. Construction of a highly selective and sensitive La(III) sensor based on *N*-(2-pyridyl)-*N′*-(4-methoxyphenyl)-thiourea for nano level monitoring of La(III) ions. Electroanalysis 2006;18:1091–6.
[22] Ganjali MR, Norouzi P, Shamsolahrari L, Ahmadi A. PPb level monitoring of lanthanium by a novel PCV membrane sensor based on 4-methyl-2-hydrazinobenzothiazole. Sens Actuators B 2006;114:713–9.
[23] Ganjali MR, Norouzi P, Yousefian N, Faridbod F, Adib M. Sub-micro molar monitoring of La^{3+} by a novel lanthanum PVC-based membrane sensor based on 3-hydroxy-*N′*-(pyridin-2-ylmethylene)-2-naphthohydrazide. Bull Korean Chem Soc 2006;27:1581–6.
[24] Singh AK, Saxena P, Mehtab S, Gupta B. A selective membrane electrode for Lanthanum(III) ion based on a hexaaza macrocycle derivative as ionophore. Anal Sci 2006;22:1339–44.
[25] Zamani HA, Ganjali MR, Norouzi P, Adib M, Aceedy M. Synthesis of *N′*-(1-pyridin-2-ylmethylene)-2-furohydrazide and its application in construction of a highly selective PVC-based membrane sensor for La(III) ions. Anal Sci 2006;22:943–8.
[26] Ganjali MR, Ganjali H, Larijani B, Norouzi P. Lanthanide recognition: an asymmetric lanthanum(iii) microsensor based on a new Schiff's base. Int J Electrochem Sci 2009;4:914–22.
[27] Ganjali MR, Aghabalazadeh S, Rezapour M, Hosseini M, Norouzi P. La^{3+} Carbon nanotube (CNT) based electrode using a new (1-({2-2-(2-hydroxy-1-naphtyl)-3-(2-{(E)-1-(2-hydroxy-1-naphtyl)methylidene amino}ethyl)-1-imidazolidinyl ethyl}imino)methyl-2-naphthol). Int J Electrochem Sci 2010;5:1743–53.

[28] Ganjali MR, Moghaddam MR, Hosseini M, Norouzi P. A nano-composite carbon paste lanthanum(iii) sensor. Int J Electrochem Sci 2011;6:1981–90.
[29] Islamnezhad A, Zanjanchi MA, Shariati S, Abri A. A fast response membrane sensor based on ethyl 1,2,3,4-tetrahydro-6-methyl-4-phenyl-2-thioxopyrimidine-5-carboxylate for detection of lanthanum (iii) ions at wide concentration range. Acta Chim Slov 2011;58:46–52.
[30] Zavar MHA, Heydari S, Rounaghi GH, Ashraf N. Graphite disk lanthanum(III)-selective electrode based on benzo-15-crown-5. J Electrochem Soc 2011;158:F142–6.
[31] Taher MA, Fayazi M, Pooramiri B, Ghanei-Motlagh M, Dorehgiraee A. Lanthanum-selective sensors based on 3-amino-2-mercapto-3H-quinazolin-4-one in PVC matrix. J Iran Chem Soc 2013;10:151–9.
[32] Yuan XJ, Chai YQ, Yuan R, Zhao Q. Improved potentiometric response of solid-contact lanthanum (III) selective electrode. Anal Chim Acta 2013;779:35–40.
[33] Shamsipur M, Yousefi M, Ganjali MR. PVC-based 1,3,5-trithiane sensor for cerium(III) ions. Anal Chem 2000;72:2391–4.
[34] Shamsipur M, Yousefi M, Hosseini M, Ganjali MR. PVC-based 1,3,5-trithiane coated graphite electrode for determination of cerium(III) ions. Anal Lett 2001;34:2249–61.
[35] Gaber AAA. A novel PVC membrane sensor for selective determination of cerium(III) ions. Anal Lett 2003;36:2585–96.
[36] Karami H, Mousavi MF, Shamsipur M, Yavari I, Alizadeh AA. A new ion-selective electrode for potentiometric determination of Ce(III). Anal Lett 2003;36:1065–78.
[37] Akhond M, Najafi MB, Tashkhourian J. A new cerium (III)-selective membrane electrode based on 2-aminobenzothiazole. Sens Actuators B 2004;99:410–5.
[38] Gupta VK, Singh AK, Gupta B. A cerium(III) selective polyvinyl chloride membrane sensor based on a Schiff base complex of *N,N* ′-bis 2-(salicylideneamino)ethyl ethane-1,2-diamine. Anal Chim Acta 2006;575:198–204.
[39] Pouretedal HR, Semnani A, Keshavarz MH. Cerium(III) ion-selective electrodes based on 1,4,7-trithiacyclononane. Turkish J Chem 2006;30:711–21.
[40] Saleh MB, Gaber AAA, Khalaf MMR, Tawfeek AM. A new ion-selective electrode for potentiometric determination of Ce(III) ions. Sens Actuators B 2006;119:275–81.
[41] Zamani HA, Ganjali MR, Adib M. Construction of a highly selective PVC-based membrane sensor for Ce(III) ions. Sens Actuators B 2007;120:545–50.
[42] Abedi MR, Zamani HA, Ganjali MR, Norouzi P. Determination of cerium(III) ions in soil and sediment samples by Ce(III) PVC-based membrane electrode based on 2,5-dioxo-4-imidazolidinyl. Int J Environ Anal Chem 2008;88:353–62.
[43] Ganjali MR, Khoshsafar H, Shirzadmehr A, Javanbakht M, Faridbod F. Improvement of carbon paste ion selective electrode response by using room temperature ionic liquids (RTILs) and multi-walled carbon nanotubes (MWCNTs). Int J Electrochem Sci 2009;4:435–43.
[44] Mittal SK, Kumar SKA, Sharma HK. Cerium(III)-selective membrane electrode based on dibenzo-24-crown-8 as a neutral carrier. E-J Chem 2010;7:849–55.
[45] Singh AK, Singh P. Determination of cerium ion by polymeric membrane and coated graphite electrode based on novel pendant armed macrocycle. Anal Chim Acta 2010;675:170–80.
[46] Rounaghi G, Kakhki RMZ, Sadeghian H. A new cerium (III) ion selective electrode based on 2,9-dihydroxy-1,10-diphenoxy-4,7-dithia decane, a novel synthetic ligand. Electrochim Acta 2011;56:9756–61.
[47] Afkhami A, Madrakian T, Shirzadmehr A, Tabatabaee M, Bagheri H. New Schiff base-carbon nanotube-nanosilica-ionic liquid as a high performance sensing material of a potentiometric sensor for nanomolar determination of cerium(III) ions. Sens Actuators B 2012;174:237–44.
[48] Bagheri H, Afkhami A, Saber-Tehrani M, Shirzadmehr A, Husain SW, Khoshsafar H, Tabatabaee M. Novel sensor fabrication for the determination of nanomolar concentrations of Ce^{3+} in aqueous solutions. Anal Methods 2012;4:1753–8.

[49] Ali TA, Mohamed GG, Azzam EMS, Abd-elaal AA. Thiol surfactant assembled on gold nanoparticles ion exchanger for screen-printed electrode fabrication. Potentiometric determination of Ce(III) in environmental polluted samples. Sens Actuators B 2014;191:192–203.
[50] Ganjali MR, Mirnaghi F, Norouzi P, Adib M. Novel Pr(III)-selective membrane sensor based on a new hydrazide derivative. Sens Actuators B 2006;115:374–8.
[51] Ganjali MR, Norouzi P, Mirnaghi FS, Riahi S, Faridbod F. Lanthanide recognition: monitoring of praseodymium(III) by a novel praseodymium(III) microsensor based on *N*′-(pyridin-2-ylmethylene)benzohydrazide. IEEE Sens. J 2007;7:1138–44.
[52] Mittal SK, Sharma HK. Zirconium (IV) Antimonotungstate Based Sensor for Praseodymium (III) Ions. Res J Biotechnol 2008;470–4.
[53] Zamani HA, Ganjali MR, Norouzi P, Meghdadi S. Application of novel praseodymium (III) PVC-membrane electrode for determination of Pr(III) ions in soil and sediment samples. Anal Lett 2008;41:902–16.
[54] Ganjali MR, Khoshsafar H, Faridbod F, Shirzadmehr A, Javanbakht M, Norouzi P. room temperature ionic liquids (RTILs) and multiwalled carbon nanotubes (MWCNTs) as modifiers for improvement of carbon paste ion selective electrode response; a comparison study with PVC membrane. Electroanalysis 2009;21:2175–8.
[55] Zamani HA, Masrournia M, Sahebnasagh S, Ganjali MR. Fabrication of a praseodymium(III) PVC-membrane sensor based on *N*′1,*N*′2-Bis(2-*oxo*-1,2-diphenylethylidene) ethanedihydrazide. Anal Lett 2009;42:555–70.
[56] Chandra S, Singh DR. Development of Pr(III) selective electrode using 1,6,7,12-tetramine-2,5,8,11-tetraoxo-1(12),6(7)-di(biphenyl)dodecane (TATODBDD) as a neutral carrier. Chinese J Chem 2010;28:833–8.
[57] Faridbod F, Zamani HA, Hosseini M, Pirali-Hamedani M, Ganjali MR, Norouzi P. Praseodymium selective carbon paste electrode based on carbon nanotubes and ionic liquids. Int J Electrochem Sci 2011;6:3694–703.
[58] Harish KS, Neha V, Kapoor JK. Potentiometric sensors for Pr(III) based on *N*-*N*′ o-phenylene-bis(salicylideneimine) using epoxy resin and PVC as binding materials. E-J Chem 2011;8:S155–64.
[59] Zamani HA, Arvinfar A, Rahimi F, Imani A, Ganjali MR, Meghdadi S. Praseodymium analysis in aqueous solution by Pr^{3+}-PVC membrane sensor based on *N*,*N* ′-bis(4-hydroxysalicylidene)-1-3-phenylenediamine. Mater Sci Eng C 2011;31:307–12.
[60] Pourjavid MR, Rezaee M, Hosseini MH, Razavi T. Monitoring of praseodymium(III) ions in aqueous solutions, soil and sediment samples by a PVC membrane sensor based on a furan-triazole derivative. Quim Nova 2012;35:1973–80.
[61] Gupta AP, Agrawal H, Shukla SK, Bharduwaj VR. Studies on PVC based chelating inorganic ion exchange resin membrane sensor for neodymium(III) ion. Indian J Chem Technol 2004;11:500–3.
[62] Agarwal H, Chandra S. Studies on PVC-based chelating inorganic ion exchange resin membrane sensor for neodymium(III) ion. Indian J Chem Section A 2004;43:2361–4.
[63] Shamsipur M, Hosseini M, Alizadeh K, Mousavi MF, Garau A, Lippolis V, Yari A. PVC membrane potentiometric sensor based on 5-pyridino-2,8-dithia 9 (2,9)-1,10-phenanthroline-phane for selective determination of neodymium(III). Anal Chem 2005;77:276–83.
[64] Ganjali MR, Ahmadalinezhad A, Norouzi P, Adib M. Nd(III)-PVC membrane sensor based on 2-{ (6-aminopyridin-2-yl)imino methyl}-phenol. J Appl Electrochem 2006;36:931–6.
[65] Norouzi P, Ganjali MR, Ahmadalinezhad A, Adib M. Novel neodymium(III) membrane sensor based on *N*-(2-furylmethylene)pyridine-2,6-diamine. J Braz Chem Soc 2006;17:1309–15.
[66] Behmadi H, Zamani HA, Ganjali MR, Norouzi P. Determination of neodymium(III) ions in soil and sediment samples by a novel neodymium(III) sensor based on benzyl bisthiosemicarbazone. Electrochim Acta 2007;53:1870–6.
[67] Chandra S, Singh DR. Polymeric membrane neodymium(III)-selective electrode based on 11,13-diaza-4,7,12-trioxo-2(3),8(9)-dibenzoyl-cyclotetridecane-1,11-diene. Mater Sci Eng A 2009;502:107–10.

[68] Gupta VK, Goyal RN, Sharma RA. Comparative studies of neodymium (III)-selective PVC membrane sensors. Anal Chim Acta 2009;647:66–71.

[69] Singh AK, Singh J, Jain AK. Highly selective Nd(III) sensors: novel macrocyclic compounds for potentiometric determination of neodymium. Electroanalysis 2010;22: 2443–52.

[70] Ganjali MR, Hosseini M, Pirali-Hamedani M, Zamani HA. MWCNT-NH2/Nanosilica/RTIL/ionophore nanocomposite carbon paste electrode for Nd(III). Int J Electrochem Sci 2011;6:2808–17.

[71] Kumar R, Sharma HK, Dogra P. Potentiometric determination of neodymium (III) ions using tin (IV) antimonotungstate and zirconium (IV) antimonotungstate as electroactive materials. Int J Environ Anal Chem 2011;91:280–90.

[72] Zamani HA, Imani A, Arvinfar A, Rahimi F, Ganjali MR, Faridbod F, Meghdadi S. Neodymium(III)-PVC membrane sensor based on a new four dentate ionophore. Mater Sci Eng C 2011;31:588–92.

[73] Pourjavid MR, Yousefi SR, Hosseini MH, Rezaee M, Razavi T, Sehat AA. Determination of Neodymium(III) in aqueous and soil samples with use of a high-sensitive and selective membrane sensor. Int J Electrochem Sci 2012;7:5147–62.

[74] Zamani HA. Determination of Nd^{3+} ions in solution samples by a coated wire ion-selective sensor. E-J Chem 2012;9:83–8.

[75] Zamani HA, Zaferoni M, Meghdadi S. Neodymium(III) PVC membrane electrochemical sensor based on *N*-benzoylethylidene-2-aminobenzylamine. E-J Chem 2012;9:1941–50.

[76] Bandi KR, Singh AK, Upadhyay A. Biologically active Schiff bases as potentiometric sensor for the selective determination of Nd^{3+} ion. Electrochim Acta 2013;105:654–64.

[77] Zamani HA, Sahebnasagh S. Application of 1,2-dimaleimidobenzene as a neutral ionophore in construction of a highly selective Neodymium(III) liquid membrane electrode. Int J Electrochem Sci 2013;8:2722–33.

[78] Gupta VK, Jain R, Hamdan AJ, Agarwal S, Bharti AK. A novel ion selective sensor for promethium determination. Anal Chim Acta 2010;681:27–32.

[79] Ganjali MR, Pourjavid MR, Rezapour M, Haghgoo S. Novel samarium(III) selective membrane sensor based on glipizid. Sens Actuators B 2003;89:21–6.

[80] Shamsipur M, Hosseini M, Alizadeh K, Eskandari MM, Sharghi H, Mousavi MF, Ganjali MR. Polymeric membrane and coated graphite samarium(III)- selective electrodes based on isopropyl 2- (isopropoxycarbothioyl)disulfanyl ethanethioate. Anal Chim Acta 2003;486:93–9.

[81] Shamsipur M, Hosseini M, Alizadeh K, Talebpour Z, Mousavi MF, Ganjali MR, Arca M, Lippolis V. PVC membrane and coated graphite potentiometric sensors based on Et(4)todit for selective determination of samarium(III). Anal Chem 2003;75:5680–6.

[82] Ganjali MR, Rezapour M, Pourjavid MR, Haghgoo S. Ppt level detection of samarium(III) with a coated graphite sensor based on an antibiotic. Anal Sci 2004;20:1007–11.

[83] Mittal SK, Sharma HK, Kumar ASK. Samarium(III) selective membrane sensor based on tin(IV) boratophosphate. Sensors 2004;4:125–35.

[84] Mittal SK, Sharma HK. A samarium (III) selective electrode based on zirconium (IV) boratophosphate. J Anal Chem 2005;60:1069–72.

[85] Saleh MB, Gaber AAA, Khalaf MMR, Tawfeek AM. Novel Sm(III) ion-selective polymeric membrane sensor based on spiro oxirane-pyrazolidinedione derivative ionophore. Anal Lett 2006;39:17–31.

[86] Zamani HA, Ganjali MR, Adib M. Fabrication of a new samarium(III) ion-selective electrode based on 3-{2-*oxo*-1(2H)-acenaphthylenyliden amino}-2-thioxo-1,3-thiazolidin-4-on e. J Braz Chem Soc 2007;18:215–22.

[87] Ganjali MR, Memari Z, Faridbod F, Dinarvand R, Norouzi P. Sm^{3+} potentiometric membrane sensor as a probe for determination of some pharmaceutics. Electroanalysis 2008;20:2663–70.

[88] Ganjali MR, Memari Z, Faridbod F, Norouzi P. Samarium microsensor: an asymmetric potentiometric membrane sensor. Int J Electrochem Sci 2008;3:1169–79.

[89] Ganjali MR, Tavakoli M, Faridbod F, Riahi S, Norouzi P, Salavati-Niassari M. Interaction Study of a new bis-bidentate Schiff's base with some metal ions and its application in fabrication of Sm(III) potentiometric membrane sensor. Int J Electrochem Sci 2008;3:1559–73.
[90] Naddaf E, Zamani HA. Samarium(III)-PVC-membrane sensor based on ethoxy-1-ethoxycarbonyl-1,2-dihydroquinoline. Anal Lett 2009;42:2838–52.
[91] Zamani HA. Determination of fluoride ions in mouthwash samples by a PVC membrane Samarium(III) sensor. Anal Lett 2009;42:615–29.
[92] Agrahari SK, Srivastava AK. Graphite Electrode Coated with a 7,16-dibenzyl-1,4,10,13-tetraoxa-7,16-diazacyclooctadecane-multiwalled carbon nanotube composite as sensor for detection of samarium. Electroanalysis 2011;23:1531–5.
[93] Upadhyay A, Singh AK, Jain AK, Gupta VK, Bandi KR. Potentiometric study of coated graphite electrode and polymeric membrane electrode for the determination of sm3 + ion. Electroanalysis 2012;24:1630–8.
[94] Rawat A, Chandra S. Synthesis, characterization and applications of a hexadentate macrocyclic ligand as Sm(III) electrode. Sensor Lett 2013;11:1453–9.
[95] Zarnani HA, Naghavi-Reyabbi F, Faridbod F, Mohammadhosseini M, Ganjali MR, Tadjarodi A, Rad M. Fabrication of a PVC membrane samarium(III) sensor based on *N,N ',N"*-tris(4-pyridyl)trimesic amide as a selectophore. Mater Sci Eng C 2013;33:870–4.
[96] Ganjali MR, Rahimi-Nasrabadi M, Maddah B, Moghimi A, Borhany S. An Eu(III) sensor based on *N,N*-diethyl-*N*-(4-hydroxy-6-methylpyridin-2-yl)guanidine. Anal Sci 2004;20:1427–31.
[97] Ganjali MR, Norouzi P, Daftari A, Faridbod F, Salavati-Niasari M. Fabrication of a highly selective Eu(III) membrane sensor based on a new S-N hexadentates Schiff's base. Sens Actuators B 2007;120:673–8.
[98] Zamani HA, Rajabzadeh G, Ganjali MR. A new europium(III) PVC membrane potentiometric sensor based on 4-(2-hydroxybenzylideneamino)-6-methyl-3-thioxo-3,4-dihydro-1,2,4-triazi n-5(2H)-one. Bull Chem Soc Japan 2007;80:172–7.
[99] Ganjali MR, Davarkhah N, Ganjali H, Larijani B, Norouzi P, Hossieni M. A novel Europium(III) sensor based on 4E-4-(2-phenylviazenyl)-2-((E)-(2-aminoethylimino)methyl)phenol. Int J Electrochem Sci 2009;4:762–71.
[100] Abedi MR, Zamani HA. Construction of Eu^{3+} ion-selective electrode based on 1,2-diaminopropane-*N,N,N ',N '*-tetraacetic acid. E-J Chem 2011;8:S467–73.
[101] Norouzi P, Hosseini M, Ganjali MR, Rezapour M, Adibi M. A nano-composite modified carbon paste Europium(III) sensor. Int J Electrochem Sci 2011;6:2012–21.
[102] Zamani HA, Kamjoo R, Mohammadhosseini M, Zaferoni M, Rafati Z, Ganjali MR, Faridbod F, Meghdadi S. Europium (III) PVC membrane sensor based on *N*-pyridine-2-carboxamido-8-aminoquinoline as a sensing material. Mater Sci Eng C 2012;32:447–51.
[103] Sharma HK, Sharma N. A new Europium (III) sensor based on Titanium IV tungstophosphate as an electroactive material. Res J Chem Environ 2013;17:34–9.
[104] Upadhyay A, Singh AK, Bandi KR, Jain AK. Fabrication of coated graphite electrode for the selective determination of europium (III) ions. Talanta 2013;115:569–76.
[105] Ganjali MR, Emami M, Rezapour M, Shamsipur M, Maddah B, Salavati-Niasari M, Hosseini M, Talebpoui Z. Novel gadolinium poly(vinyl chloride) membrane sensor based on a new S-N Schiff's base. Anal Chim Acta 2003;495:51–9.
[106] Ganjali MR, Tahami M, Shamsipur M, Poursaberi T, Haghgoo S, Hosseini M. Novel gadolinium PVC-based membrane sensor based on omeprazole as an antibiotic. Electroanalysis 2003;15:1038–42.
[107] Ganjali MR, Rezapour M, Norouzi P, Salavati-Niasari M. A new pentadentate S-N Schiffs' base as a novel lonophore in construction of a novel Gd(III) membrane sensor. Electroanalysis 2005;17:2032–6.
[108] Ganjali MR, Norouzi P, Alizadeh T, Tajarodi A, Hanifehpour Y. Fabrication of a highly selective and sensitive Gd(III)-PVC membrane sensor based on *N*-(2-pyridyl)-*N'*-(4-nitrophenyl)thiourea. Sens Actuators B 2007;120:487–93.

[109] Zamani HA, Rajabzadeh G, Ganjali MR, Norouzi P. Determination of gadolinium(III) ions in soil and sediment samples by a novel gadolinium membrane sensor based on 6-methyl-4-{ 1-(2-thienyl)methylidene amino}3-thioxo-3,4-dihydro-1,2,4-t riazin-5-(2H)-one. Anal Chim Acta 2007;598:51–7.
[110] Faridbod F, Ganjali MR, Labbafi S, Norouzi P. Lanthanide recognition: development of an asymmetric gadolinium microsensor based on *N*-(2-pyridyl)-*N*′-(4-nitrophenyl) thiourea. Anal Lett 2008;41:2972–84.
[111] Sharma HK, Sharma N. Potentiometric sensor for Gadolinium(III) ion based on zirconium iv tungstophosphate as an electroactive material. E-J Chem 2009;6:1139–49.
[112] Singh AK, Jain AK, Singh J. Novel coated graphite electrode for the selective determination of Gd(III) in rocks and waste water samples. Electrochim Acta 2009;54(24):5640–7.
[113] Harish KS, Nadeem S, Pernita D. Potentiometric sensor for Gadolinium (III) ion based on Tin II tungstophosphate as an electroactive material. Res J Chem Environ 2011;15:897–902.
[114] Rezaei B, Askarpour N, Hadadzadeh H. Experimental and PM6/SPARKLE Semiempirical study of interaction between 4-methoxyphenylcyanamide and Gadolinium(III) as a fast polymeric membrane sensor. Electroanalysis 2011;23:1029–37.
[115] Zamani HA, Behmadi H. Gadolinium(III) ion-selective electrode based on 3-methyl-1H-1,2,4-triazole-5-thiol. E-J Chem 2012;9:308–12.
[116] Zamani HA, Mohammadhosseini M, Haji-Mohammadrezazadeh S, Faridbod F, Ganjali MR, Meghdadi S, Davoodnia A. Gadolinium(III) ion selective sensor using a new synthesized Schiff's base as a sensing material. Mater Sci Eng C 2012;32:712–7.
[117] GuptaVK,SinghAK,KumawatLK.Anovelgadoliniumion-selectivemembraneelectrode based on 2-(4-phenyl-1, 3-thiazol-2-yliminomethyl) phenol. Electrochim Acta 2013;95: 132–8.
[118] Vahdani M, Zamani HA, Karimian S. Quantitative determination of gadolinium ion by a gd3+-liquid membrane sensor based on 1-(4-Nitrophenyl)-3-(4-phenylazophenyl) triazene as a sensing material. Int J Electrochem Sci 2013;8:2734–45.
[119] Zamani HA, Faridbod F, Ganjali MR. A new selectophore for gadolinium selective sensor. Mater Sci Eng C 2014;43:488–93.
[120] Ganjali MR, Ghesmi A, Hosseini M, Pourjavid MR, Rezapour M, Shamsipur M, Salavati-Niasari M. Novel terbium(III) sensor based on a new bis-pyrrolidene Schiff's base. Sens Actuators B 2005;105:334–9.
[121] Zamani HA, Rajabzadeh G, Ganjali MR. Construction of a novel potentiometric terbium(III) membrane sensor and its application for the determination of terbium ion in binary mixture and fluoride ion in a mouth wash preparation. J Braz Chem Soc 2006;17:1297–303.
[122] Mittal SK, Raj K, Pernita D, Sharma HK. Potentiometric determination of terbium (III) ion using zirconium (IV) boratophosphate as an electroactive material. Res J Chem Environ 2007;11:47–51.
[123] Gupta VK, Singh AK, Gupta B. Electroanalytical performance of a terbium(III)-selective sensor based on a neutral ionophore in environmental and medicinal samples. Anal Bioanal Chem 2008;390:2171–81.
[124] Zamani HA, Ganjali MR, Norouzi P, Tadjarodi A, Shahsavani E. Determination of terbium(III) ions in phosphate rock samples by a Tb^{3+}-PVC membrane sensor based on *N*, *N*-Dimethyl-*N*′, *N*″-bis(4-methoxyphenyl)phosphoramidate. Mater Sci Eng C 2008;28:1489–94.
[125] Mohammadhossieni M, Zamani HA, Nekoei M. Terbium(III) ion-selective electrochemical sensor based on hematoporphyrin. Anal. Lett 2009;42:298–311.
[126] Zamani HA, Ganjali MR. Tb3 + PVC-membrane electrode based on 2-{2-(2-anilinocarbothioyl)hydrazino)-2-oxoacetyl}-*N*-phenyl-1-hydrazine carbothioamide. anal. lett 2009;42:1958–70.
[127] Ganjali MR, Ganjali H, Hosseini M, Norouzi P. A novel nano-composite Tb3+ carbon paste electrode. Int J Electrochem Sci 2010;5:967–77.

[128] Zamani HA, Rafati Z, Meghdadi S. Construction of Tb^{3+} PVC-membrane electrode based on *N,N'*-bis(pyrrolylmethylene)-2-aminobenzylamine. E-J Chem 2011;8:S203–10.
[129] Zamani HA, Zabihi MS, Rohani M, Zangeneh-Asadabadi A, Ganjali MR, Faridbod F, Meghdadi S. Quantitative monitoring of terbium ion by a Tb^{3+} selective electrode based on a new Schiff's base. Mater Sci Eng C 2011;31:409–13.
[130] Poorghazi-Mahmoodabadi M, Zamani HA, Pasyar S. Design of highly selective and sensitive PVC-membrane sensor for terbium ions based on Di(*N*-succinimidyl) carbonate as a neutral ion carrier. Int J Electrochem Sci 2013;8:4010–22.
[131] Zamani HA, Sahebnasagh S. Evaluation and characteristics of a Tb^{3+} electrochemical sensor based on plasticized poly(vinyl chloride) membrane. Int J Electrochem Sci 2013;8:3696–707.
[132] Gupta AP, Verma GL, Agarwal H. Poly(vinyl chloride) membrane containing zirconium(IV)selenomolybdate inorganic ion exchanger as dysprosium(III)-ion selective electrode. Indian J Chem Section A 2003;42:1910–3.
[133] Ganjali MR, Ravanshad J, Hosseini M, Salavati-Niasari M, Pourjavid MR, Baezzat MR. Novel Dy(III) sensor based on a new bis-pyrrolidene Schiff's base. Electroanalysis 2004;16:1771–6.
[134] Ganjali MR, Dodangeh M, Ghorbani H, Norouzi P, Adib M. Ppb level monitoring of Dy(III) ions by a highly sensitive and selective Dy(III) sensor based on a new asymmetrical Schiff's base. Anal Lett 2006;39:495–506.
[135] Prasad K, Kala R, Rao TP, Naidu GRK. Ion imprinted polymer based ion-selective electrode for the trace determination of dysprosium(III) ions. Anal Chim Acta 2006;566:69–74.
[136] Ganjali MR, Norouzi P, Faridbod F, Hajiabdollah N, Larijani B, Hanifehpour Y. Procaine as a sensing material for determination of Dysprosium(III) ions in presence of other rare-earth elements in biological and environmental samples. Anal Lett 2007;40:2544–61.
[137] Zamani HA, Ganjali MR, Seifi N. Dysprosium(III) ion-selective electrochemical sensor based on 6-hydrazino-1,5-diphenyl-6,7-dihydropyrazolo- 3,4-d pyrimidine-4(5H)-imine. Coll Czech Chem Commun 2007;72:1189–206.
[138] Zamani HA, Mohaddeszadeh M. Determination of Dysprosium(III) ion in soil and sediment samples by an original potentiometric Dysprosium(III) membrane sensor. Anal Lett 2008;41:2710–26.
[139] Mittal SK, Kumar R, Dogra P, Sharma HK. Potentiometric determination of dysprosium(III) ion using zirconium(IV) antimonomolybdate as an electroactive material J Anal Chem 2010;65:1045–51.
[140] Harish KS, Neha V, Kapoor JK. Dysprosium (III) selective sensor based on *N, N'*-ethylene bis(salicylideneimine) as an electro-active material. Res Chem Environ 2011;15:914–9.
[141] Pourjavid MR, Sehat AA, Rezaee M, Hosseini MH, Razavi T. Design and construction of high-sensitive and selective Dysprosium(III) electrochemical membrane sensor based on a thiourea derivative. Int J Electrochem Sci 2012;7:5133–46.
[142] Zamani HA, Faridbod F, Ganjali MR. Dysprosium selective potentiometric membrane sensor. Mater Sci Eng C 2013;33:608–12.
[143] Ganjali MR, Norouzi P, Adib M, Ahmadalinezhad A. A novel Holmium(III) membrane sensor based on *N*-(1-thien-2-ylmethylene)-1,3-benzothiazol-2-amine. Anal Lett 2006;39:1075–86.
[144] Ganjali MR, Rasoolipour S, Rezapour M, Norouzi P, Amirnasr M, Meghdadi S. A novel Ho(III) sensor based on *N,N'*-bis(2-pyridinecarboxamide)-1,2-benzene as a neutral ion carrier. J Braz Chem Soc 2006;17:1211–6.
[145] Zamani HA, Ganjali MR, Norouzi P, Meghdadi S. Fabrication of a novel holmium(III) PVC membrane sensor based on 4-chloro-1,2-bis(2-pyridinecarboxamido)benzene as a neutral ionophore. J Appl Electrochem 2007;37:853–9.
[146] Ganjali MR, Nemati R, Faridbod F, Norouzi P, Darviche F. Lanthanide recognition: a ho^{3+} potentiometric membrane sensor. Int J Electrochem Sci 2008;3(11):1288–98.

[147] Faridbod F, Ganjali MR, Norouzi P. Ho^{3+} Ion selective microelectrode: an asymmetric PVC membrane sensor. Int J Electrochem Sci 2009;4:1679–90.
[148] Ganjali MR, Motakef-Kazemi N, Norouzi P, Khoee S. A modified Ho^{3+} carbon paste electrode based on multi-walled carbon nanotubes (MWCNTs) and nanosilica. Int J Electrochem Sci 2009;4:906–13.
[149] Ganjali MR, Shams H, Faridbod F, Hajiaghababaei L, Norouzi P. Lanthanide recognition: A Ho^{3+} potentiometric membrane sensor as a probe for determination of terazosin. Mater Sci Eng C 2009;29:1380–6.
[150] Faridbod F, Ganjali MR, Larijani B, Hosseini M, Norouzi P. Ho^{3+} carbon paste sensor based on multi-walled carbon nanotubes: applied for determination of holmium content in biological and environmental samples. Mater Sci Eng C 2010;30:555–60.
[151] Hassan AZ, Fatemeh R, Fatemeh NR, Ali A, Alihossien I, Mohammad RG, Farnoush F, Masoud SN. Quantitative analysis of holmium ion by a Ho^{3+} ion-selective sensor based on a symmetric Thio-Schiff's base. Chinese J Chem 2011;29:1523–8.
[152] Zamani HA. PVC-membrane potentiometric electrochemical sensor based on 2-(4-Oxopentan-2-ylideneamino)isoindoline-1,3-dione for selective determination of Holmium(III). E-J Chem 2011;8:S97–S104.
[153] Zamani HA. Fabrication of a Ho3+-PVC membrane sensor based on *N*-phenyl-2-(thiophen-2-ylmethylene)hydrazinecarbothioamide for determination of holmium ions. chinese chem lett 2011;22:201–4.
[154] Zamani HA. *N*-1,*N*-2-Bis 1-(2-hydroxyphenyl)methylidene ethanedihydrazide as a neutral ionophore for preparation of a new Ho3+ PVC-membrane sensor. Chinese Chem Lett 2011;22:701–4.
[155] Zamani HA, Langroodi S, Meghdadi S. Application of *N*-quinoline-2-carboxamido-8-aminoquinoline in fabrication of a Ho(III)-PVC membrane sensor. E-J Chem 2011;8:S237–44.
[156] Zamani HA, Zanganeh-Asadabadi A, Rohani M, Zabihi MS, Fadaee J, Ganjali MR, Faridbod F, Meghdadi S. A Ho(III) potentiometric polymeric membrane sensor based on a new four dentate neutral ion carrier. Mater Sci Eng C 2013;33:984–8.
[157] Ganjali MR, Faridbod F, Norouzi P, Adib M. A novel Er(III) sensor based on a new hydrazone for the monitoring of Er(III) ions. Sens Actuators B 2006;120:119–24.
[158] Faridbod F, Ganjali MR, Larijani B, Norouzi P, Riahi S, Mirnaghi FS. Lanthanide recognition: an asymetric erbium microsensor based on a hydrazone derivative. Sensors 2007;7:3119–35.
[159] Faridbod F, Ganjali MR, Larijani B, Norouzi P. Multi-walled carbon nanotubes (MWCNTs) and room temperature ionic liquids (RTILs) carbon paste Er(III) sensor based on a new derivative of dansyl chloride. Electrochim Acta 2009;55:234–9.
[160] Nekoei M, Zamani HA, Mohammadhossieni M. Erbium(III) PVC membrane ion-selective sensor based on 4-(2-Thiazolylazo)resorcinal. Anal Lett 2009;42:284–97.
[161] Norouzi P, Rafiei-Sarmazdeh Z, Faridbod F, Adibi M, Ganjali MR. Er^{3+} Carbon paste electrode based on new nano-composite. Int J Electrochem Sci 2010;5:367–76.
[162] Zamani HA, Ganjali MR. Fabrication of an Er^{3+} PVC membrane sensor based on oxalic acid bis 2- (phenylamino)carbonyl hydrazide. Monatsh Chem 2010;141:1183–9.
[163] Zamani HA, Mohammadhosseini M, Nekoei M, Ganjali MR. Determination of erbium ions in water samples by a PVC membrane erbium-ion selective electrode. Sensor Lett 2010;8:303–7.
[164] Zamani HA. Erbium(III) PVC membrane sensor based on *N*-(benzyloxycarbonyloxy) succinimide as a new neutral ionophore. Chinese Chem Lett 2011;22:346–9.
[165] Zamani HA, Rohani M, Mohammadhosseini M, Ganjali MR, Faridbod F, Meghdadi S. Quantitative monitoring of erbium ion in alloy samples by a erbium selective sensor. Sensor Lett 2011;9:1745–9.
[166] Zamani AA, Parinejad M, Jamali F, Yaftian MR, Matt D. Evaluation of a wide rim phosphorylated calix 4 arene's properties as a sensory molecule in an Er(III)-PVC membrane sensor. Turkish J Chem 2012;36:907–16.

[167] Zamani HA, Faridbod F. Selective determination of erbium in the mixture of other lanthanide ions by a potentiometric sensor. Sensor Lett 2013;11:571–5.
[168] Ganjali MR, Rasoolipour S, Rezapour M, Norouzi P, Adib M. Synthesis of thiophene-2-carbaldehyde-(7-methyl-1,3-benzothiazol-2-yl)hydrazone and its application as an ionophore in the construction of a novel thulium(III) selective membrane sensor. Electrochem Commun 2005;7:989–94.
[169] Ganjali MR, Norouzi P, Tamaddon A, Husain SW. Novel Tm(III) membrane sensor based on 2,2 ′-dianiline disulfide and its application for the fluoride monitoring of mouth wash preparations. Bull Korean Chem Soc 2006;27:1418–22.
[170] Ganjali MR, Norouzi P, Akbari-Adergani B. Thulium(III) ions monitoring by a novel thulium(III) microelectrode based on a S-N Schiff base. Electroanalysis 2007;19:1145–51.
[171] Ganjali MR, Norouzi P, Farrokhi R, Faridbod F, Larijani B, Meghdadi S. Lanthanide recognition: determination of Thulium(III) ions in presence of other rare earth elements by a Thulium(III) sensor based on 4-Methyl-1,2-bis(2-pyridinecarboxamido) benzene as a sensing material. Anal Lett 2008;41:2322–43.
[172] Zamani HA, Nekoei M, Mohammadhosseini M, Ganjali MR. Construction of Tm3+-PVC membrane sensor based on 1-(2-thiazolylazo)-2-naphthol as sensing material. Mater Sci Eng C 2010;30:480–3.
[173] Ganjali MR, Ranaei-Siadat SO, Rashedi H, Rezapour M, Norouzi P. Thulium selective sensor based on nanographene/RTIL/ionophore/graphite. Int J Electrochem Sci 2011;6:3684–93.
[174] Ganjali MR, Rezapour M, Torkestani SK, Rashedi H, Norouzi P. Long-term stable fabrication of a nanocomposite Tm(III) Sensor containing nanographene/nanosilica/RTIL/ionophore. Int J Electrochem Sci 2011;6:2323–32.
[175] Zamani HA, Feizyzadeh B, Faridbod F, Ganjali MR. Application of 1-ethyl-3-(2,5-dihydro-4-(3,5-dimethyl-1H-pyrazol-4-yl)-5-*oxo*-1H-pyrazol -3-yl)thiourea as sensing material for construction of Tm3+-PVC membrane sensor. Mater Sci Eng C 2011;31:1379–82.
[176] Zamani HA, Feizyzadeh B, Faridbod F, Ganjali MR. Thulium(III) sensor based on a derivative of thiourea doped in polymeric membrane. Sensor Lett 2011;9:1767–73.
[177] Zamani HA, Naghavi-Reyabbi F, Mohammadhosseini M, Feizyzadeh B, Abedi MR, Faridbod F, Ganjali MR. Quantitative monitoring of thulium ions by a new thulium selective polymeric membrane sensor. Sensor Lett 2012;10:112–6.
[178] Hatamluyi B, Zamani HA, Behmadi H. Fabrication of a PVC-membrane electrochemical sensor based on 4-(3-nitrophenyl)-2,6-di-2-thienylpyridine as a sensing material for determination of Thulium(III) ions. Int J Electrochem Sci 2013;8:2746–55.
[179] Ganjali MR, Naji L, Poursaberi T, Shamsipur M, Haghgoo S. Ytterbium(III)-selective membrane electrode based on cefixime. Anal Chim Acta 2003;475:59–66.
[180] Ganjali MR, Rasoolipour S, Rezapour M, Norouzi P, Tajarodi A, Hanifehpour Y. Novel ytterbium(III) selective membrane sensor based on *N*-(2-pyridyl)-*N*′-(2-methoxyphenyl)-thiourea as an excellent carrier and its application to determination of fluoride in mouth wash preparation samples. Electroanalysis 2005;17:1534–9.
[181] Ganjali MR, Norouzi P, Ghorbani H, Larijani B, Tadjarodi A, Hanifehpour Y. Lanthanide recognition: fabrication of an ytterbium PVC membrane sensor based on *N*-(6-picolyl)-*N*′-(4-metoxyphenyl)thiourea(PCMPT). Can J Anal Sci Spectrosc 2006;51:279–87.
[182] Ganjali MR, Norouzi P, Tamaddon A, Adib M. Nano-level monitoring of ytterbium(III) by a novel ytterbium(III) membrane sensor based on 3-hydroxy-*N*′- (2-hydroxyphenyl) methylene -2-naphthohydrazide. Sens Actuators B 2006;114:855–60.
[183] Zamani HA, Rajabzadeh G, Ganjali MR. A new ytterbium(III) PVC membrane electrode based on 6-methy-4-{ 1-(1H-pyrrol-2-yl)methylidene amino}-3-thioxo-3,4-dihydro-1, 2,4-triazin-5(2H)-one. Talanta 2007;72:1093–9.
[184] Ebrahimi M, Zamani HA. Determination of Ytterbium(III) ions in soil and sediment samples by a new polymeric Yb^{3+}-PVC membrane sensor. Anal Lett 2009;42(7): 1041–55.

[185] Singh AK, Singh P. Nano-level monitoring of Yb(III) by fabrication of coated graphite electrode based on newly synthesized hexaaza macrocyclic ligand. Anal Chim Acta 2009;643:74–82.
[186] Zamani HA, Ganjali MR. Original potentiometric Ytterbium(III) PVC-membrane sensor based on N1,N2-Bis- 1-(2-hydroxy-1,2-diphenyl)ethylidene ethanedihydrazide. Anal Lett 2009;42:1014–28.
[187] Faridbod F, Ganjali MR, Pirali-Hamedani M, Norouzi P. mwcnts-ionic liquids-ionophore-graphite nanocomposite based sensor for selective determination of Ytterbium(III) Ion. Int J Electrochem Sci 2010;5:1103–12.
[188] Rezaei B, Hemati A, Hadadzadeh H. Investigation of Yb(III)-PVC membrane interfacial interaction by semiempirical PM6/SPARKLE method based on 1,10-phenanthroline-5,6-dione as a suitable neutral ionophore. J Colloid Interf Sci 2011;354:268–74.
[189] Zamani HA, Sahebnasagh S. Construction of a highly selective and sensitive Ytterbium(III) membrane sensor based on 4-methoxybenzyl carbazate as a new neutral ionophore. Int J Electrochem. Sci 2013;8:3429–40.
[190] Ganjali MR, Tamaddon A, Norouzi P, Adib M. Novel Lu(III) membrane sensor based on a new asymmetrically S-N Schiff's base. Sens Actuators B 2006;120:194–9.
[191] Ganjali MR, Norouzi P, Akbari-Adergani B, Riahi S, Larijani B. An asymmetric lutetium(III) microsensor based on *N*-(2-furylmethylene) pyridine-2,6-diamine for determination of lutetium(III) ions. Anal Lett 2007;40:1923–38.
[192] Ganjali MR, Norouzi P, Faridbod F, Hajiabdollah N, Dinarvand R, Meghdadi S. Lutetium(III) ions determination in biological and environmental samples by a Lutetium(III) sensor based on *N,N'*-bis(2-pyridinecarboxamide)-1,3-benzene as a sensing material. Anal Lett 2008;41:3–23.
[193] Ganjali MR, Norouzi P, Atrian A, Faridbod F, Meghdadi S, Giahi M. Neutral *N,N'*-bis(2-pyridinecarboxamide)-1,2-ethane as sensing material for determination of lutetium(III) ions in biological and environmental samples. Mater Sci Eng C 2009;29:205–10.
[194] Zamani HA, Rohani M, Zangeneh-Asadabadi A, Zabihi MS, Ganjali MR, Salavati-Niasari M. A novel lutetium(III) PVC membrane sensor based on a new symmetric S-N Schiff's base for Lu(III) analysis in real sample. Mater Sci Eng C 2010;30:917–20.
[195] Abedi MR, Zamani HA. Application of 2,2'-dithiobis(4-methylthiazole) as sensing material for construction of Lu^{3+} PVC-membrane sensor. Chinese Chem Lett 2011;22(8):977–80.
[196] Norouzi P, Pirali-Hamedani M, Ranaei-Siadat SO, Ganjali MR. Nanocomposite carbon paste sensor for selective determination of Lu(III) ion. Int J Electrochem Sci 2011;6:3704–13.
[197] Zamani HA, Ganjali MR, Faridbod F. A lutetium PVC membrane sensor based on (2-*oxo*-1,2-diphenylethylidene)-*N*-phenylhydrazinecarbothioamide. J Serb Chem Soc 2011;76:1295–305.
[198] Pourjavid MR, Razavi T. 2-Amino-4-(4-aminophenyl)thiazole application as an ionophore in the construction of a Lu(III) selective membrane sensor. Chinese Chem Lett 2012;23:343–6.
[199] Karimian S, Zamani HA, Vahdani M. Construction of a new Lutetium (III) PVC-membrane electrochemical sensor based on 4'-carboxybenzo-18-crown-6. Int J Electrochem. Sci 2013;8:2710–21.
[200] Pasyar S, Zamani HA, Poorghazi-Mahmoodabadi M. New potentiometric poly(vinyl chloride) membrane sensor for selective monitoring of Lutetium(III) based on Di(*N*-succinimidyl)oxalate. Int J Electrochem Sci 2013;8:4023–35.
[201] Tamaddon A, Amiri R, Hazini F. Novel Lu^{3+} Carbon paste electrode based on functionalized multiwalled carbon nanotubes. Electroanalysis 2014;26:612–7.
[202] Heineman, P.W. R. Laboratory techniques in electroanalytical chemistry, 3rd ed. Revised and Expanded 2nd ed. CRC, 1996, ISBN 0-8247-9445-1.
[203] Zoski, Cynthia G. handbook of electrochemistry. Elsevier Science, 2007, ISBN 0-444-51958-0.

[204] Bard, A; Larry R. F. electrochemical methods: fundamentals and applications 2nd ed. Wiley, 2000, ISBN 0-471-04372-9.
[205] Nicholson RS, Irving S. Theory of stationary electrode polarography. single scan and cyclic methods applied to reversible, irreversible, and kinetic systems. Anal Chem 1964;36:706–23.
[206] Brainina KhZ. Stripping voltammetry in chemical analysis. New York: Wiley; 1974.
[207] Iikova SB, Zazimko EA, Antonov AV. Determination of cerium in rare earth- and yttrium oxides by inversion voltammetry with a graphite electrode. Indust Lab 1983;49:105–7.
[208] Wang J, Farias PAM, Mahmond JS. Trace determination of lanthanum, cerium, and praseodymium based on adsorptive stripping voltammetry. Anal Chim Acta 1985;171: 215–23.
[209] Abollino O, Aceto M, Mentasti E, Sarzanini C, van den Berg CMG. Determination of trace europium by adsorptive cathodic stripping voltammetry after complexation with cupferron. Electroanalysis 1997;9:444–8.
[210] Khoo SB, Zhu J. Poly(catechol) film modified glassy carbon electrode for ultratrace determination of cerium(III) by differential pulse anodic stripping voltammetry. Electroanalysis 1999;11:546–52.
[211] Araiza JEA, Mendoza RN, Jimenez GG, Martinez IG. Determination of La(III) by adsorptive stripping voltammetry using carminic acid in acetate buffer. Quim Anal 2000;19:159–63.
[212] Moretto LM, Brunetti B, Chevalet J, Ugo P. Multiple square wave voltammetry of nanomolar and ubnanomolar concentrations of europium(III) at polymer-coated electrodes. Electrochem Commun 2000;2:175–9.
[213] Moretto LM, Chevalet J, Mazzocchin GA, Ugo P. Advances in multiple square wave techniques for ion-exchange voltammetry at ultratrace levels: the europium(III) case. J. Electroanal Chem 2001;498:117–26.
[214] Mlakar M. Determination of ytterbium traces by cathodic stripping voltammetry. Electroanalysis 2003;15:27–32.
[215] Liu SM, Li JN, Zhang SJ, Zhao JQ. Study on the adsorptive stripping voltammetric determination of trace cerium at a carbon paste electrode modified in situ with cetyltrimethylammonium bromide. Appl Surf Sci 2005;252:2078–84.
[216] Javanbakht M, Khoshsafar H, Ganjali MR, Norouzi P, Badei A, Hasheminasa A. Stripping voltammetry of cerium(III) with a chemically modified carbon paste electrode containing functionalized nanoporous silica gel. Electroanalysis 2008;20:203–6.
[217] Kumric KR, Trtic-Petrovic TM, Ignjatovic LM, Comor JJ. Indirect determination of lutetium by differential pulse anodic stripping voltammetry at a hanging mercury drop electrode. Cent Eur J Chem 2008;6:65–9.

CHAPTER

6

Spectrometric Determination of Lanthanides Series

OUTLINE

Spectrophotometric methods are the most popular and relatively cheaper techniques for quantitative determination of metal ions. The lanthanide contents of various geological, industrial, environmental, food, biological, and clinical samples widely have been analyzed by spectroscopic methods. Both atomic (atomic absorption spectroscopy [AAS], inductively coupled plasma–optical emission spectroscopy [ICP-OES], and inductively

Lanthanides Series Determination by Various Analytical Methods. http://dx.doi.org/10.1016/B978-0-12-804704-0.00006-2

coupled plasma–mass spectrometry [ICP-MS]) and molecular spectroscopic techniques (UV–vis and luminescence spectrophotometry) have been used for determination of these elements. Here, first the basic concepts of the methods are reviewed and then reports on the application of spectroscopic methods in the analysis of lanthanide are considered.

ATOMIC ABSORPTION SPECTROMETRY

Atomic absorption spectroscopy (AAS) is commonly used for quantitative analyses and is based on the measurement of absorption of light by atomized elements in the gaseous state, relying on the Beer–Lambert law [1].

Briefly, the electrons of the atomized elements are excited to higher orbitals for a short period of time, on the absorption of a given wavelength, the energy of which should equal that required for the transition in a particular element. The wavelengths used in the method, correspond to distinct elements, and have a very narrow width of only a few picometers. This makes the technique very selective. The radiation flux in the presence and absence of the analyte is measured by the detector, and the ratio of the two values (i.e., the absorbance) is related to the concentration or mass of the analyte based on the Beer–Lambert law. As mentioned, the sample is atomized prior to analysis, using atomizers like flame or electrothermal (graphite tube) devices. Further techniques, like glow-discharge, hydride, and cold-vapor atomization, also have been used. Next, the atoms are irradiated by radiation usually from an element-specific or a continuous radiation sources, and the radiation passes through a monochromator before entering the detector. A schematic diagram of an atomic absorption spectrometer is shown in Fig. 6.1.

The most conventional atomization technique used in AAS of liquid or dissolved sample is mainly the air-acetylene flame, which can reach

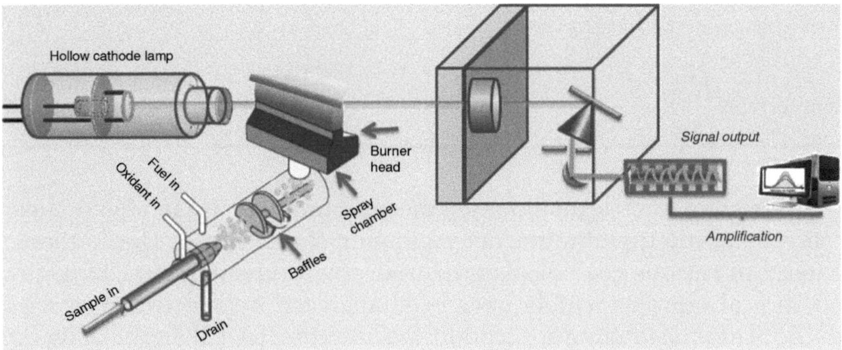

FIGURE 6.1 Schematic diagram of an atomic absorption spectrometer.

2300°C, or the N_2O-acetylene flame, which heats up to 2700°C. When using these atomizers, a pneumatic nebulizer aspirates and sprays the sample into a chamber in the form of an aerosol. There, the sample is mixed with the flame gases. The sample is treated so that only very fine droplets (<10 μm) enter the flame, so about 95% of the aspirated sample is lost in this stage, but the fine droplets guarantee reduced interference.

The burner located on top of the spray chamber produces a thin but long flame, through which the light beam passes. The gas flow and burner height can be adjusted to manipulate the concentration of atoms in the flame.

In general, after the sample enters the flame, its solvent is evaporated, leaving the dry sample particles, and next it is vaporized into gaseous molecules, before being dissociated into free atoms. The analyte atoms might even partially or wholly ionize based on the energy available in the flame (it is noteworthy that ionization is considered undesirable since it decreases the number of atoms that are available for measurement and hence reduces the sensitivity). During each of these stages, there exists the risk of interferences, if the phase transfer behavior of the analyte in the calibration standard and sample is different.

The different sources of interference in AAS include absorption of the radiation by other species; ionization of the analyte, which leads to a decrease in the population of the atomic species (this is moderated by addition of ionization suppressors); self-absorption, where the absorption is higher at the center of the line than its edges, leading to changes in shape and intensity of the line; and background absorption, where incompletely atomized particles absorb the radiation (this is also controlled by increasing the flame temperature).

Another important part of the atomic absorption spectrometer is the monochromator. Regarding the fact that each element has a distinct set of energy levels and hence very narrow absorption lines, the monochromator is very vital for obtaining a linear calibration curve since the bandwidth of the absorbing species should be wider than that of the source, which is a hard task. For measuring elements using the AAS methods, a simple calibration method, a standard addition method, and an internal standard method are commonly used.

In a more advanced and sensitive technique, called graphite furnace atomic absorption spectrometry (GFAAS), a graphite-coated furnace is used to vaporize the sample. The GFAA and flame AAS measurement principle is the same. The difference between these two techniques is the way the sample is introduced into the instrument. In GFAA analysis, an electrothermal graphite furnace is used instead (also known as electrothermal atomic absorption spectrometry [ETAAS]). The sample is heated stepwise (up to 3000°C) to dry. The biggest advantage of the graphite furnace is that the detection limit is about two orders of magnitude better than that of AAS. Furthermore, the graphite furnace has several other advantages over a flame furnace. First it accepts solutions, slurries, or solid samples. Second, it is a much more efficient atomizer than a flame furnace and it can directly accept very small absolute quantities of sample. It also provides a reducing environment for easily oxidized elements. Samples are placed directly into the graphite furnace, and the furnace is electrically heated in several steps to dry the sample (drying step; evaporation of solvents [10–100 s]), ash organic matter (pyrolysis step; removal of volatile hydroxides, sulfates, carbonates [10–100 s]), and vaporize the analyte atoms (atomization step; atomization of remaining analyte [1 s]). It accommodates smaller samples but it is a difficult operation, since the high energy that is provided to atomize the sample particles into a ground state might also excite the atomized particles into a higher energy level, thus lowering the precision.

AAS can be used to determine more than 70 different elements in solution or directly in solid samples used in pharmacology, biophysics, and toxicology research. Today these procedures are used more than ever in materials and environmental applications. However, because of the scientific advances of ICP-OES, ICP-MS, have left AAS behind. This technique, however, is excellent and has a larger specificity that ICP does not have.

In determination of lanthanides, the basic problem with the application of AAS is the formation of very stable molecular species like monoxide molecules, under flame conditions, which leads to the inefficiency of the atomization process. Some efforts have been made to overcome this situation through the use of oxygen acetylene flames in highly fuel-rich regimes, which set the conditions for the better atomization of the species. GFAAS has also been used for more sensitive single-element analyses of lanthanides, but the method still suffers pronounced matrix interferences.

The history of using AAS techniques to the measurement of lanthanides dates back to the middle of the twentieth century. In a study in 1964 on the AAS of the ethanolic solutions of perchlorate salts of lanthanides in fuel-rich oxyacetylene flames fed with more than 1000 lines were reported in the 2500–6500 °A region, and it was mentioned that most of these lines have the intensity needed for analytical determination through AAS [2]. Also, in studies dating back to 1971, lanthanides like yttrium, europium, terbium, dysprosium, holmium, erbium, thulium, ytterbium, and lutetium in zirconium and calcium rare earth silicates were studied through AAS [3]. The application of the standard addition technique to the determination of 12 lanthanides through the ASS was also reported in 1974 [4], and a similar study was made on neodymium in 1976 [5].

In the same year, Gupta [6] reported improvements in the sensitivity of atomic-absorption and flame-emission analyses of lanthanides and yttrium by a factor of 2–5, on the aspiration of an absolute ethanol solution of the perchlorate of the metal on the nitrous oxide-acetylene flame, and a method was developed for the accurate determination of the mentioned analytes, where 1% of lanthanum was used as a spectroscopic buffer for the elimination of interferences and enhancing the sensitivities in certain determinations, but it was mentioned that the element can be replaced with 2000 ppm of sodium, in case it causes interference.

Later reports mentioned the electrothermal atomization of lanthanides reaching a detection limit of 10^{-10} to 10^{-12} g [7] or the electrothermal atomization of Sm, Eu, Dy, and Er separated from uranium [8]. Later, a so-called timeless method with a graphite cell was reported for the determination of concentrations of Eu, Gd, Dy, Ho, Er, Tm, Yb, and Y, whereas La, Pr, Nd, Sm, Eu, Gd, Tb, Dy, Ho, Er, Tm, Yb, and Y were determined in rare earth ores through the flame atomic absorption method [9].

Also, a procedure was developed for the determination of traces of Y, Pr, Nd, Sm, Eu, Gd, Tb, Dy, Ho, Er, Tm, Yb, and Lu in rocks by electrothermal atomization in a pyrolytic graphite–coated furnace, reaching sensitivities in the order of 10^{-9} to 10^{-2} g at 2500°C [10]. More recent reports on the application of AAS are the determination of Er through the electrothermal method with the help of a pyrolytic graphite–coated tube and a pyrolytic graphite–coated tube lined with tantalum foil [11] or study of the enhancing effects of reagents and surfactants on the sensitivity of measuring Yb, Al, and Cr through AAS [12]. The latter study showed that a new substance other than ytterbium sodium dodecyl sulfate or ytterbium-sulfosalicylic acid was formed in the experiments, which was very easily atomized. It was also observed that the presence of surfactants lowered the surface energy of the solution and increased the effect of organic reagents.

The analysis of ytterbium with using GFAAS with the aid of a pyrolytic graphite–coated tube and a pyrolytic graphite–coated tube lined with a tungsten spiral and a tantalum foil is another example [13]. Another work focused on the effect of the presence of Sm and Eu on the electrothermal determination of ytterbium in HCl and HNO_3 media [14], and the method proved to require low atomization temperature; lead to high selectivity, sensitivity, and repeatability; as well as low detection limits and memory effect. It was also found that a variety of rare earths and non–rare earth elements do not interfere with the method.

The low sensitivity and severe memory effects observed in the determination of lanthanides during GFAAS was also studied [15], and the experiments showed that during the determination, lanthanum nitrate converted to lanthanum oxide (and to its carbide at 2300 K or above, which is an important reason for the low sensitivity and memory effects of the element. The study also suggested that the formation of a thin film of carbon of an unknown origin also led to the low sensitivity of lanthanum. In the case of samarium nitrate, after the formation of samarium oxide through the formation of SmO from Sm_2O_3. At above 2000 K, the formation of the carbide was noticed as a result of the low sensitivity and memory effects of Sm. The study suggested that the use of a tantalum probe atomization in the case of samarium led to high sensitivity because tantalum substrate meddles with the surface reactions of Sm species through the formation of $SmTaO_4$. There is no need to mention that these interactions lead to damages to the tantalum probe, which might, in turn, adversely influence the sensitivity.

In a theoretical and experimental study of the atomization process of neodymium GFAAS [16], the same fact about the formation of stable oxide and carbide and their dissociation for the atomization of Nd was observed.

In 1998, a method was proposed for the determination of ytterbium in animal feces, which was called tungsten coil electrothermal atomic absorption spectrometry [17]. Another approach to the determination of lanthanides through AAS is the indirect determination. An example is the indirect determination of the total yttrium and lanthanides by flame AAS [18], where the Y^{3+}, La^{3+}, Pr^{3+}, Ce^{3+}, Nd^{3+}, and Sm^{3+} were extracted into isobutyl methyl ketone as mixed metal complexes with purpurin and Mg^{2+}. The measurements were next based on the absorbance of magnesium.

A more recent example of the application of AAS for the analysis of lanthanides is the determination of Eu^{3+} through GFAAS by preconcentration with 2-thenoyl trifluoroacetone phenanthrolene modified electrode [19], where a tungsten filament decorated with a mixture of 2-thenoyl trifluoroacetone and phenanthrolene (phen) was used for preconcentration of europium in $(CH_2)_6N_4$–HCl buffer. The electrode was next inserted into a so-called pyrolytic coated graphite cup for atomization and measurement. The method showed a linear behavior in the range of 0.29×10^{-8} to 5.9×10^{-8} mol/L.

In 2005, Li et al. reported the AAS determination of traces of La, Ce, Pr, Nd, Sm, Gd, Tb, Dy, Yb, and Lu using carbon powder, potassium sulfate, barium sulfate, strontium sulfate, and scandium chloride as buffer [20]. They used scandium as the internal standard line and did not perform any sample separation and chemical treatment on the sample but achieved satisfactory results.

Biju and Rao in 2005 determined Nd, Eu, and Lu in diluted aqueous and synthetic seawater solutions by flame AAS after solid phase extraction and preconcentration with 5,7-dichloroquinoline-8-ol functionalized naphthalene [21].

A rather recent report made in 2011 has also focused on the determination of rare-earth elements in rare-earth magnets by polarized Zeeman flame atomic absorption spectrometry [22].

INDUCTIVELY COUPLED PLASMA–OPTICAL EMISSION SPECTROMETRY (ICP-OES)

The technique known as inductively coupled plasma atomic emission spectroscopy (ICP-AES), which is also referred to as inductively coupled plasma optical emission spectroscopy (ICP-OES) (because it is

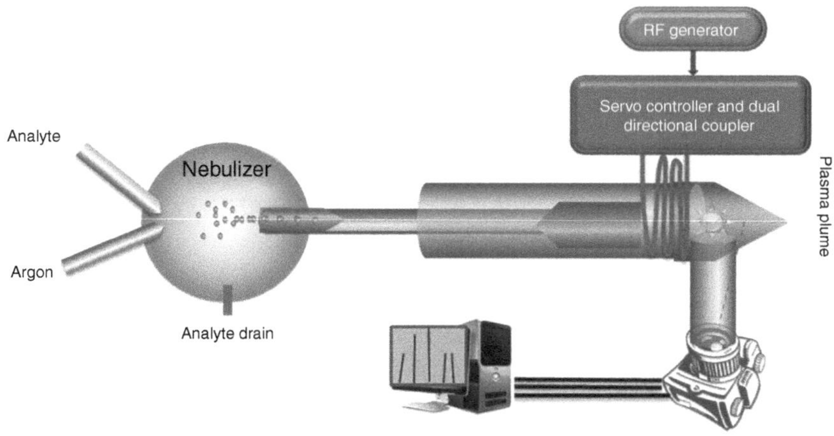

FIGURE 6.2 A schematic illustration of a typical ICP-OES instrument.

a type of emission spectroscopy that uses the inductively coupled plasma to produce both excited *atoms* and *ions*, which emit electromagnetic radiation at wavelengths characteristic of a particular element, and so it is better to use ICP-OES) is widely used for the analysis of trace concentrations of metals.

ICP-OES Instrumentation

Basically, the technique involves injection of samples into an argon plasma through a nebulizer. The plasma is induced using a radio frequency (RF). The nebulized sample loses its solvent in a very short time and vaporizes into the gas phase, where it is excited as a result of collisions with other particles at very high temperatures. The excited species emit spectral lines that are characteristic of their electron configurations, and the emissions are collected by an optical element before entering the slit of another element for separating the wavelengths, which guides the emissions into a detector that can be as simple as a photomultiplier tube (PMT). Fig. 6.2 shows a typical ICP-OES instrument. ICP-OES is capable of analyzing about 70 elements either alone or in combination with each other. Single-element measurements with instruments having a simple monochromator/PMT combination can be cost effective. However, simultaneous multielement determinations that can analyze even up to 70 elements require more sophisticated instrumentation, such as polychromators or array detectors.

The ICP torch is composed of three fused-silica tubes in a concentric arrangement, containing the outer, intermediate, and inner gas tubes. The other element of the torch is the load coil, which is a copper coil that is connected to the RF generator and surrounds the top part of the torch. The coil is also equipped with a water cooling system.

The flow of argon from the outer gas tube is about 10–15 L/min and is necessary for sustaining the plasma, while the sample is carried by another lower flow of argon of about 0.5–1.5 L/min through the inner tube. The intermediate gas tube is devised for optionally blowing another flow of Ar (0–1.5 L/min) with the aim of lifting the plasma as well as changing the concentration of the inner flow. The coil surrounding the quartz torch induces an RF with power in the range 700–1500 W, altering which induces an alternating current based on the frequency of the RF generator, which is either 27 or 40 MHz in the case of ICP devices.

The strong electromagnetic field within the coil of the torch is the result of the RF applied to the coil, by the RF generator. On turning of the torch, the Ar (or any other alternative) gas is "ignited" by a Tesla unit. This means some ionized argon is produced by a short discharge arc that is produced by the Tesla unit, which can be described as a lighter that turns off after this. The rest of the gas is ionized by the strong electromagnetic field, and the cations and electrons start flowing concurrently in a particular pattern with respect to the magnetic field, which creates a very hot plasma phase (at 7000–10,000 K) because of the collisions and friction among the atomic and ionized species present.

The sample on the other hand is pumped into the nebulizer, which sprays it in the form of a mist with very fine particles into the plasma, where it readily atomizes and ionizes and gets excited at the high temperature of the phase. The excited species naturally emit radiation, based on their electronic structures.

Some instruments use a shear gas, which is usually N_2 or pressurized dry air to cut a specific spot of the plasma. The devices next use a plurality of transfer lenses for focusing the light that is emitted on a diffraction grating. The gratings are meant to separate the light into its wavelengths in an optical spectrometer, whereas in other designs an optical surface consisting of an orifice is used. The plasma strikes the optical interface, while a constant flow of Ar arises from the orifice not only cooling and leading to the deflection of the plasma but also letting the emitted light enter an optical chamber.

The optical elements initially separate the light, which is within the UV–vis region (i.e., 160–800 nm), into different wavelengths, and next the intensity of the desired wavelength is measured by the different detection systems and arrangements possible. Yet modern arrayed detectors have the ability to measure the intensity of a range of spectral lines, which allows multielemental analyses.

The intensity of the measured emissions are then compared with the data acquired for standard samples to evaluate the concentration of the analyte in the sample, and their wavelength can also be used as an indication of the presence of specific elements in the sample.

The emission from the torch can be observed through radial, axial, or an intermediate configuration, each of which has its own advantages and disadvantages. In the radial or side view, which can be regarded as the classical ICP-OES mode, the plasma is operated in a vertical orientation, and the emissions are observed from its side. This approach reportedly limits the observation volume and hence the influence of spectral and background interferences that might occur, whereas in the axial mode, the plasma is in a horizontal position and observations are made at its end. This approach helps lower the detection limit 5- or 10-folds because it gives a longer viewing path. The approach, however, suffers drawbacks like increased spectral and matrix-induced interferences, as well as increased risk of self-absorption because of the lower temperature of the plasma at its tail. Measures like controlling the flow of the shear gas can help mitigate these to some extent, through using spectral lines with less interference or making interelement corrections as well as manipulating the shear gas.

Analytical Aspects of the ICP-OES Method

ICP-OES is used in trace analysis of many elements in a wide range of samples, including minerals, alloys, oil, soil, sediments, water, agriculture products, food, and biological samples. ICP-OES enjoys several advantages as compared to the other rivaling atomic spectroscopic techniques. Advantages of the technique include the high temperature (7000–10 000 K) and electron density (1014–1016 cm^{-3}) of the plasma, which leads to reduced chemical interferences (occurring during atomization of the analyte, the two most common being the formation of nonvolatile compounds containing the analyte and ionization of the analyte); the substantial ionization for many elements; the possibility for simultaneous multielemental analysis of 70 elements, including sulfur, phosphorus, and refractory elements; low background emission; low chemical interferences; robustness; high stability; good accuracy and precision; low limits of detection (LODs, in the range of 0.1–100 ng/mL); wide linear response range of about 4–6 orders of magnitude; and reasonable costs.

In general, it is worthy to know that ICP-OES cannot be used for the determination of analytes (elements), which are also present in the argon gas sources; elements making up the sample solvent; or those present in the environment or atmosphere. The method is also not applicable to species like halogens, which require very high energies for excitation, at least with low limits of detection (LODs); and finally it is not applicable to short-lived radioactive elements.

The elements that can be analyzed using the technique are those that have emissions in the wavelength window of the instrument. Important examples include alkaline metals that emit in wavelengths above 500 nm or some halogens (Cl, Br, etc.) or elements like N_2 and As whose wavelengths are below 190 nm, or even below 160 nm.

An important point in analytical analysis is the choice of the spectral line(s) of the elements, while avoiding spectral overlap with other elements. The lines also should be suitable for the monochromator and detection systems and further produce acceptable signal levels with respect to the concentration of the elements. Another practical consideration is that when using an internal standard procedure, the comparisons between the ionic and elemental lines of the analyte should be made with equivalent standard species.

The selectivity of ICP-OES is very much affected by the resolution of the monochromator, which reaches 5 pm in the case of the best instruments. The monochromator's selectivity plays an important role in minimizing the overlap of the spectral lines of the analyte with those of elements W, Co, Nb, Ta, and lanthanides, which have numerous spectral lines, as well as in increasing the signal-to-background ratio. Commonly, higher selectivities are achievable at the cost of lower sensitivity, slower analysis, and limited range of wavelengths.

The accuracy of the method, like any other analytic technique, decreases by approaching the LOD. It has been reported that for reaching accuracies in the range of ±10%, the minimum concentration of the analyte be five times more than the LOD, whereas for more precise tests, for example, ±2%, this ratio should be 100 times or more than the LOD.

The ICP-OES method also has the advantage of minimal interference compared to its counterparts. This can be due to the inertness and very high temperatures of the torch. The inactivity of the plasma sources in comparison with other techniques, in addition to the high temperature that breaks down almost every species in the sample, minimizes much of the chemical interferences, but such chemical interferences are not nil.

The interferences are also reduced using higher RF power, less inner argon flows, or their combination. In the case of easily ionizable elements, where high concentrations may subdue or boost the signals, the dilution of the samples can also be of help. Further methods that are used to this end are higher RF powers, mathematical corrections, or performing observations at varied conditions.

The most important source of inaccuracy in ICP-OES is the so-called background or spectral interference that arises from the high energy and excitation power of plasma, which crowds the acquired spectra, especially in the case of complex and concentrated samples. Several solutions have been proposed for overcoming this problem. One method is the use of high-resolution instrumentation, but even the best devices still show interferences. Advanced background correction techniques are used as another means for overcoming the problem.

The background problem is dealt with either by choosing another spectral line with no background interference or through correction of the results by measuring the background emission on two other points on both sides of the analytical line, where the emission of the analyte does not exist, and using the results for background correction by subtracting the average of the emissions at two points from the emission at the spectral line. In case another element has a direct spectral overlap with the analyte, the problem can be solved by eliminating the interfering species during sample preparation or through evaluation and subtraction of the magnitude of the interference signal, which is also known as the concentration ratio method. To evaluate the concentration and hence the strength of the signal of the interfering species, in the latter approach, another measurement is made, in parallel to the main measurement, at a spectral line corresponding to the interfering element only.

Sometimes, however, no correction is possible because of the complex background shift. In this case, the intensity of the background significantly varies on the correction point chosen, most probably due to the incidence of some very close spectral lines near the emission line of the analyte. In such cases, another analytical line should be chosen for an accurate analysis to become possible [23].

As mentioned above, the presence of considerable amounts of elements, like lanthanides or iron, with crowded spectra can lead to limitations in the ICP-OES analyses because of the increased possibility of overlap with the spectral lines of the analyte, a higher standard deviation of the background, and decreased signal-to-background ratios. This can be tackled by preconcentration of the analyte and separation of the interfering elements through procedures like ion exchange, solvent extraction, and coprecipitation.

ICP-OES analyses typically include preparing the samples and standards; setting the selected wavelengths, acquiring the spectra, choosing the integration time, mode of data acquisition, choosing the correction method, calibration of the instrument, and performing the analyses.

One of the sources of error in ICP-OES is the differences in the standard and sample matrices, no matter how slight these are. Such differences can lead to [24] variations in the efficiency of the introduction of the sample and standards into the system, because of their different physical properties; or alteration of the plasma power arising from the different sample matrices. Such errors are alleviated through matrix matching, standard addition, and the internal standard methods [25].

Matrix matching technique is applicable if the sample matrix is known, in which case standard samples are used to approximate the sample matrix and draw a calibration curve. Standard addition and internal standard techniques, on the other hand, are used if the sample matrices are not known and/or difficult to approximate.

The internal standard method, on the other hand, is applicable to eliminate matrix effects in ICP-OES. An assumption made in the IS method is that the effects of the matrix are identical for all elements, which is not completely true in many cases. One should note that different sample matrices can change the plasma temperature, leading to different responses for identical elements, which can be regarded as a limitation of the internal standard method. Another shortcoming of the method is its inability in the presence of elements that are easily ionizable. In this light, the standard addition method is preferred in case of complex or unknown matrices [26-29].

Standard addition is however rather laborious and demanding, which makes it unsuitable when time is a key element. Some works in the beginning of the century have reported success in terms of facilitation of the method. An instance is the work of C. Huang and Beauchemin [30] or that of Yamada et al. [31], where the flow injection technique is used for performing the standard addition. Another example is a new approach by Yousefi et al. [32], based on a fast on-line standard addition for the analysis of samples with concentrated matrices, using a T-piece, prior to the sample introduction system. The advantages of the method were shorter analysis times and increased simplicity.

Analysis of Lanthanides by the ICP-OES Method

ICP-OES has been used widely in the analysis of lanthanides either individually or together with different concentrations of interfering. Table 6.1 reveals the typical detection limits for analysis of lanthanides achieved through the application of the ICP-OES method reported by Perkin–Elmer [33]. All detection limits are reported as 3σ of blank and were obtained on a Perkin–Elmer Optima 3000 under simultaneous multielemental conditions with side-viewed plasma. Detection limits using an axially viewed plasma are typically improved 5–10 times.

ICP-OES has been used in the analysis of lanthanides in a wide range of samples. Determination of trace amounts of lanthanides in certain samples was not possible by ICP-OES alone because of spectral interference by the matrix, leading to the total or partial overlap of analyte and interfering species' lines. This is especially the case when the interfering species has a rich emission spectrum, as in the case of iron, which is a strong interfering species, during the determination of lanthanides. To overcome such problems, preconcentration and separation techniques are used to remove the interfering species prior to ICP-OES analyses. Examples include the hyphenation of chromatographic methods with ICP-OES and the development of new sorbents as well as on-line column or microcolumn techniques for this purpose. Another separation technique is the application of on-line solvent extraction, but the method suffers from the disadvantage that the organic solvents used changes the properties of the plasma, decreasing its energy and excitation capability.

TABLE 6.1 ICP-OES Limits of Detection (μg/L) for Lanthanides

La	Ce	Pr	Nd	Sm	Eu	Gd	Tb	Dy	Ho	Er	Tm	Yb	Lu
1	5	1	1	2	0.1	1	2	2	0.4	1	0.6	0.3	0.2
	Th 70		U 15										

A final point in ICP-OES is its optimization, which requires finding and setting the optimal values for a range of parameters, like the RF power, plasma and auxiliary gas flows, the pressure of the nebulizer, pump rate, PMT voltage, and generator frequency, etc.

A review of the literature revealed about 168 reports [32, 34-200] on the determination of lanthanide ions through ICP-OES, which is discussed in some detail below under five categories based on the nature of the sample.

Geological Samples

Before starting the review of the analysis of samples through ICP-OES, a short review of the sample preparation techniques used for geological sample (solid samples) preparation are presented below. The methods include

- open acid digestion
- closed vessel digestion
- microwave digestion
- alkaline fusion

Open acid digestion (OAD) is generally based on weighing a certain amount of the sample, transferring it to a proper dish (e.g., platinum dish), adding concentrated nitric acid, hydrofluoric acid, and perchloric acid (1:2:0.5) to it, and then mixing it with an inert rod (made from, e.g., polytetrafluoroethylene [PTFE]). Next the dish is placed in a bath and its content is allowed to evaporate to incipient dryness; this process is repeated twice before transferring it to a sand-bath, where perchloric acid fumes leave the mixture. As the next step, the residue is dissolved in a proper amount of HCl or HNO_3, while heating up to boiling, and next cooled, filtered with a proper filter paper, and washed with acid and diluted to a certain volume.

The closed-vessel digestion, on the other hand, is based on transferring a known amount of the sample to a Teflon pressure decomposition vessels. Next, an appropriate amount of $HClO_4$ and a mixture of HNO_3, HF, and HCl (3:6:1) is added. The mixture is stirred, before the vessel is sealed and placed in an electric oven (e.g., at 110°C for 12 h). Next the vessel and its content is cooled to ambient temperature and the fumes are allowed to leave by removing the lid. This is repeated if necessary, and eventually the residue is dissolved in a proper amount of HNO_3 by gentle heating and diluted to a certain volume.

In microwave digestion, a microwave oven is used for the dissolving the sample, which greatly reduces the time required for the digestion. As in other methods, a known amount of the sample is digested using the microwave oven for a short period of time in a certain volume of a mixture of concentrated acids (e.g., HCl/HNO_3/HF at 2:5:10) in a PTFE closed vessel. Next, the mixture is cooled down and the contents are transferred into a PTFE beaker. In order to remove HF and organic matter, 2 mL of conc. $HClO_4$ are added and evaporated to dryness. The residue is taken up in 2 mL of conc. HCl, transferred with water into a beaker, and heated on a hot plate. If the sample contains phases that are not dissolvable in acids, the solid phase is filtered and mini-fused with Na_2O_2, before being dissolved in HCl in molten form and mixed with the filtrate of the previous step. Like the previous cases, the resulting solution is diluted to a certain volume with water.

Finally, alkaline fusion is based on fusing the molten geological sample powder (ground to about 200 mesh) with a eutectic of Li meta- and tetraborate at 1000°C. This is done by heating the mixture in a furnace, quenching the product, and dissolving it in HCl, before diluting to a certain volume.

The lanthanide content of geological samples are commonly analyzed through techniques like instrumental neutron activation analysis (NAA), isotopic dilution mass spectrometry (IDMS), ICP-OES, and ICP-MS. In spite of their sensitivity for most lanthanides, NAA and IDMS are rather expensive, laborious, and time consuming. An inherent limitation of IDMS is its inability to analyze Y, Pr, Tb, Ho, and Tm. ICP-OES, on the other hand, has the ability for sensitive multielemental analysis of lanthanides, especially if the interferences are eliminated through an adequate technique. Another important step is the choice of the spectral lines used for the analysis of the analyte, especially due to the interferences from Al, Ca, Ba, Fe, K, Na, Ti, and Mg, which commonly occur in geological samples. This task is performed using software for spectral recognition.

A chronologically ordered summary of the reports on the analysis of lanthanides in geological samples and related materials including anorthosites, granite, bauxite, apatite, bastnesite, britholite, basalt, monazite, phosphate, etc., is given in Table 6.2 [32,34-109].

In general, rare earth elements (REEs) at parts per million levels in lanthanum oxide was determined without prior separation and preconcentration by use of high-resolution inductively coupled plasma–emission derivative spectrometry (ICP-EDS) [38].

A common separation technique used for eliminating the matrix or preconcentrating the sample is the cation exchange when using resins like sulfonated polystyrene [35,40,59,61,65,90,105]. The step is conducted after sample digestion, leading to the elimination of matrix elements through acid elution. Samples are usually rather concentrated to meet the volume requirements but this also leads to high matrix element salt loads, causing the undesired recovery of lanthanides during the separation with resins. A procedure described by Webster and Gilstrap [65] tends to evade using big resin and acid volumes through a calcium oxalate precipitation before the mixed-resin ion exchange. Another precipitation reaction used for matrix elimination is based on the application of oxalate anion for precipitation [56].

TABLE 6.2 Some Reports on Determination of Lanthanides in Geological Samples

Year	Analyte	Sample	Important Aspects	Ref.
1981	REE	Geological and related materials	-	[34]
1981	REE	Rock	Cation-exchange resin	[35]
1982	REE	Geological and related materials	-	[36]
1983	Sc, Y, and Selected REE	Geological and related materials	-	[37]
1983	REE	Lanthanum oxide	ICP-derivative spectroscopy	[38]
1983	REE	Rock	HPLC-ICP	[39]
1984	REE	Geological materials	Cation-exchange chromatography-ICP	[40]
1984	Sm	Apatite minerals	Solvent extraction	[41]
1985	REE	Geological materials	HPLC-ICP	[42]
1985	REE	Phosphate mineral	-	[43]
1985	REE	Geological materials	Thin-film X-ray-fluorescence and ICP	[44]
1985	REE	Sediment	-	[45]
1985	REE	Rock	-	[46]
1986	REE	Geological materials	-	[47]
1986	REE	Ore	-	[48]
1986	REE	French Geostandards	-	[49]
1986	REE	Rock sample	-	[50]
1987	REE	Rock sample	-	[51]
1987	Light REE	Hydroxide participate of apatite and baestnesite	-	[52]
1987	REE	Reference sample and glauconite	-	[53]
1987	REE	Reference mineral of the britholite apatite	-	[54]
1987	REE	Reference rock sample	-	[55]

TABLE 6.2 Some Reports on Determination of Lanthanides in Geological Samples (*cont.*)

Year	Analyte	Sample	Important Aspects	Ref.
1988	REE	Geological samples	Oxalate coprecipitation and cation-exchange column separation	[56]
1988	REE	International silicate reference materials	-	[57]
1988	Tb, Dy, Ho, Er, Yb	Ore and concentrate	-	[58]
1988	REE	Geological reference samples	-	[59]
1988	REE	Geological materials	-	[60]
1988	REE	-	Separation and determination	[61]
1989	REE	Silicate rock	-	[62]
1989	REE	Silicate rock	-	[63]
1990	REE	Geochemical reference materials	-	[64]
1990	REE	Geological matrix	Calcium content—oxalate precipitation and cation exchange	[65]
1990	REE	Rock sample	organic solvent extraction	[66]
1991	REE	Phosphate rock	Cation-exchange	[67]
1991	REE	Phosphate rock	-	[68]
1991	REE	Silicate rocks	Without ion-exchange separation	[69]
1991	REE	U.S. Geological Survey Kesterson Bottom Sediment	-	[70]
1992	REE	Phosphate samples	-	[71]
1992	REE	Ores	Ion-exchange separation	[72]
1992	REE	Sediments	cation-exchange separation	[73]
1992	REE	Ores and concentrates	With separation	[74]
1993	REE	Phosphates and related sedimentary rocks	Ultrasonic nebulization	[75]

(*Continued*)

TABLE 6.2 Some Reports on Determination of Lanthanides in Geological Samples (*cont.*)

Year	Analyte	Sample	Important Aspects	Ref.
1993	REE	Geological specimens	-	[76]
1993	REE	Geological samples	Preconcentration by countercurrent chromatography	[77]
1993	REE	Rock	-	[78]
1994	REE	Rock	-	[79]
1994	REE	Reference samples	-	[80]
1995	REE	Anorthosites and related rocks	-	[81]
1995	REE	Geological materials	Ion-exchange chromatographic separation	[82]
1995	REE	Geological samples	Preconcentration by countercurrent chromatography	[83]
1995	REE	Granites and greisens	Preconcentration by thin-layer chromatographic	[84]
1996	REE	Geological samples	Microwave decomposition	[85]
1996	REE	Geological samples	Ion-exchange chromatographic technique	[86]
1996	REE	Niobate-tantalate, carbonatite, and fergusonite samples	Acid hydrolysis and hydroxide precipitate	[87]
1997	La	Monazite sand and standard samples	Solvent extraction with crown hydroxamic acid, 5,14-*N*,*N*′-hydroxyphenyl-4,15-dioxo-1,5,14,18-tetraaza hexacosane	[88]
1997	REE	Bauxite	Precipitated with oxalic acid using calcium as a carrier	[89]
1997	REE	Geological silicate materials	Dowex 50W-X8 cation resin exchange	[90]
1998	REE	Ion-adsorption-type ores	-	[91]

TABLE 6.2 Some Reports on Determination of Lanthanides in Geological Samples (*cont.*)

Year	Analyte	Sample	Important Aspects	Ref.
1998	REE	Uranium-rich geological materials	Precipitation of the analytes as hydroxides using H_2O_2/NaOH in the presence of iron as carrier	[92]
1998	REE	Geological materials, rare earth minerals and niobate/ tantalate-type of samples	Solvent extraction and acid hydrolysis separation	[93]
1999	REE	Rock	Solvent extraction with a mixture of 2-ethylhexyl dihydrogenphosphate and bis(2-ethylhexyl) hydrogenphosphate	[94]
2000	REE	Geological sample	-	[95]
2000	REE	Iron-rich geological samples	-	[96]
2000	REE	Silicate rocks	Ultrasonic nebulization and on-line ion-exchange separation	[97]
2002	REE	Silicate rocks	Separation in flow system	[98]
2002	REE	Granitic rocks	Chromatographic separation	[99]
2002	REE	Marine sediment	ICP-OES instrument equipped with a dual-view plasma torch	[100]
2003	REE	Geological samples	Separation and microcolumn preconcentration on nanoscale TiO_2	[101]
2003	REE	Refractory minerals zircon, ilmenite, rutile, columbite-tantalite, garnet, and silliminite	Cation exchange chromatographic group separation	[102]
2003	REE	Geological samples	Flow injection liquid–liquid extraction	[103]
2004	REE	Bastnasite ores	Microwave digestion	[104]
2005	REE	Apatite selective leaching from granitic rocks	Cation exchange	[105]

(*Continued*)

TABLE 6.2 Some Reports on Determination of Lanthanides in Geological Samples (*cont.*)

Year	Analyte	Sample	Important Aspects	Ref.
2009	REE	Geological samples	Precipitated as hydroxides with sodium hydroxide using Mg as the carrier in the presence of triethanolamine as the complexing agent	[106]
2011	REE	Geological materials	preconcentration by Cu-diethyldithiocarbamate (DDTC) as coprecipitative carrier	[107]
2012	REE	Geological samples	-	[108]
2012	REE	Geological, red mud, and coal fly ash samples	Separation as oxalates precipitation	[109]
2014	REE	Geological samples	On-line standard addition technique	[32]

Premadas [109] studied the separation of oxalate salts at different pH values using calcium as carrier and observed that by using 300 mg of calcium as the carrier (pH 1), traces of lanthanides and thorium are quantitatively recovered. A further step included the removal of calcium through reaction with ammonium hydroxide. The overall method was found to be very effective in removing most matrix elements. In another work [106], Premadas studied another two-stage precipitation separation of Sc, and Th for the ICP-OES determination of the same. The method was based on the precipitation of the hydroxides of Sc and Th using Mg as the carrier and triethanolamine as a chelating agent. The precipitates were next separated and dissolved in hydrochloric acid. A further precipitation reaction was carried out using NH_4OH in the presence of Fe as the carrier. The filtered residues were next dissolved in hydrochloric acid and analyzed using ICP-OES.

Another work describes the determination of yttrium, scandium, and other rare earth elements in geological samples rich in uranium using ICP-AES. The method was baed on the separation of uranium through the selective precipitation of Y and Sc using H_2O_2/NaOH and iron as carrier, followed by redissolution of the precipitates and the analysis of the resulting solution [92].

In the case of using solvent extraction for matrix elimination, a mixture of (2-ethylhexyl) dihydrogenphosphate and bis(2-ethylhexyl) hydrogenphosphate in kerosene has been suggested for silicate rock samples [94].

In the work of Xu et al. [103], a direct sampling was made through solvent extraction followed by ICP-AES coupled with flow injection liquid–liquid extraction. They used the intensity ratio of Mg II (279.55 nm) and Mg I (285.21 nm) lines.

Some reports mention the application of chromatographic separation followed by ICP-OES detection [39,40,42,77,82-84,86,99,102]. For instance, HPLC-ICP-OES has been reportedly applied to the determination of 15 rare earth elements, where a cation exchange column and an ammonium lactate buffer were used in the separation stage. The application of ICP-OES detection in other chromatographic techniques like the counter-current chromatography has also been reported [39,40,42,77,83].

Another example of the application of ICP-OES is the case of determination of individual rare earth elements and yttrium in granites and greisens after thin-layer chromatographic separation and cleanup of the rare earth element group on a Dowex 50-X8 cation exchanger [84]. Another report is based on the application of a microcolumn packed with nanometer-sized titanium dioxide for the separation of traces of Sm, Tm, Ho, and Nd from geological samples at pH 7 and their analysis by ICP-OES [101].

In another report by Pozebon and Martin, an ICP-OES instrument was equipped with a dual-view plasma torch [100]. This technology allows the determination of trace elements in the axial view mode along with major components in the radial view [100].

Industrial Samples

An overview on ICP-OES analyses of the lanthanide content of industrial samples, including ceramics, alloys, catalysts, magnets, glass, nodule, metallurgical samples, steel, aluminum, garnet, YAG, phosphors, oxides, coal, petroleum, crude oil, nuclear fuel, paint, and pen ink samples is provided in Table 6.3 [110-154]. In general, the samples are digested through methods described in the previous section. A pre-analysis treatment of the samples is also usually performed through separation techniques for trace analysis.

A first example of such analyses is the determination of erbium in heavy metal fluoride optical fibers by ICP-OES [129] through the steps of sample dissolution through alkaline fusion with sodium carbonate, acid leaching with dilute hydrochloric acid, and measurements of emission intensities at 337.276 nm.

Another example includes the use of a hyphenated HPLC-ICP-OES system together with a UV analysis of Gd^{3+} complexes [136] and using the mixed information for the analysis of Gd concentration together with collecting information on the chromophoric groups based on the UV information. Another instance of the analysis of lanthanides through ICP-AES is the determination of La, Eu, and Yb by ICP-OES after solid-liquid extraction with microcrystalline naphthalene [142]. Based on the

TABLE 6.3 Some Reports on Determination of Lanthanides in Industrial Samples

Year	Analyte	Sample	Important Aspects	Ref.
1981	Ca, Fe, Ge, Lu, Sm, Y	Bubble garnet film	-	[110]
1981	Ce	Paint		[111]
1982	REE	High-purity yttrium oxide	-	[112]
1983	Al, Co, La, and Ru	Alumina based catalyst	-	[113]
1983	REE	Metallurgical samples	-	[114]
1983	REE	Aluminum	-	[115]
1983	REE	High purity yttrium oxide	-	[116]
1984	REE	Manganese Nodules	-	[117]
1986	Sm, Nd	Permanent magnet (Sm-Co, Nd-Fe-B)	Ion chromatography-ICP	[118]
1986	REE	Uranium	-	[119]
1986	REE	Rare earth transitional metal ultrathin magnetic films	-	[120]
1986	Dy, Eu, Gd and Sm	Nuclear-grade uranium oxide	-	[121]
1990	La, Ce, Pr, Nd	Alloy steels	-	[122]
1990	REE	In yttrium, lanthanum, gadolinium and ytterbium matrices	Second-derivative spectroscopy	[123]
1990	REE	Rare earth impurities in high-purity yttrium oxide	-	[124]
1991	Ce, Pr, Nd and Sm	High-purity La_2O_3	-	[125]
1991	Y, Sm, Eu, Gd, Dy, Ho, Er	High-purity terbium oxide	-	[126]
1991	REE	Impurities in terbium oxide	HPLC-ICP	[127]
1992	REE	Terbium	Chromatographic separation	[128]
1992	Er	Optical fibers	-	[129]

TABLE 6.3 Some Reports on Determination of Lanthanides in Industrial Samples (*cont.*)

Year	Analyte	Sample	Important Aspects	Ref.
1992	heavy rare earth elements (Eu-Lu)	High-purity yttrium oxide		[130]
1992	REE	Ferromanganese nodule reference samples	-	[131]
1992	REE	Uranium	-	[132]
1994	REE	Gd_2O_3	-	[133]
1994	REE	Trace impurities in high-purity europium oxide	Extractive chromatographic separation	[134]
1995	REE	Indian polymetallic nodule reference sample	-	[135]
1995	Gd	Analysis of gadolinium polyaminopolycarboxylic complexes' stability in heating and solution	HPLC–ultrasonic nebulizer	[136]
1996	REE	Impurities in "pure" rare earth matrices	-	[137]
1997	REE	Rare earth oxides of Ce, Pr, Nd, Sm, and Dy	-	[138]
1997	REE	Pure cerium dioxide	-	[139]
1997	REE	Pure rare-earth bases	-	[140]
2000	REE	Impurities in rare earth chloride and light rare earth oxide	-	[141]
2001	La, Eu, Yb	Certified reference materials	Preconcentration with tribromoarsenazo-cetylpyridinium bromide supported on microcrystalline naphthalene	[142]
2001	Eu, Sm	Cathodoluminophors	-	[143]
2002	La, Y	Uranium hydrometallurgical products	-	[144]
2003	REE	Rare earth impurities in high-purity europium oxide	-	[145]

(*Continued*)

TABLE 6.3 Some reports on Determination of Lanthanides in Industrial Samples (*cont.*)

Year	Analyte	Sample	Important Aspects	Ref.
2004	Eu	Blue phosphors for fluorescent lamp	-	[146]
2005	REE	Trialkylphosphine oxide	-	[147]
2007	Dy, Eu, and Sm	Impurities in a Gd matrix	With an extraction chromatographic column	[148]
2007	La, Ce, Pr, Nd	Rare earth hydrogen storage alloy (AB_5 type)	-	[149]
2008	Eu	Blue ball point pen inks/ offset printing inks	-	[150]
2010	La, Nd, Eu and Gd	High-purity ytterbium oxide	Ion-exchange fiber column	[151]
2010	REE	Uraninite samples	After solid-phase extraction on activated carbon	[152]
2010	REE	Clean steel	After removal of iron matrix with anion-exchange resin	[153]
2012	REE	Heat-resistant nickel alloys	-	[154]

procedure described, the analytes were quantitatively adsorbed on tribromoarsenazo-cetylpyridinium bromide-naphthalene (TBA-CPB-naphthalene) that was used as the packing in a column, and were next desorbed through the use of a 3 M HCl solution, so that they were determined through the ICP-AES.

In a later study, the europium and samarium contents of the red-emitting cathodoluminophors CaS:Eu, and CaS:Eu, Sm, which are of great importance because of their high efficiency and chromaticity for electroluminescence and cathodoluminescence displays, were determined through a sequential ICP-AES [143].

In another experiment [151], traces of La, Nd, Eu, and Gd were analyzed after a separation and enrichment step using a strong acid ion-exchange fiber column from high-purity Yb_2O_3 and then determined by ICP-OES. The ion-exchange fiber's breakthrough capacity for Yb was reported as 134 mg/g.

Further, uraninite samples were analyzed by ICP-OES after a separation and preconcentration step using activated charcoal [152]. The sample preparation was based on recovering rare earths from activated carbon (AC) by igniting into ashes. The presence of thorium impurity in the samples led to the focus on testing various emission lines, to find those free from interference.

In a report by Xu et al. [153], the lanthanide content of steel was evaluated. The sample was dissolved in a mixture of hydrochloric acid (HCl), nitric acid (HNO_3), and hydrogen peroxide (H_2O_2) and passed through a column containing a DOWEX 1 x 8–200 anion-exchange resin, and were analyzed using ICP-AES after proper separation.

Another report on the determination of lanthanides through ICP-AES is the mass fraction determination of microalloying additions—Zr, Ce, La, and Y (from 0.001 to 0.1 wt%)—in heat-resistant nickel alloys [154]. The presence of small amounts of lanthanides is due to the modern technology of microalloying, which tends to improve the properties of cast heat-resistant nickel. The lanthanide-containing alloys for instance have been found to have improved short- and long-term strength; ductility; impact toughness; and resistance to deformation at high temperatures, stress corrosion cracking, vacancy cavitation, and radiation-induced swelling.

Environmental Samples

Although the reports on the applications of ICP-AES in the analysis of geological and industrial samples date back to a rather distant past, the majority of reports on the analysis of environmental samples only date back to the past decade [155-186]. In most of the works reported, a suitable sorbent is used for solid-phase extraction and preconcentration of lanthanide prior to ICP-OES analysis. The sorbent range from organic [184] or inorganic compounds modified by organic molecules [176,178,181-183] to nanosorbents [161,167,170,178,186]. For the sake of simplicity, a list of the sorbents used in the cited reports is shown in Table 6.4.

During solid-phase extraction (SPE), the compounds that are dissolved or suspended in a liquid mixture are separated from other compounds in the mixture based on their physical and chemical properties, which helps to concentrate and purify analytes from different matrices. The overall procedure includes contacting the sorbent with the analyte solution for a period of time under optimal conditions and recovering the separated species by a suitable washing solvent. In the case of solid samples, a digestion step, according to the methods in the previous section, precedes the SPE.

The application of different sorbents (Table 6.4) has proven the SPE to be a promising matrix purification/preconcentration route for the determination lanthanides in various natural water samples to be analyzed by ICP-OES.

The efficiency of different zeolites (e.g., clinoptilolite, mordenite, zeolite Y, zeolite Beta), ion exchangers (Amberlite CG-120, Amberlite IR-120, Rexyn 101, Dowex 50W X18), and chelating resins (Muromac, Chelex 100, Amberlite IRC-718) for absorbing lanthanides was investigated by Pasinli et al. [168], and the optimal separation conditions (i.e. pH, stirring time and amount of sorbent) were compared. In general, most sorbents were found to absorb lanthanides in a wide pH range. Further experiments with

TABLE 6.4 Some Reports on Determination of Lanthanides in Environmental Samples

Year	Analyte	Sample	Important aspect	Ref.
1985	REE	Mussel, ark shell, and coral	-	[155]
1985	REE	Sea water	-	[156]
1991	Er	Environmental samples	-	[157]
1992	Dy, Ho, Er, and Yb	Waste water	Macroporous poly(vinylphosphoramidic acid) resin adsorbent	[158]
1994	REE	Mineral water	Dowex cation exchanger	[159]
2001	REE	Natural waters	Flow injection on-line sorption preconcentration in a knotted reactor	[160]
2001	La, Y, Yb, Eu, Dy	Environmental samples	Nanometer-sized titanium dioxide microcolumn on-line preconcentration	[161]
2002	REE	Soils and sediments	Cation-exchange separation	[162]
2002	REE	Soils and sediments	Using $Ti(OH)_4$-$Fe(OH)_3$ coprecipitation preconcentration	[163]
2002	REE	Seawater	Solid-phase microextraction on activated charcoal	[164]
2005	REE	Water sample	Using C-18-cartridge modified with 1-(2-pyridylazo) 2-naphtol (PAN) prior to simultaneous determination	[165]
2005	REE	Water sample	Preconcentration of rare earth elements on silica gel loaded with 1-phenyl-3-methyl-4-benzoylpyrazol-5-one	[166]
2005	REE	Lake and sea-water	Preconcentration with multiwalled carbon nanotubes	[167]
2005	REE	Natural water	Preconcentration with variety of sorbent	[168]
2006	REE	Lake water	Microcolumn separation and preconcentration by pyrocatechol violet complexes on anthracene-cetyltrimethyl ammonium bromide adsorbent	[169]

TABLE 6.4 Some Reports on Determination of Lanthanides in Environmental Samples (*cont.*)

Year	Analyte	Sample	Important aspect	Ref.
2007	REE	Stream sediment reference material	After preconcentration with immobilized nanometer titanium dioxide	[170]
2007	REE	Environmental water	Sequential-injection on-line preconcentration using chitosan resin functionalized with 2-amino-5-hydroxy benzoic acid	[171]
2007	Ce(IV)	Water sample	Selective extraction and preconcentration by cloud point extraction	[172]
2007	Dy	Environmental samples	A new ion-imprinted silica gel sorbent for on-line selective solid-phase extraction of dysprosium(III)	[173]
2009	REE	River water sample	High-capacity chitosan-based chelating resin for on-line collection of transition and rare-earth metals	[174]
2009	REE	Water sample	On-line collection/concentration and determination by multi-Auto-Pret system coupled	[175]
2009	Ce(III)	Water and soils	Thio-functionalized MCM-41	[176]
2009	REE	Seawater	Chelating resin having fluorinated beta-diketone immobilized on styrene divinyl benzene	[177]
2012	Dy	Water sample	Selective modified nanoporous silica as sorbent for separation and preconcentration	[178]
2012	REE	High-level radioactive liquid waste solutions	-	[179]
2013	REE	Orchards soil	-	[180]
2014	REE	Water samples	Preconcentration with 6-(2-thienyl)-2-pyridinecarboxaldehyde functionalized Amberlite	[181]

(*Continued*)

TABLE 6.4 Some Reports on Determination of Lanthanides in Environmental Samples (*cont.*)

Year	Analyte	Sample	Important aspect	Ref.
2014	REE	Tap water and seawater samples	Preconcentration using Amberlite XAD-4 modified with 2,6-pyridinedicarboxaldehyde	[182]
2014	REE	Water samples	Solid phase extraction using chemically modified Amberlite XAD-4 resin	[183]
2014	Yb, Ho	Natural water	Ultrasound-assisted emulsification microextraction	[184]
2014	REE	Aqueous sample	On-line surfactant-based extraction using ion-pair microparticles	[185]
2015	REE	Sea water and sediment samples	Graphene oxide-TiO_2 composite as a novel adsorbent	[186]

clinoptilolite, zeolite Y, and Chelex 100 showed that all three sorbents have very fast sorption kinetics (higher than 96% in 1 min). For evaluating the desorption phenomena, 2.0 M HNO_3 was used for clinoptilolite; and 0.1 M HNO_3 was used for zeolite Y and Chelex 100. Only the lower concentration range (0.01–2.0 mg/L) of matrix-matched standards were reportedly used in quantitation, but the results showed that the calibration graphs were linear at least up to 10.0 mg/L for all tested elements. The detection limits (3 s), in the absence of preconcentration, was 0.1, 1.0, and 0.2 μg/L for Eu, La, and Yb, respectively.

The application of cloud point extraction (CPE), as a preconcentration stage prior to ICP-AES analysis has been tested for cerium(IV) in aqueous solutions [172] using *n*-*p*-tolylbenzohydroxamic acid (*n*-TBHA) to form an extractable complex of Cr(IV) which is extracted into the micellar phase of Triton X-114 formed above the cloud point temperature (CPT). The surfactant-rich phase was next diluted up to 1.0 mL, after it was extracted from the solution and was used for analysis by ICP-OES. The results revealed that under optimal extraction and analysis conditions, the preconcentration of only 10.0 mL of the sample using 0.09% of Triton X-114 led to an enhancement factor of 13.8. Further investigations revealed that the calibration curve was linear from 1.5 to 1200 μg/L, with a detection limit of low as 0.4 μg/L.

More target-specific sorbents like thenoyltrifluoroacetone imprinted with dysprosium(III) and modified with silica gel [173] have also been used for separation of the analyte. The sorbent was used in a flow injection solid-phase extraction system with an ICP-AES for the determination of

Dy(III). The most interfering species were observed in the case of La(III), Nd(III), and Gd(III). The adsorption capacity and selectivity coefficient of the imprinted sorbent were higher than those of the nonimprinted sorbent. The sorbent modified with silica gel proved to have fast adsorption/desorption kinetics. It was reported that at a sample loading flow rate of 2.5 mL/min for 48 s, preconcentration and elution flow rate of 1.5 mL/min for 8 s elution, an enrichment factor of 10 and a sample frequency of 40 samples h^{-1} were obtained. The method was reported to have a detection limit of 0.2 ng/mL with a relative standard deviation (RSD) of 1.5% ($n = 8$, $c = 40$ ng/mL).

The application of a multi-auto-pret (MAP) system (with three auto-pret systems equipped with minicolumns capable of preconcentrating trace metals sequentially or simultaneously) hyphenated with ICP-OES has been reported [175]. The MAP system reduced the analysis time to a third, as well as reducing the cost of Ar gas and labor. A novel chelating resin (i.e., ethylenediamine-*N,N,N′*-triacetate-type chitosan) was also used for the collection of traces of metals prior to ICP-OES, and a very good adsorption ability and no interference from alkali and alkaline-earth metal ions were observed in an acidic media. The method was successfully used to analyze SLRS-4 river water reference materials for trace metals.

ICP-OES was also reportedly used for the determination of lanthanides in sea water [177]. After preconcentration of the analytes with a novel chelating resin based on the immobilization of fluorinated beta-diketone group on solid support styrene divinyl benzene. Sample solutions were passed through a polyethylene column packed with 250 mg of the resin, and the operating conditions were optimized. The method was found to be capable of yielding recoveries of more than 95% for La, Ce, Nd, Sm, Eu, Gd, Dy, Er, Yb, and Lu with RSDs of less than 5%.

Physically nanoporous silica (SBA-15) functionalized using *N′*-[(2-hydroxy phenyl) methylene] benzohydrazide (BBH) has also been used as a selective sorbent for Dy used prior to ICP-OES [178]. The results included an extraction recovery of 96.5, with a linear calibration curve from 0.2 to 1000 μg/L, and a detection limit of 0.05 ng/mL with a standard deviation of less than 3.2%. The high adsorption capacity and short equilibrium times for Dy(III) ions could be mentioned as the advantages of the method, which was used for recovery and determination of dysprosium in some natural water samples.

ICP-OES has also been used for analysis of traces of lanthanides in high-level liquid waste solutions of reactor fuels from the plutonium reprocessing plant and waste immobilization plant in Tarapur, India, under optimal conditions and the results were validated through the analysis of three synthetic high-level liquid waste samples of a pressurized heavy water reactor, research reactor, fast breeder reactor, and high-level radioactive liquid waste [179].

Using the microwave digestion method and a mixture of HF, HNO_3, and HCl, the spectral interference of La, Ce, Nd, Sm, Eu, Gd, Tb, Yb, Lu, and Y in soil samples during ICP-AES analysis was alleviated through a double precipitation treatment [180]. The procedure was found to lead to good results in terms of sensitivity and elimination of interferences and was further used for the analysis of soil standard reference materials (GSS-4, GSS-5). The detection limit for each element was in the range of 0.2–38.2 μg/L, with RSDs between 0.8% and 3.2%.

Karadas and Kara [181] introduced a new method for the determination of Ce, Pr, Nd, Sm, Eu, Gd, Tb, Dy, Ho, Er, Tm, Yb, and Lu in water samples. The method included a preconcentration step using a mini-column packed with 6-(2-thienyl)-2-pyridinecarboxaldehyde functionalized Amberlite XAD-4 before the analysis of the sample using ICP-AES. The optimal pH for preconcentration was evaluated and found to range from 6.0 to 8.0. Under the optimal conditions, the detection limit was found to be 0.032 and 0.963 μg/L for a 10-mL sample volume and 0.006 and 0.193 μg/L for a 50 mL sample, and the method was successfully applied to determination of analytes in water samples, with recoveries in the range of 90.1%–110.5%.

The same group [182] used a preconcentration and determination method for the analysis of Nd, Sm, Eu, Gd, Tb, Dy, Ho, Yb, Lu, and Ce in water samples, using a 2,6-pyridinedicarboxaldehyde-functionalized Amberlite XAD-4 resin at pH 7.0; elution with a 1.0 mol/L HNO_3 solution; and determination by ICP-OES. The main preconcentration parameters including sample pH, sample and eluent flow rates, and sample volume evaluated and under the optimal values (i.e., pH 7.0 and sample and eluent flow rates 1.0 and 4.0 mL min^{-1} respectively), the detection limits were found to be between 0.011 and 0.298 μg/L for a 25 mL sample volume and 0.006 and 0.149 μg/L for a 50 mL sample. The sorption capacities of the resin were found to range between the two extremes of 49.0 μmol/g (for Lu) and 66.7 μmol/g (for Sm). The validation of the method was performed using a surface water certified reference material (SPS-SW2 Batch 127) and the method was successfully used for the determination of lanthanides in tap water and seawater samples, with recovery values for the spiked water samples in the range of 90.0–101.7%.

Sereshti et al. [184] introduced an optimized ultrasound-assisted emulsification-microextraction technique that was coupled with ICP-OES for the simultaneous extraction and determination of Ho^{3+} and Yb^{3+} in water samples. The method was based on the application of 1-(2-pyridylazo)-2-naphthol as a chelating agent, which forms complexes with the target species that can be extracted more easily. The detection limits were evaluated to be 0.364 and 0.252 μg/L for Ho^{3+} and Yb^{3+}, respectively, with RSDs of 4.2%–4.5% (C = 200 μg/L, n = 6). The proposed method was successfully applied to the analysis of real water samples, with recoveries in the range of 80%–95%.

Yousefi et al. [185] also developed an on-line and in situ procedure for the extraction and preconcentration of rare earth elements prior to ICP-AES, based on the application of dodecyltrimethylammonium bromide (as a cationic absorbant) and PAN as a complexing agent. These were added to the aqueous samples, before adding hexafluorophosphate (an ion-pairing agent). Upon the addition of the ion-pairing agent, the solutions turned cloudy as a result of the formation of ion pairs between the surfactant and hexafluorophosphate. The solid microparticles next passed through a microcolumn filter and re-eluted after separation from the matrix for the ICP-OES analysis. The optimized method, and in the case of a 20 ml of sample, enhancement factors of 90–102 and detection limits of 0.02–1 μg/L were obtained.

In a rather recent report, nanosorbents were used for preconcentration before ICP-OES. Zhang et al. used a graphene oxide-TiO_2 composite as the adsorbent for preconcentrating heavy metals and rare earth elements in environmental samples [186]. The method was used for the determination of heavy metals and REEs in environmental water and sediment samples, yielding recoveries that ranged from 82.4%–115.5% and 83.8%–114.8% for the spiked water and sediment samples, respectively. The optimal conditions led to a wide linear range over 0.5–1000 ng/mL for the method.

Food Samples

Table 6.5 lists an overview of the reports on the application of ICP-AES for the determination of lanthanide in samples, including rice, tea, fruits, seeds, food, beans, and wheat [187-192].

The first such report was made using a device with tungsten coil, which was used for the electrothermal vaporization for sample and its introduction into ICP. This work, which was performed by Mei et al. in 1992 [187], focused on sample introduction as an important stage of ICP-AES analysis (ICP). This is because the conventionally used pneumatic nebulization system for introducing liquid samples into ICP, which has been accepted by most analysts today, has the disadvantages of low sample introduction efficiency (generally 1%–3%), frequent nebulizer blockage, and the need for larger sample volumes. The application of electrothermal vaporization (ETV) as the sample introduction technique for ICP-AES was evaluated for the determination of trace amounts of rare earth and other elements in rice samples, and the factors like drying and vaporization parameters, carrier gas flow rate, volume of the vaporization chamber, and matrix effects were evaluated and optimized. Under optimal conditions, the detection limits for Mg, Cu, Mn, Cr, Fe, Co, and Ni and eight rare earth elements were found to be from 10^{-9} to 10^{-11} g. The detection limits for the rare earth elements tested were comparable to, and in most instances exceeded, those for the GFAAS and conventional pneumatic nebulization-ICP-AES, and the RSDs were less than 6% in all cases.

TABLE 6.5 Some Reports on Determination of Lanthanides in Food Samples

Year	Analyte	Sample	Important aspect	Ref.
1992	REE	Rice samples	Sample introduction by tungsten-coil electrothermal vaporization	[187]
2005	REE	Tea leaves	Separation/preconcentration by microcolumn packed with immobilized 1-phenyl-3-methyl-4-bonzoil-5-pyrazone on nanometer Al_2O_3	[188]
2007	REE	Pig liver, agaric and mushroom	On-line separation and preconcentration with acetylacetone-modified silica gel using microcolumn	[189]
2010	REE	Tea, mushroom	Cloud point extraction with/without chelating agent on-line coupled	[190]
2010	REE	*Tricholoma giganteum*	Separation/preconcentration by microcolumn with nanometer Al_2O_3	[191]
2010	REE	Pure tea	-	[192]

The sample preparation technique included washing the grain from the Jiangxi REE-mineral area once with tap water and then with double-distilled water, and dried under an infrared lamp. Next the sample was wrapped up with a plastic film and pressed to crush. The resulting husk and rice powders were separately stored in polyethylene bottles. One to 2 g of the sample were soaked for about 20 h with 30 mL 65% HNO_3 in a quartz beaker for pre-decomposition. Next, 10 mL of a 3:1 $HNO_3/HClO_4$ mixture was added to the samples and the resulting mixture was heated on a hot plate, to gradually dry before being transferred into a calibrated test tube and made up to 5 mL with HC1 (1:1).

Using a microcolumn packed with immobilized 1-phenyl-3-methyl-4-bonzoil-5-pyrazone (PMBP) on nanometer-size Al_2O_3 powder as the adsorption material, the adsorption and elution behaviors of rare earth ions (Sc^{3+},Y^{3+}, and La^{3+}) on the above material under dynamic conditions were studied with ICP-OES [188]. The conditions for preconcentration of the lanthanide ions were optimized, and the results show that the analyte ions can be quantitatively absorbed on the absorbing material at pH 4.5, and the analytes adsorbed on the column can be eluted with 0.5 mol/L HCl solution. The detection limits of the method for Sc, Y, and La were 0.16, 0.19, and 0.39 μg/L, respectively, with RSDs of 2.7%, 3.2%, and 1.6% ($n = 9$, $C = 0.5$ mg/L), respectively. The proposed method was used for the determination of trace amounts of Sc, Y, and La in GBW 07605 tea leaves, and the determined values were in good agreement with the certified values. Later, an identical method was reportedly used for determination of these three REEs in *Tricholoma giganteum* [191].

Another study of food samples was based on the use of on-line microcolumn separation and preconcentration coupled to ICP-OES [189]. The column was packed with acetylacetone-modified silica gel. The RSDs obtained for the determination of 250-ng/mL ($n = 8$) REE standard solutions ranged from 1.7% for Y to 4.4% for Sm. The method was successfully used for the determination of trace REEs in pig liver, agaric, and mushroom. The tested samples (i.e., pig liver, agaric, and mushroom samples) were chosen since they may accumulate and transfer REEs to the human body through the food chains. Pig liver, agaric, and mushroom were frozen in a refrigerator at 4°C before analysis. For analysis, the samples were washed with deionized water and aired; next, they were dried at 80°C for 3 h. The liver sample was cut into pieces, and the agaric and mushroom were dried at 105°C till their weights were constant. A 0.5000-g portion of the samples was weighed and transferred into a PTFE digestion vessel.

Another method used for the analysis of lanthanides is the on-line incorporation of cloud point extraction (CPE) [190] with or without 8-hydroxyquinoline (8-Ox). 8-Hydroxyquinoline was used as a chelating agent in the flow injection system, and an ICP-OES detection system was used.

Biological Samples

In the case of monitoring lanthanides in samples with biological origins (e.g., serum, urine, fecal, and biological fluids) through ICP-AES, the works were reviewed and the summary is given in Table 6.6 [193-200].

In one work, sprodiamide injection (S-043 Injection, Nycomed Salutar; WIN 59080, Sterling Winthrop), which is a magnetic susceptibility–based MRI contrast agent containing 500 mM of dysprosium diethylenetriaminepentaacetic acid bis(methylamide) (DyDTPA-BMA), and 25 mM of caldiamide sodium (CaNaDTPA-BMA) were administered on cynomolgus monkeys [194]. According to the report made by the researchers, 18 cynomolgus monkeys, were divided into three groups of 6, and the sprodiamide injection was intravenously administered in three doses of 0.25, 0.5, and 2.5 mmol/kg, respectively. Next, the dysprosium content of the blood serum of the animals was determined in a serum–hydrochloric acid matrix using ICP-AES. The method showed a dynamic linear range from 0 to 50 μg/mL, and the limit of quantification (LOQ) was 24 ng/mL.

In another work, ICP-OES was used for the determination of gadodiamide in serum, peritoneal dialysate, and feces [195]. The LOQ in serum and peritoneal dialysate were found to be 6.5 and 1.6 μM Gd, respectively, and the within-day and between-day precision determination of gadolinium in feces were found to be 1.0–5.3 and 2.2%–7.9%, respectively, with an

TABLE 6.6 Some Reports on Determination of Lanthanides in Some Biological Samples

Year	Analyte	Sample	Important aspects	Ref.
1988	Yb	Biological origin	-	[193]
1993	Dy	Monkey serum after the administration of sprodiamide injection (MRI drug)	-	[194]
2000	Gd	Human serum and peritoneal dialysate and feces	-	[195]
2001	Dy, Eu, Yb	Animal fecal markers	-	[196]
2002	Gd	Urine	On-line cloud point preconcentration and using flow injection	[197]
2003	Dy	Urine	On-line complexation/ cloud point preconcentration	[198]
2006	Gd	Urine	Using flow injection on-line sorption preconcentration in a knotted reactor	[199]
2009	Gd	Water and human serum	HPLC	[200]

accuracy of 104%–116%. The LOQ for Gd was 11 nmol Gd/g. A comparative study of gadodiamide in serum and peritoneal dialysate from severely impaired renal patients by ICP-AES and HPLC revealed no chelator metabolism or transmetallation of gadolinium, even when the samples had been obtained 7 days after dosing.

In another report, the matrix effect was evaluated during the determination of Dy, Eu, and Yb used as sheep fecal markers using ICP-OES, through axial and radial viewing configurations [196]. Sample preparation procedure was based on microwave-assisted acid digestion using an oven with closed vessels. The comparison of the performances of the two ICP-OES instruments was performed through using the Mg II/Mg I intensity ratios under different operating conditions. The results showed that regardless of the operating conditions and the nature of the elements, the detection limits of the axial configuration were at least threefold better, but the results were more influenced by incremental background emission increases at increased plasma powers and lower nebulization gas flow rates. The accuracy of the results was verified using the INAA technique, and a paired *t* test revealed the results to be in agreement at a 95% confidence level.

Another on-line cloud point extraction (CPE) incorporated into the flow injection system has been reported as a sample preconcentration system for ICP-OES analysis [197], using the complexation reaction of Gd^{3+} with 2-(5-bromo-2-pyridylazo)-5-diethylaminophenol in the presence of non-ionic micelles of PONPE-7.5. The temperature of the micellar system containing the complex was fixed at 25°C, and it was loaded into the FI system at 5 mL/min. The surfactant rich-phase was retained in a microcolumn packed with cotton, at pH 9.2. A 4 mol/L nitric acid solution was used to elute the retained surfactant-rich phase, at a flow rate of 1.5 mL/min, and the eluted phase was directly fed into the nebulizer of the ICP-OES system. The results showed an enhancement factor of 20 for 10 mL of the sample solution, and the detection limit of the method was reported to be 40 ng/L for Gd^{3+}. The applicability of the method was successfully tested for the determination of Gd^{3+} in urine samples. An identical procedure was used for the on-line preconcentration and determination of Dy^{3+} [198].

A highly sensitive and robust method for the determination of six gadolinium species (i.e., Gd-DTPA(2-), Gd-DOTA(1-), Gd-HP-DO3A, Gd-DTPA-BMA, and Gd-DTPA-BMEA) by reversed-phase liquid chromatography coupled with inductively coupled plasma–optical emission spectrometry (HPLC-ICP-OES) has been reported [200].

INDUCTIVELY COUPLED PLASMA–MASS SPECTROMETRY (ICP-MS)

Inductively coupled plasma–mass spectrometry is a very useful and attractive technique used for quantitative and qualitative analyses. Besides unique advantages of hyphenating two very powerful techniques of ICP-OES and MS, its major disadvantage is the high costs. The method can be used for analysis of metallic and many nonmetallic elements, at low concentrations. It is much faster, more sensitive, and precise than the AAS, ICP-OES, because of the highly sensitive mass spectrometer used as the detector. One should, however, consider the fact that if compared with pure mass spectrometric results, the spectra acquired in this case includes some noise, which is due to the presence of some interfering species originating from the ICP ionizer, that is, argon ions. Table 6.7 compares ICP-OES and ICP-MS techniques.

The main figures of merit of ICP-MS are the very low detection limits for most elements (even down to parts per trillion [ppt]); wide linear response range (from 2 to 9 orders of magnitude); the ability to perform fast isotopic analysis, and numerous qualitative and quantitative analyses [201].

TABLE 6.7 Comparison Between ICP-OES and ICP-MS

Parameter	ICP-OES	ICP-MS
Amount of sample required for analysis	0.1–1 g	0.01–0.1 g
Optimal linear range	Up to 6 orders of magnitude	Up to 9 orders of magnitude
Measurement uncertainly	0.5%–3% relative to reported value	3%–5% relative to reported value
Limit of detection (3σ)	0.01 ppb–100 ppb	0.1 ppt–10 ppb
Element	From Li to U, except atmospheric species (C, H, O, N) and Nobel gases	From Li to U, except S, atmospheric species (C, H, O, N) and Nobel gases
Sample throughput	5–30 elements/min	All elements 2–6 min
Isotope analysis	No	Yes
Capital cost	High	Very high

Instrumentation

An ICP-MS instrument consists of two components of an ICP and a mass spectrometer (MS), as well as an interface that allows the hyphenation of an ICP device with an MS, which works under vacuum. The inductively coupled plasma has the same setup as that of an ICP-OES device. The sample introduction includes a nebulizer and spray chamber that feed the sample into the ICP torch. The sample is introduced to the MS through the interface, which links the ICP ion source to the high-vacuum MS. The vacuum is created by the vacuum system of the MS, which creates a high vacuum in the area where the optics, quadrupole, and detector are placed.

The collision or reaction cell is a chamber located before the MS, and should remove interferences to avoid adverse effects on the detection limit. Some cells can act as either of the collision and the reaction cells, which are called universal cells. The so-called ion optics tends to guide the ionic species of desire into the quadrupole, while filtering the neutral species and photons. MS, which is the main part of the detecting system of the instrument, separates the ions based on the mass-to-charge ratio (m/z) of each ion. The detector of the MS counts the ions leaving the quadrupole and reaching it and sends the signal to the data computer system.

Mechanism of Operation

The sample introduction in ICP-MS is like the one in ICP-OES, but the output is very different in the light of the fact that the ICP-OES results

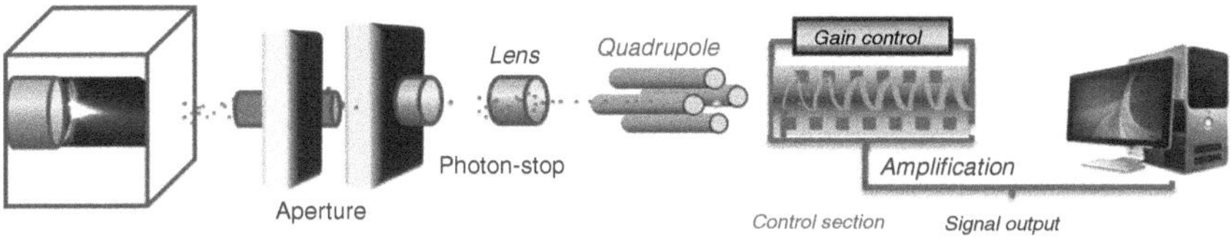

FIGURE 6.3 The schematic illustration of an ICP-MS instrument.

in optical spectra in the range of 165–800 nm, which is acquired either sequentially or simultaneously, the latter being faster and yet more expensive. Although in many cases, simultaneous ICP-OES tends to be faster, the method is rather cost intensive in comparison to the sequential mode. A factor influencing this is the number and concentration of the elements. A wide range of recent ICP-OES spectrometers have the ability to conduct experiments at 120 nm, which provides them with the ability to analyze chlorine.

After the sample enters the plasma and ionizes there, the instrument leads the ions into the interface. This section, which consists of a sampler and a skimmer cone, has the ability to avoid sudden pressure drop and instead it differentially reduces the pressure from the atmospheric value in the ICP to that of the MS, which is around 10^{-5}–10^{-7} Torr. After entering the vacuum of the MS detector, the ions pass through the ion optics, which is designed to filter off the neutral species and photons, and enter the quadrupole or other types of mass filters (i.e., double-focusing magnetic-electrostatic sector systems or as time-of-flight systems) before reaching the detector. The output of the ICP-MS, as opposed to the ICP-OES, is in terms of atomic mass unit or Daltons (in the range of 3–250 Da).

The quantitative analyses can be performed using calibration techniques by standard reference material. Fig. 6.3 shows a schematic illustration of an ICP-MS instrument with a quadrupole filter.

Applications of ICP-MS in the Analysis of Lanthanides

As a result of its potentials, ICP-MS can handle the analysis of a broad range of elements over a wide concentration range with detection limits as low as parts-per-trillion levels, in a quite rapid manner (<5 min per sample), and hence the method has been applied to the analysis of different geological, industrial, environmental, and biological samples [202-424]. On the other hand, ICP-MS can also conduct isotopic analyses, and although the technique has a much more robust rival (i.e., isotope dilution thermal ionization mass spectrometry [ID-TIMS]) in terms of precision and accuracy, the ICP-MS analyses are much easier and win the competition in case the number of samples is considerable. An overview of the works conducted on the analysis of lanthanides in different samples using

ICP-MS is provided in the following lines, where works are divided into five groups based on the nature of the samples.

Geological Samples

ICP-MS plays an intense role in the analysis of geological samples. This can be attributed to its capability to determine traces of a wide range of elements. The ICP-MS technique, as well as the rather unique capability of isotope analysis, is, for example, used in the case of Re-Os chronometer pairs in ore genesis studies. Another advantage of the technique that makes it suitable for the analysis of rather complex geological samples, in comparison with ICP-AES, is the relative simplicity of its spectra and its lower detection limits.

A list of the reports [202-270] on the analysis of geological samples through ICP-MS is provided in Table 6.8. The sample preparation techniques are identical to those explained before in the ICP-OES section.

A very old method is the work of Hirata et al. [206], in which very accurate determination of lanthanides was performed through ICP-MS, through the application of internal standardization, matrix matching, or elimination of the lanthanides from the matrix elements. The results obtained were in excellent agreement (within 5%) with those obtained from the isotope dilution method using ID-IMS. The best data were reportedly acquired from sample solutions that were free from matrix elements.

In another report, the lanthanide content of international reference silicate standards was determined through ICP-MS [215], because of the high sensitivity of the method and its rather high selectivity. This allowed for the direct estimation of the concentration of the elements in 0.1% rock solutions without ion-exchange separation or preconcentration with precisions in the range of 1%–7%.

The impact of ICP-MS in exploration geochemistry has been reviewed by Hall in 1992 [218], together with a description of the instrumentation, the principles of the technique, and the procedures suitable for its optimal use. To develop a realistic insight, one should, however, consider the shortcomings of ICP-MS, including some degree of interferences, which is of course less than that of ICP-OES; or its intolerance to high concentrations of dissolved solids (>0.2%) in sample solutions. Some have tried to overcome the latter through changing the mechanism of sample introduction from the classic nebulization to methods like electrothermal vaporization, laser ablation, slurry nebulization, and flow injection.

In another work, ICP-MS and ICP-OES were used for the determination of lanthanides in ores like xenotime and monazite [222], leading to analytical sensitivities of about 2–70 ppt and 1–50 ppb, respectively. ICP-MS did not show serious interferences from other elements present in xenotime, but the Gd and Tb intensities in monazite were altered because of the presence of praseodymium or neodymium monoxides. Such

TABLE 6.8 Some Reports on Determination of Lanthanides in Geological Samples by ICP-MS

Year	Analyte	Sample	Important aspect	Ref.
1985	REE	Geological materials	-	[202]
1987	REE	Geological materials	-	[203]
1987	REE	Geological materials	-	[204]
1988	REE	Geological materials	-	[205]
1988	REE	Geological standard rocks	-	[206]
1988	REE	Rocks and minerals	-	[207]
1988	REE	Geological materials	-	[208]
1989	REE	Geological materials	-	[209]
1989	REE	Geological samples	-	[210]
1989	REE	Silicate rock	-	[211]
1990	REE	Rock sample	-	[212]
1990	REE	Rock sample	-	[213]
1990	REE	Geological samples	-	[214]
1991	REE	Geological samples	-	[215]
1991	REE	Greek bauxites	-	[216]
1992	REE	Rock standards	-	[217]
1992	REE	Geological materials	-	[218]
1992	REE	Minerals and ores	-	[219]
1993	REE	Standard reference samples of oceanic deposits	-	[220]
1993	REE	Rock and sediment standard reference samples	-	[221]
1993	REE	Phosphate minerals	-	[222]
1993	REE	Fluid inclusions	-	[223]
1994	REE	Geological samples	-	[224]
1994	REE	Rock	Using potassium-based fluxes for sample decomposition	[225]
1995	REE	Rock samples	-	[226]
1995	Gd	Geological samples	-	[227]

(*Continued*)

TABLE 6.8 Some Reports on Determination of Lanthanides in Geological Samples by ICP-MS (*cont.*)

Year	Analyte	Sample	Important aspect	Ref.
1995	REE	Geological samples	Liquid chromatography	[228]
1995	Sm, Nd	Geological materials	-	[229]
1995	REE	Ultramafic rocks	-	[230]
1996	REE	Silicate reference samples	Ion exchange separation	[231]
1996	REE	Apatite minerals	-	[232]
1996	REE	Rocks	-	[233]
1997	REE	Chondritic meteorites	-	[234]
1998	REE	Chondritic meteorites	-	[235]
1999	REE	Geochemical samples	Liquid chromatography	[236]
1999	REE	Geological reference materials	-	[237]
2000	Y/Ho ratio	Late Cenozoic basalts	-	[238]
2000	REE	Rock reference samples	With cation exchange pretreatment	[239]
2000	REE	Rock reference samples	With cation-exchange pretreatment	[240]
2000	REE	Geochemistry reference materials	-	[241]
2000	REE	Granites	-	[242]
2001	REE	Rock samples	-	[243]
2001	REE	High silicate archeological specimens	-	[244]
2002	REE	Geological reference materials	-	[245]
2003	REE	Limestone geological standard reference material	Following solvent extraction	[246]
2003	REE	Felsic rocks	After lithium metaborate fusion	[247]
2003	REE	Rock samples	-	[248]

TABLE 6.8 Some Reports on Determination of Lanthanides in Geological Samples by ICP-MS (*cont.*)

Year	Analyte	Sample	Important aspect	Ref.
2003	REE	Sulfide minerals	With ion-exchange preconcentration	[249]
2004	REE	Rare earth impurities in high purity neodymium oxide	-	[250]
2005	REE	Ultramafic rocks	After preconcentration	[251]
2006	Eu, Gd	Geological materials	capillary electrophoresis	[252]
2006	REE	Monazite mineral	Solvent extraction separation	[253]
2006	REE	Garnet minerals, geological materials	-	[254]
2007	REE	Marine sediment reference materials	-	[255]
2007	REE	Ultramafic rocks	-	[256]
2008	REE	Basaltic and related rocks	By microwave oven digestion	[257]
2008	REE	Geologic reference materials	-	[258]
2008	REE	Indian coastal monazite	-	[259]
2009	REE	Geological reference materials	-	[260]
2009	REE	Ultramafic and sulfide samples	-	[261]
2010	REE	Authigenic fraction of Andaman Sea sediments	-	[262]
2010	REE	Certified reference materials (CRMs) and marine sediment samples	-	[263]
2012	REE	Rock samples	After sorption preconcentration using Pol-DETATA sorbent	[264]
2012	REE	Rock samples	-	[265]
2012	REE	Indian kimberlite	-	[266]

(*Continued*)

TABLE 6.8 Some Reports on Determination of Lanthanides in Geological Samples by ICP-MS (*cont.*)

Year	Analyte	Sample	Important aspect	Ref.
2012	REE	Romanian meteoritic material	-	[267]
2013	REE	Monazite	-	[268]
2013	REE	Ultramafic samples	Combined with chemical separation and pre-concentration	[269]
2014	REE	Oil accumulation water samples	Solid-phase extraction by the use of surface-functionalized nanoparticles	[270]

corrections were reported to be easy. The samples were digested using hot concentrated sulfuric acid, and the resulting solutions were directly injected into the ICP.

One study reported the results of 13 rare earth elements in 16 geochemical reference samples [231]. The samples were prepared through the dissolution of the rock or mineral separate, spiking with Tm, and separation of the REEs through an ion-exchange chromatography step. The resulting REE solutions were eventually analyzed through ICP-MS, and the results were tested against ID-TIMS results. The precision of the technique was reported to be better than 5% (2 sigma) for all the REE.

ICP-MS has also been used for determining traces of REEs (i.e., Th and U) in chondritic meteorites [234], in the light of the fact that chondrites has low contents of the elements (10^{-2} to 10^{-4} × crustal rock averages). The procedure was designed to be conducted in a minimal scale, so as to minimize the procedural blank. To eliminate the ion suppression effects due to the presence of Fe concentrations (20–35 wt%), internal standards (Rh for Y, In and Tl for lanthanides, and Bi for Th and U), and dilution factors (10^4 for Y and 10^3 for the rest of the elements).

Another report based on the application of LC-ICP-MS in the determination of REEs in geochemical standard reference samples [236] is based on the use of liquid chromatography, where a strong ion-exchange resin and ammonium lactate were used as the stationary and mobile phases, respectively. ICP-MS detection was used for multielement detection using the peak hopping mode. The determination of REEs in the standard rock sample and the standard sediment samples was performed by the system without any correction of the matrix effect and polyatomic ion interferences.

Lanthanoids (Ln) and yttrium (Y) in 11 GSJ (Geological Survey of Japan, Tsukuba) rock reference materials have also been determined through

ICP-MS [239,240]. The process was started with digesting the samples using a conventional acid digestion technique. The resulting solution underwent a cation exchange process in a column to remove the matrix. The isobaric interferences from metal oxide were also corrected based on the experimental results.

In another report, the REE content of a limestone standard was determined through ICP-MS, using phosphoric acid 2-ethylhexyl ester solvent extraction [246], after fractionating the limestone samples into the two acetic acid soluble (or carbonate) and insoluble (or residue) fractions. The method was reportedly modified to achieve quantitative recovery of most REEs present in the carbonate fraction, except for the heavy Tm, Th, and Lu. The poor recovery of the three elements was not clarified by the researchers. The mass-spectroscopic interference of BaO with Eu lead to the accurate and simultaneous determination of both lighter REEs and Eu with precision of 20% or better.

ICP-OES was also reportedly used in monitoring traces of heavy rare earth elements (HREEs) in Indian monazite mineral. This analysis is challenging owing to the spectral interferences and also the changes in the plasma excitation conditions in the presence of high amounts of light [253]. The HREEs were separated from LREEs and Th using a 0.22 M bis-(2-ethyl hexyl) hydrogen phosphate (HDEHP) solution for liquid–liquid extraction, followed by an ICP-MS detection.

One of the sources of interference in ICP-MS is the formation of oxide and hydroxide ions (e.g., $^{135}BaO^+$ on $^{151}Eu^+$, $^{144}NdO^+$ on $^{160}Gd^+$, $^{144}SmO^+$ on $^{160}Gd^+$, and $^{143}NdO^+$ on $^{159}Tb^+$), in the ICP segment [224,263]. Different approaches taken to overcome this problem include the measurement of bivalent ions, mathematical corrections, adjustment of instrumental factors, or the application of desolvation methods [263].

The sorption preconcentration of REEs, using Pol-DETATA as the sorbent, prior to ICP-MS analysis has also been reported [264]. In more detail, the proposed procedure is based on lithium metaborate fusion, preconcentration on Pol-DETATA sorbent, elution with nitric acid, and flow-injection sample introduction to the ICP-MS spectrometer. 5-Sulfosalicylic acid was used as a masking reagent, and the flow-injection sample introduction of 50 μL of the highly acidic eluate was used to avoid corrosion of the ICP-MS instrument. The results were double-checked by the analysis of the certified reference materials of rock samples, and the REE recoveries were reported to fall within an 85%–100% interval.

Rather recently, the determination of traces of REEs and thorium has been performed on ultramafic samples using ICP-MS [269]. Because the conventional digestion of ultramafic rocks with HNO_3 and HF mixtures leaves considerable amounts of insoluble fluorides (arising from the high Mg content [generally up to 24% m/m] of the rocks), H_3BO_3 was used as a complexing agent that breaks down the insoluble fluorides, and separates the REEs from Fe and Mg major and Ba, Ca, and Cr minor matrices through anion exchange and coprecipitation, respectively. The detection limits were between the two values of 0.51 ng/g for Ce and 0.03 ng/g for Lu, and the REE recovery was reported to range from 92.9% for Y to 102.0% for Er with ±3%.

Because of the role of REEs in identifying and indexing the origin of historical and geological samples, some have tried new approaches to the analysis of these element. In one work, the characterization of REEs in aqueous sample solutions with high salinity was conducted through ICP-AES [270], after treatment of the samples with surface-functionalized nanoparticles, to separate the analytes. The method was described as performing a combined matrix separation and internal standard correction. The method also enjoyed the advantage of complete elimination of the intrinsic limitations of SPE such as column clogging or incomplete analyte elution. The method also helped avoid time-consuming approaches for signal quantification, like matrix-matched calibration, because the internal standard created during the process made it possible to make corrections on the matrix-induced deviations in sample extraction and ICP-MS processes. The results included detection limits of less than 1 ng/L for the lanthanides tested, with RSDs of 3%–26% for unspiked and of less than 1%–2% for spiked samples. To validate the applicability of the method to highly saline samples, the method was used for recovery experiments with spiked oil accumulation water samples.

Industrial Samples

Table 6.9 illustrates the critical information in reports on the determination of lanthanides in industrial samples by ICP-MS [271-311].

An application of the very sensitive ICP-MS in industries is the monitoring of traces of the elements as impurities in high-purity oxides [272,274-277,280-282,289,293,294,296,302,306]. One example of such applications is the determination of very low levels of rare earth impurities in high-purity Y_2O_3 and Gd_2O_3, through high-resolution inductively coupled plasma–mass spectrometry (HR-ICP-MS) [272]. To overcome interferences, GdO, GdH, and GdOH on Tb, Yb, Tm, and Lu bivalent ions were detected. Further, the mass resolution of the method was set at 400, which is far beyond the conventional values of 80–90 applied in quadrupole mass filter and much lower than the standard setting for ICP-MS in the high-resolution mode. Consequently, the anticipated transmission decrease in quadrupole mass filter on increasing the resolution did not cause any problem. The sole factor controlling the sensitivity was the rate of the production bivalent species, and hence the optimization was performed to maximize the proportion of such species. The method yielded detection limits of 0.05–3 ng/L in solutions, in case of bivalent and 5–30 pg/L for monovalent species.

TABLE 6.9 Some Reports on Determination of Lanthanides in Industrial Samples by ICP-MS

Year	Analyte	Sample	Important aspect	Ref.
1992	REE	Iron-formation reference samples	-	[271]
1993	REE	Impurity in high-purity rare earth element samples	-	[272]
1994	REE	In hydrothermal fluids and plume particulates	-	[273]
1995	REE	Impurities in high-purity scandium	Liquid–liquid extraction of the matrix	[274]
1995	REE	Impurities in high-purity scandium	Liquid–liquid extraction of the matrix	[275]
1995	REE	Rare earth impurities in ultrapure europium oxide	-	[276]
1997	REE	REEs impurities in high-purity cerium oxide	Solvent extraction	[277]
1998	REE	Fluid inclusions	-	[278]
1999	REE	Impurities in high-purity gadolinium oxide	After 2-ethylhexylhydrogen-ethylhexy phosphonate extraction chromatographic separation	[279]
1999	REE	Impurities in high-purity ytterbium oxide	-	[280]
1999	REE	Rare-earth element oxides	-	[281]
1999	REE	Impurities in high-purity Pr_6O_{11}	-	[282]
2000	REE	Rare earth impurities in high-purity La_2O_3	HPLC combined using 2-ethylhexyl hydrogen-2-ethylhexylphosphonate resin as a stationary phase	[283]
2000	REE	Impurities in ultra-pure rare-earth compounds	HPLC combined	[284]

(*Continued*)

TABLE 6.9 Some Reports on Determination of Lanthanides in Industrial Samples by ICP-MS (*cont.*)

Year	Analyte	Sample	Important aspect	Ref.
2001	REE	Certified reference materials	Cation-exchange chromatographic separation	[285]
2001	REE	Certified reference materials	Algebraic correction with chromatographic separation	[286]
2001	REE	Industrial waste incineration fly	After $LiBO_2$ fusion	[287]
2002	REE	Elemental and isotopic analysis of lanthanides and actinides	Laser ablation (LA) sample introduction system and the use of high-pressure liquid chromatography (HPLC)	[288]
2003	REE	Rare earth impurities in high-purity europium oxide	-	[289]
2004	Lu	Isotopic determination	Multiple collector-ICP-MS	[290]
2004	REE	Seepage waters caused by acid mine drainage	-	[291]
2004	REE	Teeth enamel	-	[292]
2004	REE	Rare earth impurities in high-purity europium oxide	-	[293]
2005	REE	Trace amounts of rare earth impurities in various rare earth oxides	-	[294]
2005	REE	Petroleum	-	[295]
2006	REE	Impurities in high-purity rare earth erbium oxide	-	[296]
2007	REE	Petroleum refining catalysts and ambient fine aerosols	Microwave-assisted extraction	[297]
2007	Ce, St, Ti	Ceramic materials used in radiologic dispersal devices (RDDs)	-	[298]

TABLE 6.9 Some Reports on Determination of Lanthanides in Industrial Samples by ICP-MS (*cont.*)

Year	Analyte	Sample	Important aspect	Ref.
2007	REE	Rare earth impurities in high-purity samarium oxide	After extraction chromatographic separation	[299]
2008	REE	Niger Delta crude oils	-	[300]
2008	REE	Fission products in nuclear samples	-	[301]
2009	REE	Impurities in high-purity neodymium oxide	-	[302]
2010	REE	Conversion of rare earth elements to molecular oxide ions in a dynamic reaction cell	-	[303]
2010	REE	Zirconia and zirconium alloys	Microwave-assisted	[304]
2010	REE	Uranium-bearing materials	-	[305]
2012	REE	Impurities in nuclear-grade uranium oxide	Solid-phase separation	[306]
2012	REE	Zircon reference materials	-	[307]
2012	REE	Nuclear fuel sample	Separation by isotachophoresis	[308]
2013	Nd	Nuclear fuel	-	[309]
2014	REE	Gadolinium aluminate	Time of flight mass	[310]
2014	REE	Spent catalyst samples from oil refinery	-	[311]

In another experiment, the REE content of high-purity Gd_2O_3 was analyzed through ICP-MS [297], in a procedure where the matrix suppression effect of Gd_2O_3 on the signals corresponding to the trace rare earth impurities was compensated for by Re internal standardization. The interferences due to spectral overlap from GdH, GdO, GdOH (n = 1–3) on Tb, Tm, Yb, and Lu were eliminated through 2-thylhexylhydrogen-2-ethylhexy phosphonate (EHEHP) extraction chromatographic separation, and the results showed detection limits of 0.005–0.017 ng/mL for solution and 0.002–0.05 μg/g for solid samples, as well as recoveries ranging from 88% to 121%, with RSDs of 1.0%–7.5%, for the lanthanide species tested.

In another experiment, the spectral interference from polyatomic barium ions (i.e., BaO^+, $BaOH^+$) during the ICP-MS analysis of REEs was limited through chromatographic separation by AG50W-x8 cation exchanger [285]. According to the method, Ba, Ca, Mg, etc. were eluted with 2 mol/L HNO_3 solution, whereas the lanthanides of desire were retained. In a next step, the retained species were eluted with a 4 mol/L HNO_3 solution. The process led to the elimination of greater than 99.5% of Ba content of the sample, yielding REE recoveries in the range of 96% to 110%.

Klinkhammer et al. [273] described the a procedure for the measurement of the rare earth elements in fluids from deep-sea hydrothermal vents and in the particles resulting from the mixture of these fluids mixed with sea water. The extraction of the REE is due to their coprecipitation with hydrated iron(III), and the elements are finally isolated from other elements on miniature cation exchange columns before the ICP-MS. The column separations help enhance the sensitivity through the elimination of the high levels of solids dissolved and minimize isobaric interferences from barium. The interlaboratory calibrations were made with ID-TIMS and proved that the ICP-MS results are accurate at a 6% (2 sigma) level.

The application of chromatographic methods prior to ICP-MS detection has also been of interest [283-286,288,299], because chromatographic methods can eliminate matrix interferences as well as the severe overlapping of oxides and hydroxides peaks.

Analysis of traces of REE impurities in high-purity La_2O_3 has been reported through HPLC-ICP-MS [283], using 2-ethylhexyl-hydrogen-2-ethylhexylphosphonate (P507) resin as the stationary and an HNO_3 solution as the mobile phases. The effects of pH and flow rate of the mobile phase, and the column oven temperature on the separation of La and Ce, were investigated and optimized, to reach a desirable La and Ce separation. In comparison with the open column method, the proposed HPLC-ICP-MS was improved in terms of separation time, reagent consumption, and operation complexity.

Microwave-assisted acid digestion has also been used prior to ICP-MS determination of REEs (La, Ce, Pr, Nd, Sm, Gd, Eu, and Dy) in fluidized-bed catalytic cracking (FCC) catalysts and atmospheric fine particulate matter (PM2.5 from an industrial area of Houston) [297]. The harsh conditions of the digestion procedures, including the high temperature (200 degrees C), high pressure (200 psig), and strong acid mixture (i.e., HNO_3, HF, and H_3BO_3), led to the effective solubilization of REEs present in the samples (fresh and spent catalysts, as well as PM2.5). Nonspectral interferences were effectively eliminated through internal standardization using In-115.

Another report on the application of ICP-MS has been the determination of fission products and of their isotopic composition in nuclear fuels. This is achieved through the radiometric and mass spectrometric methods; the former being applicable to some of the fission products. The mass spectrometric methods, on the other hand, face limitations because of the isobaric interferences, which necessitate a prior chemical separation, through a separation technique. In this case, a capillary electrophoresis step is coupled with ICP-QMS and ICP-SFMS detection systems [301]. The detection limits were 6 ng/mL and 4 pg/mL for cesium as well as 8 ng/mL and 7 pg/mL through CE-ICP-QMS and CE-ICP-SFMS, respectively. The sample and injection volumes were in the microliter and nanoliter ranges.

Another application of ICP-MS is the quantification of traces of REEs in ZrO_2, zircaloy-2, and zirconium–1% niobium [304], after matrix separation through precipitation using Mandelic acid followed by solvent extraction.

Vio et al. reported the development of a new electrolyte system for the separation of lanthanides by isotachophoresis [308], based on a leading electrolyte incorporating 2-hydroxy-2-methylbutyric acid as a complexing agent, after optimizing the separation conditions (i.e., concentration of the complexing agent, pH, capillary dimensions, injection conditions, and current intensity) using a commercial capillary instrument with contactless conductivity detection. The method was used for the separation of lanthanides (except Pm and Ho) and successive determination using an ICP-MS.

Another application of the ICP-MS detection has been the report on the precise quantification of La, Ce, Pr, Nd, Sm, Eu, Tb, Dy, Ho, and Tm, in gadolinium aluminate ($GdAlO_3$), through nebulization in an ultrasonic desolvating device and subsequent ICP-MS detection [310]. The optimized microwave digestion procedure was used for digesting 100 mg of the refractory oxide using a mixture of H_2SO_4, H_3PO_4, and water at 1400 W power (10 min ramp and 60 min hold time). An ultrasonic desolvating sample introduction system was employed, which was found to enhance the analyte sensitivities through minimizing the formation of oxide ions in the plasma. Indium was evaluated as an internal standard and was successfully applied to matrix correction and variation of the sensitivity of the method for the analyte. During this study, the individual oxide ion formation yields were evaluated in the matrix matched solution, and the results were applied for correcting polyatomic interferences of light REE (LREE) oxide ions on the intensities of middle and heavy rare earth elements (MREEs and HREEs). The results showed recoveries of more than 90%, and the detection limits were in the range of 1–5 ng/L and 7–64 μg/kg, respectively.

In another work, REEs were determined in spent oil refinery catalyst samples [311], using three sample treatment procedures. The treatments included one without HF and two using different amounts of HF and boric acid, and a microwave evaporation system was applied to promote the dissolution. The results showed that the systems containing 1.0 mL of HF and 8.0 mL of boric acid led to complete digestion of the samples. During the studies, a reaction cell was used to overcome the interferences firm polyatomic species on four elements during ICP-MS analysis.

Environmental Samples

Despite its low detection limits, as compared to ICP-OES, and although ICP-MS is applicable for the direct determination of REEs in geological and some industrial samples, its applications for the analysis of REEs in environmental samples is limited. This is mainly due to the extremely low levels of these elements in such samples, which fall below the detection limits of the method. Another limitation is due to the high concentration matrices of environmental samples, which makes them incompatible with the sampling system of ICP-MS. Consequently, the analysis of REEs in environmental samples, as in the case of ICP-OES, requires preconcentration and matrix separation stages prior to analysis. As with other samples, the commonest preconcentration technique is the coprecipitation of REEs with Fe and subsequent cation exchange. This technique, however, has a number of limitations, including nonquantitative recovery and further obligatory measures for elimination of Mg, Ca, and Fe. These increase sample manipulation and the chances of contamination of blanks. Table 6.10 summarizes some of the reports on the determination of lanthanides in environmental samples through ICP-MS detection [312-384].

One such instance is the report on the preconcentration and determination of REEs in seawater samples through ICP-MS [313], based on the solvent extraction and back-extraction preconcentration of REEs using a mixture of bis(2-ethylhexyl) hydrogen phosphate (HDEHP) and 2-ethylhexyl dihydrogen phosphate (H_2MEHP) absorbed on C18 cartridges. Passing the seawater samples through the C18 cartridge led to extensive complexation of the lanthanides with HDEHP and H_2MEHP, whereas the rest of the elements present in the matrix were eluted with the mobile phase. In a next step, the REEs were desorbed using an acidic eluent. The results showed 200–1000-fold enrichment factors. Another example of the analysis of lanthanides in sea water samples through ICP-MS [317] was based on the coprecipitation of the REEs and heavy metals with gallium hydroxide for matrix elimination. To this end the trace elements were quantitatively coprecipitated at pH 10. The method was tested for the analysis of Al, Ti, Cr, Mn, Fe, Co, Ni, Cu, Zn, As, Y, Zr, Sn, La, Pb, Th, and U. A similar group of analytes in sea water samples was determined through ICP-MS after preconcentration using chelating resin in a batch method [318], where Na, K, Mg, and Ca were removed to minimize the matrix effects in ICP-MS.

TABLE 6.10 Some Reports on Determination of Lanthanides in Environmental Samples by ICP-MS

Year	Analyte	Sample	Important aspect	Ref.
1992	REE	Seawater	Ion exchange chromatography	[312]
1992	REE	Seawater	Preconcentration of by complexation with bis(2-ethylhexyl) hydrogen phosphate and 2-ethylhexyl dihydrogen phosphate adsorbed on a C18 cartridge	[313]
1993	REE	Water samples	On-line preconcentration	[314]
1993	REE	Natural fresh water	-	[315]
1995	REE	Ground water	-	[316]
1995	REE	Seawater	Gallium coprecipitation	[317]
1995	REE	Coastal sea water	Preconcentration using chelating resin	[318]
1995	REE	Sediment	Using both In and Bi as internal standards	[319]
1995	REE	Aqueous samples	Flow injection system	[320]
1996	REE	Drinking water and its sources	-	[321]
1996	REE	Highly saline brines	On-line method	[322]
1996	REE	Environmental materials	Liquid-liquid extraction	[323]
1996	REE	Seawater	-	[324]
1997	REE	Soil	-	[325]
1998	REE	Fertilizer, animal fodder and manure samples	-	[326]
1998	REE	Algae	-	[327]
1998	REE	Sea-water	With an ion chromatographic separation system	[328]
1998	REE	Arima spring water	With ion-exchange separation	[329]

(*Continued*)

TABLE 6.10 Some Reports on Determination of Lanthanides in Environmental Samples by ICP-MS (*cont.*)

Year	Analyte	Sample	Important aspect	Ref.
1998	REE	Seawater	Using flow injection preconcentration	[330]
1998	REE	Seawater	With poly(acrylaminophosphonic dithiocarbamate) chelating fiber prior to determination	[331]
1999	REE	Soils	After microwave digestion	[332]
1999	REE	Seawater	Preconcentration with 8-hydroxyquinoline immobilized polyacrylonitrile hollow fiber membrane	[333]
1999	REE	Environmental and geological samples	Flow injection on-line group preconcentration	[334]
2000	REE	Plant (hot pepper) and soil samples	After microwave digestion	[335]
2001	REE	Saline water samples	Chelating resins	[336]
2001	REE	Soils and sediments	After microwave digestion	[337]
2001	REE	Plant certified reference materials	-	[338]
2001	REE	Seawater	With 8-hydroxyquinoline-5-sulfonic acid-cellulose microcolumn	[339]
2001	REE	Seawater	TOF	[340]
2001	REE	Seawater	After tandem preconcentration with cooperation of chelating resin adsorption and lanthanum coprecipitation	[341]
2002	REE	Ground water	With preconcentration 2-ethylhexyl hydrogen 2-ethylhexylphosphonate (P-507) resin as the stationary phase and HCl/EDTA as a mobile phased	[342]
2002	REE	Seawater	By on-line column preconcentration	[343]

TABLE 6.10 Some Reports on Determination of Lanthanides in Environmental Samples by ICP-MS (*cont.*)

Year	Analyte	Sample	Important aspect	Ref.
2002	REE	Seawater	After pretreatment with a chelating resin disk by an on-line flow injection method	[344]
2002	REE	Environmental water	Using chelating resin disk separation	[345]
2002	REE	Natural water	With solvent extraction	[346]
2003	Th and light REE	Soil water	On-line-coupled size-exclusion chromatography	[347]
2003	REE	Seawater	On-line column preconcentration using metal alkoxide glass immobilized 8-quinolinol	[348]
2003	REE	River water	By fully automated on-line column inductively coupled plasma mass spectrometry using iminodiacetate chelate resin as a column	[349]
2003	REE	Carbonaceous black shales	-	[350]
2003	REE	Carbonaceous black shales	-	[351]
2004	Gd	River water	SPE preconcentration	[352]
2004	REE	Seawater	On-line column preconcentration using 8-quinolinole-immobilized fluorinated metal alkoxide glass	[353]
2004	REE	Peach leaves and pond sediment	Solid-phase preconcentration with chelating resin fibers and anion exchanger filters	[354]
2005	REE	Iron-rich water samples	Extraction chromatography	[355]
2005	La, Eu, Yb	Water sample	After solid-phase extraction of their 1-phenyl-3-methyl-4-benzoylpyrazol-5-one complexes on silica gel column	[356]

(*Continued*)

TABLE 6.10 Some Reports on Determination of Lanthanides in Environmental Samples by ICP-MS (*cont.*)

Year	Analyte	Sample	Important aspect	Ref.
2005	REE	Sea and natural water	Sorbent extraction technique using a chelating agent–impregnated porous PTFE filter tube	[357]
2006	REE	Antarctic ice	Using a desolvation system	[358]
2006	REE	Plant reference materials	-	[359]
2006	REE	Sediment, soil, and seawater samples	Flow injection on-line solid-phase extraction using maleic acid–grafted polytetrafluoroethylene fibers as sorbent	[360]
2006	REE	Seawater	After preconcentration with a cheresin-packed minicolumn	[361]
2007	REE	Seawater	On-line preconcentration with a novel alkyl phosphinic acid extraction resin	[362]
2007	REE	Seawater	After APDC chelate coprecipitation combined with iron	[363]
2007	REE	Mangrove tree rings	-	[364]
2008	REE	Antarctic ice	-	[365]
2008	REE	Seaweed samples	-	[366]
2009	REE	Certified reference material, which combined branch and leaves of a shrub	On-line column preconcentration and separation with single-wall carbon nanotubes	[367]
2009	REE	Tomato plants	-	[368]
2009	REE	Estuarine and coastal waters	Preconcentration with NOBIAS-chelate resin	[369]
2010	Gd	Environmental water samples	Hydrophilic interaction chromatography hyphenated	[370]
2010	REE	Multimineral phase soil	-	[371]

TABLE 6.10 Some Reports on Determination of Lanthanides in Environmental Samples by ICP-MS (*cont.*)

Year	Analyte	Sample	Important aspect	Ref.
2011	REE	Environmental samples	-	[372]
2011	REE	Seawater	With off-line column preconcentration using 2,6-diacetylpyridine functionalized Amberlite XAD-4	[373]
2011	REE	Seawater	Separation and preconcentration by polyhydroxamic acid cartridge	[374]
2012	REE	Plant materials	-	[375]
2012	REE	Groundwater	Dispersive liquid–liquid microextraction for simultaneous preconcentration	[376]
2012	REE	Seawater	Preconcentration using TSK (TM)-HD-MW-CNTs (highly dispersive multiwalled carbon nanotubes)	[377]
2013	Gd	Surface waters	Hydrophilic interaction liquid chromatography	[378]
2013	La, Eu, and Yb	Certified reference material	Solidified floating organic drop microextraction and electrothermal vaporization	[379]
2013	Nd, Tb, and Ho	Water samples	After column preconcentration and separation with titanium dioxide nanotubes	[380]
2013	Gd	Surface water samples and plants	Hydrophilic interaction chromatography	[381]
2013	REE	Dust deposited on tree leaves	-	[382]
2013	REE	Seawater and estuarine water	With 4-(2-thiazolylazo) resorcinol immobilized Chromosorb 106	[383]
2014	REE	Tea leaves and water samples	After magnetic solid-phase extraction with Fe_3O_4–SiO_2–polyaniline-graphene oxide composite	[384]

Another ICP-MS analysis was used for the determination of REEs at concentrations as low as 0.1 ppt (0.5 pmol/kg) in aqueous samples using flow injection coupled with an inductively coupled plasma–mass spectrometer [320]. The method was rather rapid because the samples proceeded to the purification stage in less than 90 min, whereas more than 99.5% of the Ba impurity was removed, ensuring minimizing the isobaric interference of BaO on the REE to less than 2% of the REE signal. The detection limits of the method were improved about 30 times, because of the application of the flow injection technique, which allows for the samples to be concentrated in a smaller volume for analysis. The applicability of the method was tested through its use for determination of REE in Icelandic hydrothermal fluids. Another experiment [340] based on timed flow-injection (FI) system, using a column containing iminodiacetate-based resin, was also used for the analysis of the lanthanide content of sea water samples. The detection limits (3 s) ranged from 20 to 50 pg/L, with RSDs of about 5%–6%.

The REE content of sea water samples was also determined after a preconcentration using a nitrilotriacetate (NTA)-type chelating resin [328], which retained the REES, whereas the alkaline earth metals were eluted from the column by elution with 0.5 mM nitric acid. The results showed a linear calibration, and repeatability of the signal intensity of ICP-MS was obtained in the determination of trace metals at the <1 μg/L level.

Another analysis of REE in sea-water and highly saline brines through ICP-MS [322] applied the separation of the analytes from the saline water, using a microcolumn of silica-immobilized 8-hydroxyquinoline for absorbing the REEs at pH 9, followed by their desorption using a small volume of mixed acids, in less than 10 min. The detection limits achieved ranged between 0.06 ng/L for Tb and Ho to 0.6 ng/L for Ce and Nd.

An inductively coupled plasma sector field mass spectrometry coupled with a microflow nebulizer and a desolvation system was also reported for the direct determination of La, Ce, Pr, Nd, Sm, Eu, Gd, Tb, Dy, Ho, Er, Tm, Yb, and Lu at concentrations down to the subpicogram per gram levels in 1 mL of molten Antarctic ice [358]. The interference and contamination issues were carefully considered through the use of an ultra-clean procedure during the sample pretreatment, and the application of the desolvation system for sample introduction also greatly reduced spectral interferences from oxide formation. The method also considered residual interfering contributions through calculation and and subtraction whenever necessary, reaching to instrumental detection limits from 0.001 pg/g for Ho, Tm, and Lu to 0.03 pg/g for Gd.

In another work, Chen et al. in 2013 [379] applied a so-called solidified floating organic drop microextraction using 1-(2-pyridylazo)-2-naphthol (PAN) as a chelating agent together with an electrothermal vaporization (ETV) as the matrix elimination/preconcentration stage prior to ICP-MS.

According to the descriptions of the work, PAN was also found to act as a modifier in ETV. The influencing parameters including pH, amount of PAN, extraction time, stirring rate, volume of sample solution, and temperature program were optimized to get detection limits of 2.1, 0.65, and 0.91 pg/mL for La, Eu, and Yb, respectively.

In rather recent work, a novel Fe_3O_4–SiO_2–polyaniline-graphene oxide composite (MPANI-GO), prepared by a noncovalent method, was used in the magnetic solid-phase extraction (MSPE) of traces of rare earth elements in tea leaves and environmental water samples, and an ICP-MS detection was conducted as the final step [384]. The best detection limits (DLs, 3 sigma) were from 0.04 to 1.49 ng/L with RSDs ($c = 20$ ng/L, $n = 7$) of 1.7%–6.5%. The MPANI-GO MSPE-ICP-MS method was described as simple, rapid, and highly sensitive, with a high enrichment factor and suitable for the analysis of traces of REEs in complex matrices.

Food Samples

The extensive application of lanthanides in fertilizers, food additives, etc. has increased their occurrence in environmental and biological samples and finally can naturally influence human health, and hence the determination of even traces of these elements has found considerable significance. Table 6.11 illustrates a list of the reports on the ICP-MS determination of lanthanides in some food samples such as wine, tea, feed, seed, and fruits. It is clearly evident that the focus of the reports is on the past decade [385-404].

Because the presence of lanthanides in the irrigation system of rice can increase both the production capacity and the resistance to disease, this has become commonplace in some countries. So the analysis of different parts of such plants, including their root stalk and grain, becomes important. In one experiment, the parts of rice plants were analyzed by both INAA and ICP-MS [386], and the results prove the presence of some 40 elements in the plant tissues. It was also shown that the presence of lanthanides was focused in the roots but did not reach the grains.

In another study, the accumulation profile of REEs in *Camellia oleifera* seeds from farms near REE mines was studied by ICP-MS [393]; and detection limits mostly lower than 3.00 ng/L with RSDs of 1.97%–11.7% were achieved for 10 REEs. THE authors concluded that the REE content of the *C. oleifera* seeds from such farms was too low to cause harm to human health.

In one study, walnut shell was used as an adsorbent for the on-line preconcentration and detection of REEs in food and agricultural products, prior to ICP-MS analysis of the analytes [397]. Given the porous surface and abundance of –COO– groups on the surface of this adsorbant, the walnut shell–packed microcolumns were found to provide high adsorption efficiencies and high tolerable capacities for coexisting ions. The

TABLE 6.11 Some Reports on Determination of Lanthanides in Food Samples by ICP-MS

Year	Analyte	Sample	Important aspect	Ref.
1996	REE	Wine	-	[385]
1997	REE	Rice	-	[386]
1997	REE	Plant foods	Calibration with indium as the internal standard	[387]
1998	REE	Tea	With microwave digestion and AG50W-x8 cation exchange chromatography	[388]
2004	REE	German white wines	-	[389]
2007	REE	Certified reference materials of tea leaves and mussel	Carbon nanofibers as solid-phase extraction adsorbent for the preconcentration	[390]
2007	REE	White wines	-	[391]
2007	La, Ce	Feed	-	[392]
2011	REE	*Camellia oleifera* seeds	-	[393]
2011	REE	Lotus seed	-	[394]
2011	REE	Corn seeds	-	[395]
2012	REE	Indian tea	-	[396]
2012	REE	Food and agricultural products	Flow injection walnut shell packed microcolumn preconcentration	[397]
2013	REE	Gannan navel orange fruit	-	[398]
2013	REE	Certified reference material of tea leaves	Titanium dioxide nanotubes as solid-phase extraction adsorbent for on-line preconcentration	[399]
2013	REE	Macedonian wines	-	[400]
2013	REE	Rice	-	[401]
2013	Sm, Eu	Wheat	On-line separation and preconcentration	[402]
2014	REE	Navel oranges	-	[403]
2014	REE	Fruit samples	-	[404]

authors reported that under optimal preconcentration conditions, the adsorption efficiency of the microcolumns was as high as 98.9% for 15 REEs. At the same time, the tolerable concentration ratios ranged between 2000 and 80 000 000 for 37 coexisting interfering ions. Enhancement factors of 79–102 were achieved for 15 REEs at sample loading volumes of 4.7 mL, and the detection limits of the overall method were from 2 to 34 pg/g. The method was successfully tested to the selective determination of REEs in 4 environmental and biological certified reference materials and 18 locally collected food and agricultural products.

The REE content of the navel oranges from the three main production sites in China (i.e. Southern Jiangxi Province, Southern Hunan Province, and the Three Gorges District of the Yangtze River) were also studied through ICP-MS [403]. Because the Southern Jiangxi and Southern Hunan are known for presence of ionic REEs in their soil, which makes it easily transferable to the plants and fruits, an ICP-MS analysis was conducted after the microwave digestion of the samples. The results proved that the concentrations of La, Ce, Pr, and Nd in the samples from Southern Jiangxi and Southern Hunan (1.26–1.86 μg/g) were much higher than those from the Three Gorges (0.23–0.46 μg/g). Another conclusion was that yttrium is relatively accumulated (0.25–0.29 μg/g) in navel orange peels from Southern Jiangxi at almost twice that of samples from Southern Hunan (0.15 μg/g). The authors declared that the navel orange peels from Southern Jiangxi posed no risk to consumers, based on the maximum allowed level (0.7 μg/g, wet weight) of rare earth elements in food issued in China (GB 2762-2005).

Biological Samples

Table 6.12 tends to provide an overview on the reports concerned with the ICP-MS determination of lanthanides in biological samples [405-424]. What follows tries to provide some detail about these reports.

Regarding the applications of La and Lu texaphyrins as photosensitizers in photodynamic therapy and identification of malignant tumors, a number of studies have been conducted on the determination of these elements. One report describes the determination of La and Lu in liver homogenates through ICP-AES, after the in vitro addition or in vivo treatment with these metal-containing texaphyrins [406], reaching a detection limit of 0.5 ng/mL for elements. The report introduces the ICP-MS method, as a means for facilitating the study of concentration profile of La and Lu texaphyrins in tissues that can be adapted for the determination of traces of other metals in metal-incorporated texaphyrins, such as those containing manganese, cadmium, gadolinium, or terbium.

Another example is the determination of rare earth elements in a blood serum reference samples through ICP-MS, after sample digestion and concentration pretreatment of the dry-freezed samples [407]. The procedure

TABLE 6.12 Some Reports on Determination of Lanthanides in Biological Samples by ICP-MS

Year	Analyte	Sample	Important aspect	Ref.
1990	REE	Plasma and urine	-	[405]
1995	La, Lu	In metal-containing texaphyrins (Anticancer drug)	-	[406]
1996	REE	Blood serum reference sample	By chelating resin preconcentration	[407]
1997	REE	Human blood serum	Preconcentrated by using the chelating resin prior to determination	[408]
1998	Gd	Tissue samples	-	[409]
1998	REE	Human hair and wheat flour reference materials	-	[410]
1998	REE	Urinary calculi	-	[411]
1999	REE	Biological samples	-	[412]
1999	REE	Bovine whole blood reference material	After coprecipitation preconcentration with heme-iron as coprecipitant	[413]
1999	REE	Standard biological samples	-	[414]
2000	REE	Human plasma	Indium as an internal standard element is used to compensate for matrix suppression effect and sensitivity drift	[415]
2000	REE	Human blood serum	After chelating resin preconcentration	[416]
2004	REE	Chloroplast pigments of hyperaccumulator *Dicranopteris dichotoma*	Off-line separation by normal-phase liquid chromatography	[417]
2006	REE	Chinese herbal medicine	-	[418]
2007	REE	Reference material of human hair	Microcolumn packed with modified carbon nanofibers on-line preconcentration	[419]

TABLE 6.12 Some Reports on Determination of Lanthanides in Biological Samples by ICP-MS (*cont.*)

Year	Analyte	Sample	Important aspect	Ref.
2008	Gd	Hair and fingernail	-	[420]
2010	REE	Marine biological tissues	-	[421]
2011	La	Serum	-	[422]
2012	REE	*Sophora japonica* L–related herbal medicine	-	[423]
2013	REE	Human urine and serum samples	Polymer monolithic capillary microextraction combined on-line	[424]

included digestion of about 0.8 g of the samples (equivalent to 10 mL of blood serum) with HNO_3 while heating on a hot plate. The digested sample was next diluted with 100 mL of 0.1 M HNO_3. Finally, the REEs content of this solution was adsorbed by a chelating resin (Chelex(R) 100), eluted from the adsorbate using 10 mL of 2 M HNO_3 aqueous solution and analyzed by ICP-MS. The reports indicated recovery values of 70%–80% and RSDs less than 5% ($n = 3$).

Off-line normal-phase liquid chromatography-ICP-MS analysis of REEs in chloroplast pigments of *Dicranopteris dichotoma* has also been reported [417]. During the study, no destruction of REE-bound pigments was observed. In the next step, the free REE ions accumulated on the microcrystalline cellulose column were eluted with a 5 mmol/L 2-ethylhexyl hydrogen 2-ethylhexyl-phosphonate (P507) mobile phase, to avoid their exchange with metals from the pigments. After taking these precautions, the REE-bound pigments in *D. dichotoma* were studied through ICP-MS, the results of which study revealed that REES were present in chlorophylls and lutein, although REE concentrations in carotene and pheophytin were both below procedural blank levels. The careful study of the eluate fractions containing chlorophyll revealed that the REE-bound chlorophyll a in *D. dichotoma* was slightly enriched in the fractions with relatively short retention times. The results also indicated that the retention time of REE-bound chlorophyll a is a little less than that of magnesium chlorophyll a. It was also concluded that the REE-bound chlorophylls might be more polar than magnesium-bound chlorophylls, which is explained by the special double-decker sandwich structure of REE-bound chlorophylls. The researchers finally concluded that REE can replace magnesium in chlorophyll a of *D. dichotoma*, and the role of REE-bound chlorophylls in photosynthesis is rather considerable.

Because in patients with stage V chronic kidney disease, lanthanum carbonate is used as a phosphate binder, and although no toxicity has been reported for the bone and liver metabolisms as well as on cognitive functions by this compound, there are still concerns over its toxicity. Hence, a simple method was developed for the measurement of lanthanum content of serum through ICP-MS [422]. To validate the results, internal quality control solutions containing La^{3+} (0.20, 0.70, and 4.00 μg/L) were used.

The measurements were conducted on samples from 20 patients with chronic kidney disease administering lanthanum carbonate (500–1500 mg/d) and a control group of patients undergoing hemodialysis who did not use lanthanum carbonate, and RSDs of 5% were reported. The response of the method was reported to be linear from 0.1–5 μg/L (R^2 = 0.99), and the lower limit of quantification (LLOQ) was evaluated to be 0.1 μg/L. In the case of the control group patients, La content of the serum was below the LLOQ, while among those using the medicine, the amounts of La was measureable in the case of 12 and reached 0.11–0.60 μg/L. An interesting observation was that no considerable relationship was observed between the dose and serum content of lanthanum in these patients, which was regarded as a sign of the patients' poor adherence to the treatment regimen.

Finally, a rapid and sensitive method based on polymer monolithic capillary microextraction combined on-line with microconcentric nebulization ICP-MS was reported. The method was used for the monitoring of trace/ultratraces of REEs in biological samples [424]. The researchers used a iminodiacetic acid modified poly(glycidyl methacrylate-trimethylolpropane trimethacrylate) monolithic, and optimized the extraction efficiency factors like sample pH, flow rate, sample/eluent volume, as well as the behavior of the coexisting ions. The optimal LODs ranged from 0.08 for Er to 0.97 ng/L for Nd with a sampling frequency of 8.5 h^{-1}. The RSDs were between 1.5% for Sm and 7.4% for Nd (c = 20 ng/L, n = 7), and the method was successfully used for monitoring of REEs in human urine and serum samples, with recoveries in the range of 82%–105% for spike samples.

UV–VIS SPECTROPHOTOMETRY

Ultraviolet–visible (UV–vis) spectroscopy or spectrophotometry technique refers to the methods performed in the UV–vis spectral region and hence correspond to the electronic transitions of molecules. The use of the absorption of the UV or visible light is a very conventional laboratory practice, offering merits of time and cost efficiency as well as ease of use, apart from the technical abilities. UV–vis techniques can be used as complementary to fluorescence ones, because of the fact that the latter also deals with electronic transitions, and the only difference might be in that in the case of fluorescence techniques, the transitions studied are those from the excited state to the ground state, which can be generally considered as the reverse route as to that of UV–vis

According to the electronic structure of molecules, the π or even nonbonding electrons can absorb UV or visible wavelengths and move to higher orbitals. Naturally, the energy of the wavelengths is dependent on the energy difference between the ground-state and excited-state orbital that is the highest occupied molecular orbital (HOMO) and the lowest unoccupied molecular orbital (LUMO), but the absorption wavelength is also influenced by external factors like the nature of the solvent and its properties like pH, and temperature, ionic power, and the presence of interfering species, which can alter the energy levels of HOMO and LUMO to some extent. Operational variables like the effective bandwidth of the excitation radiation (that is determined by the width of the slit) can further influence the spectrum and hence should be considered in practical work.

In the spectrometer, the intensity of light in the absence and presence of the sample (symbolized by I0 and *I*) is measured to yield so-called transmittance (*T*), which is the I/I_0 ratio and is expressed as a percentage. Next, the absorbance *A* of the sample is calculated by

$$A = -\log(\%T / 100\%) \tag{6.1}$$

In some applications, another parameter, namely, the reflectance (*R*) of the sample is used, in which mode the instrument measures the intensity of light reflected by the sample (*I*), and compares it to that of a reference material (I_0), and the I/I_0 ratio is expressed as a percentage (%*R*).

In the case of absorbance based on the Beer–Lambert law, which related the absorbance of a solution to the concentration of the absorbing species in it, as well as the path length in which the light travels through the sample, can be used for the evaluation of the concentration of the sample. The proportionality constant of the Beer–Lambert law is the extinction coefficient, which is a measure of how the absorbance changes with concentration and can be found in references or determined using a calibration curve [425]. The equation is then written as follows:

$$A = \log_{10}(I / I_0) = \varepsilon \mathrm{bc} \tag{6.2}$$

where A is the absorbance (expressed in absorbance units (AU)); I and I0 are the intensities of the light in the presence and absence of the sample, respectively; b is the length of the path that light travels through the sample; c is the concentration of the sample; and ε is the so-called extinction or molar absorptivity coefficient, which is a fundamental molecular property in a given solvent, at a particular temperature and pressure for a given wavelength and has units of 1/Mcm or AU/Mcm.

It should be noted that although the Beer–Lambert law is applicable to many compounds, it is not a universal relationship that describes the behavior of all substances and in the case of large and complex molecules like xylenol orange or neutral red, the absorbance-concentration relationship is a second-order polynomial [425]. This might, in part, be due to the fact that the Beer–Lambert law's assumptions are not met as a result of experimental factors like the physicochemical nature and conditions of the sample, which can influence the extinction coefficient.

Another condition that might lead to deviations from the law is the concentration of the sample. Above certain concentrations (which depend on the nature of the sample), the absorption also deviates from the linear behavior, which is due to reaching the maximum absorption, which cannot increase even by the increase in the concentration of the sample leading to flattened curves. Since the concentration and path length both have direct effects on the absorption, varying the path length can help determine whether the deviations are due to the high concentration of the sample. A further cause of flattening is the inhomogeneity of the solutions, which can lead to a phenomenon such as surrounding of the analyte with the suspended particles, leading to deviations from the Beer–Lambert law even at low concentrations.

Instrumentation

A UV–vis spectrophotometer necessarily includes a light source, a sample holder, a monochromator for separating the different wavelengths of light, and a detector. The source of radiation is typically a Tungsten filament or a deuterium or xenon arc lamp, which produces continuums of wavelengths from 300 to 2500, 190 to 400 nm, and 160 to 2000 nm, respectively. More recent visible sources include light-emitting diodes (LEDs).

As for the detector, photomultiplier tubes, single or arrayed photodiodes or charge-coupled devices (CCD) are used. If the instrument has a scanning monochromator which separates the wavelengths sequentially, single photodiodes photomultiplier tubes can serve as the detector. In devices with fixed monochromators, however, CCDs or photodiode arrays are used. These detectors are actually a combination of several single detectors forming a dimensional surface of arrays, which can collect different wavelengths on different pixels or groups of pixels at the same time.

The UV–vis instruments are also classified based on being single or double beam (Fig. 6.4), which naturally changes their instrumental requirements.

In a single-beam device, the radiation produced by the source passes through the sample cell, and to measure the I_0, the sample is removed in another step. The double-beam devices, however, split the radiation produced by the source into two beams, directing one toward the sample

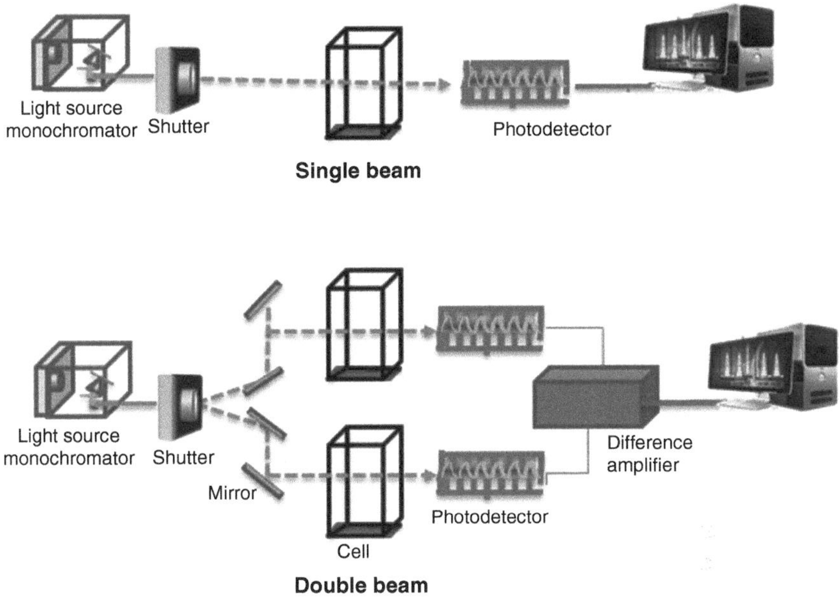

FIGURE 6.4 Two types of UV–vis spectrophotometer.

cell and the other through a void to act as the reference beam (I_0), the intensity of which is considered as 100% transmission or zero absorbance. In addition to the optical elements used for the splitting of the light, the double-beam devices can also have two detectors, which allows for the simultaneous measurement of I and I_0, while in single-detector devices, the two beams are blocked sequentially using a chopper, in a way that only one of the two reaches the detector at a certain time.

Since the UV–vis techniques are conventionally used for liquid samples (although gases and even of solids can be tested under certain conditions), the sample container is a transparent cell (cuvette). These transparent cells are usually rectangular with internal widths of 1 cm to simplify the calculation of path length (b) in the Beer–Lambert law. To be transparent in the UV–vis spectral region, the cells are mostly made of high-quality fused silica or quartz, which are transparent throughout the UV–vis and even in the near-infrared regions. Because of the high cost of quartz cuvettes, glass and plastic cells are also available for analyses in which the UV region is not under study.

Derivative UV–Vis Spectrophotometry

Because the UV–vis spectra show the sum of the absorbance of analyte and matrix, which might have spectral overlaps, one of the disadvantages of this technique its low selectivity, which is caused by the nature of the signal under study. Solutions to this problem include performing measurements versus a blank solution (where a blank solution with a similar composition as the matrix is used), or the elimination of matrix. The former method is applicable to samples with simple and known matrixes. There is no need to mention that no matter how effective these methods might be, in overcoming the selectivity issues, they certainly increase the required time and costs, as well as increasing the chances of errors due to contamination of the sample.

A simpler way to overcome this problem is the mathematical technique of derivativing the spectra, which allows for eliminating the spectral interferences. The technique is based on derivative spectra of basic, zero-order spectra. The resulting curve is called the derivative spectrum, which can be expressed as

$$nD_{X,\lambda} = d^n A / d\lambda^n = f_\lambda \tag{6.3}$$

or

$$nD_{x,v} = d^n A / dv^n = f_{(v)} \tag{6.4}$$

in which A is the absorbance, and $nD_{x,\lambda}$ and $nD_{x,v}$ represent the nth-order derivative of analyte x at the wavelength λ or at wavelength number v.

In derivative spectrophotometry, the Beer–Lambert law is still observed, but it is expressed in its differential form as shown below:

$$nD = \frac{d^n A}{d\lambda^n} = \frac{d^n \varepsilon}{d\lambda^n} \cdot c \cdot b \tag{6.5}$$

in which ε is the extinction coefficient, c is the sample concentration, and b is the path length.

Also, as in the case of normal UV–vis spectroscopy, the derivative spectra of a mixture is the sum of the derivative spectra of each component:

$$nD_{mix} = nD_1 + nD_2 + \ldots + nD_n \tag{6.6}$$

Modern software for control and data acquisition from spectrophotometers also offer the ability to add, subtract, multiply, or derivatize the acquired data, which can help obtain the derivative data required in this approach. A further approach used in spectroscopic studies is finding the best polynomial equation that can illustrate the curve. For further information on the mathematics of derivative, one can consult the literature [426].

UV–Vis Spectrophotometric Determination of Lanthanides

Characteristic absorption bands of rare earth in the UV–Vis region, corresponding to the transitions of the 4f shell electrons, have been used

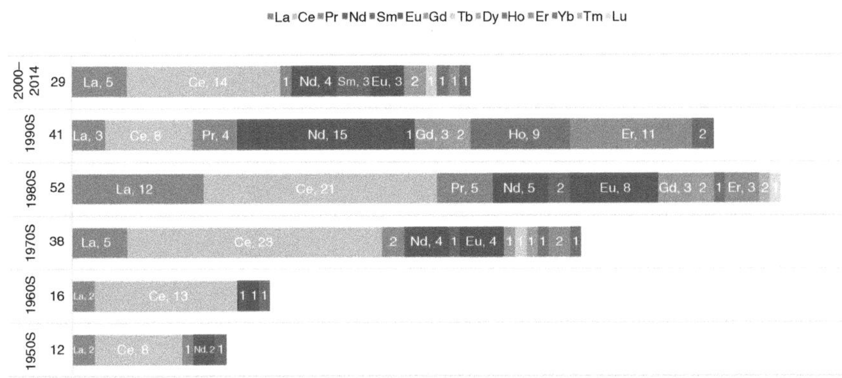

FIGURE 6.5 Number of reports on determination of lanthanides by UV–vis spectrophotometric method.

for the quantitative determination of these elements. A survey in the literature reveals that there are about 188 reports on the spectrophotometric determination of lanthanide species [427-614]. Fig. 6.5 shows the historical trend of reports on the application of the technique during the past decades. As it can be seen, since the 1970s the determination of lanthanide-based spectrophotometric methods grew [455-492], reaching its peak in the 1980s and 1990s [493-585]. After the year 2000, the number of reports gradually reduced, which may be due to the development of new, more sensitive techniques for such analyses. Thus, the role of spectrophotometric methods (colorimetry) in monitoring of lanthanides became less pronounced. Instead, in the recent years, UV–vis optical sensors, which shall be discussed later in this chapter, were devised for measuring the concentration of lanthanides.

Fig. 6.6 shows the number of reports for each lanthanide, and as can be observed, the largest body of research has been focused on the determination of cerium, which, as mentioned before, is among the most abundant lanthanides with a wide variety of applications. Hence, the methods for the determination of this element, which can be customized for the other lanthanides, has been considered of great importance by researchers. The second and third most studied lanthanides are La and Nd.

Generally, the UV–vis spectrophotometric determination of lanthanides, sometimes referred to as colorimetry, falls into two categories. In some reports, the lanthanide intrinsic absorbances (especially in case of cerium) have been directly used for the measurements. And in others, suitable reagents have been applied to form a color complex with one or more lanthanide ions, and then absorbance of the formed complex was considered as a base for determinations. The concentration is then indirectly determined. In the second approach, the reagent should rapidly and stoichiometrically form a stable complex with lanthanides.

FIGURE 6.6 Number of spectrophotometric determination reports on each lanthanide members.

Some of studies have used derivative spectrophotometry (2nd, 3rd, or 4th derivatives) to achieve more selective and sensitive response [532-534,537,539,555,566,567,569,570,572-574,577,578,584-587,596,604]. The majority of the available colorimetric determination techniques used for cerium depend on the oxidizing properties or color of ceric solutions.

In 1951, the yellow color of the ceric ion in acidic media was used as the basis for the colorimetric procedure, where cerous ions were oxidized by persulfate in a hot acidic solution in the presence of catalytic amounts of silver ion [428]. The yellow color of the resulting ceric solution was sufficiently intense for the determination of milligram amounts of cerium. Ce^{4+} was found to show an absorption maximum at approximately 320 nm, with a molar extinction coefficient of 5.58×10^8, in 1 N sulfuric acid. The determination of Ce^{3+} was also performed through the measurement of its ultraviolet absorption peak at 253.6 nm in 1 N sulfuric acid [429].

In a report in 1955 [431], Gordon and Feibush described an indirect method for the determination of cerium, where ceric ion was formed through the oxidation of cerous ions with lead dioxide, was treated with excess amounts of a ferrous sulphate solution, and the excess ferrous ions were spectrophotometrically determined with *o*-phenanthroline. Testing the method using microgram amounts of cerium led to rather inconsistent results. This was attributed to the difficulty of complete removal of traces of suspending lead dioxide, and the oxidation of ferrous ion before addition of the complexing agent.

Cerium also could be estimated by treating the sample with excess amounts of the solution of a complex. The decrease in the optical density of the solution, caused by the oxidation of the ferrous complex to the faintly colored ferric-phenanthroline complex, would be an indirect measure of the presence and concentration of cerium. In this method, the ferrous ions added were stabilized by phenanthroline and would therefore be less likely to be oxidized by oxygen, which is abundant in the solution. The ferrous solution was separately added as described in the literature [431]. The ferrous-phenanthroline complex is completely stable over the pH range of 2–9. At pH values below 2, the complex is partly decomposed and its color is much weaker. When the ferrous-phenanthroline reagent is added to a blank solution treated with persulfate, the low pH of the resultant solution (<1) causes rapid fading of the color. Because cerous hydroxide precipitates at pH 2.5, the applicable pH range for the determination of cerium is between 2.0 and 2.4. It was found that if the solution was buffered to pH 2.3, the color was stable for about 1 h. The sensitivity of the method was claimed to be doubled by substituting phenanthroline with sulfonated bathophenanthroline in the reagent.

More sensitive procedures have been reported based on using the yellow color formed, when cerous salt is treated with potassium carbonate in the presence of air or hydrogen peroxide [433,441].

In another method, the ceric ion formed through oxidation with persulfate, in the presence of catalytic amounts of silver, was reduced to the cerous ions by treatment with excess of a ferrous-phenanthroline reagent. The decrease in the optical density of the ferrous-phenanthroline compared with the reagent blank was used as a measure of the amount of cerium present. The only interfering ions were reported to be Mn^{2+} and Cr^{3+} [442].

Generally, in acid solution, ceric ion can oxidize a number of organic compounds, such as sulfanilic acid [451], brucine, benzidine, and the

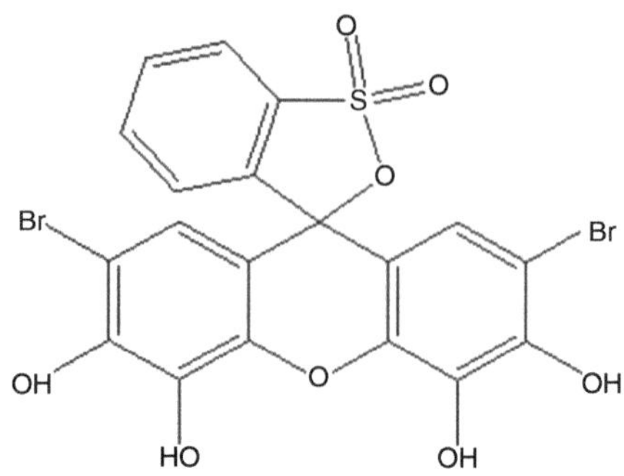

FIGURE 6.7 Bromopyrogallol red chemical structure.

leucobase of malachite green, to colored compounds that can be used for the photometric determination of the element.

Several other methods also used oxidation reactions to determine lanthanides, but interferences from the matrix and other cationic species and the instability of the solution during the measurements. Therefore, gradually, chromogenic reagents as complexing agents were used in the determination of rare earth elements.

One conventional chromogenic reagent is Bromopyrogallol red (Fig. 6.7) [439], which forms soluble blue to violet-blue complexes with many metal ions and rare earths (i.e., yttrium, cerous, and cerium) in pH >4.5. This reagent is also an acid–base indicator, with orange yellow in strong acidic media, claret red in nearly neutral solutions, and blue in alkaline solution.

Another complexing agent is 2-methyl-8-quinolinol (Fig. 6.8) [444], which forms a complex with cerium at pH 10 using maleic acid as an auxiliary chelating agent. The colored Ce^{4+} complex is completely soluble and remarkably stable in CCl_4. The carbon tetrachloride extract containing this complex has a maximum absorbance at 485 nm.

Another method, for the spectrophotometric determination of rare earth elements, is based on xylenol orange (Fig. 6.9), which is a metallochromic indicator. The complexes of Ce^{3+}, La^{3+}, Nd^{3+}, and Yb^{3+} with this

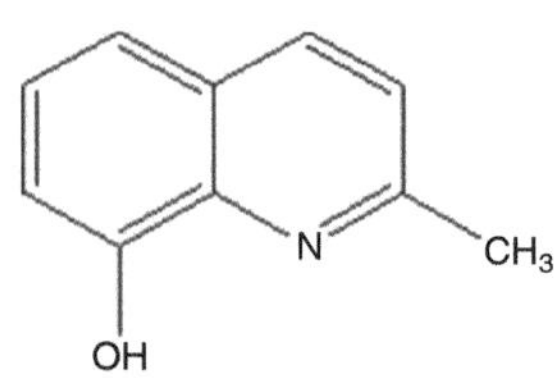

FIGURE 6.8 2-Methyl-8-quinolinol chemical structure.

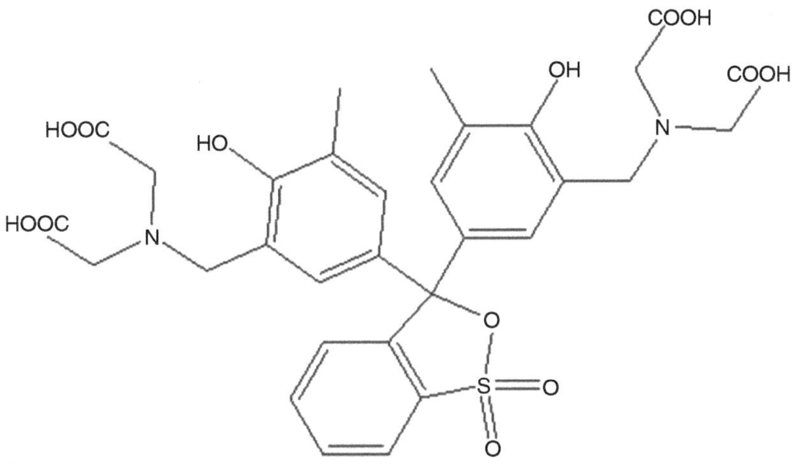

FIGURE 6.9 Xylenol orange chemical structure.

reagent show maximum absorptions from 575 to 578 nm. All complexes are 1:1 and have formation constants around 105 under the conditions of the method [445].

Later, a mixture of xylenol orange and cetylpyridinium bromide was proposed for the spectrophotometric determination of traces of La^{3+} [452], based on the formation of a dark blue $La[(XO)(CP)_2]_2$ complex with a molar absorption of 92 000 at 625 nm.

Another work described the use of 3-thianaphthenoyltrifluoroacetone (Fig. 6.10) for the analysis of trace amounts of cerium(IV) in slightly acidic media, through the formation of an orange-red species which can be extracted into benzene [455].

The analogous compound, 2-thenoyltrifluoroacetone, was also used in the spectrophotometric determination of cerium [440], but 3-thianaphthenoyltrifluoroacetone was reported to have the distinct advantages of higher sensitivity, higher selectivity due to the different absorption wavelength (424 nm), and lack of interference by tartrate.

o-(*p*-Tolylsulfonamido)aniline [456], methyl xylenol blue [457], and chromotrope 2R (Fig. 6.11) [458] were also used as chelating agents, in the determination of cerium as well as Pr, Nd, Sm, Eu [463], and Yb ions [467].

During the 1970s, several other chromogenic reagents (Table 6.13), were suggested for the determination of lanthanides.

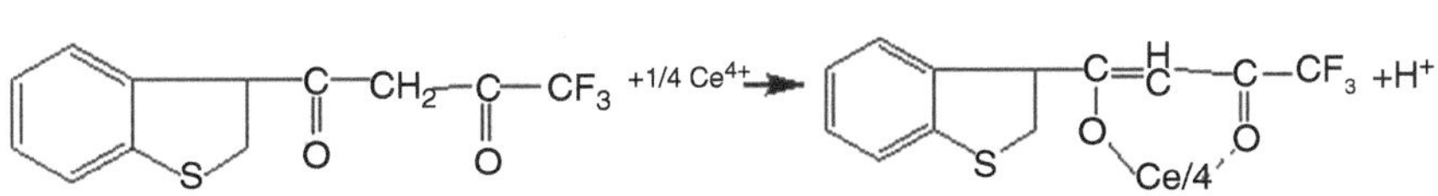

FIGURE 6.10 3-Thianaphthenoyltrifluoroacetone chemical structure.

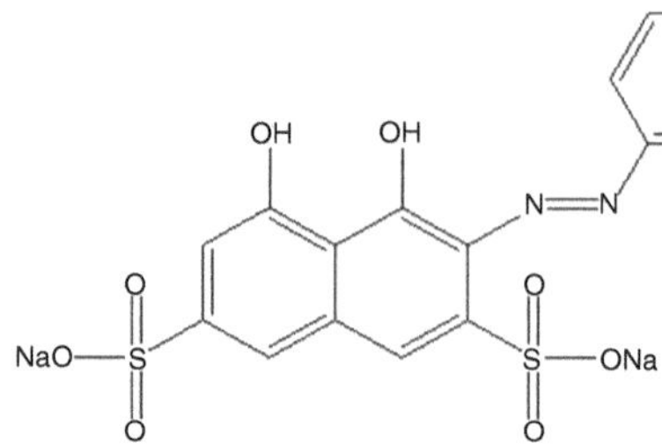

FIGURE 6.11 Chemical structure of chromotrope 2R.

UV–vis spectrophotometry has been used for the analysis of lanthanides in various industrial, environmental, geological, and biological samples or for detecting the impurities in minerals (Table 6.14).

Some reports cover the application of UV–vis spectrophotometric techniques after the separation or preconcentration of the samples [562, 579]. One simple method developed to determine trace amounts of Nd^{3+}, Th^{3+}, and U^{4+} in brine [562] involves the separation of the cations from brine through extraction using chlorophosphonazo-III as the extractant in 1-butanol as the solvent. Next, the spectrophotometric determination was performed and LODs of 5×10^{-7} M for Nd^{3+} and 2×10^{-7} M for Th^{3+} and U^{4+}.

Extraction using cloud point phase separation of nonionic surfactants was also reported for the preconcentration of parts-per-billion amounts of gadolinium in urine, before its determination through the UV–vis absorption procedure [579]. Another extraction method has been based on the formation of Gd^{3+} complexes with 2-(3,5-dichloro-2-pyridylazo)-5-dimethylaminophenol in a phase rich with the nonionic polyethyleneglycolmono-*p*-nonylphenylether (PONPE 7.5). The variables influencing the method were evaluated and optimized, and the optimized procedure was applied to the determination of total and free Gd^{3+}contents in real urine samples.

Separation and preconcentration of trace amounts of Ce^{3+} on octadecyl silica membrane discs, which were modified with 1,3,5-trithiacyclohexane (TTCH) prior to spectrometric determination of the analyte has also been reported [592]. The highest capacity of the modified membrane discs containing 15 mg of TTCH was 547 mg of cerium, and the LOD was reported to be 2 ng/mL.

Reports on the application of UV–vis detection in normal or automated flow injection systems for the analysis of lanthanides [582,614,503,509,540] also exist. A fully automated flow-injection analyzer (FIA) was used for simultaneous UV–vis determination of lanthanum, and the total amount of two ions using 2-(2-arsenophenylazo)-7-(2,4,6,-trichlorophenylazo)-1,8-dihydroxynaphthalene-3,6-disulfonic acid and 2-(2-phosphono-4-chlorophenylazo)-7-(4-fluorophenylaxe)-1,8-dihydroxy-3,6-disulfonaphthalene, respectively [540].

TABLE 6.13 The Chemical Structures of Some Chromogenic Reagents Used for Spectrophotometric Determination of Lanthanide Ions

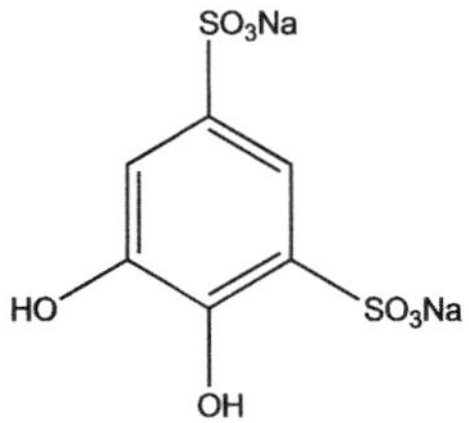 **Tiron (4,5-dihydroxy-1,3-benzenedisulfonic acid disodium salt)** **Nd, Ho, and Er [459]; Eu [464]; Ho [548]**	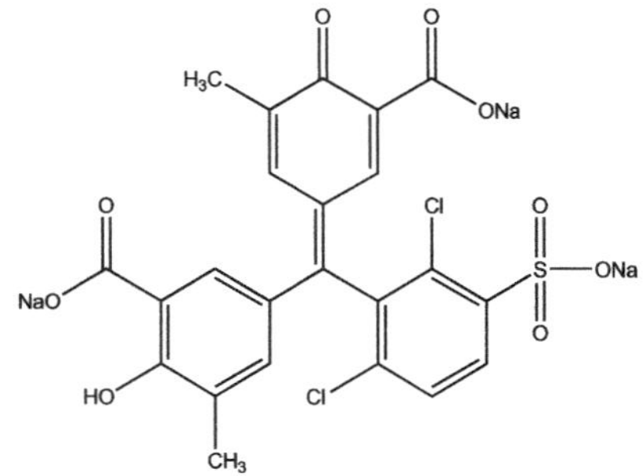 Chrome azurol-s Eu [460]	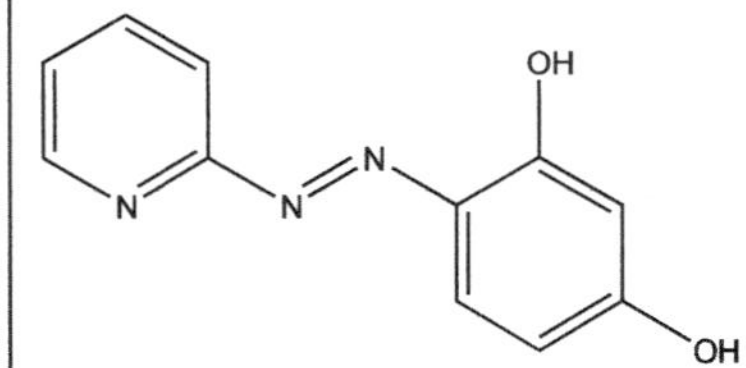 4-(2-Pyridylazo)resorcinol Gd [465]; Dy [553]
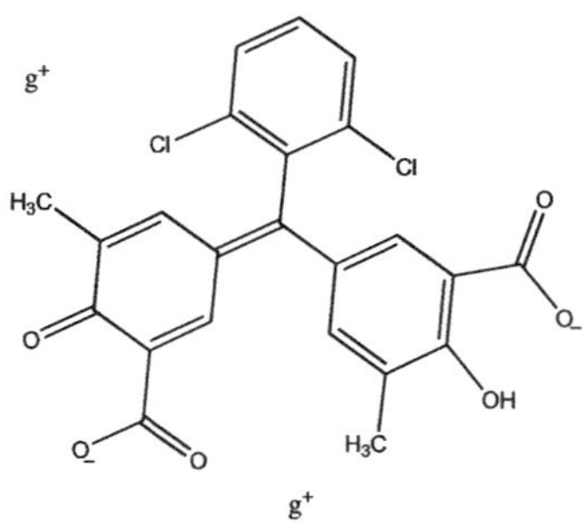 **Solochrome azurine bs** **Ce(IV) [468]**	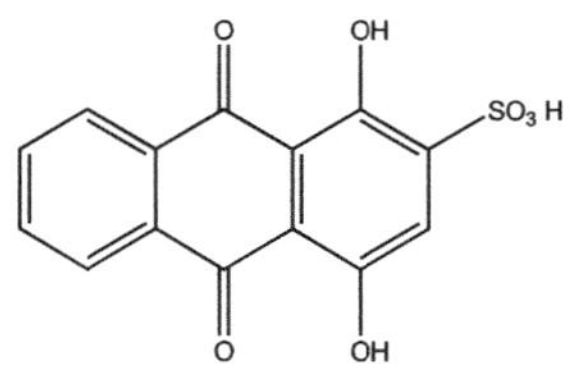 Quinizarin sulfonic-acid Ce(III) [469]	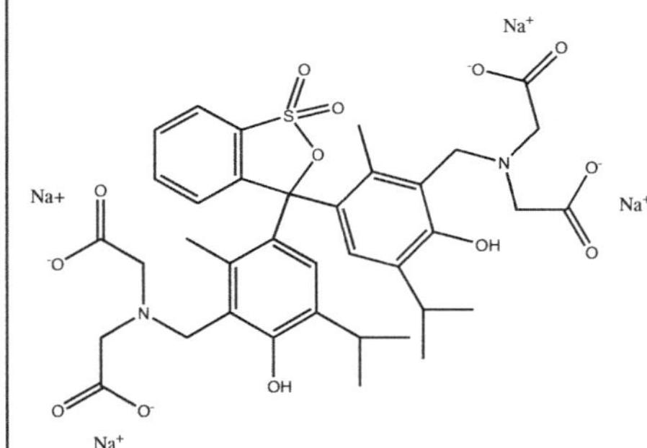 Methylthymol blue Er, La [470]; Ce [506]; Nd [566]

(Continued)

TABLE 6.13 The Chemical Structures of Some Chromogenic Reagents Used for Spectrophotometric Determination of Lanthanide Ions (*cont.*)

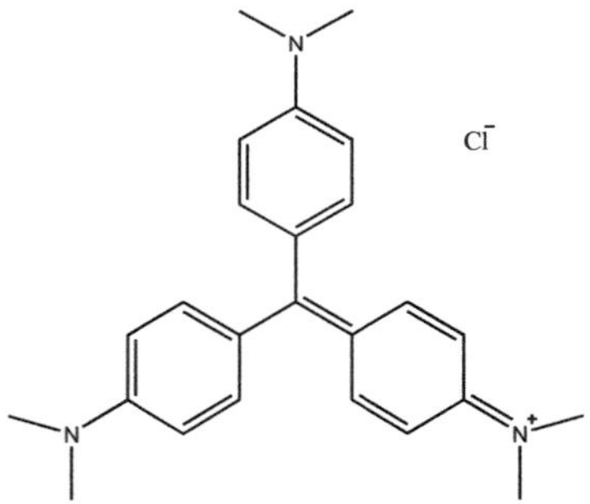	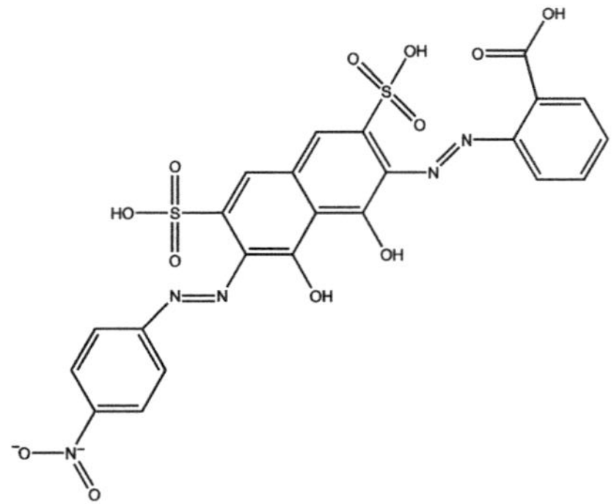	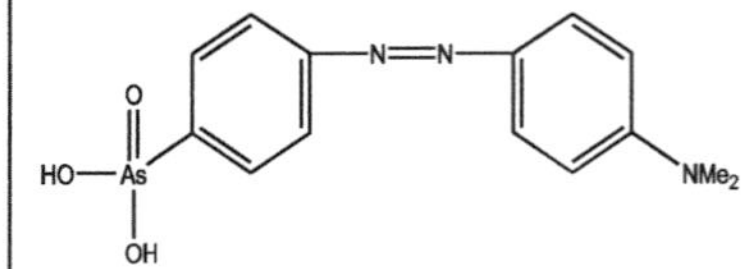
Crystal violet	Carboxynitrazo	4-Dimethylamino-azobenzene-4′-arsonic acid
Ce(IV) [472]	La, Ce [474]	Ce(IV) [475]
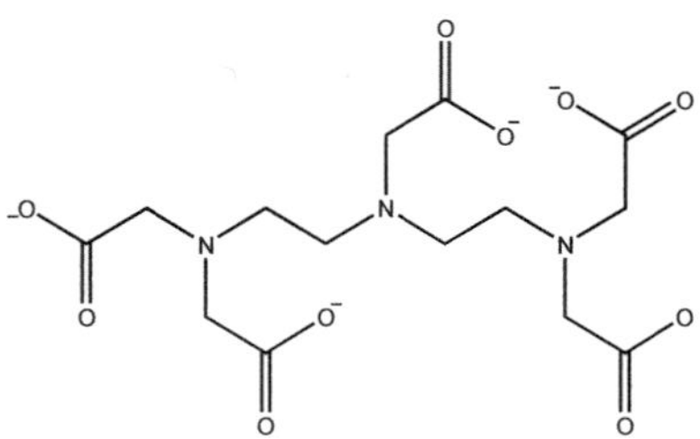	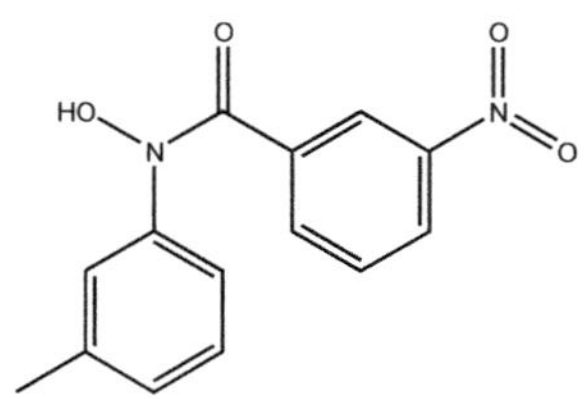	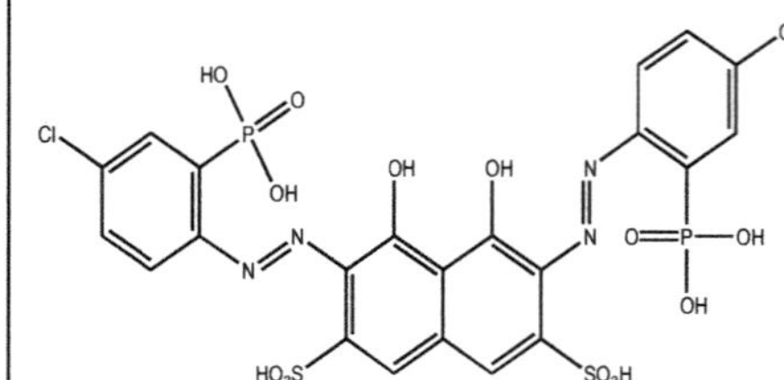
Diethylenetriaminepentaacetate	*N*-Meta-tolyl-meta-nitrobenzohydroxamic acid	Chlorphosphonazo
Pr [476]	Ce [478]	La [480]

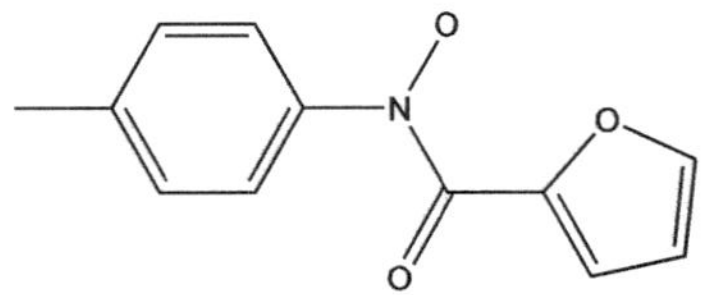 **N-(p-Tolyl)-2-furohydroxamic acid** **Ce(IV) [481]**	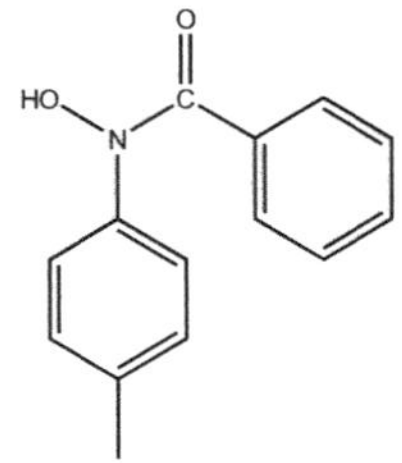 *N-para*-Tolylbenzohydroxamic acid Ce(IV) [482]	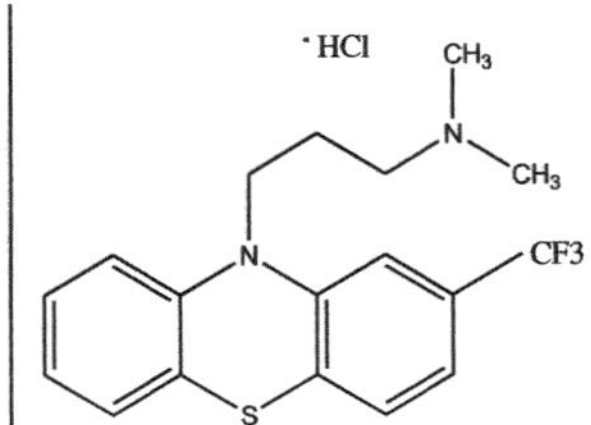 Triflupromazine hydrochloride Ce(IV) [483]
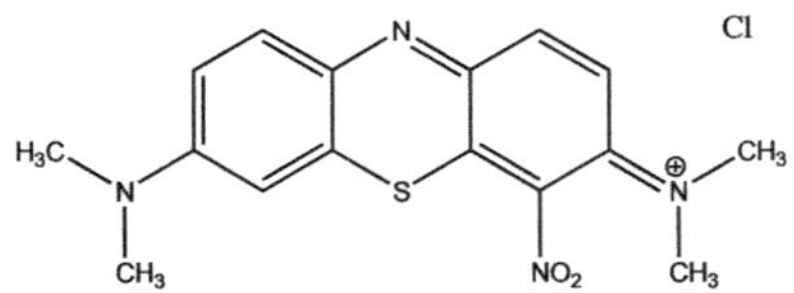 **Methyl green** **Ce(IV) [484]**	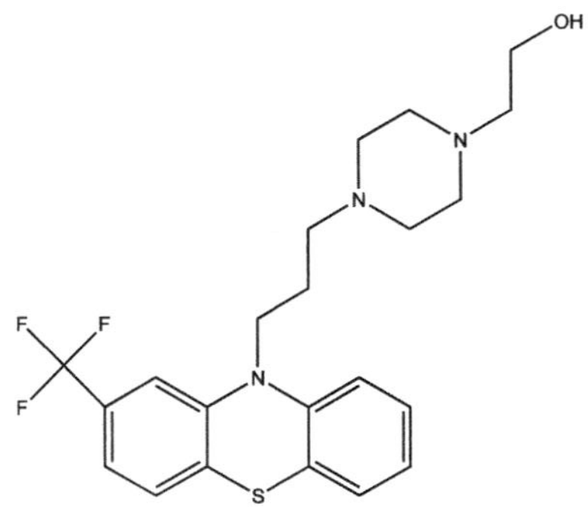 Fluphenazine hydrochloride Ce(IV), Pd(II), As(III), Nitrite [485]	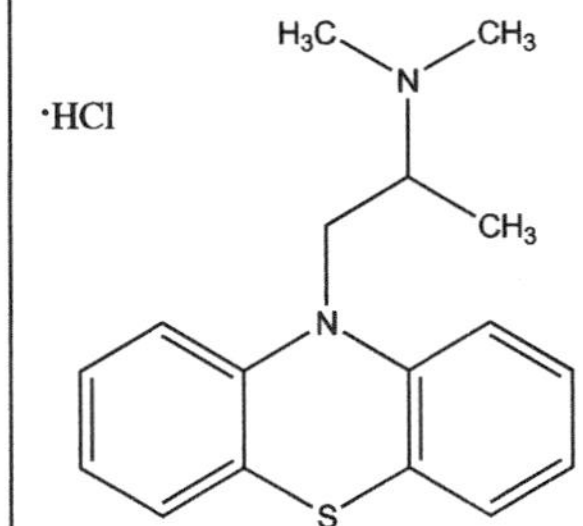 Promethazine hydrochloride Ce(IV), As(III), Nitrite [486]

(Continued)

TABLE 6.13 The Chemical Structures of Some Chromogenic Reagents Used for Spectrophotometric Determination of Lanthanide Ions (*cont.*)

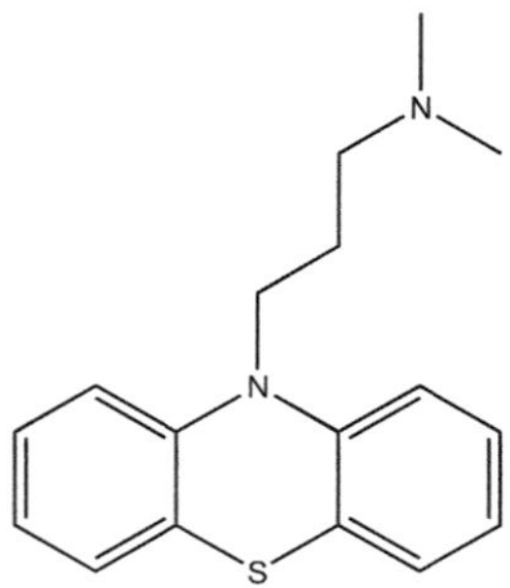	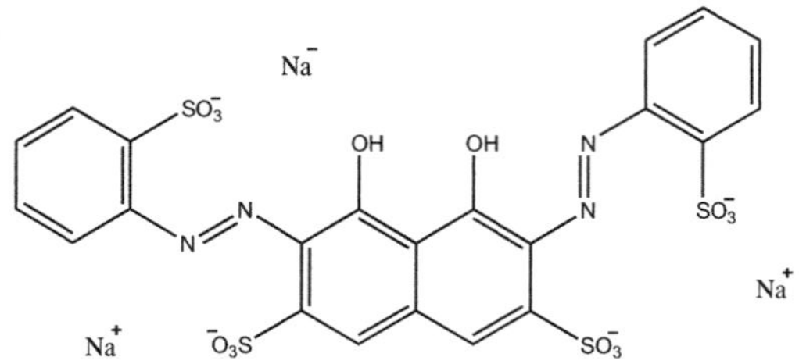	
Promazine hydrochloride	Sulfonazo-II	
Ce(IV), As(III), Nitrite [487]	La, Ce [490]	
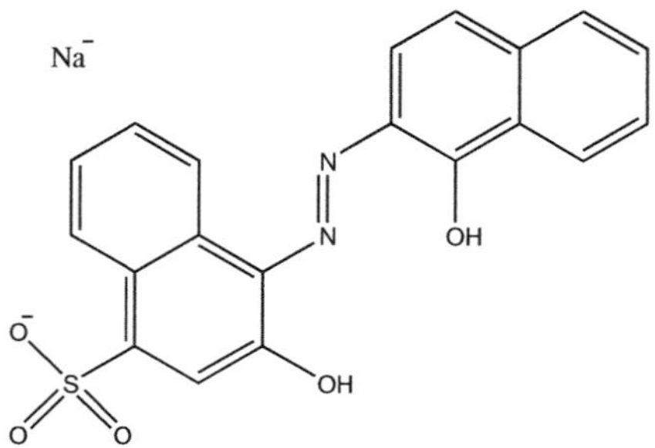	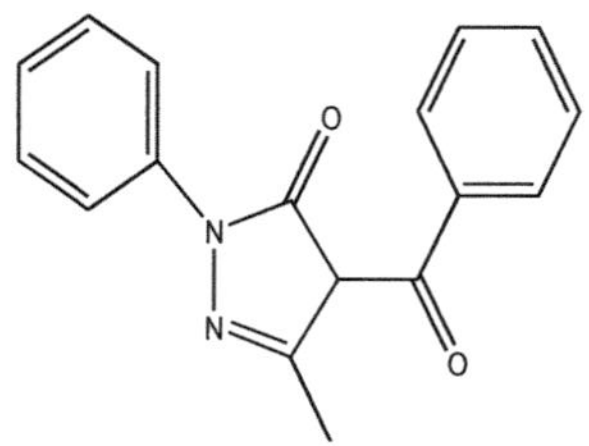	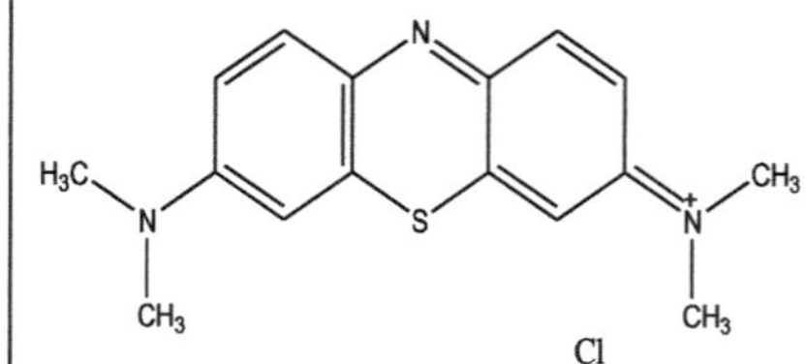
Solochrome black 6B	4-Benzoyl-2,4-dihydro-5-methyl-2-phenyl-3*H*-pyrazol-3-one	Methylene blue
Ce(IV) [492]	Ce(IV) [493]	La [494]

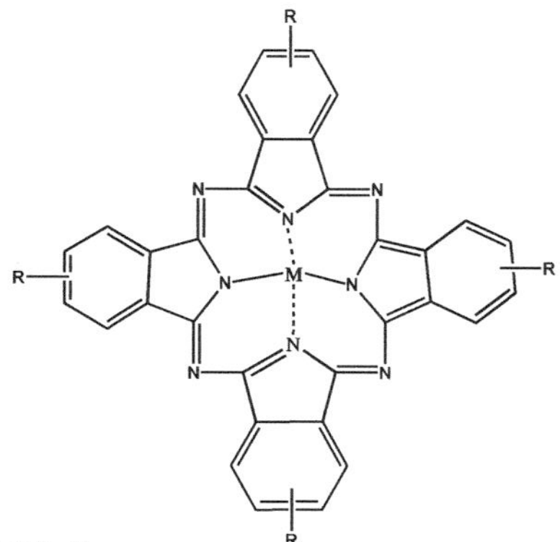 R=SO3 or H **Tetrasulfonated metal phthalocyanine** **Ce(IV), As(III) [495]**	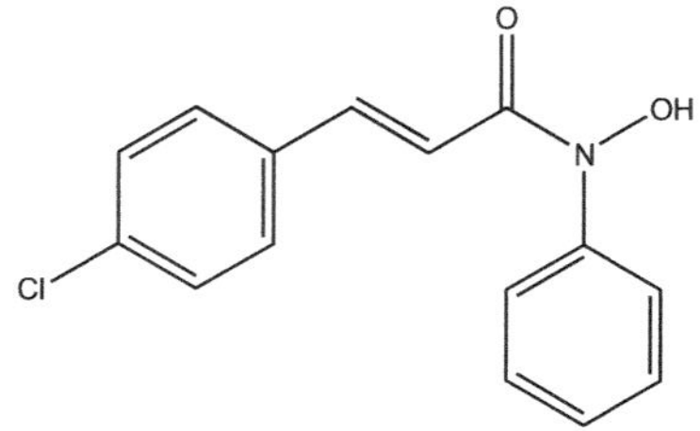 *N*-Phenyl-*para*-chlorocinnamohydroxamic Ce [496]	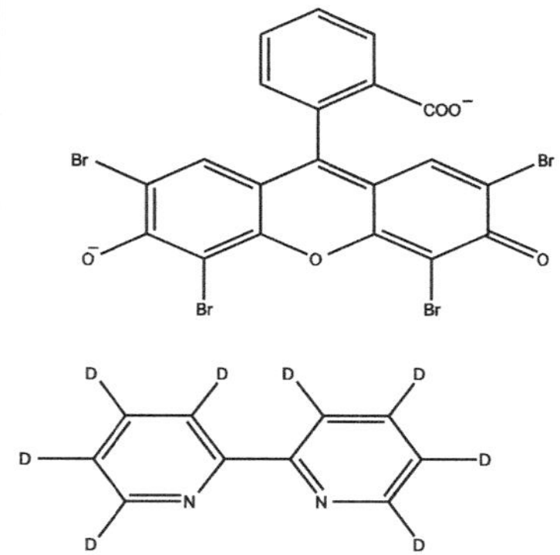 Eosin and 2,2′-dipyridyl Nd [497]
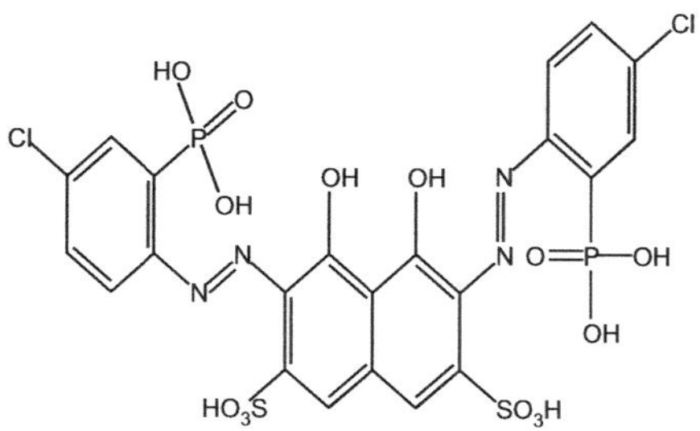 **Chlorophosphonazo-meta** **Ce [500]**	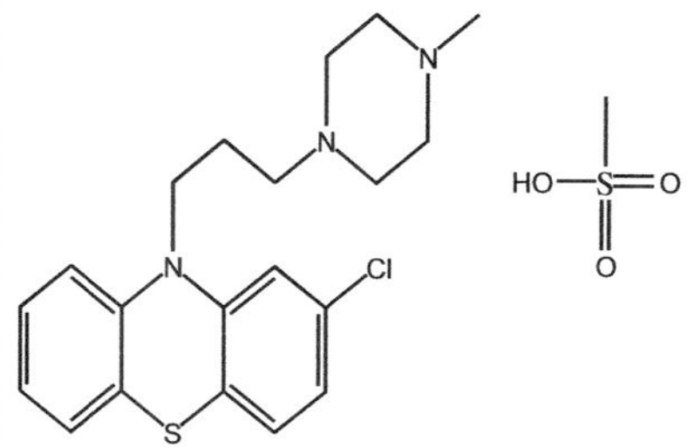 Prochlorperazine bismethanesulfonate Ce(IV), As(III), Nitrite [507]	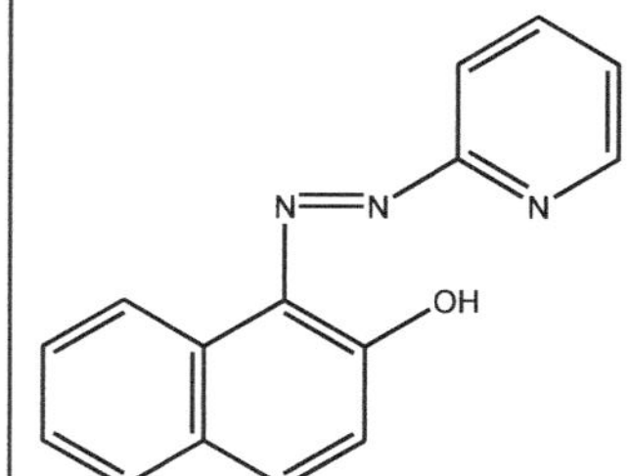 1-(2-Pyridylazo)-2-naphthol La [511]

(*Continued*)

TABLE 6.13 The Chemical Structures of Some Chromogenic Reagents Used for Spectrophotometric Determination of Lanthanide Ions (*cont.*)

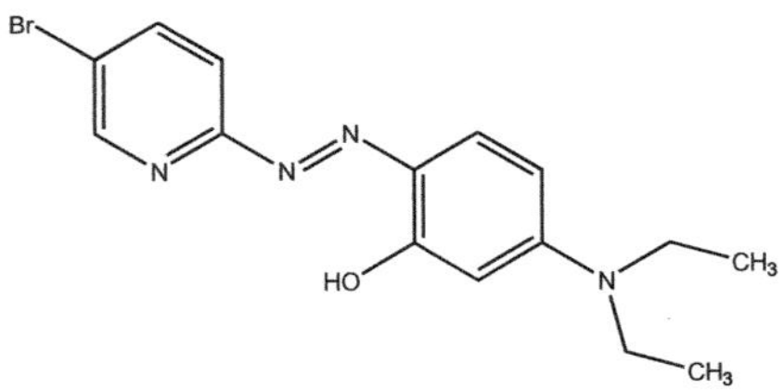	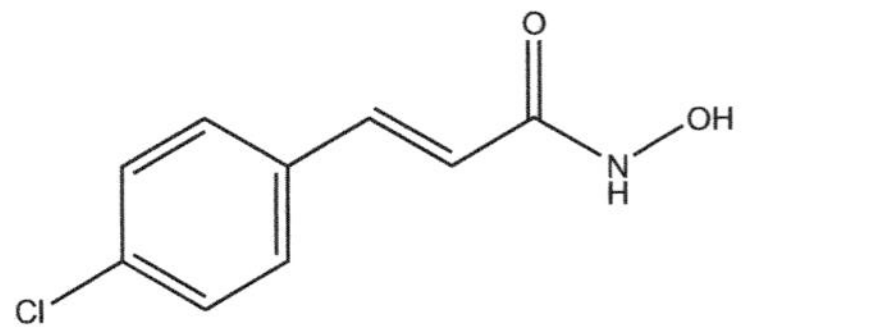	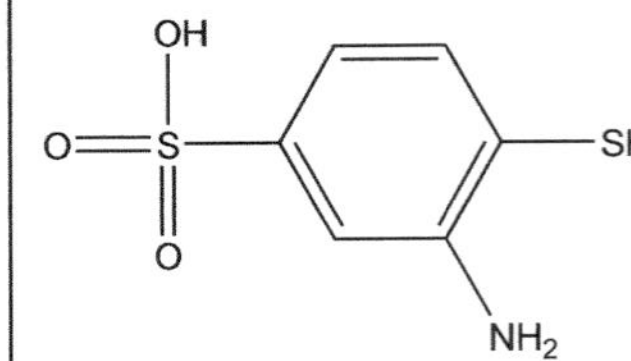
2-(5-Bromo-2-pyridylazo)-5-diethylaminophenol	*N-para*-Chlorophenylcinnamohydroxamic acid	4-Sulfo-2-aminobenzenethiol
La, Ce, Y [512]; Yb [551, 580]; Gd [559]; Nd [571]; Er [581]; Ho [583]	Ce(IV) [514]	Ce(IV) [521]
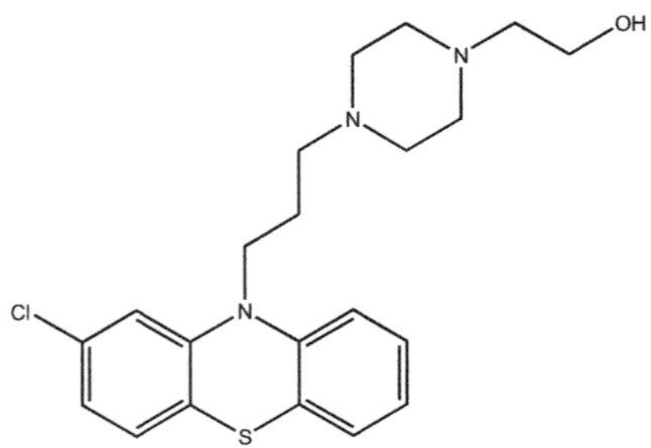	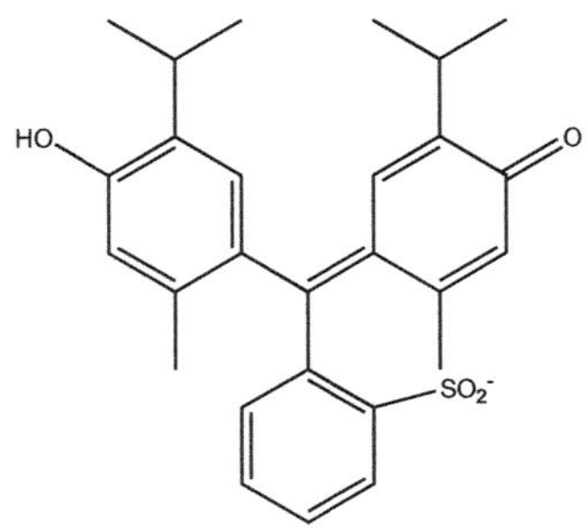	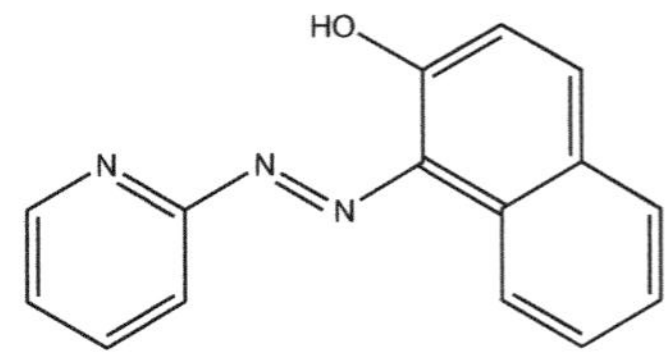
Perphenazine	Thymol blue	1-(2-Pyridylazo)-2-naphthol
Ce(IV), As(III), Nitrite [522]	Sm [534]	Er, Pr [535]; Nd [542]; Pr [547]; Nd, Sm, Gd, Tb, Dy, Ho [613]

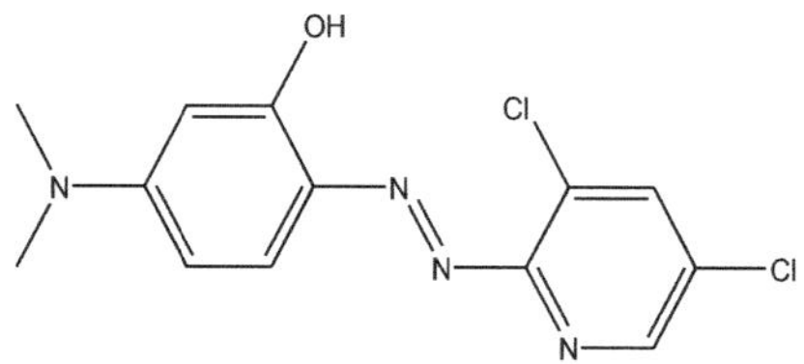	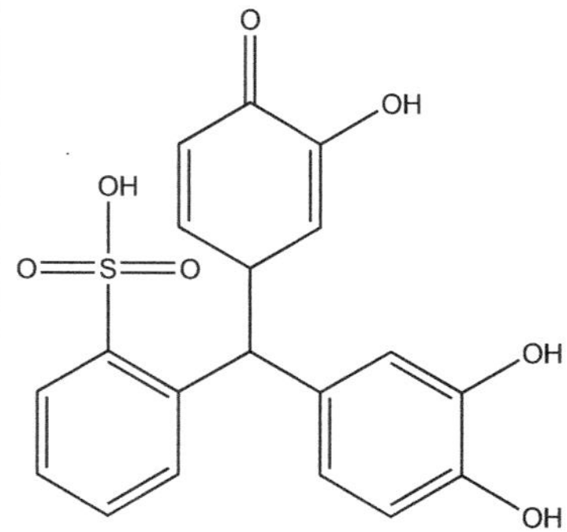	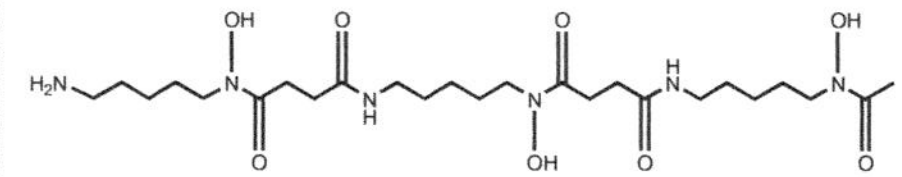
2-(3,5-Dichloro-2-pyridylazo)-5-dimethylaminophenol	Pyrocatechol violet	Desferrioxamine-B
La [545]; Er [576]; Gd [579]	Nd [546]	Ce(IV) and Fe(III) [552]
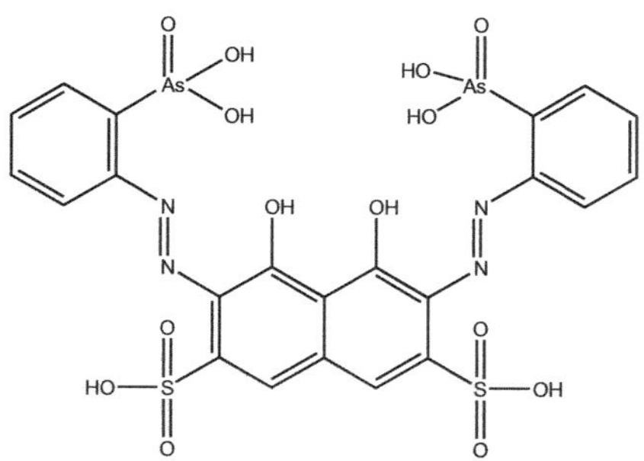	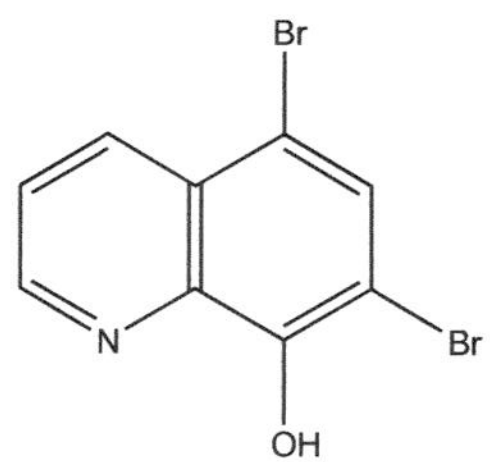	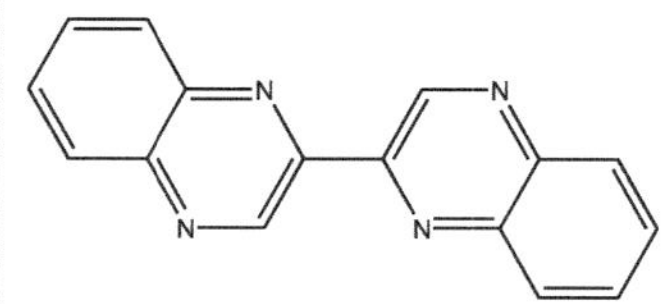
Arsenazo-III	5,7-Dibromo-8-hydroxyquinoline	2,2′-Biquinoxalyl
La, Ce, Nd, Pr [558]	Nd, Er [560]	Eu [561]

(Continued)

TABLE 6.13 The Chemical Structures of Some Chromogenic Reagents Used for Spectrophotometric Determination of Lanthanide Ions (*cont.*)

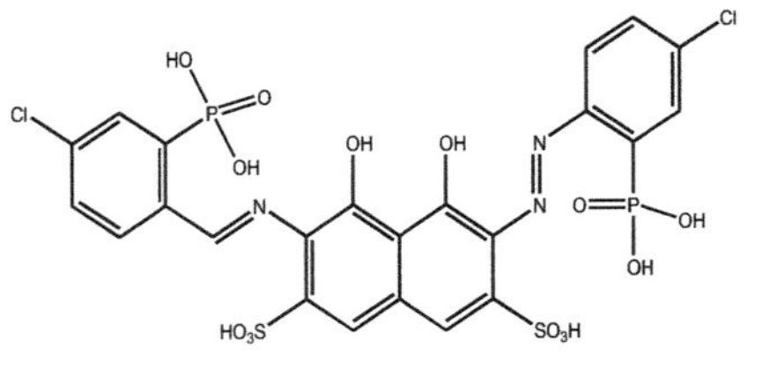 **Chlorophosphonazo-III** **Nd, Th(IV), U(VI) [562]**	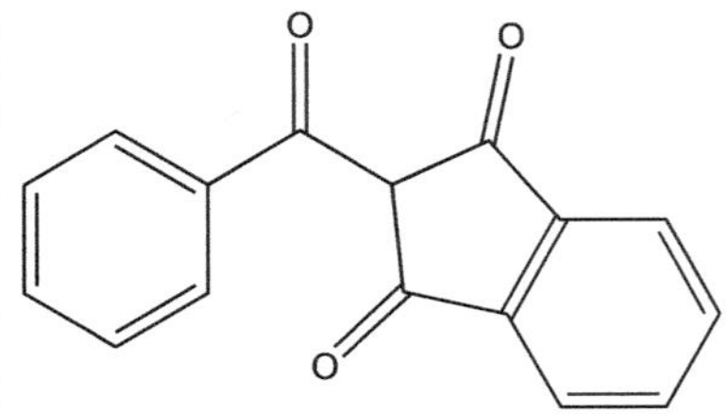 Benzoyl-indan-1,3-dione Ho [573]; Nd, Er [578]	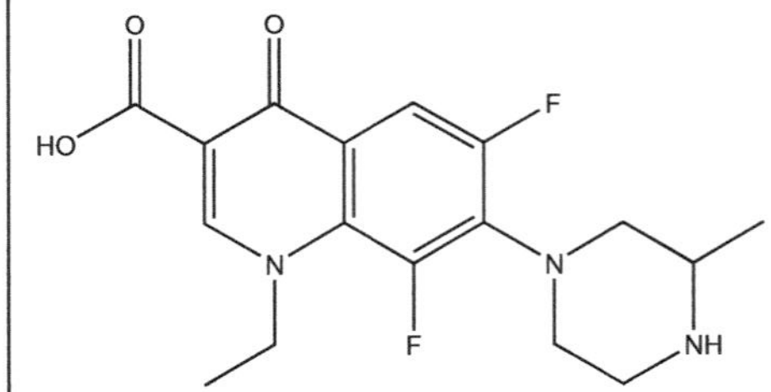 1-Ethyl-6,8-difluoro-7-(3-methyl-1-piperazinyl)-4-*oxo*-1,4-dihydro-3-quinoline carboxylic acid Nd, Ho, Er [585]
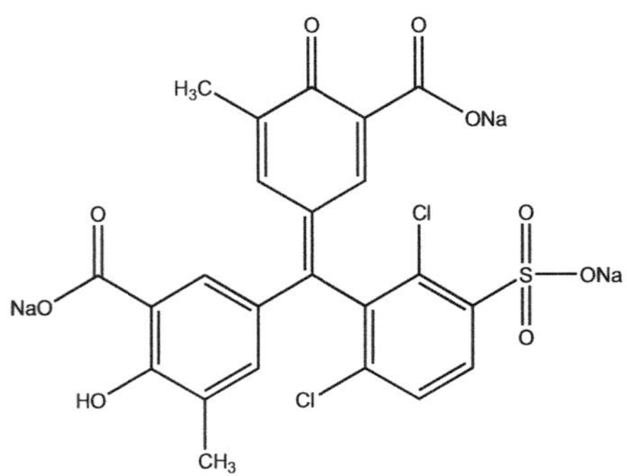 **Chrome azurol S** **Sm [586]**	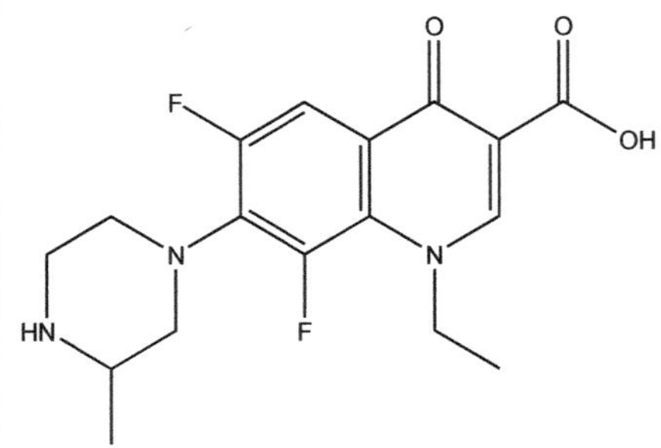 Lomefloxacin Pr [587]	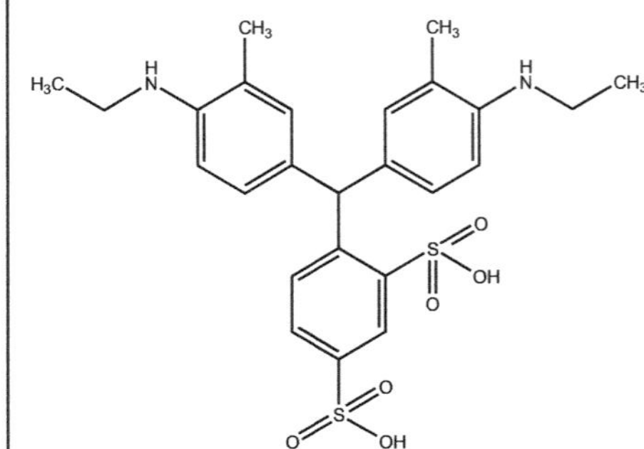 Leuco xylene cyanol FF Ce [591]

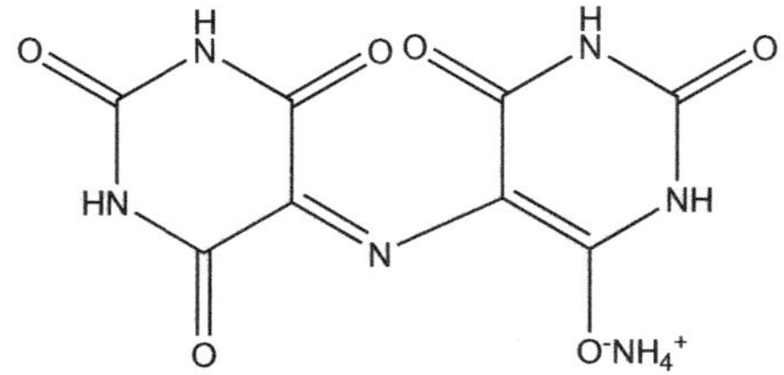

Murexide

Er [593]

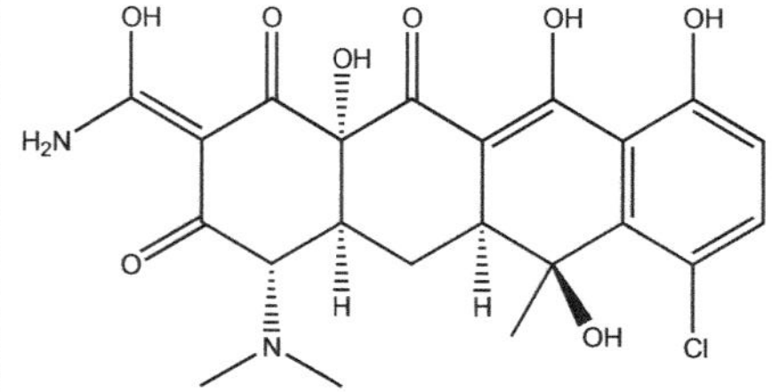

Chlortetracycline

Eu [594]

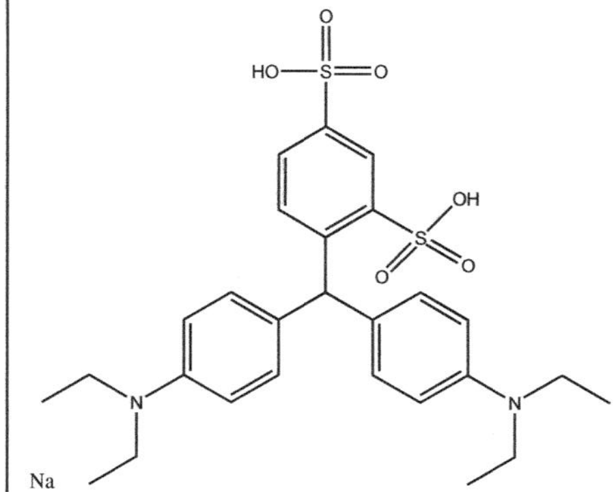

Leuco disulfine blue

Ce [595]

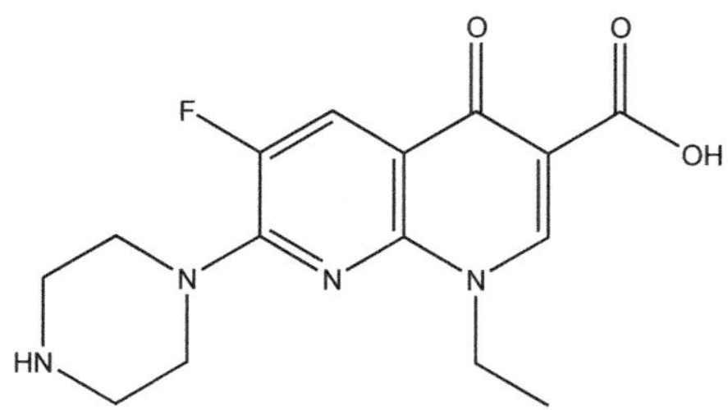

Enoxacin

Nd [596]

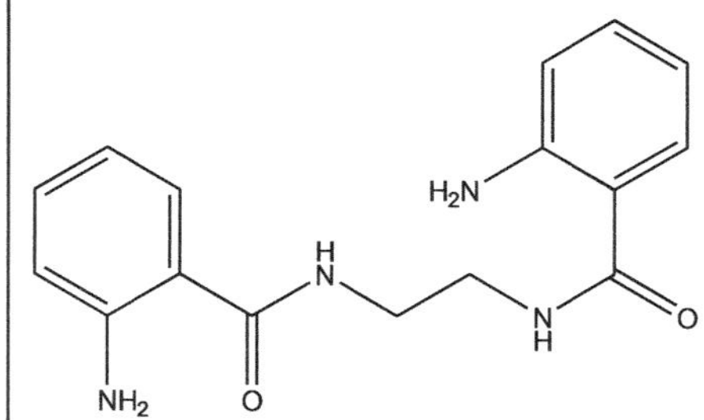

N,N′-Bis(2-aminobenzoyl)ethylenediamine

Nd [611]

TABLE 6.14 Some of the Chelating Reagents Reported for the Spectrophotometric Determination of Lanthanides in Various Samples

Year	Reagent	Ions	Sample	Ref.
1983	Molybdophosphoric acid	Eu	Lanthanide oxides and the minerals bastnasite and monazite	[509]
1989	1-(2-Pyridylazo)-2-naphthol (PAN)	Nd	Glasses	[542]
1993	5,7-Dichloroquinolin-8-ol and rhodamine 6g	Pr	Praseodymium in aluminium metal and aluminium salts	[556]
1994	Chlorophosphonazo III	Nd, Th(IV), U(VI)	Brine	[562]
1994	*p*-Acetylchlorophosphonazo (CPApA)	Ce(IV)	Nickel-base alloys	[563]
1996	2-(5-Bromo-2-pyridylazo)-5-diethylaminophenol	Nd	Rock	[571]
1996	2-Isobutylformyl-1,3-dione-indan and octylphenol poly-(ethyleneglycol)ether	Ho	Rare earth mixtures	[572]
1997	2-(Diphenylacetyl)indan-1,3-dione and octylphenyl poly(ethyleneglycol) ether	Ho	Rare earth mixture	[577]
1998	2-(3,5-Dichloro-2-pyridylazo)-5-dimethylaminophenol	Gd	Monitoring the elimination of gadolinium-based pharmaceuticals in urine	[579]
1998	2-(5-Bromo-2-pyridylazo)5-diethylaminophenol	Yb	Geological samples and bimetallic catalysts	[580]
1999	2-(5-Bromo-2-pyridylazo)-5-diethylaminophenol	Er	Geological materials and glasses	[581]
1999	2-(5-Bromo-2-pyridylazo)-5-diethylaminophenol	Ho	Geological and lasers samples	[583]
1999	1-Ethyl-6,8-difluoro-7-(3-methyl-1-piperazinyl)-4-*oxo*-1,4-dihydro-3-quinoline carboxylic acid	Nd, Ho, Er	Derivative spectrophotometric determination in rare earth mixtures	[585]
2000	lomefloxacin	Pr	Derivative spectrophotometric determination in rare earth mixtures	[587]
2001	Rhodamine G in sulfuric acid medium	Ce(IV)	In rare earth oxides and sea water	[588]

TABLE 6.14 Some of the Chelating Reagents Reported for the Spectrophotometric Determination of Lanthanides in Various Samples (*cont.*)

Year	Reagent	Ions	Sample	Ref.
2002	2,6-Dibromo-4-chlorocarboxyarsenazo	Ce(IV)	Biological samples 2,6-dibromo-4-chlorocarboxyarsenazo. In 1.0 M H_3PO_4, Ce(III) reacts with 2,6-dibromo-4-chlorocarboxyarsenazo to form a 1 : 2 blue complex with an absorption maximum at 630 nm.	[589]
2003	Enoxacin	Nd	Derivative spectrophotometric determination	[596]
2005	Xylenol orange	La	Human blood serum zero and first derivative spectrophotometry in micellar media	[599]
2009	Malachite green iodide	Ce	Geological and environmental samples	[606]
2009	Naphthol green B	Ce(IV)	Hair sample	[607]
2010	Tribromoarsenazo (TB-ASA)	Ce	Human blood and chicken liver samples	[608]
2011	*N,N'*-Bis(2-aminobenzoyl) ethylenediamine	Nd	Natural samples and industrial effluents	[611]
2014	Arsenazo-III	Eu	Environmental water samples	[614]

Another simple and accurate FIA coupled with UV–vis spectrophotometry was used for the preconcentration and determination of Eu^{3+} in aqueous samples [614]. The method included the preconcentration of the analyte in a column packed with Ambulate XAD-4 resin impregnated with nalidixic acid at pH 7.0. Next the analyte was desorbed from the column using 0.1 mol/L HCl and mixed with arsenazo-III solution and studied with a UV–vis spectrometer at 661 nm. The system showed good precision and accuracy, with an RSD of 1.5%.

The combination of liquid–liquid extraction with an automatic method for determination of europium in the presence of other lanthanides, yttrium, and scandium has been reported [509]. Eu^{3+} is selectively reduced to Eu^{2+}, and the divalent species was extracted into isoamyl alcohol for spectrophotometric determination. A rather different technique was based on the catalytic spectrophotometric determination of cerium through ion-exchange separation coupled to a flow injection [582].

Some have also used color fading techniques [598,600,609] for the same purpose. Basically, the presence of the lanthanides (Eu^{3+}) inhibits the fading reaction [609]. This reaction helps determine traces of Eu^{3+}. The maximum absorption was observed at 600 nm, and the linear range was 0–0.18 μg/mL, with a detection limit of 19 ng/mL.

There is another report on the so-called flotation spectrophotometry [556], which has been claimed to provide a sensitive and simple approach for the analysis of traces of elements. The method involves shaking an aqueous pseudo-solution that contains a sparingly soluble hydrophobic compound with an appropriate less dense solvent. It was suggested that the sparingly soluble ion associate contained a multivalent anionic complex of the target metal with more than one hydrophobic monovalent cation of the dye. It was proposed that if the floatable compound is not a simple ion associate, an adduct with salt molecules of the dye was used. The precipitate was separated by flotation and washed to remove the contaminants before being dissolved in a polar organic solvent for the UV–vis measurements.

In a report by Bhagavathy et al. in 1993 [556], Pr^{3+} was determined through its reaction with 5,7-dichloroquinolin-8-01 and rhodamine 6G in a weakly alkaline medium, which was used for separating the analyte through flotation in hexane. The molar absorptivity at 530 nm, was 1.83×10^5 L/(mol cm) and the response was found to be linear in the range of 8–240 pg/L. The mixture of praseodymium and aluminum metal or salts was studied based on the method above using a Dowex 5OW-X8 cation-exchange resin.

Optical UV–Vis Sensors for Lanthanide Series

Optical sensors form another group of chemical sensors, and the signal used for detection of the analyte is the changes in the absorption or fluorescence of the sample. These so-called optical electrodes or optodes are hence devices used for measuring specific substances through the use of optical signals usually with the aid of a chemical transducer. The devices are generally composed of some major parts, including a substance with the tendency to respond to the analyte, a polymeric matrix for immobilization of the chemical transducer, an optical fiber, a light source, a detector, and other electronic parts. These devices can be used through different optical modes such as reflection, absorption, evanescent wave, luminescence (fluorescence, phosphorescence, chemiluminescence), and surface plasmon resonance.

Because the measurement of lanthanides is the major focus of the current document, it is necessary to mention that in the case of these elements, the most popular modes used for the measurement are UV–vis and luminescence.

These devices, which can be used as alternatives for other analytical instruments, are rather popular for their low cost, low power requirements, and long lives. The mechanism of action in the case of lanthanide analytes is generally based on the selective interaction of a chelating agent which has been immobilized in an organic or inorganic matrix, with the lanthanide cations, which results in the formation of complexes that produces an optical signal, which is an indicator of the amount of the analyte. The common supporting materials include PVC or cellulose derivatives, which in addition to the desirable physical properties, are transparent in the spectral area under study. The use of these supports offers several further advantages, such as freedom from electrical noise, possibility of remote sensing, and low cost of manufacturing.

Among the different classes of optodes, the colorimetric sensors suffer the disadvantage of lower sensitivities than electrochemical and potentiometric sensors, but the fluorescent or luminescentdevices can lead to comparatively lower detection limits, which make these devices interesting. The lines below shall cover an overview of the UV–vis optodes reported for the lanthanide ions.

Zare-Dorabei et al., in 2009 [615], developed a selective and sensitive Gd(III) optode through the immobilization of (Z)-*N*′-((pyridine-2-yl) methylene)-thiophene-2-carbohydrazide (PMTC) (Fig. 6.12) on a triacetylcellulose membrane and observed linear responses in the range of 5.0×10^{-8} to 2.0×10^{-5} M, reaching detection limits down to 1.1×10^{-8} M. Other characteristics of the device included its short response time of 1–2 min

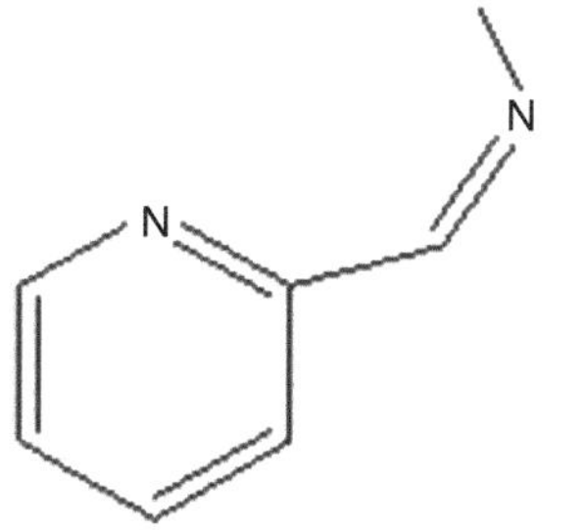

FIGURE 6.12 Structure of (Z)-*N*′-((pyridine-2-yl) methylene) thiophene-2-carbohydrazide (PMTC).

(depending on the Gd(III) concentration) and rather wide range of pH independent responses in the pH range of 2.0–9.0. The optode reportedly showed very good selectivity in the presence of La, Ce, Sm, and Eu ions, and was successfully regenerated using thiourea solutions.

The transparent triacetylcellulose membranes used for immobilizing the ligand were prepared from photographic film tapes, through their chronic treatment with sodium hypochlorite (for removing the colored gelatinous layers), before treating them with a clear solution of 0.02% (W = V) PMTC in 10 mL ethylene diamine for 2 min and washing them with water (for removing ethylene diamine and the loosely trapped indicator). The films were stored under water before use.

Bis(thiophenal) pyridine-2,6-diamine (BPD) (Fig. 6.13) [616] has also been used for the construction of Gd^{3+} optodes, which show linear behaviors in the range of 2.5×10^{-6} mol/L to 0.93×10^{-8} mol/L, with the identical response time and pH behaviors and good selectivity toward the analyte as opposed to La, Ce, Sm, Tm, Ho, and Eu ions.

The same group devised Dy^{3+} optodes through the immobilization of *N*′-[(2-hydroxyphenyl)methylene] benzohydrazide (BBH) (Fig. 6.14), on a triacetylcellulose membrane [617], and achieved linear responses in the range of 5.0×10^{-7} to 8.0×10^{-6} M (but rather shorter response times as compared to the mentioned Gd^{3+} sensors (i.e., 30–50 s). The pH range in which the sensor could be applied without considerable interferences from H^{+} ions was however rather narrower (i.e., 3.0–5.0), and the sensor had good selectivity behaviors in the presence of other lanthanide ions such as La^{3+}, Ce^{3+}, Sm^{3+}, Gd^{3+}, Tm^{3+}, Ho^{3+}, and Eu^{3+}.

Absorption spectra of the optode film in presence of the Dy(III) ion changed significantly (Fig. 6.15), which is due to the complex formation between dysprosium ion and BBH ligand used as a selectophore in the membrane. This complexation is the basis of determination of this ion.

The transparent triacetylcellulose membrane films were prepared through conventional methods, treated with BBH solution (0.003 g) in 10 mL ethylene diamine for 2 min at ambient temperature. Afterwards, they were washed with water for the removal of ethylene diamine and the loosely trapped indicator (used ligand).

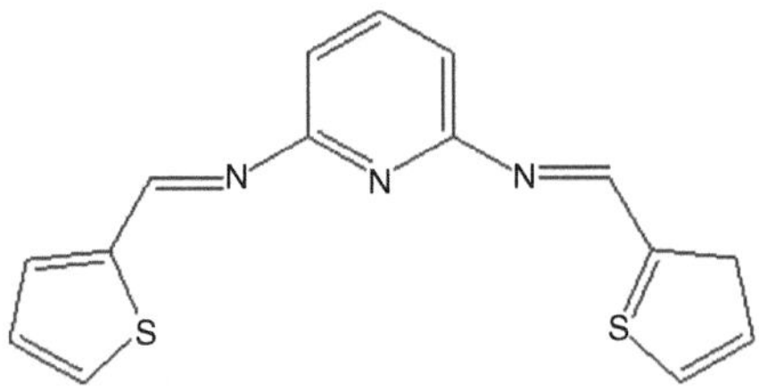

FIGURE 6.13 Structure of bis(thiophenal) pyridine-2,6-diamine (BPD).

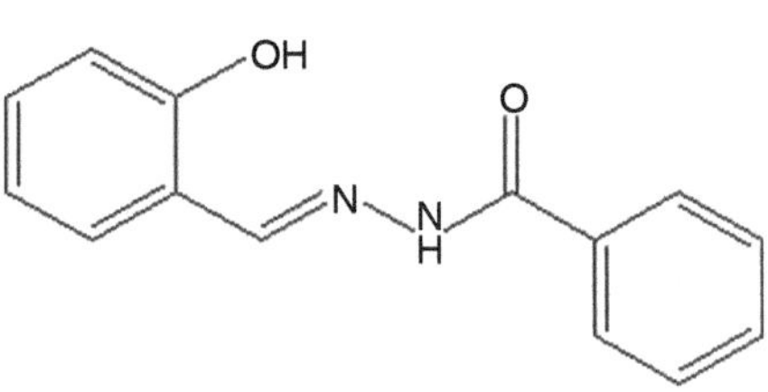

FIGURE 6.14 *N*′-[(2-Hydroxyphenyl) methylene] benzohydrazide (BBH).

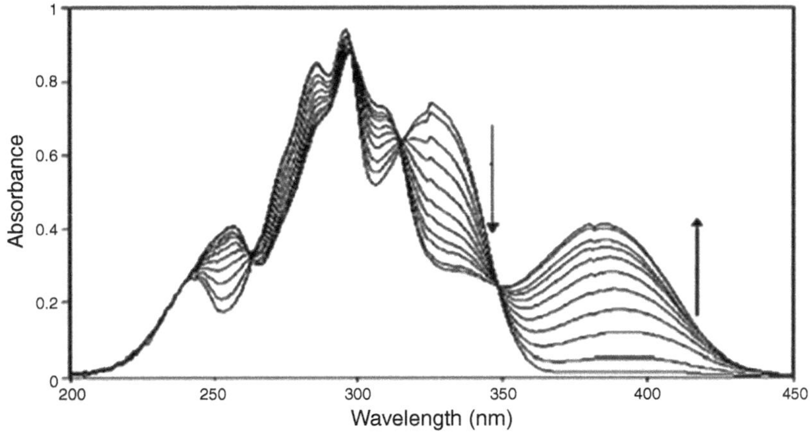

FIGURE 6.15 Absorption spectra of optode film response to Dy(III) in the range of 5.0×10^{-7} to 8×10^{-6} mol/L at pH 5.

The first optical sensor for Sm^{3+} was reported by Fouladgar and Ensafi [618], who used 4-(5-bromo-2-pyridylazo)-*N*,*N*-diethyl-3-hydroxyanilin (Br-PADAP) as the ionophore in plasticized PVC membranes. They reported that the interactions between Sm^{3+} and Br-PADAP give rise to a new peak at 525 nm, which corresponds to the resulting complex. The device was reported to have a linear response range from 2.6×10^{-6} to $2\ 6 \times 10^{-4}$ mol/L and a lower detection limit of 1.9×10^{-6}, respectively, and rather longer response times of 200 to 300 s depending on the analyte concentration.

Another optical sensor was devised for La^{3+} by Ensafi and Fooladgar [619] based on 4-hydroxysalophen (Fig. 6.16) which was immobilized on a hydrolyzed triacetylcellulose membrane through covalent bonding (Fig. 6.17). The resulting membrane, on contact with La^{3+} at pH 4.0, was reported to change color from white yellow (323–433 nm) to orange and

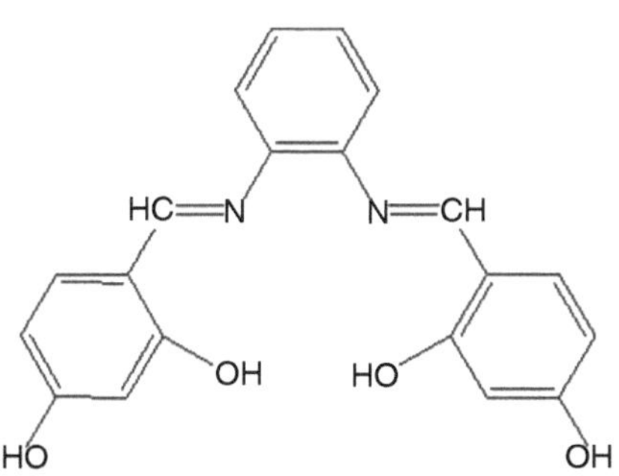

FIGURE 6.16 Structure of 4-hydroxysalophen.

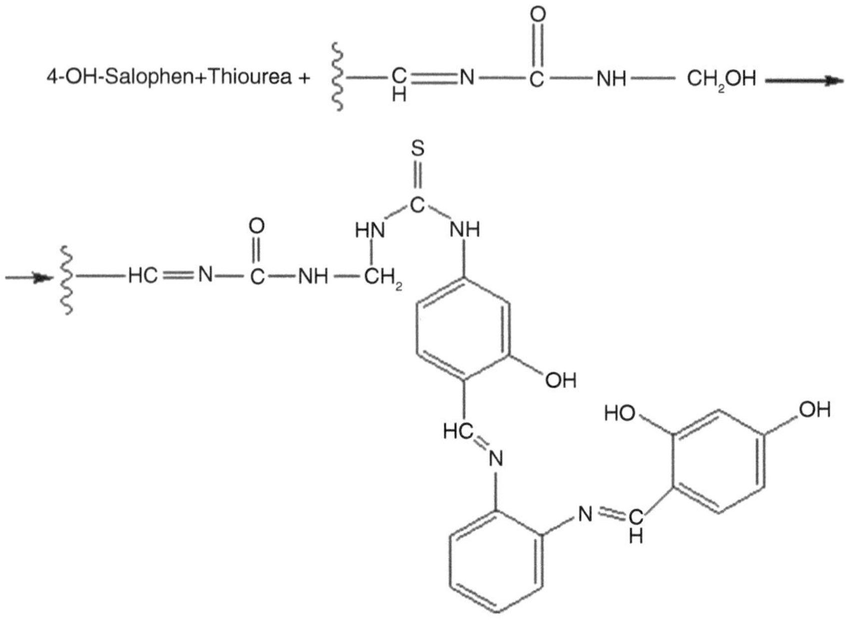

FIGURE 6.17 The possible reactions between 4-hydroxy-salophen and the membrane.

had a linear response from 1.0×10^{-6} to 1.0×10^{-2} mol/L and rather long response times of 5–6 min [619].

The researchers hydrolyzed the triacetylcellulose film by hydrolyzing it with a 0.1 mol/L KOH solution for 24 h and next activated through treatment with a polyvinyl alcohol (0.50% w/v) and thiourea (0.60% w/v) solution at 25°C for 24 h and next with a 5.7×10^{-4} mol/L 4-hydroxysalophen solution in citrate buffer (pH 6.2) for 6 h at 25°C.

The first Eu^{3+} sensor was developed in 2012 [620], using chromogenic calix[4]arene derivatives (Fig. 6.18) and was used for UV–vis absorption spectroscopy measurements. The device was reported to have good selectivity over Pb^{2+}, Cd^{2+}, and Mg^{2+} at pH 6.8.

The process for preparing the membranes included dissolving 10 mg of the ionophore in 1 mL of chloroform. The adhesion of the film to the glass plates was improved through activating the glass through successive ultrasonic treatments in acetone and methanol for 20 min before spin-coating the films on the plates at a controlled speed of 2300 rpm and their annealing at 80°C for an hour.

Most of the optical membrane sensors are prepared as described above. And a typical procedure for optical measurement of the amount of analyte by an optical sensor can be explained as follows.

The prepared optode membrane was soaked into the ions test solution in a suitable buffer solution for a certain amount of time. Next, the

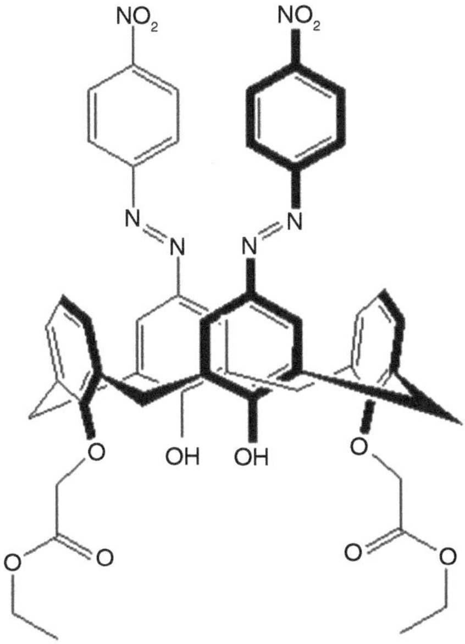

FIGURE 6.18 Chemical structure of azo-calix[4]arene: 5,17-bis(4-nitrophenylazo)-26,28-dihydroxy-25,27-di(ethoxycarbonylmethoxy)-calix[4]arene.

membrane was washed with distilled water and dried. Then, it was placed vertically inside the sample cuvette containing 3 mL of buffer solution for about 15 min to reach equilibrium. The absorbance-measuring system for the optode films is shown in Fig. 6.19. Then, the spectrum of the membrane is taken. Finally, the sample cell was titrated with ion solution,

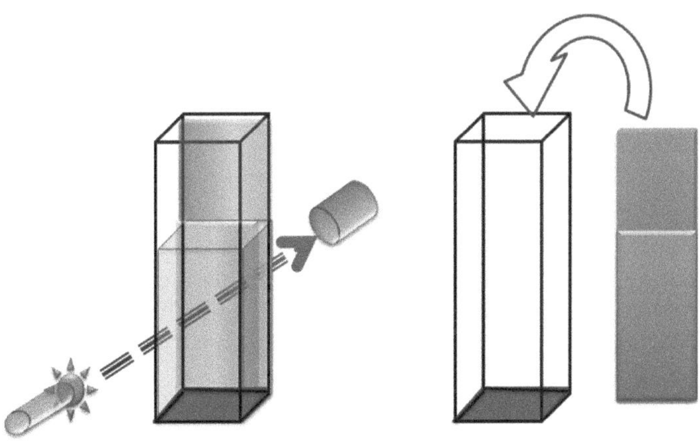

FIGURE 6.19 Absorbance-measuring system for the film optode.

and the absorbance of the system was measured over the indicated wavelength range. Then the absorption changes correlated with the concentration of the analyte are determined using a calibration graph.

LUMINESCENCE OF LANTHANIDE SERIES

By definition, luminescence, as opposed to incandescence, is used to refer to emissions of light by a substance that are not induced by heating, which changes such radiations to cold body type. The driving force behind such radiations can be chemical reactions, electrical energy, subatomic motions, or stress on crystals, based on which [621] the phenomena have been classified into photoluminescence, chemiluminescence, crystalloluminescence, electroluminescence, mechanoluminescence, and radioluminescence (Fig. 6.20).

Photoluminescence, which can be defined as the emission of light as a consequence of absorbing photons, can be subdivided into subclasses like fluorescence (where a singlet–singlet (no spin change) electronic relaxation causes the emission and has lifetimes in the order of nanoseconds), and phosphorescence (in which a triplet–singlet [spin change] electronic relaxation causes the radiation and has rather long lifetimes of a few milliseconds to hours). Chemiluminescence, where the emission is due to a chemical reaction, can be divided into groups of bioluminescence and electrochemiluminescence, where the chemical reaction is caused by a living organism or is of an electrochemical nature, respectively. A subdivision of electroluminescence is cathodoluminescence, where the emission is due to the bombardment of a luminescent material by electrons.

Crystalloluminescence phenomena are those in which the emission occurs due to crystallization, and mechanoluminescence is caused by mechanical work on solids. The latter can be subdivided into triboluminescence, or emissions due to the breaking of bonds in a material on exertion of mechanical forces; fractoluminescence, caused by breaking bonds in certain crystals; piezoluminescence, which caused by exerting pressure on the solid; or sonoluminescence, caused by sound energy. Finally, radioluminescence is referred to incidents where light is produced as a result of bombarding the arterials with ionizing radiations.

One of the similarities of lanthanides that results from their electronic arrangements is their spectroscopic properties, including the characteristic narrow-visible or near-infrared emissions, which has not only been the major reason behind the discovery of many of these elements but has led to their application in lighting and light conversion technologies [622-625].

Fluorescence sensing of chemical and biochemical analytes is an active area of research. Because the aim of this work is considering the various measurement techniques for determination of lanthanide series, the ability of the lanthanide and their complexes as fluorescent materials is not

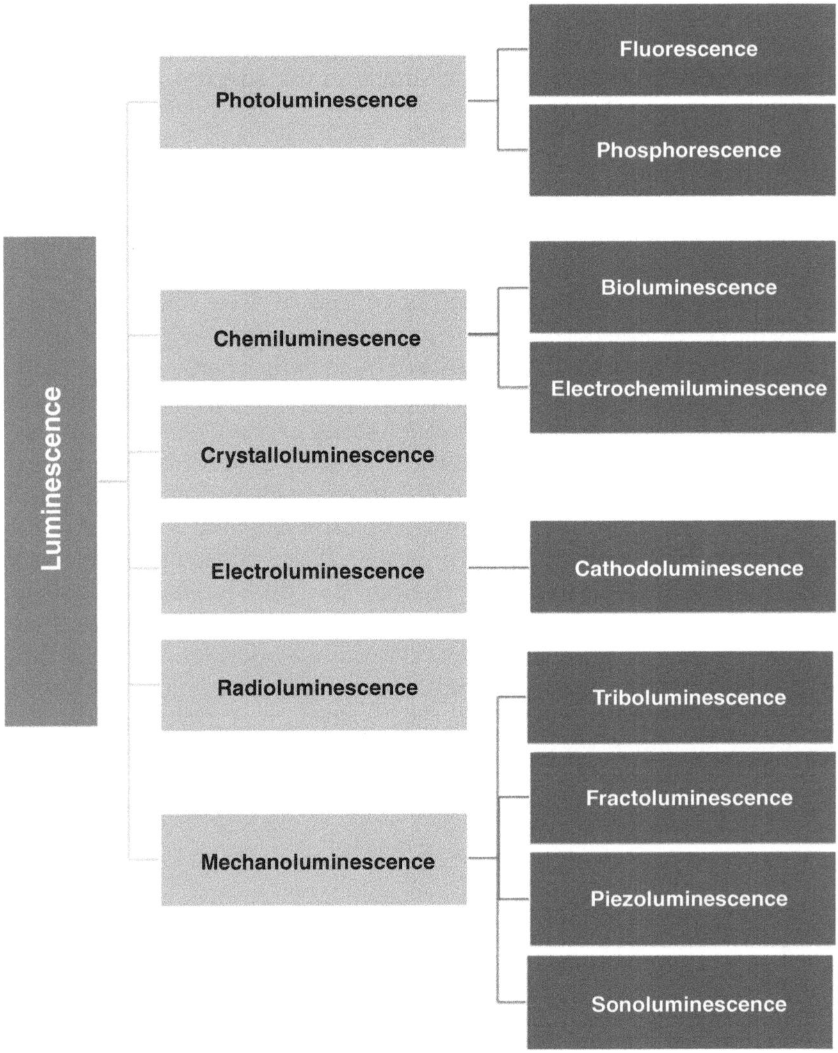

FIGURE 6.20 Various types of luminescence phenomena.

considered [626]. We focus only on the luminescence methods that determine lanthanides.

As previously discussed in the Chapter 5, finding a selective organic ligand (suitable selectophore) for one of the members of this family that is able to select this ion from other members of the family is rather difficult. Moreover, having a selectophore with fluorescence properties or even change fluorescence behavior after interaction with lanthanides causes the issue gets more difficult. Because of the shielded f orbitals, the lanthanides are rather insensitive to their local environment [625]. However, lanthanide-sensitive chemosensors have been developed.

Fluorescence sensing requires changes in the spectral properties of a suitable fluorophore in response to the analyte. Changes can occur in the intensity, excitation spectrum, emission spectrum, anisotropy, or lifetime of the sensing probe.

The most direct sensing method is when the fluorescence intensity of the probe directly changes in response to the analyte (Fig. 6.21). Quenching and in few cases enhancing of the intrinsic fluorescence of the fluorophore can be accomplished by addition of analyte. Then, the concentration of the analyte can be correlated by suitable equations (e.g., Stern-Volmer in case of quenching) to the amounts of quenching or enhancing. Another type of interactions shows wavelength-ratiometric spectra. In this case, wavelength-ratiometric probes display shifts in the absorption or emission spectra on binding of the analyte. Wavelength-ratiometric probes are desirable because the ratios are independent of the probe concentration. The analyte concentration can then be determined from the ratio of fluorescent intensities measured at two excitation and emission wavelengths.

In fact, any effect that can change the fluorescence intensity or wavelength of a sample can be utilized as a means for building sensing. Typically, there is equilibrium between the fluorophore free in solution and the fluorophore bound to analyte. One form can be nonfluorescent, in which case emission is only seen in the absence or presence of analyte, depending on which form is fluorescent. Probes that act in this manner

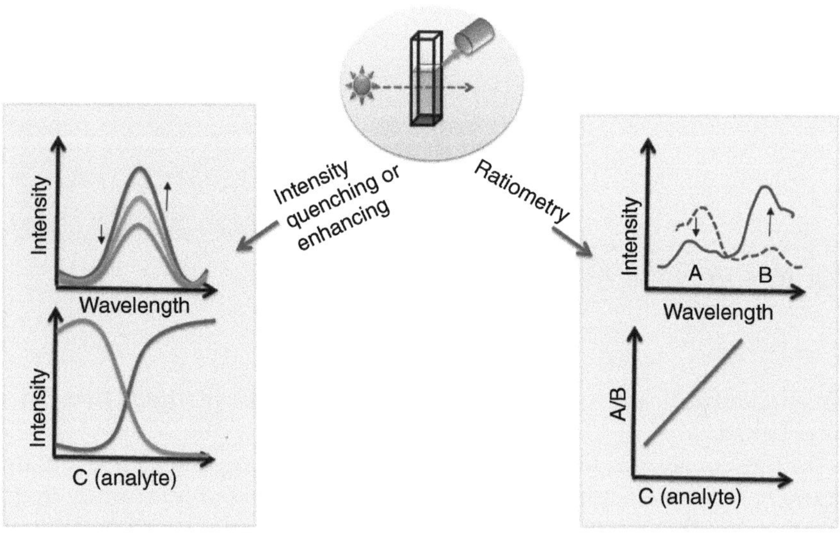

FIGURE 6.21 Spectral observable for fluorescence sensing. From left to right, sensing is performed using intensity quenching or enhancing and wavelength ratios measurements.

are not wavelength-ratiometric probes. Alternatively, both forms may be fluorescent but display different quantum yields or emission spectra. This type of behavior is often seen for pH probes, where ionization of the probe results in distinct absorption and/or emission spectra. Spectral shifts are also seen for probes that bind specific cations such as calcium. Such probes allow wavelength-ratiometric measurements. In this case, the change in intensity or shift in the emission spectrum is used to determine the analyte concentration. Probes that bind specific analytes are often referred to as probes of analyte recognition.

As in other techniques, instrumentation plays a crucial role in the success of fluorescence experiments, especially in light of the fact that many phenomena can influence the data.

The overall structure of a general-purpose spectrofluorometer is shown in Fig. 6.22. The xenon acts as an excitation source, and the monochromator is used for selecting the excitation and emission wavelengths [625]. Photomultiplier tubes are used for the detection of the signal.

Two general types of instruments exist:

- Filter fluorometers use filters to isolate the incident light and fluorescent light.
- Spectrofluorometers use diffraction grating monochromators to isolate the incident and fluorescent light.

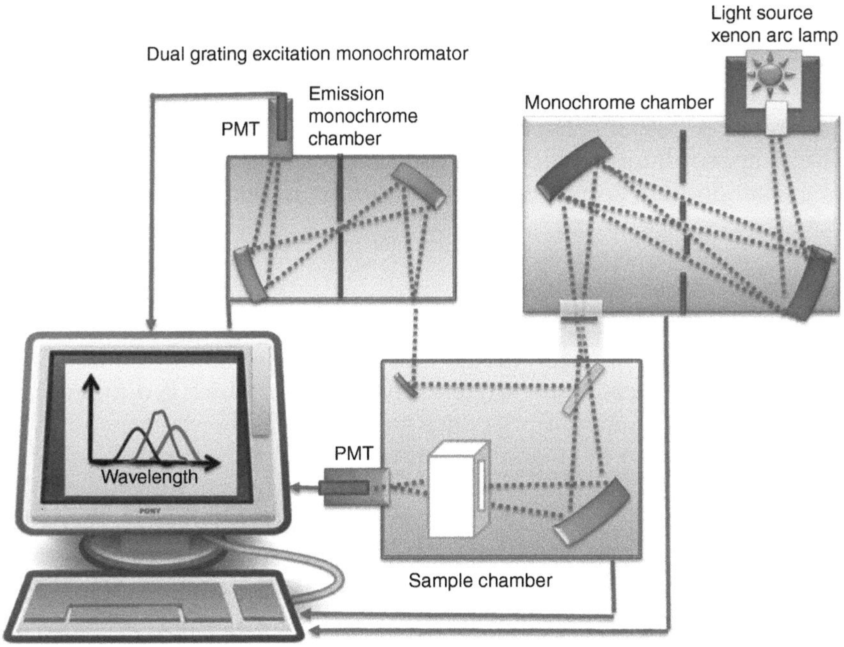

FIGURE 6.22 Schematic diagram of a spectrofluorometer.

In both types, the light from a source passes through a filter or monochromator, and interact with sample. A proportion of the incident light is absorbed by the sample and causes some of the molecules in the sample to fluoresce. The fluorescent light is emitted in all directions. Some of this fluorescent light passes through a second filter or monochromator and reaches a detector, which is usually placed at 90° to the incident light beam to minimize the risk of transmitted or reflected incident light reaching the detector.

Several light sources may be used as excitation sources, including lasers, LEDs, xenon arcs, and mercury-vapor lamps in particular. A laser only emits light of high irradiance at a very narrow wavelength interval, typically under 0.01 nm, without needing the excitation monochromator or filter. The disadvantage of this method is that the wavelength of a laser cannot be changed by much. A xenon arc has a continuous emission spectrum with nearly constant intensity in the range from 300 to 800 nm and a sufficient irradiance for measurements down to just above 200 nm.

Another important part of the fluorimeters is filters and/or monochromators. A monochromator transmits light of an adjustable wavelength with an adjustable tolerance. The most common type of monochromator utilizes a diffraction grating; that is, collimated light illuminates a grating and exits with a different angle depending on the wavelength. The monochromator can then be adjusted to select which wavelengths to transmit. The detector can either be single-channeled or multichanneled. The single-channeled detector can only detect the intensity of one wavelength at a time, whereas the multichanneled one detects the intensity of all wavelengths simultaneously, making the emission monochromator or filter unnecessary. The different types of detectors have both advantages and disadvantages.

Most spectrofluorometers with dual monochromators and a continuous excitation light source can record both an excitation spectrum and a fluorescence spectrum [627].

Fluorescence Optical Sensors

Compared with other methods, fluorescence detection techniques are more sensitive, specially due to detection systems, which is differential in absorption and direct in fluorescence techniques. For instance, in the case of a 10^{-10} M solution of an analyte with a hypothetical a molar extinction coefficient of 10^5 M^{-1} cm^{-1}, the absorbance shall equal very low (about 10^{-5} per cm) and shall equal a transmission of 99.9977%, which is very difficult to detect even with today's technology. However, in the case of fluorescence detection, such measurements are rather common, specifically because the measurement is conducted relative to a dark background, which greatly decreases the chance of error, because even single photons can be detected with common photomultiplier tubes.

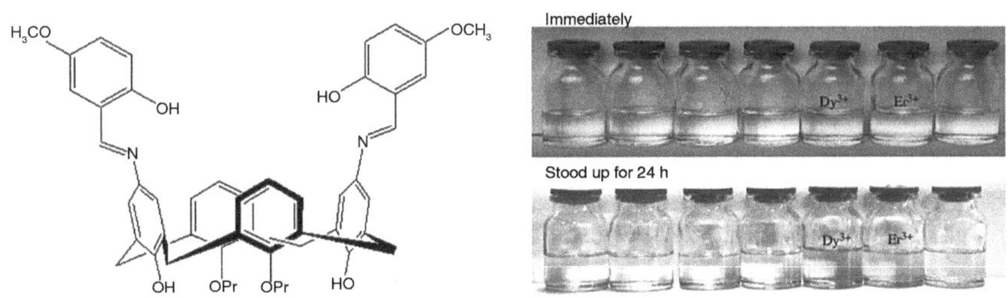

FIGURE 6.23 Color changes immediately of compound 4 in dichloromethane induced by addition of Dy^{3+} (top); color changes of compound 4 in dichloromethane induced by addition of Er^{3+} (bottom) stood up for 24 h in the dark condition (from left to right: addition of La^{3+}, Pr^{3+}, Eu^{3+}, Gd^{3+}, Dy^{3+}, Er^{3+}, and Yb^{3+}). *Reprinted with permission from [628].*

In general, there have been a few reports in the past 5 years on the determination of lanthanide ion using a fluorescence technique. Two general approaches have been used in designing the method. In the first one, the fluorescence of the probe (selectophore) has been studied and then its fluorescent behavior in interaction with the lanthanide ions has been considered. Quenching, enhancing, and ratiometric properties of the fluorophore is used as detection method and called fluorescent turn off, turn on, and ratiometric chemosensors, respectively. In others, a plasticized PVC membrane has been used to immobilize the fluorophore and then the extraction of the ions from the bulk and interaction with the fluorophore, which causes the changes in basic fluorescence of the optode, has been considered.

Liang et al. reported one of the first fluorescent optodes for Dy^{3+} and Er^{3+} ion [628] using photochromic 5,17-*N*,*N*′-(5,5′-di-methoxyl-benzlaldehyde)diimine-25,27-dipropoxycalix[4]arene (Fig. 6.23) and observed clear color changes in the system in the presence of the two analytes, which were attributed to the decrease in the intensity of the band at 450 nm and the emergence of one at 550 nm (Fig. 6.24) due to the increase in the degree of π-conjugation as well as donor–acceptor charge transfer in the resulting complexes. The researchers also reported a considerable increase in the fluorescence and a red shift from 550 to 598 nm, in the presence of the analyte in aqueous media.

Also, Hosseini et al. [629] reported a fluorescent chemosensor for La^{3+} based on (1-[({2-[2-(2-hydroxy-1-naphthyl)-3-(2-{[(*E*)-1-(2-hydroxy-1-naphtyl)methylidene]amino}ethyl)-1-imidazolidinyl]ethyl}-imino) methyl]-2-naphthol) (L) (Fig. 6.25). They reported that the complexation between the compound L and La(III) gives rise to a huge red shift from 430 to 522 nm in the emission spectra, because of the formation of a 1:1 metal–ligand complex (Fig. 6.26).

One of the initial reports on a selective fluorometric optode membrane belongs to an optode for Er^{3+} [630], based on 5-(dimethylamino) naphthalene-1-sulfonyl-4-phenylsemicarbazide (L) (Fig. 6.27) as the Er^{3+}

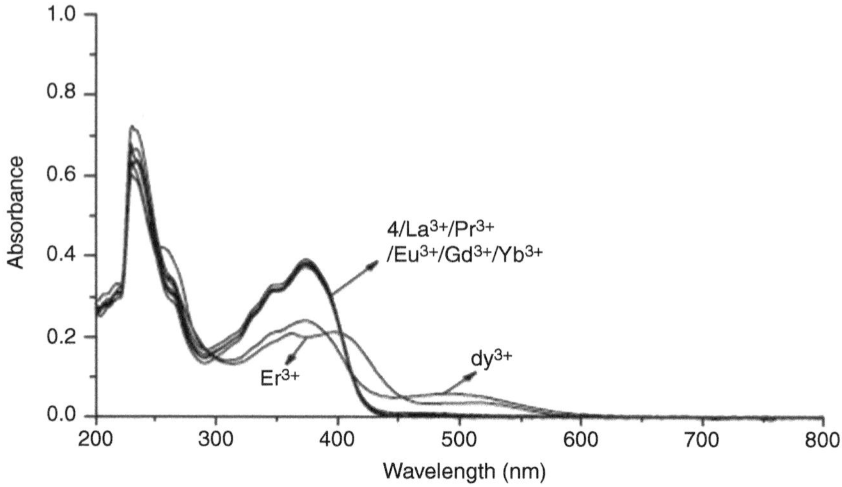

FIGURE 6.24 Fluorescence spectra of the ligand (1×10^{-5} mol/L) and its complexes in CH_2Cl_2; $\lambda_{ex} = 375$ nm.

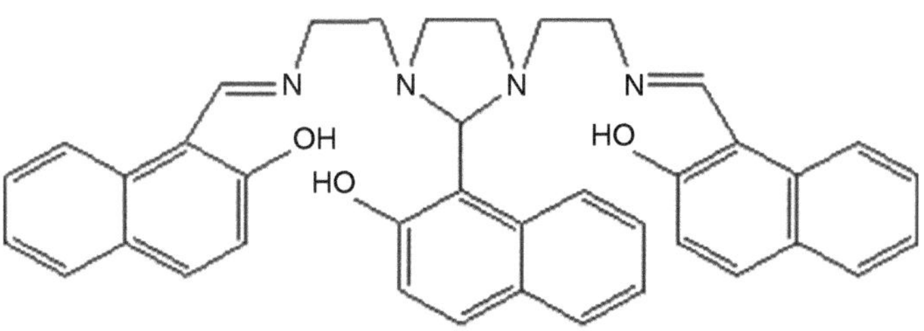

FIGURE 6.25 Chemical structure of compound L.

fluoroionophore, in a plasticized PVC membrane, which further contained tetraphenyl borate as a lipophilic anionic additive. At a pH value of 5, the optode was reported to show a strong fluorescence response in the concentration range of 1.0×10^{-10} to 1.0×10^{-2} mol/L. The responses were reported to need less than 50 s to reach equilibrium.

The response of the sensor was attributed to the formation of a 1:2 complex between the fluoroionophore (L) and Er^{3+} in the organic phase as below:

$$2L(org) + 3Na^{+}(org) + 3TPB^{-}(org) + Er^{3+}(aq)$$
$$= ErL_2^{3+}(org) + 3TPB^{-}(org) + 3Na^{+}(aq)$$

Under optimal conditions, the fluorescence excitation and emission spectra were observed at $\lambda_{em} = 510$ nm and $\lambda_{ex} = 364$ nm, respectively (Fig. 6.28). The spectra in the figure show decreases in the intensities as a

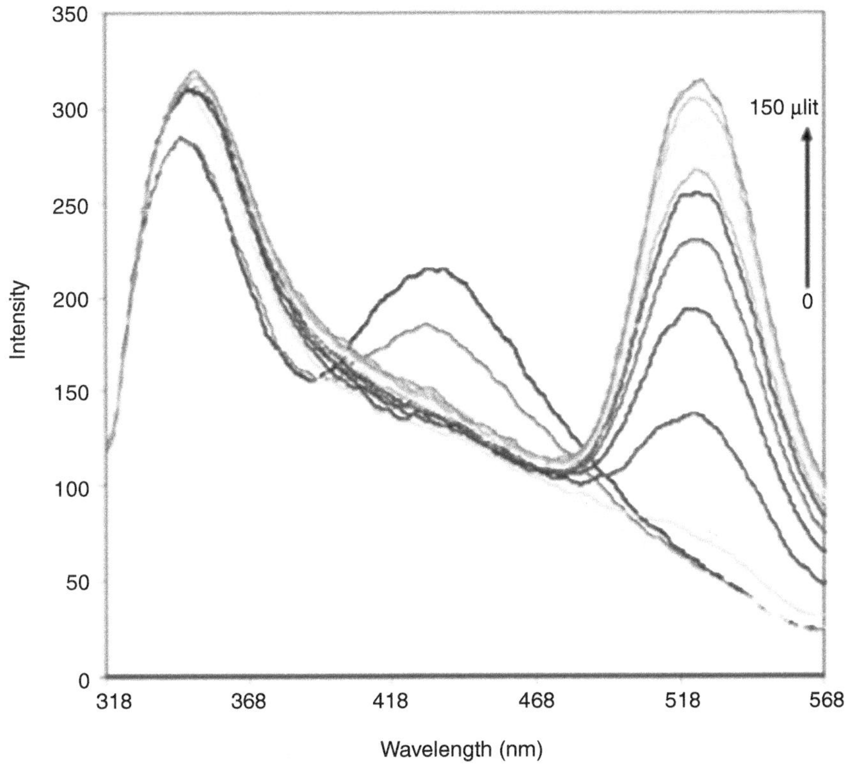

FIGURE 6.26 Effect of concentration of La (1×10^{-4} mol/L) on the fluorescence spectrum of receptor L (5×10^{-6} mol/L) (2 mL CH_3CN). *Reprinted with permission from [629].*

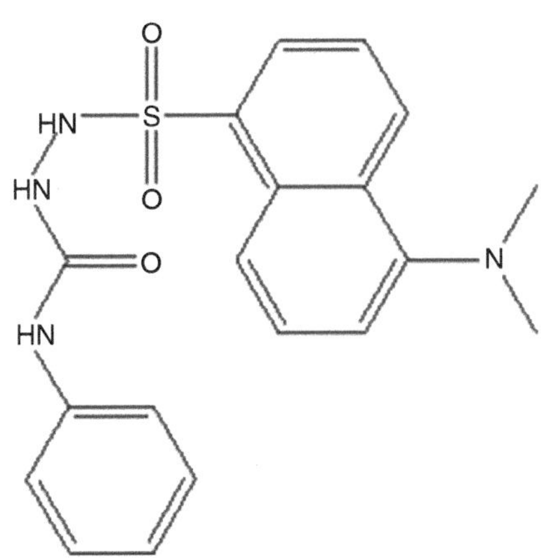

FIGURE 6.27 Chemical structure of 5-(dimethylamino) naphthalene-1-sulfonyl-4-phenylsemicarbazide (L).

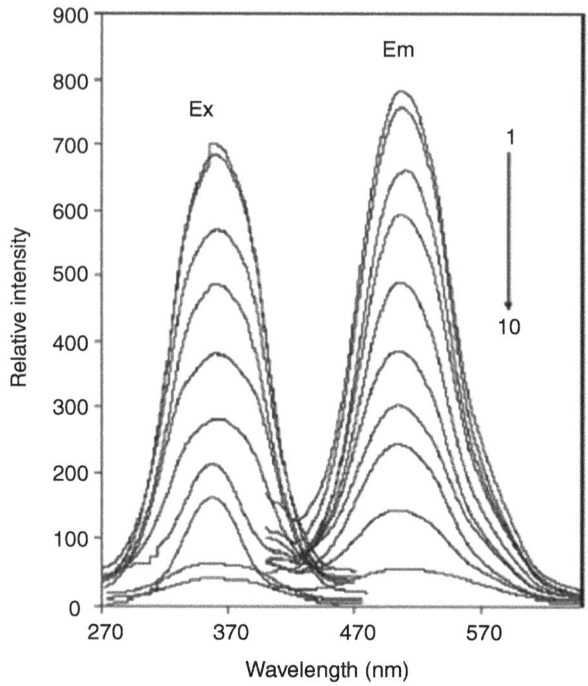

FIGURE 6.28 Excitation (E_x) and emission (E_m) spectra of the erbium-selective optical sensor (the PVC membrane composition used is that given for membrane no. 4) in the presence of varying concentration of Er^{3+} ion in acetate buffer of pH 5.0: (1) 0.0 mol/L, (2) 1.0×10^{-10} mol/L, (3) 1.0×10^{-9} mol/L, (4) 1.0×10^{-8} mol/L, (5) 1.0×10^{-7} mol/L, (6) 1.0×10^{-6} mol/L, (7) 1.0×10^{-5} mol/L, (8) 1.0×10^{-4} mol/L, (9) 1.0×10^{-3} mol/L, and (10) 1.0×10^{-2} mol/L. *Reprinted with permission from [630].*

result of the addition of Er^{3+}, whereas no shifts are observed in the location of the spectrum.

In the same year, a novel fluorescent bulk chemosensors was designed for Tb^{3+} ions using a dansyl derivative [631], namely, 2-(5-(dimethylamino) naphthalen-1-ylsulfonyl)-*N*-phenylhydrazinecarbothioamide (Fig. 6.29).

The excitation and emission were observed at 360 nm and 530 nm. The emission intensity quenched in the presence of to Tb^{3+} because of the formation of a selective 1:1 complex formation between the ligand and the ions, and led to linear responses in the range 4.0×10^{-7} to 1.0×10^{-5} mol/L of Tb^{3+}, with a detection limit of 1.4×10^{-7} mol/L.

The response was studied for a range of interfering cations (i.e., Nd^{3+}, Sm^{3+}, Yb^{3+}, Er^{3+}, La^{3+}, Ce^{3+}, Lu^{3+}, Eu^{3+}, Gd^{3+}, Ho^{3+}, Tm^{3+}, Pr^{3+}, Dy^{3+}, Cu^{2+}, Hg^{2+}, Pb^{2+}, Ag^{+}, Cs^{+}, and K^{+}) and was found to be very low (Fig. 6.30). A typical interfering study curve is presented in Fig. 6.30.

The same fluorophore was later used in an optical membrane sensor [632], in a plasticized PVC membrane that further composed on sodium

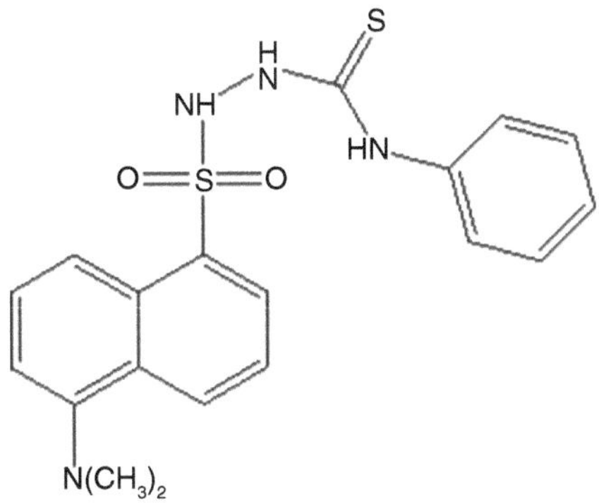

FIGURE 6.29 Chemical structure of compound L.

tetraphenyl borate, based on the strong and sensitive quenching observed in its fluorescence at pH 4.7, in the range of 1.0×10^{-2} to 1.0×10^{-8} mol/L.

The lipophilic anionic TPB^- were reported to enhance the ion-exchange properties of the membrane, and the interaction between the ligand and Tb^{3+} ions was described as

$$L(org) + 3Na^+(org) + 3TPB^-(org) + Tb^{3+}(aq) \\ = TbL^{3+}(org) + 3TPB^-(org) + 3Na^+(aq)$$

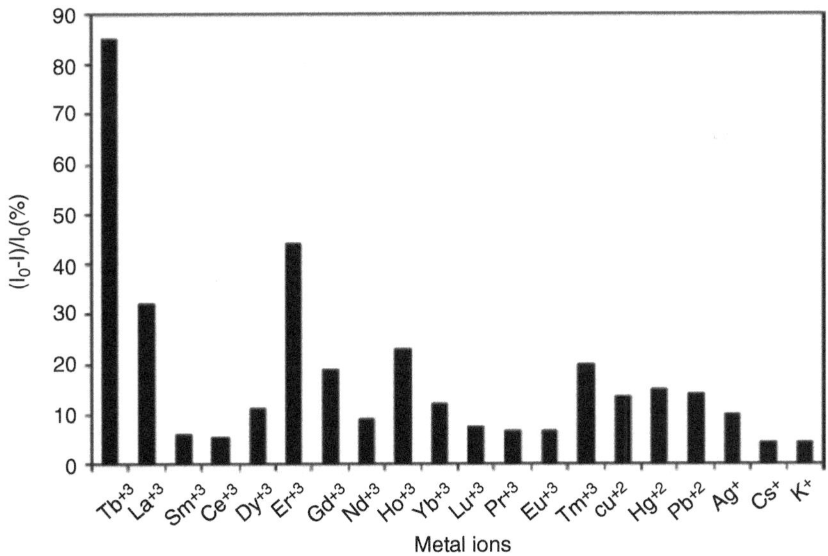

FIGURE 6.30 Quench ratio $(I_o - I)/I_o$ of fluorescence intensity of L (5×10^{-6} mol/L) on the addition of two equivalent metal ions in MeCN solution.

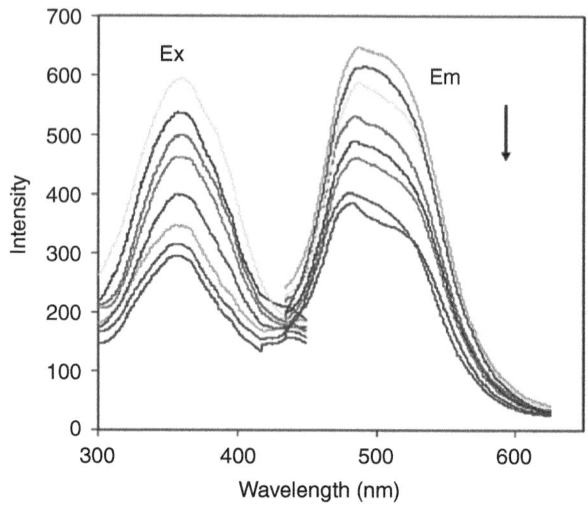

FIGURE 6.31 Relative fluorescence intensity versus pTb^{3+}. *Reprinted with permission from [632].*

The fluorescence excitation (at λ_{em} = 487 nm) and emission (at λ_{ex} = 360 nm) spectra of the proposed fluorescent membrane sensor in a buffer of pH 4.7 and in the presence of increasing concentrations of terbium ions are shown in Fig. 6.31. As seen from Fig. 6.31, both the intensities are decreased considerably on increasing terbium ion concentration, with no measurable spectral position shifts as a result of the strong and selective quenching effect of Tb^{3+} ions.

A typical response mechanism of the membrane optodes in which the fluorescence of the fluorophore (selectophore) immobilized in a PVC membrane quenches by addition of the analyte, as explained above, is explained below.

As a result of the complexation between Tb^{3+} ion in the aqueous sample solution (A) and the fluoroionophore L in the organic membrane phase, an A_aL_n complex can be formed based on a following equilibrium:

$$aA_{(aq)} + nL_{(org)} \overset{K_e}{\leftrightarrow} A_aL_{n(org)}$$
$$K_e = \frac{[A_aL_n]_{(org)}}{[A]^a_{(org)}[L]^n_{(org)}} \tag{6.7}$$

where a and n indicate the complexing ratio between Tb^{3+} and L, and K_e is the equilibrium constant of the overall reaction. It should be noted that the extent of ion-pairing of terbium nitrate in aqueous solutions over the low concentration ranges studied is negligible.

By definition, the relative fluorescence intensity is the ratio of uncomplexed L in the membrane phase, $[L]_{(org)}$, to its total amount, $[Lt]_{(org)}$, so that

$$\alpha = \frac{[L]_{(org)}}{[L_t]_{(org)}} \tag{6.8}$$

$$[A_a L_n]_{(org)} = [L_t]_{(org)} \frac{1-\alpha}{n} \tag{6.9}$$

In practice, value during titration with Tb^{3+} ions can be determined by the fluorescence intensity measurement of the optical sensor, *FI*:

$$\alpha = \frac{Fl - Fl_0}{Fl_\alpha - Fl_0} \tag{6.10}$$

where FI_o and FI_α are the limiting fluorescence intensities of the fluorescent intensities of the fluorescent membrane sensor at $\alpha = 0$ (i.e., totally complexed L) and $\alpha = 1$ (i.e., uncomplexed L), respectively. The relationship between the α-value and the concentration of Tb^{3+} ion in aqueous sample solution, $[A]_{(aq)}$, can be obtained by combining Eqs. (6.8) and (6.9), as follows:

$$\frac{a^n}{1-\alpha} = \frac{1}{nK_e [L_t]_{(org)}^{n-1} [A]_{(org)}^a} \tag{6.11}$$

Eq. (6.11) can then be used as a basis for the quantitative determination of terbium ions, using the proposed optical membrane (Fig. 6.32).

A third selective Tb(III) sensor was constructed using 5-(dimethylamino)-*N′*-(2-hydroxy-1-naphthoyl) naphthalene-1-sulfonohydrazine (Fig. 6.33) in a plasticized PVC membrane containing sodium tetraphenyl borate [633].

Like previous optodes, the response of the optode is based on the strong fluorescence quenching of L by Tb^{3+} ions. At a pH value of 5.0, the optode displays a wide concentration range of 1.0×10^{-7} to 1.0×10^{-2} mol/L, with a relatively fast response time of less than 45 s. In addition to high stability and reproducibility, the sensor shows a unique selectivity toward Tb^{3+} ion with respect to common cations. The optode was applied successfully to the trace determination of terbium ion in binary mixture and water samples and the determination of Tb^{3+} in phosphate rock samples.

Another optode for lanthanide ions was reported by Das et al. in 2010 [634], who developed a colorimetric sensor for Nd^{3+} ions with a selection limit in the range of parts per million while having a good selectivity in

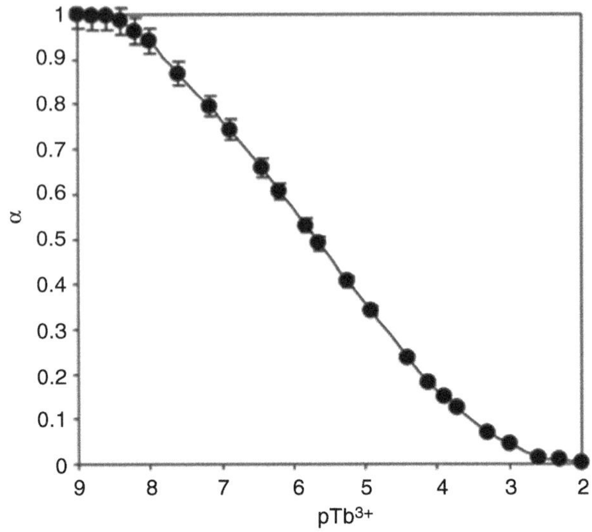

FIGURE 6.32 Excitation (E_x) and emission (E_m) spectra of the terbium-selective optical sensor in the presence of varying concentration of Tb^{3+} ion: from top to bottom 0.0, 1.0×10^{-8}, 1.0×10^{-7}, 1.0×10^{-6}, 1.0×10^{-5}, 1.0×10^{-4}, 1.0×10^{-3}, and 1.0×10^{-2} mol/L. *Reprinted with permission from [632].*

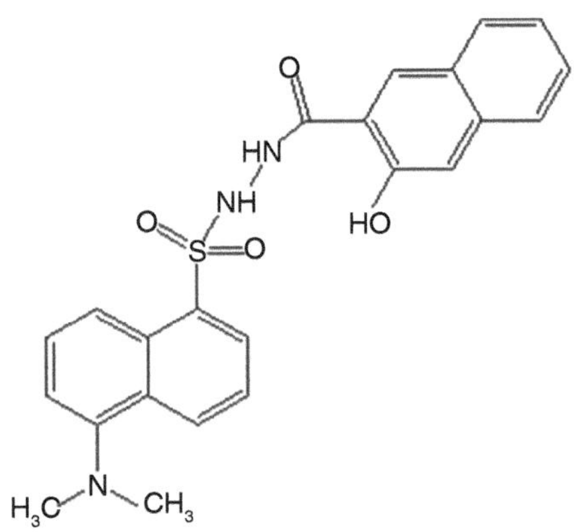

FIGURE 6.33 Structure of *N'*-(2-naphthoyl)-8-(dimethylamino)naphthalene-2-sulfonohydrazide (L).

the presence of La^{3+}, Pr^{3+}, Yb^{3+}, Eu^{3+}, Er^{3+}, Sm^{3+}, Tb^{3+}, Ce^{3+}, and Dy^{3+}. The presence of Nd^{3+} ions was found to induce a red shift of 68 nm in the emission spectrum of L (Fig. 6.34) from 440 to 508 nm. The authors reported a 2:1 complex (i.e., two ligands for one cation), in which the presence of a second Nd^{3+} was suggested to induce the quenching observed due to an intramolecular charge transfer between the donating section of the ligand (i.e., dimethyl azobenzene) and the acceptor fragment (i.e., phthalimide)

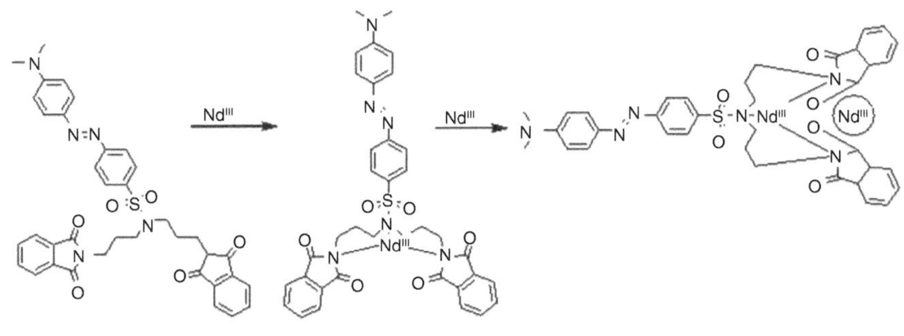

FIGURE 6.34 Probable coordination for L_1 to Nd^{3+} ion. *Reprinted with permission from [634].*

Fig. 6.35 shows absorption spectra of L (2.3×10^{-5} mol/L) in CH_3CN in the presence of varying concentrations of Nd^{3+} ($[Nd^{3+}] = 0–1.202 \times 10^{-2}$ mol/L). Color changes observed for L_1 (1.062×10^{-4} mol/L) in the presence of different lanthanide ions (from left to right: L_1, Eu, Er, Nd, Pr, Sm, Yb, and Tb).

In **2010**, Ganjali et al. synthesized (4*E*)-4-(2-phenyldiazenyl)-2-((*E*)-(2-aminoethylimino) methyl)phenol (PAMP) and used it as a ratiometric fluorescent chemosensor for recognition of Eu^{3+} ions in acetonitrile solution [635]. PAMP shows a fluorescent emission at 536 nm. When it forms a complex with a Eu^{3+} ion, two new fluorescent enhancements at 418 and

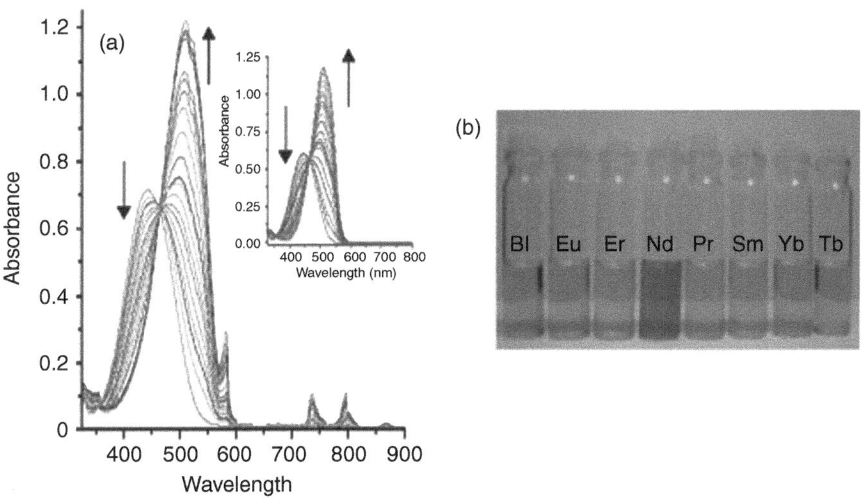

FIGURE 6.35 (a) Absorption spectra of L (2.3×10^{-5} mol/L) in CH_3CN in the presence of varying concentrations of Nd^{3+} ($[Nd^{3+}] = 0–1.202 \times 10^{-2}$ mol/L). Inset: absorption titration as shown in (a) after appropriate correction for respective $[Nd^{3+}]$. (b) Color changes observed for L_1 (1.062×10^{-4} mol/L) in the presence of different lanthanide ions (from left to right: L, Eu, Er, Nd, Pr, Sm, Yb, and Tb [tripositive charge of each metal ions is not shown for clarity], whereas $[M^{n+}] = 2.0 \times 10^{-2}$ mol/L (M^{n+} stands for different lanthanide ions). *Reprinted with permission from [634].*

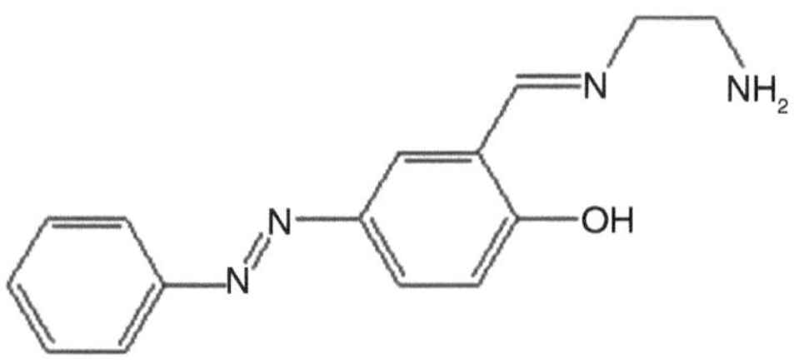

FIGURE 6.36 Structure of compound PAMP.

496 nm appear (Fig. 6.36). In acetonitrile solution, the blue shift of fluorescent emission upon europium binding is due to the formation of a 1:1 metal ligand complex. The chemosensor shows a linear response toward Eu^{3+} in the range of 8.3 x 10^{-8} M to 8.3 x 10^{-6} M. The new fluorescent probe exhibits high selectivity toward Eu^{3+} ions over a large number of interfering cations (Fig. 6.37).

In the same year, Zapata et al. reported a selective Yb^{3+} optode using an aza-substituted anthracene ionophore (Fig. 6.38) [636], in which the aza–butadiene bridge acted as the binding, the interaction of which with the cation led to the quenching of the fluorescence emission intensity and the anthracene segment acted as the fluorophore. The color of the molecule changed from yellow to dark orange upon interaction with solutions of Yb^{3+} in acetonitrile, which was indicated by the emergence of a new band at 503 nm, through UV spectroscopy. The reaction was found to be of a selective nature and did not repeat in the case of other metal ions.

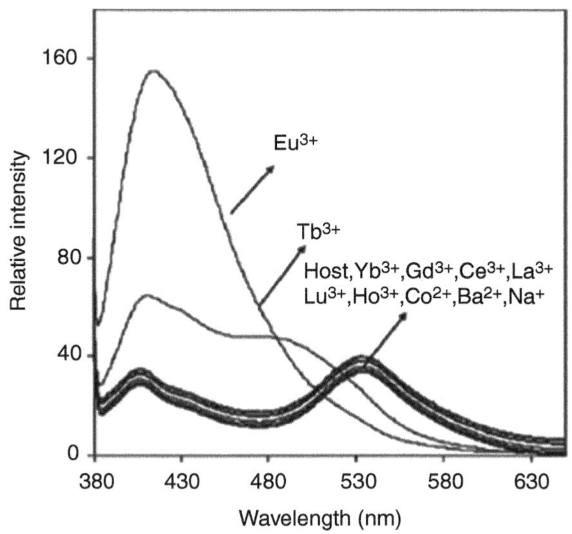

FIGURE 6.37 Fluorescence emission spectra of PAMP (3 mL 5 × 10^{-6} mol/L) in acetonitrile solution in the presence of different metal ions (1.6 × 10^{-6} mol/L). Excitation was performed at 370 nm. *Reprinted with permission from [635].*

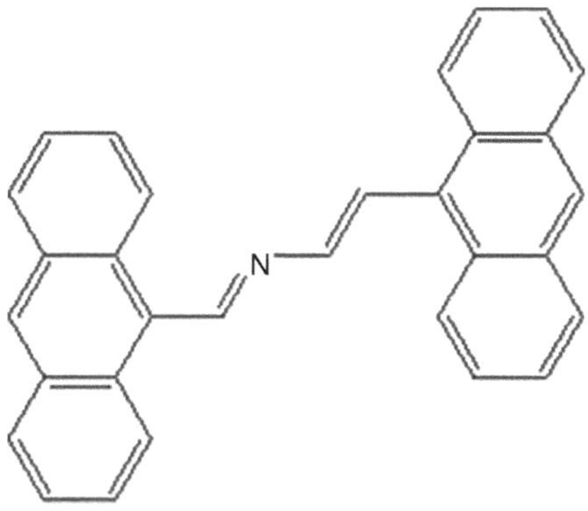

FIGURE 6.38 Structure of compound used as a fluorescent selectophore for Yb^{3+}.

The interaction of Yb^{3+} with the ligand was found to greatly increase the quantum yield of the ligand from $\phi = 6.2 \times 10^{-4}$ to 2.1×10^{-2}, which was regarded as an indicator of the turning off of the internal electron transfers from the aza-butadiene, which led to the quenching of the anthracene fluorescence that could happen in the absence of the cation.

In the same year, Hosseini et al. synthesized and used *N*′-(1-oxoacenaphthylen-2(1*H*)-ylidene)furan-2-carbohydrazide (Fig. 6.39) in ratiometric fluorescent chemosensor Yb^{3+} ions in acetonitrile [637] and reported observing a quenching phenomenon for the fluorescence of the ligand at 420 nm, while a new fluorescent band emerged and enhanced at 516 nm (Fig. 6.40). The complexation ratio was reported to be 2:1 metal–ligand complex, and a red shift of fluorescent emission spectrum was also observed. The device had a linear range in the range of 3.3×10^{-7} mol/L to 1.0×10^{-4} mol/L with a detection limit of 1.2×10^{-7} mol/L.

In 2012, Ganjali et al. developed a Dy^{3+} fluorescence optode [638] using 5-(dimethylamino)-*N*-(furan-2-yl-methylene)-naphthalene-1-sulfonohydrazide (Fig. 6.41) in a plasticized PVC membrane containing sodium

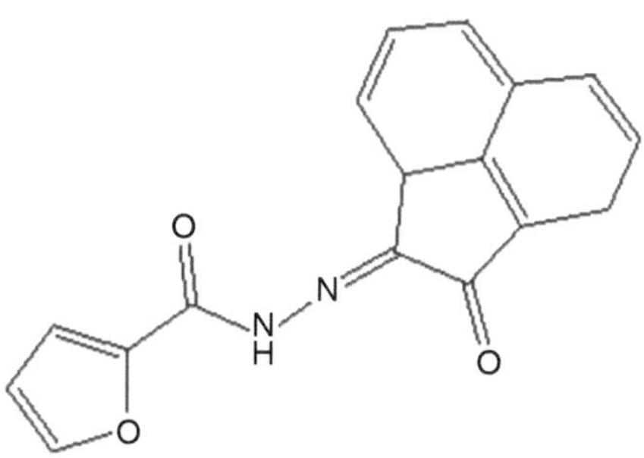

FIGURE 6.39 Structure of compound *N*′-(1-oxoacenaphthylen-2(1*H*)-ylidene)furan-2-carbohydrazide.

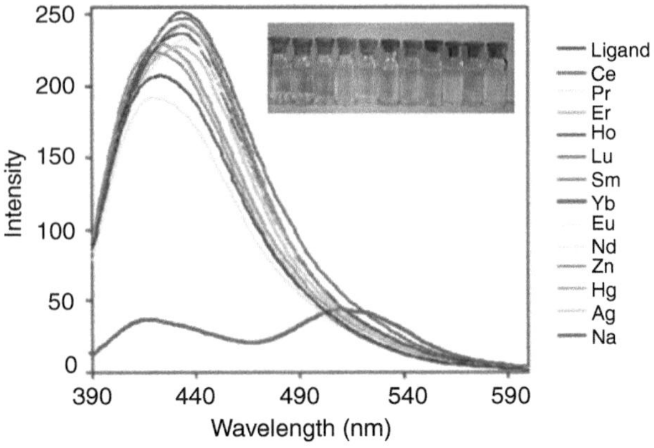

FIGURE 6.40 Color response of the receptor L in CH_3CN (4.0×10^{-5} mol/L) to the addition of ions (4.0×10^{-4} mol/L in CH_3CN) from the right to the left: L only; L + Pr, L + Nd, L + Eu, L + Er, L + Tb, L + Ce, L + Lu, L + Gd, L + Sm, L + Yb. *Reprinted with permission from [637].*

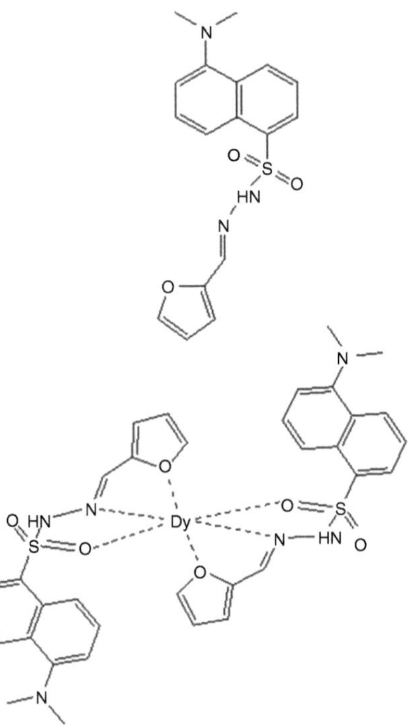

FIGURE 6.41 Chemical structure of 5-(dimethylamino)-*N*-(furan-2-yl-methylene)naphthalene-1-sulfonohydrazide (L) without and with Dy^{3+} ion.

tetraphenyl borate and observed very short response times of below 35 s, good selectivity behavior, and a linear response in the range of 1.0×10^{-10} to 1.0×10^{-2} mol/L.

The experimental results showed that the fluoroionophore has unique binding properties to Dy^{3+}, and as a result of interacting with the ion its fluorescence quenches. The complexation ratio was reported to be 1:2 (ligand/metal) in the organic phase and happened through the following ion-exchange mechanism:

$$L_{(org)} + 3Na_{(org)^+} + 3TPB_{(org)^-} + Dy_{(aq^{3+})} \leftrightarrow DyL_{2(org)^{3+}} + 3TPB_{(org)^-} + 3Na_{(aq)^+}$$

The excitation and emission bands were observed at 592 and 467 nm, respectively, and the spectral behavior in the presence of different concentrations of Dy^{3+} could be observed in Fig. 6.42.

Ganjali et al. also reported a fluorescent turn on (enhancement) chemosensor based on [639]. (*E*)-2-(1-(4-hydroxy-2-*oxo*-2*H*-chromen-3-yl) ethylidene)hydrazinecarbothioamide as a Pr^{3+} selective fluoroionophore

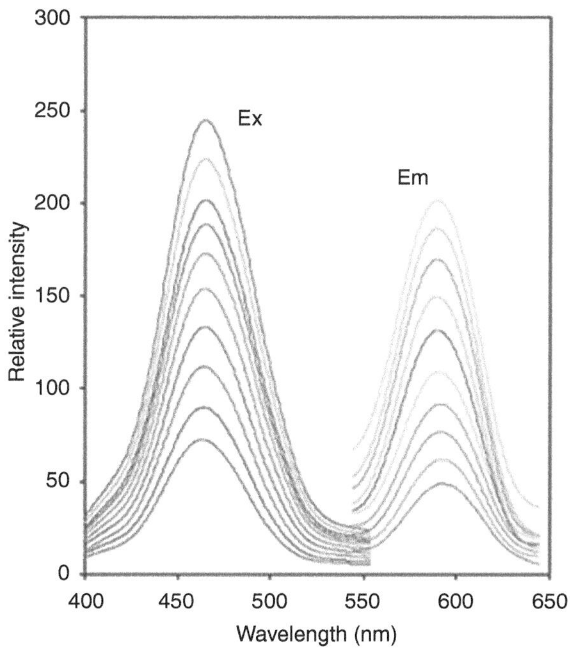

FIGURE 6.42 Excitation (E_x) and emission (E_m) spectra of the dysprosium-selective optical sensor (the PVC membrane composition used is that given for membrane no. 3) in the presence of varying concentration of Dy^{3+} ion in acetate buffer of pH 5.0: (1) 0.0, (2) 1.0×10^{-10}, (3) 1.0×10^{-9}, (4) 1.0×10^{-8}, (5) 1.0×10^{-7}, (6) 1.0×10^{-6}, (7) 1.0×10^{-5}, (8) 1.0×10^{-4}, (9) 1.0×10^{-3}, (10) 1.0×10^{-2} mol/L. *Reprinted with permission from [638].*

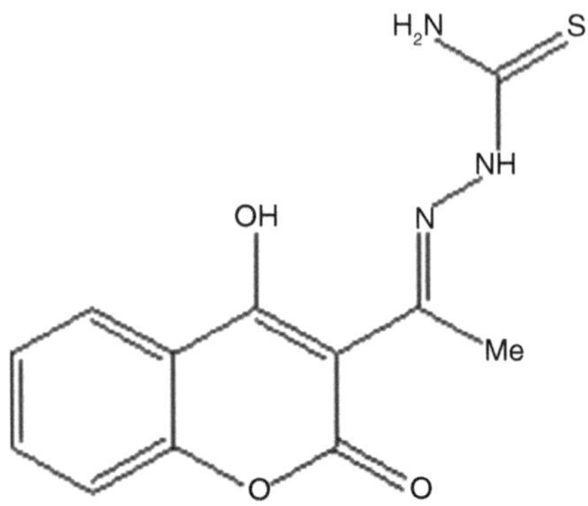

FIGURE 6.43 Chemical structure of (*E*)-2-(1-(4-hydroxy-2-*oxo*-2*H*-chromen-3-yl) ethylidene) hydrazinecarbothioamide as a Pr^{3+}-selective fluoroionophore.

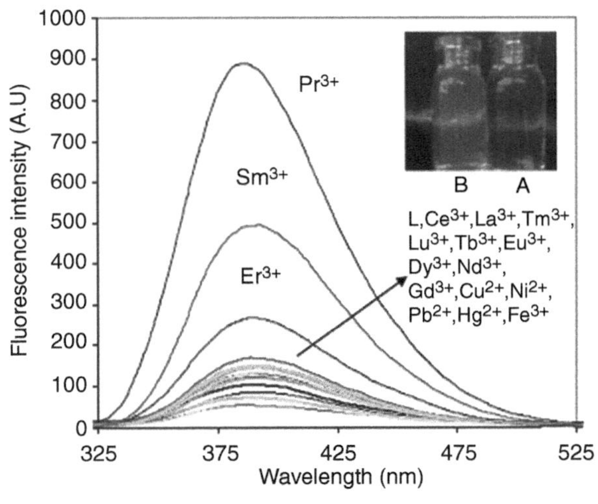

FIGURE 6.44 Fluorescence responses of L (5×10^{-6} mol/L) MeOH/H_2O (9/1:v/v) solution upon addition of cations (5×10^{-4} mol/L) (λ_{ex}= 310 nm). *Reprinted with permission from [639].*

(Fig. 6.43) and used the compound in a fluorescent probe for the detection of the same in aqueous solutions.

The fluorescence band of the ligand was found to be enhanced as a result of the presence of Pr^{3+} ions due to the formation of a 1:1 complex with Pr^{3+}. This behavior was found to linearly correlate with the Pr^{3+} concentration in the range of 1.6×10^{-7} to 1.0×10^{-5} mol/L, with a detection limit of 8.3×10^{-8} mol/L (Fig. 6.44), and was very selective toward the presence of this cation as opposed to the rest of the lanthanide elements tested.

In the same year, Ganjali et al. introduced a Eu^{3+}-sensitive fluorescent enhancement chemosensor [640], utilizing magnetic core–shell silica nanoparticle functionalized with cinchonidine (CD-Fe_3O_4@SiO_2) (Fig. 6.45). These

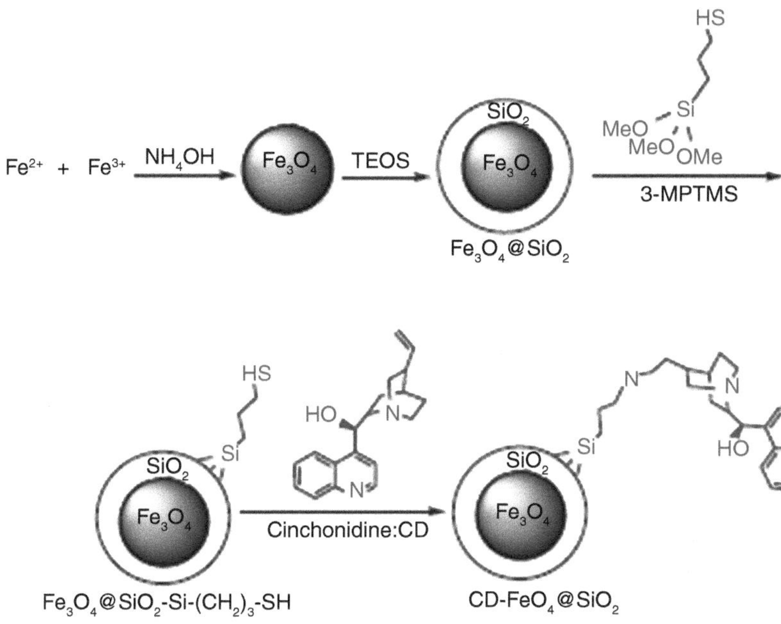

FIGURE 6.45 Synthesis route of CD-Fe_3O_4–SiO_2. *Reprinted with permission from [640].*

fluorescent nanochemosensors were found to have selective interactions with Eu^{3+} ions, and their emission intensity (at 375 nm) was significantly enhanced (Fig. 6.46) in the presence of various concentrations of Eu^{3+} (due to the photo-induced electron transfer [PET] mechanism), while negligible changes were reported for the interfering mono, di, and other trivalent cations tested.

Another report has been the development of a Lu^{3+}-sensitive fluorescent chemosensor using 8-hydroxyquinoline functionalized mesoporous silica with ordered hexagonal array structure (LUS-SPS-Q) (Fig. 6.47) [641]. During the fluorescence measurements (at an excitation wavelength of 330 nm), the emission intensity was observed to significantly increase in the presence of Lu^{3+}, which was not the case with the other ions tested as the interfering ones.

8-Hydroxyquinoline functionalized mesoporous silica with another ordered hexagonal-array structure (SBA-SPS-Q) also can be used as Lu^{3+}-sensitive fluorescent chemosensor [642]. This chemosensor demonstrated a selective interaction with the Lu^{3+} ion, most probably because of the presence of the fluorophore part at its surface. Fluorescence studies revealed that the emission intensity of the Lu^{3+}-bound mesoporous material increases significantly after addition of various concentrations of Lu^{3+}, whereas the mono-, di-, and trivalent cations result in either no changes or weak changes in

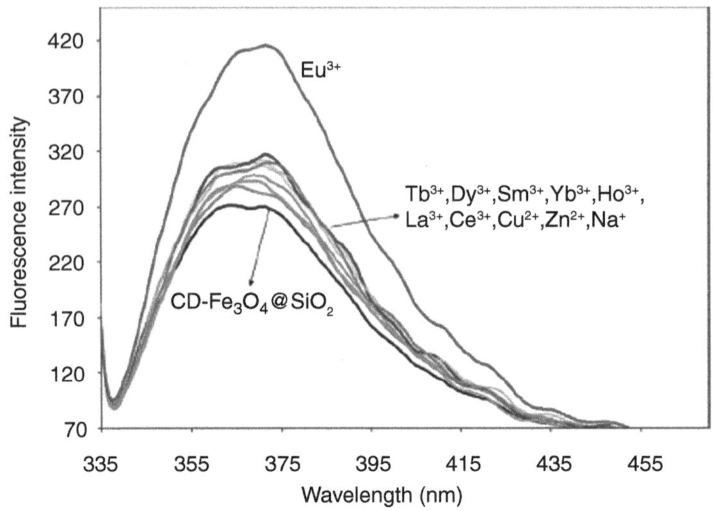

FIGURE 6.46 Fluorescence emission spectra of CD-Fe_3O_4–SiO_2 (3 mL, 0.8 mg/L) suspended in aqueous solution in the presence of different metal ions (1.0×10^{-4} mol/L). Excitation was performed at 360 nm. *Reprinted with permission from [640].*

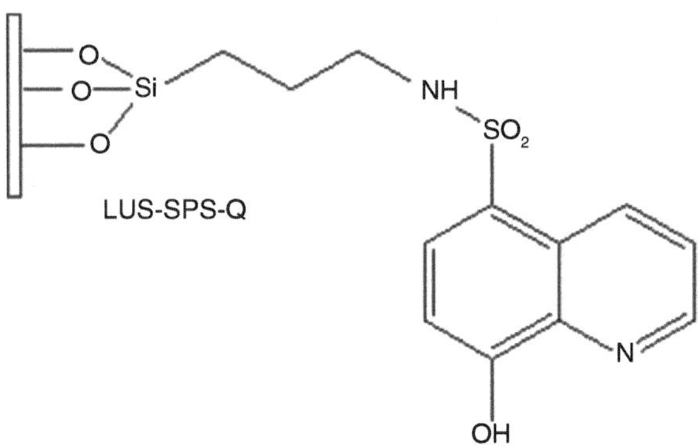

FIGURE 6.47 Lu^{3+}-sensitive fluorescent chemosensor using 8-hydroxyquinoline functionalized mesoporous silica.

fluorescent intensities. The enhancement of fluorescence is attributed to the strong covalent binding of the Lu^{3+} ion, evident from the large binding constant value (2 × 106 L/mol). The linear range of the fluorescent chemosensor covers a Lu^{3+} concentration from 1.6×10^{-7} to 2.6×10^{-6} mol/L, with a detection limit of 4.0×10^{-8} mol/L. This chemosensor has been applied to determination of Lu^{3+} ion in some soil samples where domestic devices were stored. SBA-SPS-Q showed a very weak fluorescence because of the PET from N atom of the pyridine to naphthalene (Fig. 6.48).

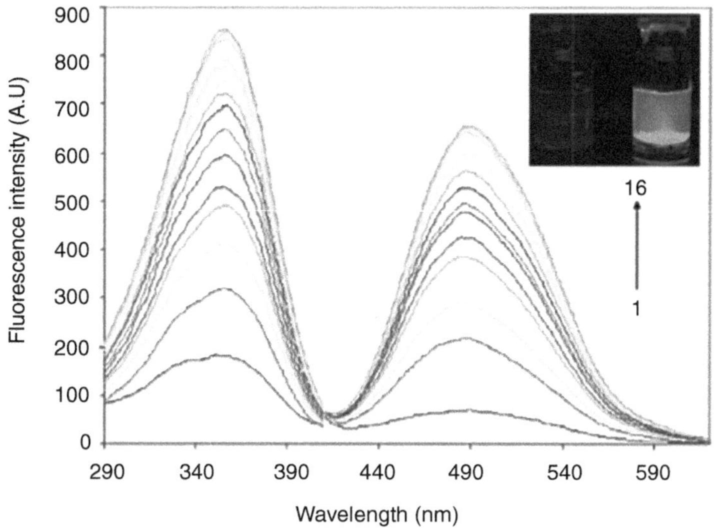

FIGURE 6.48 Excitation (A) and emission (B) spectra of the proposed chemosensor in the presence of varying concentrations of lu^{3+} ions: (1) 0, (2) 1.6×10^{-7} mol/L, (3) 3.3×10^{-7} mol/L, (4) 6.6×10^{-7} mol/L, (5) 1.0×10^{-6} mol/L, (6) 1.3×10^{-6} mol/L, (7) 1.6×10^{-6} mol/L, (8) 2.0×10^{-6} mol/L, (9) 2.3×10^{-6} mol/L, (10) 2.6×10^{-6} mol/L, (11) 3.0×10^{-6} mol/L, (12) 3.3×10^{-6} mol/L, (13) 4.0×10^{-6} mol/L, (14) 5.0×10^{-6} mol/L, (15) 6.0×10^{-6} mol/L, (16) 6.6×10^{-6} mol/L, $\lambda_{ex} = 360$ nm. (b) Visual fluorescence changes of sensor SBA-SPS-Q (100 mg/L) in the absence and excess presence of lu^{3+} ions (1.0×10^{-2} mol/L). The photo was taken under a handheld UV (365 nm) lamp. *Reprinted with permission from [642].*

After increasing the amounts of Lu^{3+}, SBA-SPS-Q showed a large CHEF (chelation-enhanced fluorescence) effect in the emission spectra, resulting from the blocking of the PET process.

Another fluorescent optode La^{3+} was developed by Shamsipur et al. in 2014 [643], where *N,N′*-bis(dansylamidoethyl)-*N,N′*-bis(2-pyridylmethyl) propylenediamine (Fig. 6.49) was incorporated into a plasticized PVC membrane containing *o*-NPOE as plasticizer, and the results showed a linear fluorescence–concentration behavior in the range of 5.0×10^{-9} to 5.0×10^{-4} mol/L, with a lower detection limit of 2.0×10^{-9} mol/L in pH values of 6.0. This equilibrium response was reached in less than 2 min.

The incorporation of lipophilic anionic TPB − , additives provide the optode with the necessary ion-exchanging properties. And under the experimental conditions, a 2:1 complex was reported to form the ligand and La^{3+} ions in the membrane phase according to a response mechanism as shown below:

$$La^{3+}_{(aq)} + 2L_{(org)} + 3NaTPB_{(org)} = \left[La(L)_2\right]^{3+}_{(org)} + 3TBP^{-}_{(org)} + 3Na^{+}(aq)$$

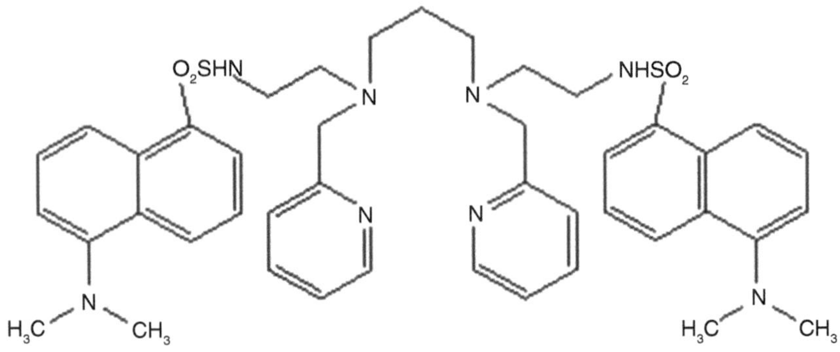

FIGURE 6.49 The chemical structure of ligand *N,N'*-bis(dansylamidoethyl)-*N,N'*-bis(2-pyridylmethyl)propylenediamine.

Very recently, Ganjali et al reported a very selective and sensitive Ho^{3+} optode [644]. Using *N,N*-bis(salicylidene)-naphthylene-1,8-diamine (Fig. 6.50) as the selective fluoroionophore, in a plasticized PVC membrane containing an anionic additive and at a pH of 5.4, observed a linear behavior in the range of 1.0×10^{-10} to 1.0×10^{-3} mol/L, at rather short response times of <1 min.

This work was followed by the development of a selective recognition system for dysprosium(III) ions using an enhanced chemiluminescence method by Hosseini et al. [645]. They studied the intensity of the light emitted by CdSe quantum dots (QDs)-H_2O_2 as a novel chemiluminescence mechanism for the determination of Dy^{3+}, based on the catalytic effect of Dy^{3+} ions, which led to considerable increase in the light emission during the reaction of the quantum dots with hydrogen peroxide. The response behavior was found to be linear from 8.3×10^{-7} mol/L to 5.0×10^{-6} mol/L, and a detection limit of 6.0×10^{-8} mol/L was achieved. A major experimental advantage of the method was reported to be its high selectivity to the analyte as compared to the rest of the coexisting cationic species (Fig. 6.51).

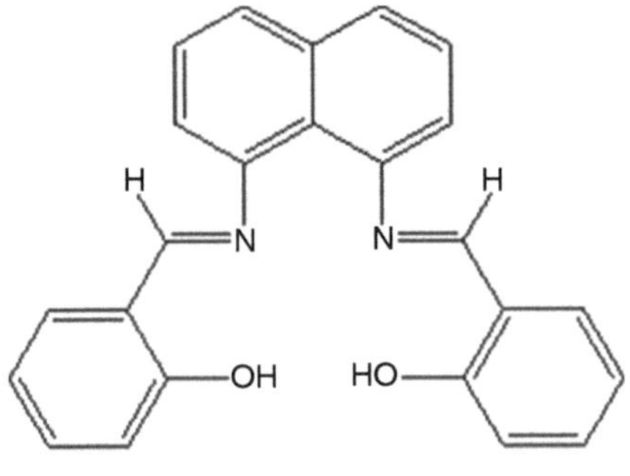

FIGURE 6.50 Chemical structure of *N,N*-bis(salicylidene)-naphthylene-1,8-diamine.

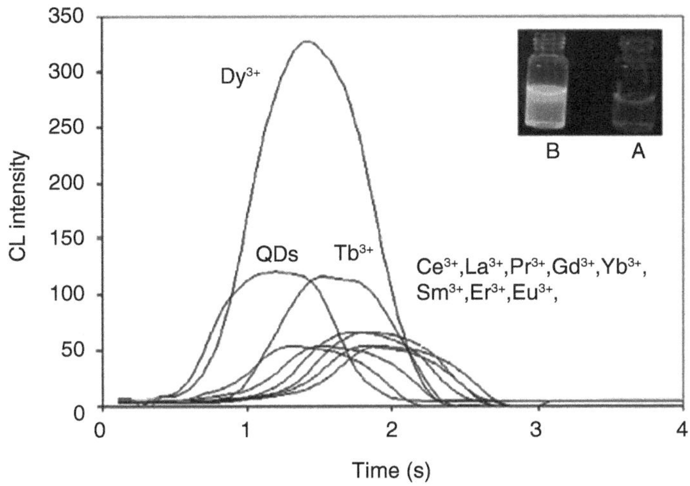

FIGURE 6.51 Effect of lanthanide ions 50 mL (1.0×10^{-5} M) on CL intensity H_2O_2–QDs (CdSe) system in optimal conditions. CL intensity H_2O_2–QDs (CdSe) system in absence (A) and presence of Dy^{3+} ions (B). *Reprinted with permission from [645].*

The overall mechanism is based on the fact that most CL reactions suffer from weak luminescence because of their low quantum efficiencies, but this can be overcome by enhancing the luminescence through the use of sensitizers, which are commonly fluorescent compounds with high quantum efficiency.

A turn-on fluorescent chemosensor also has been introduced by Faridbod et al. for the detection of Lu^{3+} ions using *N*-[3-methyl]-2-[pyridine-2-amido] phenyl] pyridine-2-carboxamide (Fig. 6.52) [646], the fluorescent emission intensity of which reportedly enhances in the presence of Lu^{3+} ions in ethanol–water solution, most probably as a result of a strong covalent binding between Lu^{3+} ion and the ligand. The linear response of the sensor has been reported to range from 3.3×10^{-7} to 1.0×10^{-5} mol/L, and a detection

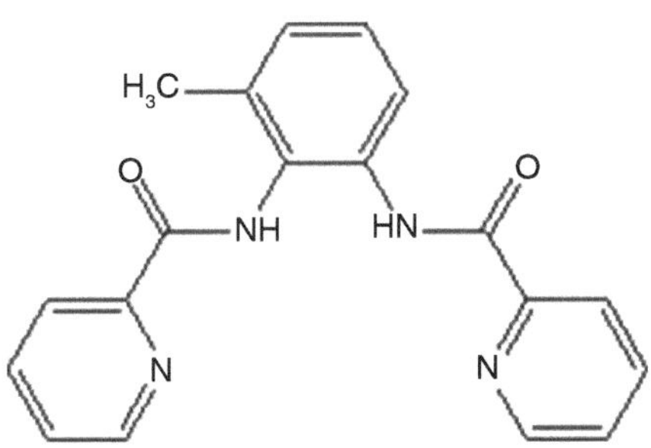

FIGURE 6.52 Chemical structure of *N*-[3-methyl]-2-[pyridine-2-amido] phenyl] pyridine-2-carboxamide.

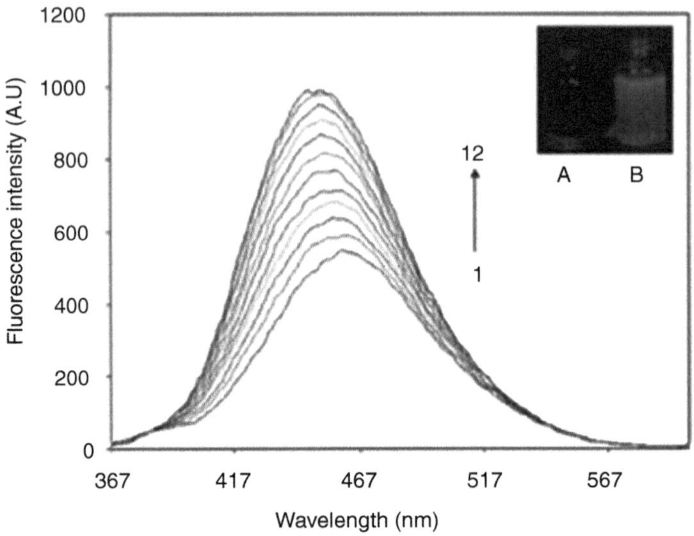

FIGURE 6.53 Fluorescence titration of L (5×10^{-6} mol/L) in EtOH/H_2O (70:30, v/v) solution in the presence of varying concentrations of Lu^{3+} ion: (1) 0, (2) 3.3×10^{-7} mol/L, (3) 5.0×10^{-7} mol/L, (4) 8.3×10^{-7} mol/L, (5) 1.6×10^{-6} mol/L, (6) 2.6×10^{-6} mol/L, (7) 3.6×10^{-6} mol/L, (8) 4.6×10^{-6} mol/L, (9) 5.6×10^{-6} mol/L, (10) 6.6×10^{-6} mol/L, (11) 1.0×10^{-5} mol/L, (12) 2.0×10^{-5} mol/L, λ_{ex} = 344 nm. (Inset) Visual fluorescence changes of sensor L (1.0×10^{-2} mol/L) in the absence (A) and excess presence of mercury ion (1.0×10^{-3} mol/L) (B). The photo was taken under a handheld UV (365 nm) lamp. *Reprinted with permission from [646].*

limit of 8.6×10^{-7} mol/L has been reported for it. The probe has high selectivity toward Lu^{3+} as compared to the other cations tested. Fluorescence titration of L (5×10^{-6} mol/L) in EtOH/H_2O (70:30, v/v) solution in the presence of varying concentrations of Lu^{3+} ion is shown in Fig. 6.53.

References

[1] García R, Báez AP. Atomic Absorption Spectrometry (AAS), Atomic Absorption Spectroscopy, Dr. Muhammad Akhyar Farrukh (Ed.) 2012; ISBN: 978-953-307-817-5, InTech, DOI: 10.5772/25925.

[2] Mossotti VG, Fassel VA. The atomic absorption spectra of the lanthanide elements. Spectrochim. Acta 1964;20:1117–27.

[3] Van Loon JC, Galbraith JH, Aarden HM. The Determination of Yttrium, Europium, Terbium, Dysprosium, Holmium, Erbium, Thulium, Ytterbium and Lutetium in Minerals by Atomic-absorption Spectrophotometry. Analyst 1971;96:47–50.

[4] Ooghe W, Verbeek F. Atomic Absorption Spectrometry of the Lanthanides in Minerals and Ores Anal Chim Acta 1974;73:87–95.

[5] Ooghe W, Verbeek F. Spectral Line Interference in the Atomic Absorption Spectrometry of Lanthanidrs Anal Chim Acta 1976;79:286–91.

[6] Sen Gupta JG. Determination of Lanthanides and Yttrium in Rocks and Mninerals by Atomic-Absorption and Flame-Emission Spectrometry. Talanta 1976;23:343–8.

[7] Montaser A, Mehrabzadeh AA. Determination of Ca, Er, Eu, Ga, In, K, Na, Mo and W by Atomic Absorption Spectrometry with an Electrothermal Graphite Braid Atomizer. Anal Chim Acta 1979;111:297–300.
[8] Sastry MKB, Savitri K, Babu Y, Joshi BD. Electrothermal AAS Studies of Sm, Eu, Dy, and Er Separated from Uranium. Fresenius Z. Anal Chem 1979;298:367–72.
[9] Sicifiska P, Michalewska M. Application of Atomic Absorption Speetrophotometry for the Determination of Lanthanides in Ores and Rare Earth Concentrates. Fresenius Z. Anal Chem 1982;312:530–2.
[10] Sen Gupta JG. Determination of Yttrium and Rare-Earth Elements in Rocks by Graohite-Furnance Atomic-Absorption Spectrometry. Talanta 1981;28:31–6.
[11] Ma YZ, Sun Di-jun BJ. Determination of Erbium by Electrothermal Atomic Absorption Spectrometry Using a Pyrolytic Graphite Coated Graphite Tube and a Pyrolytic Graphite Coated Graphite Tube Lined With Tantalum Foil, Held in Place by a Tungsten Spiral J Anal At Spectrom 1992;7:35–42.
[12] Wei J, Mu H, Wang X, Shi H, Lin F. Sensitivity Enhancement Effects of Organic Reagents on Ytterbium, Aluminium and Chromium in Atomic Absorption Spectrometry. J Anal At Spectrom 1992;7:175.
[13] Ma Y, Sun DJ. The determination of ytterbium with graphite furnace atomic absorption spectrometry using a pyrolytic graphite coated tube and a pyrolytic graphite coated tube lined with a tungsten spiral and a tantalum foil. Spectrochim Acta 1992;47B:45–471.
[14] Zhang J, Guo S. Determination of Ytterbium Using Electrothermal Atomic Absorption Spectrometry with Europium as Chemical Modifier. Analyst 1995;120:1661–4.
[15] Gao Y, Deng B. Experimental study on the reasons for low sensitivity and memory effects in the determination of lanthanum and samarium in graphite furnace atomic absorption spectrometry. Spectrochim Acta B 1996;51:1147–53.
[16] Byrne JP, Carambassis AL. Vaporization and atomization of neodymium in graphite furnace atomic absorption spectrometry. Spectrochim Acta B 1996;51:87–96.
[17] Lima EC, Krug FJ, Nóbrega JA, Nogueira ARA. Determination of ytterbium in animal faeces by tungsten coil electrothermal atomic absorption spectrometry. Talanta 1998;47:613–23.
[18] Arrebola A, Bagur MG, Sanchezvinas M, Gazquez D, Romero R, Camino M. Indirect determination of total yttrium and lanthanides by flame atomic absorption measurement of magnesium after extraction of their ternary complexes with purpurin (1,2,4-trihydroxyanthraquinone) and magnesium into isobutyl methyl ketone. J Anal At Spectrom 1998;13:765–8.
[19] He H, Zhang W, Ma G, Shen H. Determination of europium (III) by graphite furnace atomic absorption spectrometry with preconcentration by 2-thenoyl trifluoroacetone phenanthrolene modified electrode. Fenxi Huaxue 2001;29:1128.
[20] Li HZ, Zhai DT, Zhang J, Pei MS. Determination of ten trace rare earth elements in the sample by atomic emission spectrometry. Spectrosc Spect Anal 2005;25:1500–2.
[21] Biju VM, Rao TP. FAAS determination of selected rare earth elements coupled with multielement solid phase extractive preconcentration. Chemia Analityczna 2005;50:935–44.
[22] Amamoto K, Miura K, Yonetani A, Sakamoto H, Shirasaki T. Determination of rare-earth elements in rare-earth magnets by polarized zeeman flame atomic absorption spectrometry. Bunseki Kagaku 2011;60:873–6.
[23] Hou X, Jones BT. Inductively Coupled Plasma/Optical Emission Spectrometry. In: Meyers RA, editor. Encyclopedia of Analytical Chemistry. Chichester: Ó John Wiley & Sons Ltd 2000. p. 9468–85.
[24] Dean JR. Atomic Absorption and Plasma Spectroscopy. London: Wiley & Sons; 1997.
[25] Arruda MAZ. Trends in Sample Preparation. New York: Nova Publishers 2006. pp. 15–18.
[26] Koscielniakn P. Nonlinear calibration by the standard addition method. Chemom Intell Lab Syst 1999;47:275–87.

[27] Bings NH, Bogaerts A, Broekaert JAC. Atomic spectroscopy Anal Chem 2004;76:3313–36.
[28] Shirdama R, Modarres-Tehrani Z, Dastgoshadeh F. Microwave assisted digestion of soil, sludge and sediment for determination of heavy metals with ICP-OES and FAAS. Chem Rasayan J 2008;1:920–8.
[29] Armid A, Ohde S, Shinjo R, Toki T. Determination of uranium in pore water from coastal sediment by standard addition ICP-MS analysis. J Radioanal Nucl Chem 2008;27:233–7.
[30] Huang SDC, Beauchemin D. Direct multielemental analysis of human serum by ICP-MS with on-line standard addition using flow injection. J Anal At Spectrom 2003;18:951–2.
[31] Yamada K, McLeod CW, Kujirai O, Okochi H. On-line enrichment and determination of trace sulfur in high-purity iron samples by flow injection and inductively coupled plasma atomic emission spectrometry. J Anal At Spectrom 1992;7:661–5.
[32] Yousefi SR, Zolfonoun E, Ahmadi SJ. On-line standard addition technique for the determination of elements in the samples with high dissolved solids by ICP-OES: Application to determination of rare earth elements in geological samples. Anal Lett 2014;47:146–54.
[33] Boss CB, Fredeen KJ, Concepts, Instrumentation. Techniques in Inductively Coupled Plasma Optical Emission Spectrometry. 3rd edition USA: PerkinElmer, Inc.; 2004.
[34] Brenner IB, Watson AE, Steele TW, Jones EA, Goncalves M. Application of an argon-nitrogen inductively-coupled radiofrequency plasma (ICP) to the analysis of geological and related materials for their rare earth contents. Spectrochim. Acta Part B 1981;36:785–97.
[35] Walsh JN, Buckley F, Barker J. The simultaneous determination of the rare-earth elements in rocks using inductively coupled plasma source spectrometry. Chem Geol 1981;33:141–53.
[36] Crock JG, Lichte FE. Determination of rare earth elements in geological materials by inductively coupled argon plasma/atomic emission spectrometry. Anal Chem 1982;54:1329–32.
[37] Bolton A, Hwang J, Vandervoet A. The determination of scandium, yttrium, and selected rare earth elements in geological materials by inductively coupled plasma optical emission spectrometry. Spectrochim Acta Part B 1983;38:165–74.
[38] Ishii H, Satoh K. Determination of rare earths in lanthanum oxide by inductively-coupled plasma emission derivative spectrometer. Talanta 1983;30:111–5.
[39] Yoshida K, Fuwa K, Haraguchi H. Determination of 15 rare earth elements by liquid chromatography with inductively coupled plasma atomic emission spectrometric detection. Chem Lett 1983;12:1879–82.
[40] Brenner IB, Jones EA, Watson AE, Steele TW. The application of a N2-Ar medium-power ICP and cation-exchange chromatography for the spectrometric determination of the rare-earth elements in geological materials. Chem Geol 1984;45:135–48.
[41] Fujino O, Kometani Y, Sugiyama M, Matsui M. Determination of samarium in apatite minerals by solvent extraction-inductively coupled plasma atomic emission spectrometry. Bunseki Kagaku 1984;33:465–71.
[42] Aulis R, Bolton A, Doherty W, Vandervoet A, Wong P. Determination of yttrium and selected rare earth elements in geological materials using high performance liquid chromatographic separation and ICP spectrometric detection. Spectrochim. Acta Part B 1985;40:377–87.
[43] Fujino O, Araki K, Kometani Y, Sugiyama M, Matsui M. Determination of rare earth elements in phosphate minerals by inductively coupled plasma atomic emission spectrometry. Bunseki Kagaku 1985;34:386–91.
[44] Hou QL, Hughes TC, Haukka M, Hannaker P. Determination of the rare-earth elements in geological materials by thin-film X-ray fluorescence and inductively-coupled plasma atomic-emission spectrometry. Talanta 1985;32:495–9.

[45] Jarvis I, Jarvis KE. Rare-earth element geochemistry of standard sediments: a study using inductively coupled plasma spectrometry. Chem Geol 1985;53:335–44.
[46] Toyoda K, Haraguchi H. Determination of rare-earth elements in geological standard rock samples by inductively coupled plasma atomic emission spectrometry. Chem Lett 1985;7:981–4.
[47] Buchanan SJ, Dale LS. Inductively coupled plasma-atomic emission spectrometric determination of rare earth elements in geological materials. Spectrochim. Acta Part B 1986;41:237–42.
[48] Iwasaki K, Fuwa K, Haraguchi H. Simultaneous determination of 14 lanthanides and yttrium in rare earth ores by inductively-coupled plasma atomic emission spectrometry. Anal Chim Acta 1986;183:239–49.
[49] Roelandts I, Michel G. Sequential Inductively Coupled Plasma Determination of Some Rare-Earth Elements in Five French Geostandards. Geostand Newsl 1986;10:135–54.
[50] Toyoda K, Haraguchi H. Rare earth elements in six new. GSJ standard rock samples as determined by inductively coupled plasma atomic emission spectrometry. Geostand Newsl 1986;10:173–5.
[51] Govindaraju K, Mevelle G. Fully automated dissolution and separation methods for inductively coupled plasma atomic emission spectrometry rock analysis. Application to the determination of rare earth elements Plenary lecture. J Anal At Spectrom 1987;2:615–21.
[52] Kanicky V, Toman J, Malecek M, Povolny A. Determination of light rare-earth elements in hydroxide precipitates of apatite and bastnesite using inductively coupled argon plasma atomic emission-spectrometry. Chem Listy 1987;81:1304–14.
[53] Roelandts I. Determination of nine rare earth elements in eight USGS reference samples and in the ANRT Glauconite, GL-O, by sequential inductively coupled argon plasma spectrometry. Geostand Newsl 1987;11:49–74.
[54] Sengupta JG. Analysis of the CCRMP Oka-2 rare-earth reference mineral of the britholite-apatite series by electrothermal atomic-absorption and inductively-coupled plasma atomic-emission spectrometry. Talanta 1987;34:1043–7.
[55] Toyoda K, Haraguchi H. Geochemical characterization of geological standard rock samples from rare earth element distribution patterns measured by inductively coupled plasma atomic emission spectrometry. Bull Chem Soc Japan 1987;60:933–9.
[56] Iwasaki K, Haraguchi H. Determination of rare earth elements in geological samples by inductively-coupled plasma atomic emission spectrometry after oxalate coprecipitation and cation-exchange column separation. Anal Chim Acta 1988;208:163–72.
[57] Jarvis KE, Jarvis I. Determination of the Rare-Earth Elements and Yttrium In 37 International Silicate Reference Materials By Inductively Coupled Plasma-Atomic Emission Spectrometry. Geostand Newsl 1988;12:1–12.
[58] Kanicky V, Toman J, Malecek M. Determination of Tb, Dy, Ho, Er and Yb in ores and concentrates using inductively coupled argon plasma atomic emission-spectrometry. Chem Listy 1988;82:630–42.
[59] Roelandts I. Application of inductively coupled plasma spectrometry to the determination of nine rare earth elements in nine new United States geological survey reference samples. At Spectrosc 1988;9:49–54.
[60] Roelandts I. Comparison of inductively coupled plasma and neutron activation analysis for precise and accurate determination of nine rare-earth elements in geological materials. Chem Geol 1988;67:171–80.
[61] Zachmann DW. Matrix effects in the separation of rare earth elements, scandium, and yttrium and their determination by inductively coupled plasma optical emission spectrometry. Anal Chem 1988;60:420–7.
[62] Roychowdhury P, Roy NK, Das DK, Das AK. Determination of rare-earth elements and yttrium in silicate rocks by sequential inductively-coupled plasma emission spectrometry. Talanta 1989;36:1183–6.

[63] Sulcek Z, Rubeska I, Sixta V, Paukert T. Determination of rare-earth elements and yttrium in rocks using the plasma-II ICP emission spectrometer. At Spectrosc 1989;10:4–9.
[64] Watkins PJ, Nolan J. Determination of Rare-Earth Elements, Scandium, Yttrium And Hafnium in 32 Geochemical Reference Materials Using Inductively Coupled Plasma-Atomic Emission Spectrometry. Geostand Newsl 1990;14:11–20.
[65] Webster JR, Gilstrap MS. Matrix-independent separation of rare-earth elements and yttrium from geological materials using constant calcium content-oxalate precipitation and cation exchange for determination by high-resolution inductively coupled plasma atomic emission spectrometry (ICP-AES). Chem Geol 1990;85:287–94.
[66] Weiss D, Paukert T, Rubeska I. Determination of rare earth elements and yttrium in rocks by inductively coupled plasma atomic emission spectrometry after separation by organic solvent extraction. J Anal At Spectrom 1990;5:371–5.
[67] Almerey R, Bowen HJM. Determination of the rare-earth elements in phosphate rocks using both inductively coupled plasma atomic emission spectrometry and instrumental neutron activation analysis. J Radioan Nucl Chem Lett 1991;153:221–34.
[68] Gumgum B, Hamamci C. determination of some rare elements in phosphate rocks by inductively coupled plasma atomic emission-spectrometry. Spectrosc Lett 1991;24:1229–36.
[69] Rathi MS, Khanna PP, Mukherjee PK. Determination of ten rare-earth elements and yttrium in silicate rocks by ICP-AES without separation and preconcentration. Talanta 1991;38:329–32.
[70] Roelandts I. Inductively coupled plasma determination of nine rare-earth elements in the U.S. Geological Survey Kesterson Bottom Sediment, BSK-1. Spectrochim Acta Part B 1991;46:79–84.
[71] Eid MA, Broekaert JAC, Tschopel P. Application of ICP-AES to the determination of rare earth elements in phosphate samples. Fresenius J Anal Chem 1992;342:107–12.
[72] Roelandts I. Determination of 10 rare earth elements in rare earth ores by inductively coupled plasma atomic emission spectrometry. At Spectrosc 1992;13:193–8.
[73] Roelandts I, Deblond A. Rare-earth element composition of devonian sediments from southern belgium: application of an inductively coupled plasma-atomic emission spectrometry method. Chem Geol 1992;95:167–76.
[74] Rucandio MI. Determination of lanthanides and yttrium in rare earth ores and concentrates by inductively coupled plasma atomic emission spectrometry. Anal Chim Acta 1992;264:333–44.
[75] Brenner IB, Dorfman E. Application of ultrasonic nebulization for the determination of rare earth elements in phosphates and related sedimentary rocks using inductively coupled plasma atomic emission spectrometry with comments on dissolution procedures. J Anal Atom Spectrom 1993;8:833–8.
[76] Kuzmin NM, Pukhovskaya VM, Varshal GM, Spivakov BY, Maryutina TA, Volynets MP, Ryabukhin VA, Chkhetiya NN, Grebneva ON, Pavlutskaya VI. Determination of rare earth elements geological specimens by atomic-emission spectroscopy with inductively coupled plasma. J Anal Chem 1993;48:635–44.
[77] Pukhovskaya VM, Maryutina TA, Grebneva ON, Kuzmin NM, Spivakov BY. Inductively coupled plasma atomic emission spectroscopic determination of rare earth elements in geological samples after preconcentration by countercurrent chromatography-I. Spectrochim. Acta Part B 1993;48:1365–70.
[78] Voldet P. From neutron activation to inductively coupled plasma-atomic emission spectrometry in the determination of rare-earth elements in rocks. Trac-Trend Anal Chem 1993;12:339–44.
[79] Cantagrel F, Pin C. Major, minor and rare-earth element determinations in 25 rocks standards by ICP-atomic emission spectrometry. Geostand Newsl 1994;18:123–38.
[80] Germanique JC. Major, trace and rare-earth elements in fourteen GSJ reference samples. Determination by X-ray fluorescence spectrometry and inductively coupled plasma optical emission spectrometory. Geostandard Newslett 1994;18:91–100.

[81] Balaram V, Ramesh SL, Anjaiah KV. Comparative study of the sample decomposition procedures in the determination of trace and rare earth elements in anorthosites and related rocks by ICP-MS. Fresenius J Anal Chem 1995;353:176–82.
[82] Farinas JC, Cabrera HP, Larrea MT. Improvement in the ion exchange chromatographic separation of rare earth elements in geological materials for their determination by inductively coupled plasma atomic emission spectrometry. J Anal At Spectrom 1995;10:511–6.
[83] Pukhovskaya VM, Grebneva ON, Maryutina TA, Kuzmin NM, Spivakov BY. Inductively coupled plasma atomic emission spectroscopic determination of rare earth elements in geological samples after preconcentration by countercurrent chromatography-Part II. Spectrochim Acta Part B 1995;50:5–12.
[84] Safronova NS, Matveeva SS, Fabelinsky YI, Ryabukhin VA. Inductively coupled plasma atomic emission spectrometric determination of rare earth elements in granites and greisens using thin-layer chromatographic preconcentration. Analyst 1995;120:1427–32.
[85] Grebneva ON, Bayunov PA, Kubrakova LV, Kuzmin NM. Inductively coupled plasma atomic emission determination of 13 rare-earth elements and yttrium in geological samples after microwave decomposition. J Anal Chem 1996;51:927–30.
[86] Rathi MS, Khanna PP, Mukherjee PK, Saini NK. Evaluation of REE determination in geological samples by inductively coupled plasma emission spectrometry. J Geol Soc India 1996;48:325–30.
[87] Satyanarayana K. Determination of rare earth elements, yttrium, scandium, and thorium in niobate-tantalate, carbonatite, and fergusonite samples using ICP-AES. At Spectrosc 1996;17:69–76.
[88] Agrawal YK, Shrivastav P. Solvent extraction, spectrophotometric and inductively coupled plasma atomic emission spectroscopic (ICP-AES) determination of lanthanum(III) with crown hydroxamic acid. Talanta 1997;44:1307–12.
[89] Das DK, Roychowdhury P. Determination of rare earth elements in bauxite by inductively coupled plasma atomic emission spectrometry. At Spectrosc 1997;18:80–3.
[90] Rucandio MI. Cation-exchange isolation and ICP-AES determination of rare earth elements in geological silicate materials. Fresenius J Anal Chem 1997;357:661–9.
[91] Ito K, Masukawa F, Mao QL, Hirokawa T. Inductively coupled plasma-atomic emission spectral fitting analysis of rare-earth elements in ion-adsorption type ores. Anal Chim Acta 1998;362:241–5.
[92] Ramanaiah GV. Determination of yttrium, scandium and other rare earth elements in uranium-rich geological materials by ICP-AES. Talanta 1998;46:533–40.
[93] Satyanarayana K, Durani S, Ramanaiah GV. Determination of scandium in geological materials, rare earth minerals and niobate/tantalate-type of samples by inductively coupled plasma atomic emission spectrometry after solvent extraction acid hydrolysis separation. Anal Chim Acta 1998;376:273–81.
[94] Srivastava PK, Premadas A. Determination of rare earth elements and yttrium in rocks by inductively coupled plasma atomic emission spectrometry after solvent extraction with a mixture of 2-ethylhexyl dihydrogenphosphate and bis(2-ethylhexyl) hydrogenphosphate. J Anal At Spectrom 1999;14:1087–91.
[95] Jaron I, Kudowska B, Bulska E. Determination of rare earth elements in geological samples by ICP-OES. At Spectrosc 2000;21:105–10.
[96] Khorge CR, Chakraborty P, Saran R. Determination of rare earth elements in iron-rich geological samples by ICP-OES. At Spectrosc 2000;21:220–4.
[97] Lara R, Olsina RA, Marchevsky E, Gasquez JA, Martinez LD. ICP-OES determination of rare earth elements in silicate rocks using ultrasonic nebulization and on-line ion-exchange tron separation. At Spectrosc 2000;21:172–8.
[98] Boaventura G. R.; De Oliveira, R. C.; Santelli, R. E. Off-line and on-line determination of eleven rare earth elements in silicate rocks by ICP-AES using Dowex minicolumns for separation of interfering elements in continuous flow systems. Geostand Newsl 2002;26:63–73.

[99] Navarro MS, Ulbrich H, Andrade S, Janasi VA. Adaptation of ICP-OES routine determination techniques for the analysis of rare earth elements by chromatographic separation in geologic materials: tests with reference materials and granitic rocks. J Alloy Compd 2002;344:40–5.

[100] Pozebon D, Martins P. Marine sediment analysis using inductively coupled plasma optical emission spectrometry. At Spectrosc 2002;23:111–8.

[101] Hang YP, Qin YC, Shen J. Separation and microcolumn preconcentration of traces of rare earth elements on nanoscale TiO_2 and their determination in geological samples by ICP-AES. J Sep Sci 2003;26:957–60.

[102] Premadas A. Cation exchange chromatographic group separation and ICP-AES determination of rare earth elements and yttrium in refractory minerals zircon, ilmenite, rutile, columbite-tantalite, garnet, and silliminite. At Spectrosc 2003;24:149–58.

[103] Xu ZF, Liu CQ, Zhang HX, Ma YJ, Lin SL. Determination of rare earth elements in geological samples by inductively coupled plasma atomic emission spectrometry with flow injection liquid-liquid extraction. Anal Sci 2003;19:1625–9.

[104] Yenisoy-Karakas S, Gaga EO, Dogangun A, Tuncel SG. Determination of major and rare earth elements in bastnasite ores by ICP-AES. Anal Lett 2004;37:2701–9.

[105] Gasquez JA, DeLima E, Olsina RA, Martinez L. D.; de la Guardia, M. A fast method for apatite selective leaching from granitic rocks followed through rare earth elements and phosphorus determination by inductively coupled plasma optical emission spectrometry. Talanta 2005;67:824–8.

[106] Premadas A, Cyriac B, Satyanarayana K, Precipitative Separation. ICP-AES Determination of Rare Earth Elements, Yttrium, Scandium, and Thorium in Different Types of Geological Samples Including Iron- and Uranium-Rich Materials. At Spectrosc 2009;30:65–74.

[107] Krishnakumar M, Kumar M. Determination of Rare Earth Elements and Yttrium in Geological Materials by Inductively Coupled Plasma Atomic Emission Spectrometry Using Cu-DDTC as Coprecipitative Carrier. At Spectrosc 2011;32:223–7.

[108] Guimaraes-Silva AK, de Lena JC, Froes RES, Costa LM, Nascentes CC. Evaluation of Signal-to-Background and Mg II/Mg I Ratios as Response for the Optimization of Rare Earth Elements Determination by Inductively Coupled Plasma Optical Emission Spectrometry. J Braz Chem Soc 2012;23:753–62.

[109] Premadas A. Determination of Trace Level Lanthanides and Thorium by Inductively Coupled Plasma Atomic Emission Spectrometry in Different Types of Geological, Red Mud, and Coal Fly Ash Samples After Separation as Oxalates Using Calcium as Carrier. At Spectrosc 2012;33:14–23.

[110] Kojima H, Kitazume E, Nagata F, Ezawa M. Analysis of elements (Ca, Fe, Ge, Lu, Sm, Y) in bubble garnet films by inductively coupled plasma atomic emission spectrometry. Bunseki Kagaku 1981;30:667–72.

[111] Wong KL. Determination of trace amounts of cerium in paint by inductively coupled plasma atomic emission spectrometry. Anal Chem 1981;53:2148–9.

[112] Tanaka T, Yamada T, Johnokuchi T, Yamada J, Kumamoto K, Hattori K. Determination of trace rare earth elements in high-purity yttrium oxide by inductively coupled plasma-atomic emission spectroscopy. Bunseki Kagaku 1982;31:385–90.

[113] Fabec JL, Ruschak ML. Determination of aluminum, cobalt, lanthanum, and ruthenium on alumina-based catalysts by inductively coupled plasma atomic emission spectrometry. Anal Chem 1983;55:2241–6.

[114] Fischer PT, Ellgren AJ. Analysis of rare earth-containing metallurgical samples by inductively coupled plasma-atomic emission spectrometry. Spectrochim Acta Part B 1983;38:309–16.

[115] Mahanti HS, Barnes RM. Determination of rare earth elements in aluminum by inductively coupled plasma-atomic emission spectroscopy. Appl Spectrosc 1983;37:261–3.

[116] Ru NC, Chang WM, Jiang ZC, Zeug YE. The determination of trace rare earth elements in high-purity yttrium oxide by inductively coupled plasma-atomic emission spectroscopy. Spectrochim. Acta Part B 1983;38:175–82.
[117] Fries T, Lamothe PJ, Pesek JJ. Determination of Rare Earth Elements, Yttrium, and Scandium in Manganese Nodules by Inductively Coupled Argon Plasma-Atomic Emission Spectrometry. Anal Chim Acta 1984;159:329–36.
[118] Fujimoto K, Okano T, Matsumura Y. Analysis of rare-earth permanent magnets (Sm-Co, Nd-Fe-B) by ion chromatography and inductively coupled plasma atomic emission spectroscopy. T Iron Steel I Japan 1986;26:B248–B1248.
[119] Murty PS, Barnes RM. Determination of trace rare earth elements in uranium by inductively coupled plasma atomic emission spectrometry. J Anal At Spectrom 1986;1:145–8.
[120] Sanzmedel A, Uria JES, Gancedo MT, Suarez AF. Inductively coupled plasma emission spectrometry in the analysis of rare earth-transition metal ultra-thin magnetic films. J Anal At Spectrom 1986;1:149–52.
[121] Wu CI, Chang FC, Wang SJ. The Simultaneous Determination of Trace Dy, Eu, Gd and Sm in Nuclear Grade Uranium Oxide by Inductively Coupled Plasma-Atomic Emission Spectrometry (ICP-AES). J Chin Chem Soc 1986;33:35–40.
[122] Grossman AM, Ciba J, Jurczyk J, Spiewok W. Determination of lanthanum, cerium, praseodymium and neodymium in alloy steels by inductively coupled plasma atomic-emission spectrometry. Talanta 1990;37:815–8.
[123] Nakamura Y, Takahashi K, Kujirai O, Okochi. Determination of rare earth elements in yttrium, lanthanum, gadolinium and ytterbium matrices using an échelle-type spectrometer and inductively coupled plasma atomic emission. J Anal At Spectrom 1990;5:501–8.
[124] Yang JF, Zeng XJ, Huang BL. Determination of trace amount of 14 rare earth impurities in high purity yttrium oxide with ICP-AES. Acta Chim Sinica 1990;48:903–8.
[125] Biswas SS, Kaimal R, Sethumadhavan A, Murty PS. Determination of Ce, Pr, Nd and Sm in High Purity La_2O_3 by Inductively Coupled Plasma Atomic Emission Spectrometry. Anal Lett 1991;24:1885–95.
[126] Biswas SS, Sethumadhavan A, Murty PS. Determination of Y, Sm, Eu, Gd, Dy, Ho, Er in high purity terbium oxide by ICP-AES. Mikrochim Acta 1991;1:71–7.
[127] Nakamura Y, Murai Y, Ni DZ, Liu YM. Determination of rare-earth impurities in terbium oxide by HPLC ICP-AES. Bunseki Kagaku 1991;40:T125–30.
[128] Kobayashi S, Wakui Y, Kanesato M, Matsunaga H, Suzuki TM. Chromatographic separation and inductively coupled plasma atomic emission spectrometric determination of the rare earth metals contained in terbiu. Anal Chim Acta 1992;262:161–6.
[129] Lopezmolinero A, Villareal A, Anzano J. Erbium determination in preforms of optical fibres by inductively coupled plasma-atomic emission spectrometry. Talanta 1992;39:191–3.
[130] Marathe SM, Biswas SS, Patil PB, Murty PS. An ICP-AES method for the determination of heavy rare earth elements (Eu-Lu) in high purity yttrium oxide. Microchim Acta 1992;109:261–8.
[131] Roelandts I. Inductively coupled plasma determination of nine rare-earth elements in five ferro-manganese nodule reference samples. Spectrochim Acta Part B 1992;47:947–55.
[132] Vanveen EH, Deloosvollebregt MTC, Wassink AP, Kalter H. Determination of Trace Elements in Uranium by Inductively. Coupled Plasma-Atomic Emission Spectrometry Using Kalman Filtering. Anal Chem 1992;64:1643–9.
[133] Darbha L, Gangadharan S. Determination of rare earth elements in Gd_2O_3 by ICP-OES. Fresenius J Anal Chem 1994;348:284–6.
[134] Yang XJ. Extractive chromatographic separation and inductively coupled plasma atomic emission spectrometric determination of trace impurities in high purity europium oxide. Talanta 1994;41:1807–13.

[135] Balaram V, Anjaiah KV, Reddy MRP. Comparative study on the trace and rare earth element analysis of an Indian polymetallic nodule reference sample by inductively coupled plasma atomic emission spectrometry and inductively coupled plasma mass spectrometry. Analyst 1995;120:1401–6.

[136] Mazzucotelli A, Bavastello V, Magi E, Rivaro P, Tomba C. Analysis of gadolinium polyaminopolycarboxylic complexes by HPLC–ultrasonic nebulizer–ICP-AES hyphenated technique. Anal Proceed 1995;32:165–7.

[137] Daskalova N, Velichkov S, Slavova P. Spectral interferences in the determination of traces of scandium, yttrium and rare earth elements in "pure" rare earth matrices by inductively coupled plasma atomic emission spectrometry .3. Europium Spectrochim Acta Part B 1996;51:733–68.

[138] Gu S, Ying H, Zhang ZG, Zhuang ZX, Yang PY, Wang XR, Huang BL, Li B. High-resolution spectra of selected rare earth elements and spectral interferences studied by inductively coupled plasma-atomic emission spectroscopy (ICP-AES). Spectrochim Acta Part B 1997;52:1567–74.

[139] Shmanenkova GI, Teselkina AE, Karpov YA, Kolesova LV, Nedler VV, Shchelkova VP. Determination of minor rare-earth elements in pure cerium dioxide by inductively coupled plasma emission spectrometry. Ind Lab 1997;63:473–4.

[140] Shmanenkova GI, Teselkina AE, Karpov YA, Kolesova LV, Nedler VV, Shchelkova VP. Determination of admixtures of rare-earth elements in pure rare-earth bases by atomic emission spectrometry with inductively coupled plasma. Ind Lab 1997;63:214–7.

[141] Bian QZ, Peng SB, He BL, Zhong ZG. Direct determination of rare earth elements in rare earth chloride and light rare earth oxide by ICP-AES. Spectrosc Spect Anal 2000;20:357–60.

[142] Cai B, Hu B, Xiong HC, Liao ZH, Mao LS, Jiang ZC. Preconcentration of lanthanum, europium and ytterbium with tribromoarsenazo-cetylpyridinium bromide supported on microcrystalline naphthalene and their determinations by inductively coupled plasma atomic emission spectrometry. Talanta 2001;55:85–91.

[143] Paama L, Parnoja E, Must M, Peramaki P. Optimal conditions for europium and samarium determination in cathodoluminophors by inductively coupled plasma atomic emission spectrometry. J Anal At Spectrom 2001;16:1333–6.

[144] Premadas A, Srivastava PK. Inductively coupled plasma atomic emission spectrometric determination of lanthanides and Y in various uranium hydrometallurgical products. J Radioanal Nucl Chem 2002;251:233–9.

[145] Zhao JQ, Feng YH, Cheng W, Lu BY. Determination of fuorteen trace rare earth and twenty non-rare earth impurities in high purity europium oxide by inductively coupled plasma-atomic emission spectrometry. Spectrosc Spect Anal 2003;23:972–5.

[146] Mitsumata H, Aoki I. Quantitative determination of barium, magnesium, aluminum and europium in blue phosphors for fluorescent lamp using ICP-AES after alkali fusion. J Ceram Soc Japan 2004;112:608–11.

[147] Zhang L, Wang JC, Xin RX. Determination of rare earth elements in trialkyl phoshine oxide (TRPO)-kerosene by ICP-AES. Spectrosc Spect Anal 2005;25:1684–7.

[148] Choi KS, Lee CH, Kim JG, Kim WH, Kang JG. Separating Ag, B, Cd, Dy, Eu, and Sm in a Gd matrix using 2-ethylhexyl phosphonic acid mono-2-ethylhexyl ester extraction chromatography for ICP-AES analysis. Talanta 2007;71:662–7.

[149] Liu PY. The determination of La, Ce, Pr, Nd, Ni, Co and Mn in rare earth hydrogen storage alloy (AB(5) Type) by inductively coupled plasma atomic emission spectrographic method. J Ceram Process. Res 2007;8:161–3.

[150] Maind SD, Chattopadhyay N, Gandhi C, Kumar SC, Sudersanan M. Quantitative evaluation of europium in blue ballpoint pen inks/offset printing inks tagged with europium thenoyltrifluoroacetonate by spectrofluorometry and ICP-AES. Sci. Justice 2008;48:61–6.

[151] Gong Q, Chen J, Ji RW, Pan XZ, Wu J. Separation with Ion Exchange Fiber Column and Determination of La, Nd, Eu and Gd in High Purity Ytterbium Oxide by ICP-AES. J Hazard Mater Spectrosc Spect Anal 2010;30:523–7.
[152] Mahanta PL, Chakrapani G, Radhamani R. Determination of Rare Earth Elements in Uraninite Samples by ICP-OES After Solid Phase Extraction on Activated Carbon. At Spectrosc 2010;31:21–6.
[153] Xu YY, Wang H, Wang GX, Wu C, Zhou JF. Determination of Trace Amounts of Rare-Earth Elements in Clean Steel by Inductively Coupled Plasma Optical Emission Spectrometry after Removal of Iron Matrix with Anion-Exchange Resin. Appl Spectrosc 2010;64:543–6.
[154] Romanova NB, Pechishcheva NV, Shunyaev KY, Titov VI, Gundobin NV. Determining low concentrations of Zr, Ce, La, and Y in heat-resistant nickel alloys by inductively coupled plasma atomic emission. Inorg Mater 2012;48:1315–9.
[155] Daidoji H, Matsubara M. Determination of rare earth elements, thorium, and uranium in mussel, ark shell, and coral by inductively coupled plasma atomic emission spectrometry. Bunseki Kagaku 1985;34:T123–7.
[156] Daidoji H, Tamura S, Matsubara M. Determination of rare earth elements and thorium in sea water by inductively coupled plasma-atomic emission spectrometry. Bunseki Kagaku 1985;34:340–5.
[157] Mazzucotelli A, Frache R, Depaz F. Inductively Coupled Plasma Atomic Emission and Electrothermal Atomic Absorption Spectroscopy of Trace Amounts of Erbium in Environmental Samples. Appl Spectrosc 1991;45:507–71.
[158] Zhan GY, Su ZX, Lou XY, Chang XJ. Efficiency of Macroporous Poly(Vinylphosphoramidic Acid) Resin Adsorbing of Selected Elements and Determination of Trace Dysprosium Holmium Erbiom and Ytterbium in Waste Water by Inductively Coupled Plasma Optical Emission Spectrometry. Anal Lett 1992;25:561–72.
[159] Kubova J, Nevoral V, Stresko V. Determination of rare earth elements in mineral waters by inductively coupled plasma atomic emission spectrometry. J Anal At Spectrom 1994;9:241–3.
[160] Benkhedda K, Infante HG, Ivanova E, Adams FC. Determination of sub-parts-per-trillion levels of rare earth elements in natural waters by inductively coupled plasma time-of-flight mass spectrometry after flow injection on-line sorption preconcentration in a knotted reactor. J Anal At Spectrom 2001;16:995–1001.
[161] Liang P, Hu B, Jiang ZC, Qin YC, Peng TY. Nanometer-sized titanium dioxide micro-column on-line preconcentration of La, Y, Yb, Eu, Dy and their determination by inductively coupled plasma atomic emission spectrometry. J Anal At Spectrom 2001;16:863–6.
[162] Djingova R, Ivanova J. Determination of rare earth elements in soils and sediments by inductively coupled plasma atomic emission spectrometry after cation-exchange separation. Talanta 2002;57:821–9.
[163] Duan TC, Chen HT, Zeng XJ. Determination of rare and rare earth elements in soils and sediments by ICP-MS using Ti(OH)(4)-Fe(OH)(3) co-precipitation preconcentration. J Anal At Spectrom 2002;17:410–3.
[164] Murty DSR, Mohanta PL, Radhamani R. Determination of rare earth elements in different geological matrices by ICP-AES after solid phase micro extraction on activated charcoal. Atom Spectrosc 2002;23:65–74.
[165] Bahramifar N, Yamini Y. On-line preconcentration of some rare earth elements in water samples using C-18-cartridge modified with 1-(2-pyridylazo) 2-naphtol (PAN) prior to simultaneous determination by inductively coupled plasma optical emission spectrometry (ICP-OES). Anal Chim Acta 2005;540:325–32.
[166] Liang P, Chen XG. Preconcentration of rare earth elements on silica gel loaded with 1-phenyl-3-methyl-4-benzoylpyrazol-5-one prior to their determination by ICP-AES. Anal Sci 2005;21:1185–8.

[167] Liang P, Liu Y, Guo L. Determination of trace rare earth elements by inductively coupled plasma atomic emission spectrometry after preconcentration with multiwalled carbon nanotubes. Spectrochim Acta Part B 2005;60:125–9.
[168] Pasinli T, Eroglu AE, Shahwan T. Preconcentration and atomic spectrometric determination of rare earth elements (REEs) in natural water samples by inductively coupled plasma atomic emission spectrometry. Anal Chim Acta 2005;547:42–9.
[169] Suleiman JS, Huang CZ, Pu XL, Zhang N, Hu B. Microcolumn separation and preconcentration of rare earth elements as pyrocatechol violet complexes on anthracene-cetyltrimethyl ammonium bromide adsorbent and their determination in environmental samples by ICP-OES. At Spectrosc 2006;27:200–6.
[170] Liang P, Cao J, Liu R, Liu Y. Determination of trace rare earth elements by inductively coupled plasma optical emission spectrometry after preconcentration with immobilized nanometer titanium dioxide. Microchim Acta 2007;159:35–40.
[171] Sabarudin A, Lenghor N, Oshima M, Hakim L, Takayanagi T, Gao YH, Motomizu S. Sequential-injection on-line preconcentration using chitosan resin functionalized with 2-amino-5-hydroxy benzoic acid for the determination of trace elements in environmental water samples by inductively coupled plasma-atomic emission spectrometry. Talanta 2007;72:1609–17.
[172] Shemirani F, Yousefi SR. Selective extraction and preconcentration of cerium(IV) in water samples by cloud point extraction and determination by inductively coupled plasma optical emission spectrometry. Mikrochim Acta 2007;157:223–7.
[173] Zhang N, Hu B, Huang CZ. A new ion-imprinted silica gel sorbent for on-line selective solid-phase extraction of dysprosium(III) with detection by inductively coupled plasma-atomic emission spectrometry. Anal Chim Acta 2007;597:12–8.
[174] Katarina RK, Oshima M, Motomizu S. High-capacity chitosan-based chelating resin for on-line collection of transition and rare-earth metals prior to inductively coupled plasma-atomic emission spectrometry measurement. Talanta 2009;79:1252–9.
[175] Katarina RK, Oshima M, Motomizu S. On-line collection/concentration and determination of transition and rare-earth metals in water samples using Multi-Auto-Pret system coupled with inductively coupled plasma-atomic emission spectrometry. Talanta 2009;78:1043–50.
[176] Saberyan K. Determination of Ultratrace Amounts of Ce^{3+} in the Presence of Sm^{3+}, La^{3+}, Dy^{3+}, and Yb^{3+} in Water and Soils by Inductively Coupled Plasma-Optical Emission Spectroscopy After Preconcentration with Thio-Functionalized MCM-41 at Controlled pH. J AOAC Int 2009;92:555–60.
[177] Waqar F, Jan S, Mohammad B, Hakim M, Alam S, Yawar W. Preconcentration of Rare Earth Elements in Seawater with Chelating Resin Having Fluorinated beta-Diketone Immobilized on Styrene Divinyl Benzene for their Determination by ICP-OES. J Chin Chem Soc 2009;56:335–40.
[178] Berijani S, Ganjali MR, Sereshti H, Norouzi P. A selective modified nanoporous silica as sorbent for separation and preconcentration of dysprosium in water samples prior to ICP-OES determination. Int J Environ Anal Chem 2012;92:355–65.
[179] Thulasidas SK, Adya VC, Seshagiri TK, Kumar M, Bhide MK, Godbole SV. Development of ICP-AES Method for the Determination of Common Metallic and Rare Earth Elements in High-Level Radioactive Liquid Waste Solutions. At Spectrosc 2012;33:158–65.
[180] Xue J, Zhong LS, Fan YL, Li X. Determination of Rare Earth Elements in Orchards Soil by Inductively Coupled Plasma-Atomic Emission Spectroscopy. Asian J Chem 2013;25:752–4.
[181] Karadas C, Kara D. Determination of rare earth elements in waters by inductively coupled plasma optical emission spectrometry after preconcentration with 6-(2-thienyl)-2-pyridinecarboxaldehyde functionalized Amberlite XAD-4 resin. Water Sci Technol 2014;69:312–9.

[182] Karadas C, Kara D. Preconcentration of Rare Earth Elements Using Amberlite XAD-4 Modified With 2,6-Pyridinedicarboxaldehyde and Their Determination by Inductively Coupled Plasma Optical Emission Spectrometry. Water Air Soil Pollut 2014;225:1972.
[183] Karadas C, Kara D. Determination of Rare Earth Elements by Solid Phase Extraction Using Chemically Modified Amberlite XAD-4 Resin and Inductively Coupled Plasma-Optical Emission Spectrometry. Water Air Soil Pollut 2014;225:2192.
[184] Sereshti H, Kermani M, Karimi M, Samadi S. Optimized Ultrasound-Assisted Emulsification-Microextraction Followed by ICP-OES for Simultaneous Determination of Ytterbium and Holmium in Natural Water. Clean Soil Air Water 2014;42:1089–97.
[185] Yousefi SR, Zolfonoun E, Pourjavid MR, Ahmadi SJ. On-line surfactant-based extraction using ion-pair microparticles combined with ICP-OES for simultaneous preconcentration and determination of rare earth elements in aqueous samples. Anal Method 2014;6:3694–9.
[186] Zhang YA, Zhong C, Zhang QY, Chen BB, He M, Hu B. Graphene oxide-TiO_2 composite as a novel adsorbent for the preconcentration of heavy metals and rare earth elements in environmental samples followed by on-line inductively coupled plasma optical emission spectrometry detection. Rsc Adv 2015;5:5996–6005.
[187] Mei EW, Jiang ZC, Liao ZH. Determination of trace amounts of rare earth and other elements in rice samples by ICP-AES with sample introduction by tungsten-coil electrothermal vaporization. Fresenius J Anal Chem 1992;344:54–8.
[188] Jon GG, Pu XL, Jiang ZC, Hu B. Separation/preconcentration of trace rare earth elements by micro-column packed with immobilized 1-phenyl-3-methyl-4-bonzoil-5-pyrazone on nanometer Al_2O_3 and their determination by inductively coupled plasma-atomic emission spectrometry. Chinese J Anal Chem 2005;33:207–10.
[189] Zhang N, Huang C, Hu B. ICP-AES determination of trace rare earth elements in environmental and food samples by on-line separation and preconcentration with acetylacetone-modified silica gel using microcolumn. Anal Sci 2007;23:997–1002.
[190] Li YJ, Hu B. Cloud point extraction with/without chelating agent on-line coupled with inductively coupled plasma optical emission spectrometry for the determination of trace rare earth elements in biological samples. J Hazard. Mater 2010;174:534–40.
[191] Liu HG, Wang YZ. Separation/Preconcentration of Trace Rare Earth Elements in Tricholoma Giganteum by Micro-Column with Nanometer Al_2O_3 and Their Determination by ICP-AES. Spectrosc. Spect Anal 2010;30:259–61.
[192] Ning PB, Gong CM, Zhang YM, Guo KK. Analysis of Rare Earth Elements in Pu'er Tea of Yunnan by ICP-AES. Spectrosc. Spect Anal 2010;30:2830–3.
[193] Molinero AL, Castillo JR, Devega A. Determination of ytterbium by AES-ICP. Application to samples of biological origin. Fresenius Z. Anal Chem 1988;331:721–4.
[194] Lai JJ, Jamieson GC. Determination of dysprosium in monkey serum by inductively-coupled plasma atomic emission spectrometry (ICP-AES) after the administration of Sprodiamide Injection, a new contrast medium for magnetic resonance imaging. J Pharm Biomed Anal 1993;11:1129–34.
[195] Normann PT, Joffe P, Martinsen I, Thomsen HS. Quantification of gadodiamide as Gd in serum, peritoneal dialysate and faeces by inductively coupled plasma atomic emission spectroscopy and comparative analysis by high-performance liquid chromatography. J Pharm Biomed Anal 2000;22:939–47.
[196] Garcia EE, Nogueira ARA, Nobrega JA. Matrix effects on the determination of dysprosium, europium and ytterbium used as animal faecal markers by inductively coupled plasma optical emission spectrometry with axially- and radially-viewed configurations. J Anal At Spectrom 2001;16:825–30.
[197] Ortega C, Gomez MR, Olsina RA, Silva MF, Martinez LD. On-line cloud point preconcentration and determination of gadolinium in urine using flow injection inductively coupled plasma optical emission spectrometry. J Anal At Spectrom 2002;17:530–3.

[198] Ortega C, Cerutti S, Olsina RA, Silva MF, Martinez LD. On-line complexation/cloud point preconcentration for the sensitive determination of dysprosium in urine by flow injection inductively coupled plasma-optical emission spectrometry. Anal Bioanal Chem 2003;375:270–4.
[199] Salonia JA, Gasquez JA, Martinez LD, Cerutti S, Kaplan M, Olsina RA. Inductively coupled plasma optical emission spectrometric determination of gadolinium in urine using flow injection on-line sorption preconcentration in a knotted reactor. Instrum Sci Technol 2006;34:305–16.
[200] Kahakachchi CL, Moore DA. Speciation of gadolinium in gadolinium-based magnetic resonance imaging agents by high performance liquid chromatography inductively coupled plasma optical emission spectrometry. J Anal At Spectrom 2009;24:1389–96.
[201] [online] Atomic Spectroscopy - A Guide to Selecting the Appropriate Technique and System; available at: www.perkinelmer.com/atomicspectroscopy.
[202] Doherty W, Vandervoet A. The application of inductively coupled plasma mass spectrometry to the determination of rare earth elements in geological materials. Can J Spectrosc 1985;30:135–41.
[203] Date AR, Hutchison D. Determination of Rare Earth Elements in Geological Samples by Inductively Coupled Plasma Source Mass Spectrometry. J Anal At Spectrom 1987;2:269–76.
[204] Lichte FE, Meier AL, Crock JG. Determination of the rare-earth elements in geological materials by inductively coupled plasma mass spectrometry. Anal Chem 1987;59:1150–7.
[205] Doherty W. A multiple internal standardization procedure for the for the determination of yttrium and the rare earth elements in geological materials by ICP mass spectrometry. Abstr Pap Am Chem S 1988;195:128–30.
[206] Hirata T, Shimizu H, Akagi T, Sawatari H, Masuda A. Precise determination of rare earth elements in geological standard rocks by inductively coupled plasma source mass spectrometry. Anal Sci 1988;4:637–43.
[207] Jarvis KE. Low level determination of the rare earth elements in rocks and minerals by inductively coupled plasma mass spectrometry. Chem Geol 1988;70:175–1175.
[208] Jarvis KE. Inductively coupled plasma mass spectrometry: A new technique for the rapid or ultra-trace level determination of the rare-earth elements in geological materials. Chem Geol 1988;68:31–9.
[209] Doherty W. An internal standardization procedure for the determination of yttrium and the rare earth elements in geological materials by inductively coupled plasma-mass spectrometry. Spectrochim Acta Part B 1989;44:263–80.
[210] Jarvis KE. Determination of rare earth elements in geological samples by inductively coupled plasma mass spectrometry. J Anal At Spectrom 1989;4:563–70.
[211] Mochizuki T, Sakashita A, Iwata H, Ishibashi Y, Gunji N. Slurry nebulization technique for direct determination of rare earth elements in silicate rocks by inductively coupled plasma mass spectrometry. Anal Sci 1989;5:311–7.
[212] Balaram V, Manikyamba C, Ramesh SL, Saxena VK. Determination of rare earth elememts in japanese rock standards by inductively coupled plasma mass spectrometry. At Spectrosc 1990;11:19–23.
[213] Casetta B, Giaretta A, Mezzacasa G. Determination of rare earth and other trace-elements in rock samples by ICP-mass spectrometer. At Spectrosc 1990;11:222–8.
[214] Roelandts I. Inductively Coupled Plasma Determination Of Nine Rare-Earth Elements In Sixty International Geochemical Reference Samples. Geostand Newsl 1990;14:137–47.
[215] Balaram V. Determination of the rare-earth elements in geological samples by inductively coupled plasma mass-spectrometry. J Indian Chem Soc 1991;68:600–3.
[216] Ochsenkuhnpetropoulou M, Ochsenkuhn K, Luck J. Comparison of inductively coupled plasma mass spectrometry with inductively coupled plasma atomic emission spectrometry and instrumental neutron activation analysis for the determination of rare earth elements in Greek bauxites. Spectrochim Acta Part B 1991;46:51–65.

[217] Alaimo R, Censi P. Quantitative determination of major, minor and trace elements on U.S.G.S. rock standards by inductively coupled plasma mass spectrometry. At Spectrosc 1992;13:113–7.
[218] Hall GEM. Inductively coupled plasma mass spectrometry in geoanalysis. J Geochem Exp 1992;44:201–49.
[219] Marabini AM, Passariello B, Barbaro M. Determination of rare-earth elements in minerals and ores by inductively coupled plasma-mass spectrometry (ICP-MS). Mater Chem Phys 1992;31:101–6.
[220] Dubinin A. V. inductively-coupled plasma mass spectrometry determination of REE in standard reference samples of oceanic deposits. Geokhimiya 1993;11:1605–19.
[221] Dubinin AV. Determination of rare earth elements in rock and sediment standard reference samples by inductively-coupled plasma source mass spectrometry. Doklady Akademii Nauk 1993;329:342–6.
[222] Fujino O, Okuda T. Determination of Rare Earth Elements in Rare Earth Phosphate Minerals by inductively-coupled Plasma- Mass Spectrometry and Plasma-Atomic Emission Spectrometry. Nippon Kagaku Kaishi 1993;5:561–8.
[223] Ghazi AM, Vanko DA, Roedder E, Seeley RC. Determination of rare earth elements in fluid inclusions by inductively coupled plasma-mass spectrometry (ICP-MS). Geochim. Cosmochim. Acta 1993;57:4513–6.
[224] Dulski P. Interferences of oxide, hydroxide and chloride analyte species in the determination of rare earth elements in geological samples by inductively coupled plasma-mass spectrometry. Fresenius J Anal Chem 1994;350:194–203.
[225] Rivoldini A, Fadda S. Inductively coupled plasma mass spectrometric determination of low-level rare earth elements in rocks using potassium-based fluxes for sample decomposition. J Anal At Spectrom 1994;9:519–24.
[226] Barbaro M, Passariello B, Quaresima S, Casciello A, Marabini A. Analysis of Rare Earth Elements in Rock Samples by Inductively Coupled Plasma-Mass Spectrometry (ICP-MS). Microchem J 1995;51:312–8.
[227] Behrens A. Determination of gadolinium in geological samples by inductively coupled plasma mass spectrometry and multivariate calibration. Spectrochim. Acta Part B 1995;50:1521–30.
[228] Hamanaka T, Itoh A, Itoh S, Sawatari H, Haraguchi H. Multielement Determination of Rare Earth Elements in Geochemical Samples by Liquid Chromatography/Inductively Coupled Plasma Mass Spectrometry. Chem Lett 1995;5:363–4.
[229] Pin C, Telouk P, Imbert JL. Direct determination of the samarium: neodymium ratio in geological materials by inductively coupled plasma quadrupole mass spectrometry with cryogenic desolvation. Comparison with isotope dilution thermal ionization mass spectrometry. J Anal At Spectrom 1995;10:93–8.
[230] Stroh A, Bea F, Montero PG. Ultratrace-level determination of rare earth elements, thorium, and uranium in ultramafic rocks by ICP-MS. At. Spectrosc 1995;16:7–11.
[231] Barrat JA, Keller F, Amosse J, Taylor RN, Nesbitt RW, Hirata T. Determination of rare earth elements in sixteen silicate reference samples by ICP-MS after Tm addition and ion exchange separation. Geostand Newsl 1996;20:133–9.
[232] Fujino O, Umetani S, Matsui M, Nishida S, Sanada K, Orimi K. Determination of rare earth elements, uranium and thorium in apatite minerals by inductively coupled plasma-mass spectrometry. Nippon Kagaku Kaishi 1996;7:650–5.
[233] Rautiainen I, HagelBrunnstrom ML, Heiskanen L, Kallio E. Production oriented method for the determination of rare earth and other trace elements in rocks by ICP-MS. Fresenius J Anal Chem 1996;355:393–6.
[234] Shinotsuka K, Ebihara M. Precise determination of rare earth elements, thorium and uranium in chondritic meteorites by inductively coupled plasma mass spectrometry - A comparative study with radiochemical neutron activation analysis. Anal Chim Acta 1997;338:237–46.

[235] Liu HC, Liu Y, Zhang ZX. Determination of ultra-trace rare earth elements in chondritic meteorites by inductively coupled plasma mass spectrometry. Spectrochim Acta Part B 1998;53:1399–404.

[236] Itoh A, Hamanaka T, Rong W, Ikeda K, Sawatari H, Chiba K, Haraguchi H. Multielement determination of rare earth elements in geochemical samples by liquid chromatography/inductively coupled plasma mass spectrometry. Anal Sci 1999;15:17–22.

[237] Kin FD, Prudencio MI, Gouveia MA, Magnusson E. Determination of rare earth elements in geological reference materials: A comparative study by INAA and ICP-MS. Geostand Newsl 1999;23:47–58.

[238] Ivanov AV, Rasskazov SV, Chebykin EP, Markova ME, Saranina EV. Y/Ho ratios in the Late Cenozoic basalts from the Eastern Tuva, Russia: An ICP-MS study with enhanced data quality. Geostand Newsl 2000;24:197–204.

[239] Lee KH, Oshima M, Takayanagi T, Motomizu S. Simultaneous determination of lanthanoids and yttrium in rock reference samples by inductively coupled plasma-mass spectrometry coupled with cation exchange pretreatment. Bull Chem Soc Japan 2000;73:615–22.

[240] Lee KH, Shishido S, Kusachi I, Motomizu S. Determination of lanthanoids and yttrium in JGb2 and JR3 by inductively coupled plasma-mass spectrometry after cation-exchange pretreatment. Geochem J 2000;34:383–93.

[241] Liang Q, Gregoire DC. Determination of trace elements in twenty six Chinese geochemistry reference materials by inductively coupled plasma-mass spectrometry. Geostand Newsl 2000;24:51–63.

[242] Liang Q, Jing H, Gregoire DC. Determination of trace elements in granites by inductively coupled plasma mass spectrometry. Talanta 2000;51:507–13.

[243] Olive V, Ellam RM, Wilson L. A protocol for the determination of the rare earth elements at picomole level in rocks by ICP-MS: Results on geological reference materials USGS PCC-1 and DTS-1. Geostand Newsl 2001;25:219–28.

[244] Pillay AE, Punyadeera C. An enhanced procedure for the rapid digestion of high silicate archeological specimens followed by ICP-MS determination of traces of rare earth elements (REE's). J Trace Microprobe Tech 2001;19:225–41.

[245] Meisel T, Schoner N, Paliulionyte V, Kahr E. Determination of rare earth elements, Y, Th, Zr, Hf, Nb and Ta in geological reference materials G-2, G-3, SCo-1 and WGB-1 by sodium peroxide sintering and inductively coupled plasma-mass spectrometry. Geostand Newsl 2002;26:53–61.

[246] Igarashi K, Akagi T, Fu FF, Yabuki S. Determination of rare-earth elements in a limestone geological standard reference material by ICP-MS following solvent extraction. Anal Sci 2003;19:441–5.

[247] Panteeva SV, Gladkochoub DP, Donskaya TV, Markova VV, Sandimirova GP. Determination of 24 trace elements in felsic rocks by inductively coupled plasma mass spectrometry after lithium metaborate fusion. Spectrochim Acta Part B 2003;58:341–50.

[248] Raut NM, Huang LS, Aggarwal SK, Lin KC. Determination of lanthanides in rock samples by inductively coupled plasma mass spectrometry using thorium as oxide and hydroxide correction standard. Spectrochim. Acta Part B 2003;58:809–22.

[249] Rimskaya-Korsakova MN, Dubinin AV, Ivanov VM. Determination of rare-earth elements in sulfide minerals by inductively coupled plasma mass spectrometry with ion-exchange preconcentration. J Anal Chem 2003;58:870–4.

[250] He M, Hu B, Jiang ZC, Zeng Y. Development and validation method for the determination of rare earth impurities in high purity neodymium oxide by ICP-MS. Atomi Spectrosc 2004;25:13–20.

[251] Qi L, Zhou MF, Malpas J, Sun M. Determination of rare earth elements and Y in ultramafic rocks by ICP-MS after preconcentration using Fe(OH)(3) and Mg(OH)(2) coprecipitation. Geostanda Geoanal Res 2005;29:131–41.

[252] Kautenburger R, Nowotka K, Beck HP. Online analysis of europium and gadolinium species complexed or uncomplexed with humic acid by capillary electrophoresis-inductively coupled plasma mass spectrometry. Anal Bioanal Chem 2006;384:1416–22.
[253] Premadas A, Khorge CR. Solvent extraction separation of heavy rare earth elements from light rare earth elements and thorium: ICP-AES determination of REEs and yttrium in monazite mineral. At Spectrosc 2006;27:170–7.
[254] Rajendran J, Thampi PK, Balasubramanian G. Determination of rare earth elements in garnet minerals, geological materials by inductively coupled plasma-atomic emission spectral and mass spectral analysis. Anal Lett 2006;39:2297–306.
[255] Begum Z, Balaram V, Ahmad SM, Satyanarayanan M, Rao TG. Determination of trace and rare earth elements in marine sediment reference materials by ICP-MS: Comparison of open and closed acid digestion methods. At Spectrosc 2007;28:41–50.
[256] Nakamura K, Chang Q. Precise determination of ultra-low (sub-ng g(-1)) level rare earth elements in ultramafic rocks by quadrupole ICP-MS. Geostand Geoanal Res 2007;31:185–97.
[257] Navarro MS, Andrade S, Ulbrich H, Gomes CB, Girardi VAV. The direct determination of rare earth elements in basaltic and related rocks using ICP-MS: Testing the efficiency of microwave oven sample decomposition procedures. Geostand Geoanal Res 2008;32:167–80.
[258] Nikolaeva IV, Palesskii SV, Koz'menko OA, Anoshin GN. Analysis of Geologic Reference Materials for REE and HFSE by Inductively Coupled Plasma-Mass Spectrometry (ICP-MS). Geochem Int 2008;46:1016–22.
[259] Rajendran J, Balasubramanian G, Thampi PK. Determination of rare earth elements in Indian coastal monazite by ICP-AES and ICP-MS analysis and their geochemical significance. Curr Sci 2008;94:1296–302.
[260] Bayon G, Barrat JA, Etoubleau J, Benoit M, Bollinger C, Revillon S. Determination of Rare Earth Elements, Sc, Y, Zr, Ba, Hf and Th in Geological Samples by ICP-MS after Tm Addition and Alkaline Fusion. Geostand Geoanal Res 2009;33:51–62.
[261] Sun YL, Sun M. Determination of ultra-trace rare earth elements in ultramafic and sulfide samples by quadrupole inductively coupled plasma-mass spectrometry. J Anal At Spectrom 2009;24:232–6.
[262] Alagarsamy R, You CF, Nath BN, Kumar AVS. Determination of rare earth, major and trace elements in authigenic fraction of Andaman Sea (Northeastern Indian Ocean) sediments by Inductively Coupled Plasma-Mass Spectrometry. Microchem J 2010;94:90–7.
[263] Ardini F, Soggia F, Rugi F, Udisti R, Grotti M. Comparison of inductively coupled plasma spectrometry techniques for the direct determination of rare earth elements in digests from geological samples. Anal Chim Acta 2010;678:18–25.
[264] Fedyunina NN, Ossipov KB, Seregina IF, Bolshov MA, Statkus MA, Tsysin GI. Determination of rare earth elements in rock samples by inductively coupled plasma mass-spectrometry after sorption preconcentration using Pol-DETATA sorbent. Talanta 2012;102:128–31.
[265] Fedyunina NN, Seregina IF, Bolshov MA, Okina OI, Lyapunov SM. Investigation of the efficiency of the sample pretreatment stage for the determination of the Rare Earth Elements in rock samples by inductively coupled plasma mass spectrometry technique. Anal Chim Acta 2012;713:97–102.
[266] Kumar SA, Pandey SP, Kumar SD. Determination of rare earth elements in Indian kimberlite using inductively coupled plasma mass spectrometer (ICP-MS). J Radioanal Nucl Chem 2012;294:419–24.
[267] Tanaselia C, Miclean M, Roman C, Pop D. An assessment of rare earth elements composition of Romanian meteoritic material using an inductively coupled plasma mass spectrometry method. Stud Univ Babes-Bolyai, Chem 2012;57:145–9.

[268] Padmasubashini V, Satyanarayana K. Determination of Rare Earth Elements, Yttrium, Thorium, and Other Trace Elements in Monazite Samples by Inductively Coupled Plasma Mass Spectrometry. At Spectrosc 2013;34:6–14.
[269] Sun YL, Sun SL, Wang CY, Xu P. Determination of Rare Earth Elements and Thorium at Nanogram Levels in Ultramafic Samples by Inductively Coupled Plasma-Mass Spectrometry Combined with Chemical Separation and Pre-concentration. Geostand Geoanal Res 2013;37:65–76.
[270] Janski R, Neouze MA, Limbeck A. Determination of rare earth elements in saline matrices using dispersed particle extraction and inductively coupled plasma mass spectrometry. Rapid Commun Mass Spectrom 2014;28:1329–37.
[271] Balaram V, Manikyamba C, Ramesh SL, Anjaiah KV. Rare earth and trace element determination in iron-formation reference samples by ICP-MS. At Spectrosc 1992;13:19–25.
[272] Takaku Y, Masuda K, Takahashi T, Shimamura T. Determination of trace impurity rare earth elements in high-purity rare earth element samples using high-resolution inductively coupled plasma mass spectrometry. J Anal At Spectrom 1993;8:687–90.
[273] Klinkhammer G, German CR, Elderfield H, Greaves MJ, Mitra A. Rare earth elements in hydrothermal fluids and plume particulates by inductively coupled plasma mass spectrometry. Marine Chem 1994;45:179–86.
[274] Panday VK, Becker JS, Dietze HJ. Trace analysis of rare earth elements and other impurities in high purity scandium by inductively coupled plasma mass spectrometry after liquid-liquid extraction of the matrix. Fresenius J Anal Chem 1995;352:327–34.
[275] Panday VK, Becker JS, Dietze HJ. Trace impurities in zircaloys by inductively-coupled plasma-mass spectrometry after removal of the matrix by liquid–liquid-extraction. At Spectrosc 1995;16:97–101.
[276] Zhang SX, Murachi S, Imasaka T, Watanabe M. Determination of rare earth impurities in ultrapure europium oxide by inductively-coupled plasma mass spectrometry. Anal Chim Acta 1995;314:193–201.
[277] Li B, Zhang Y, Yin M. Determination of trace amounts of rare earth elements in high-purity cerium oxide by inductively coupled plasma mass spectrometry after separation by solvent extraction. Analyst 1997;122:543–7.
[278] Su WC, Qi L, Hu RZ, Zhang GP. Analysis of rare-earth elements in fluid inclusions by inductively coupled plasma-mass spectrometry (ICP-MS). Chinese Sci Bull 1998;43:1922–7.
[279] Cao XD, Yin M, Li B. Determination of rare earth impurities in high purity gadolinium oxide by inductively coupled plasma mass spectrometry after 2-ethylhexyl-hydrogen-ethylhexy phosphonate extraction chromatographic separation. Talanta 1999;48:517–25.
[280] Liu XS, Cai SQ, An P, Wu X, Chen JY, Zhang AD. Determination of high purity ytterbium oxide by inductively coupled plasma-mass spectrometry. Chinese J Anal Chem 1999;27:782–5.
[281] Ramendik GI, Stepanov AI, Kharitonov PS, Karmannikov VP. An inductively coupled plasma mass-spectrometic study of the formation of polyatomic ions in trace analysis of rare-earth element oxides. Ind Lab 1999;65:501–3.
[282] Yin M, Li B, Cao XD, Zhang Y. Determination of rare earth element impurities in high purity Pr6O11 by inductively coupled plasma mass spectrometry. Chinese J Analy Chem 1999;27:304–8.
[283] Qin S, Jiang ZC, Hu B, Qin YC, Hu SS. HPLC combined with ICP-MS for the determination of trace amounts of rare earth impurities in high-purity La_2O_3 by using 2-ethylhexyl hydrogen-2-ethylhexylphosphonate resin as a stationary phase. Fresenius J Anal Chem 2000;367:250–3.
[284] Wu HZ, Watanabe N, Tsukahara R, Itoh T, Gohshi Y, Owari M. Highly sensitive determination of trace rare-earth elements in ultra pure rare-earth compounds by combining ICP-MS with HPLC and electrothermal vaporization. Bunseki Kagaku 2000;49:91–7.

[285] Cao XD, Yin M, Wang XR. Elimination of the spectral interference from barium polyatomic ions on rare earth elements in inductively coupled plasma mass spectrometry by AG50W-x8 cation exchange chromatographic separation. Chinese J Anal Chem 2001;29:890–3.
[286] Cao XD, Yin M, Wang XR. Elimination of the spectral interference from polyatomic ions with rare earth elements in inductively coupled plasma mass spectrometry by combining algebraic correction with chromatographic separation. Spectrochim. Acta Part B 2001;56:431–41.
[287] Iwata S, Minamoto K, Fujimori E, Chiba K, Haraguchi H. Determination of rare earth elements in industrial waste incineration fly ash by ICP-MS and the characteristics of the rare earth element distribution pattern. Bunseki Kagaku 2001;50:419–25.
[288] Balaram V. Inductively coupled plasma mass spectrometric techniques in the elemental and isotopic analysis of lanthanides and actinides. J Indian Chem Soc 2002;79:479–85.
[289] Yan Z, Bin H, Man H, Jiang ZC. Direct determination of rare earth impurities in high purity europium oxide by ICP-MS. J Rare Earth 2003;21:225–9.
[290] Goolaerts A, Mattielli N, de Jong J, Weis D, Scoates JS. Hf and Lu isotopic reference values for the zircon standard 91500 by MC-ICP-MS. Chem Geol 2004;206:1–9.
[291] Merten D, Buchel G. Determination of rare earth elements in acid mine drainage by inductively coupled plasma mass spectrometry. Microchim Acta 2004;148:163–70.
[292] Wang SJ, Chang P, Su WN, Sun CH, Wang LJ, Zhang GY, Hou TP. Determination of rare-earth and trace elements in teeth enamel by inductively coupled plasma mass spectroscopy. Spectrosc Spect Anal 2004;24:1113–6.
[293] Zhang AD, Liu XS, Zhang W. Determination of rare earth impurities in high purity europium oxide by inductively coupled plasma-mass spectrometry and evaluation of concentration values for europium oxide standard material. Eur J Mass Spectrom 2004;10:589–98.
[294] He M, Hu B, Zeng Y, Jiang ZC. ICP-MS direct determination of trace amounts of rare earth impurities in various rare earth oxides with only one standard series. J Alloy Comp 2005;390:168–74.
[295] Takeda K, Arikawa Y. Determination of rare earth elements in petroleum by ICP-MS. Bunseki Kagaku 2005;54:939–43.
[296] Zhang XQ, Yi Y, Liu YL, Li X, Liu JL, Jiang YM, Su YQ. Direct determination of impurities in high purity rare earth erbium oxide dissolved in nitric acid by inductively coupled plasma mass spectrometry. Anal Chim Acta 2006;555:57–62.
[297] Kulkarni P, Chellam S, Mittlefehldt DW. Microwave-assisted extraction of rare earth elements from petroleum refining catalysts and ambient fine aerosols prior to inductively coupled plasma-mass spectrometry. Anal Chim Acta 2007;581:247–59.
[298] Packer AP, Lariviere D, Li CS, Chen M, Fawcett A, Nielsen K, Mattson K, Chatt A, Scriver C, Erhardt LS. Validation of an inductively coupled plasma mass spectrometry (ICP-MS) method for the determination of cerium, strontium, and titanium in ceramic materials used in radiological dispersal devices (RDDs). Anal Chim Acta 2007;588:166–72.
[299] Zhang X, Liu JL, Yia Y, Liu YL, Li X, Su YQ, Lin P. Determination of rare earth impurities in high purity samarium oxide using inductively coupled plasma mass spectrometry after extraction chromatographic separation. Int J Mass Spectrom 2007;260:57–66.
[300] Akinlua A, Torto N, Ajayi TR. Determination of rare earth elements in Niger Delta crude oils by inductively coupled plasma-mass spectrometry. Fuel 2008;87:1469–77.
[301] Pitois. A.; de las Heras, L. A.; Betti, M. Determination of fission products in nuclear samples by capillary electrophoresis-inductively coupled plasma mass spectrometry (CE-ICP-MS). Int J Mass Spectrom 2008;270:118–26.
[302] Song XJ, Liu XL, Duan TC, Chen HT. Determination of Impurities in High Pure Neodymium Oxide with Inductively Coupled Plasma Mass Spectrometry. Chinese J Anal Chem 2009;37:1743–8.

[303] Ardini F, Soggia F, Rugi F, Udisti R, Grotti M. Conversion of rare earth elements to molecular oxide ions in a dynamic reaction cell and consequences on their determination by inductively coupled plasma mass spectrometry. J Anal At Spectrom 2010;25:1588–97.

[304] Deb SB, Nagar BK, Saxena MK, Ramakumar KL. Microwave-Assisted Quantitative Determination of Trace Rare Earth Elements in Zirconia and Zirconium Alloys by Inductively Coupled Plasma Mass Spectrometry and Validation of Separation Procedure by Standard Addition and Tracer Techniques. At Spectrosc 2010;31:122–33.

[305] Varga Z, Katona R, Stefanka Z, Wallenius M, Mayer K, Nicholl A. Determination of rare-earth elements in uranium-bearing materials by inductively coupled plasma mass spectrometry. Talanta 2010;80:1744–9.

[306] Durani S, Krishnakumar M, Satyanarayana K. Solid phase separation and ICP-OES/ICP-MS determination of rare earth impurities in nuclear grade uranium oxide. J Radioanal Nucl Chem 2012;294:215–20.

[307] Singh AK, Padmasubashini V, Gopal L. Determination of uranium, thorium and rare-earth elements in zircon samples using ICP-MS. J. Radioanal. Nucl Chem 2012;294:19–25.

[308] Vio L, Cretier G, Chartier F, Geertsen V, Gourgiotis A, Isnard H, Rocca JL. Separation and analysis of lanthanides by isotachophoresis coupled with inductively coupled plasma mass spectrometry. Talanta 2012;99:586–93.

[309] Krachler M, Alvarez-Sarandes. R.; Van Winckel, S. Cross-validation of analytical procedures for the reliable determination of Nd concentrations in nuclear fuel using ICP-OES and sector field ICP-MS. J Anal At Spectrom 2013;28:114–20.

[310] Saha A, Deb SB, Nagar BK, Saxena MK. Determination of trace rare earth elements in gadolinium aluminate by inductively coupled plasma time of flight mass spectrometry. Spectrochim. Acta, Part B 2014;94-95:14–21.

[311] Silva JSA, Maranhao TD, de Oliveira FJS, Curtius AJ, Frescura VLA. Determination of Rare Earth Elements in Spent Catalyst Samples from Oil Refinery by Dynamic Reaction Cell Inductively Coupled Plasma Mass Spectrometry. J Braz Chem Soc 2014;25:1062–70.

[312] Moller P, Dulski P, Luck J. Determination of rare earth elements in seawater by inductively coupled plasma mass spectrometry. Spectrochim Acta Part B 1992;47:1379–87.

[313] Shabani MB, Akagi T, Masuda A. Preconcentration of trace rare-earth elements in seawater by complexation with bis(2-ethylhexyl) hydrogen phosphate and 2-ethylhexyl dihydrogen phosphate adsorbed on a C18 cartridge and determination by inductively coupled plasma mass spectrometry. Anal Chem 1992;64:737–43.

[314] Akatsuka K, McLaren JW, Berman SS. Determination of 7 trace elements in environmental water samples by inductively coupled plasma mass spectrometry with online preconcentration. Bunseki Kagaku 1993;42:423–8.

[315] Ihnat M, Gamble DS, Gilchrist GFR. Determination of trace element levels in natural fresh water by inductively coupled plasma mass spectrometry. Int J Environ Anal Chem 1993;53:63–78.

[316] Probst T, Zeh P, Kim JI. Multielement determinations in ground water ultrafiltrates using inductively coupled plasma mass spectrometry and monostandard neutron activation analysis. Fresenius J Anal Chem 1995;351:745–51.

[317] Sawatari H, Fujimori E, Haraguchi H. Multielement determination of trace elements in seawater by gallium coprecipitation and inductively coupled plasma mass spectrometry. Anal Sci 1995;11:369–74.

[318] Sawatari H, Toda T, Saizuka T, Kimata C, Itoh A, Haraguchi H. Multielement determination of rare earth elements in coastal seawater by inductively coupled plasma mass spectrometry after preconcentration using chelating resin. Bull Chem Soc Japan 1995;68:3065–70.

[319] Uchino T, Ebihara M, Furuta N. Determination of rare earth elements in Precambrian sediments at Isua by inductively coupled plasma mass spectrometry. J Anal Atom Spectrom 1995;10:25–30.

[320] Aggarwal JK, Shabani MB, Palmer MR, Ragnarsdottir KV. Determination of the rare earth elements in aqueous samples at sub-ppt levels by inductively coupled plasma mass spectrometry and flow injection ICPMS. Anal Chem 1996;68:4418–23.
[321] deBoer JLM, Verweij. W.; vanderVeldeKoerts, T.; Mennes, W. Levels of rare earth elements in Dutch drinking water and its sources. Determination by inductively coupled plasma mass spectrometry and toxicological implications. A pilot study. Water Res 1996;30:190–8.
[322] Halicz L, Gavrieli I, Dorfman E. On-line method for inductively coupled plasma mass spectrometric determination of rare earth elements in highly saline brines. J Anal At Spectrom 1996;11:811–4.
[323] Panday VK, Hoppstock K, Becker JS, Dietze HJ. Determination of rare earth elements in environmental materials by ICP-MS after liquid-liquid extraction. At Spectrosc 1996;17:98–105.
[324] Zhang J, Nozaki Y. Rare earth elements and yttrium in seawater: ICP-MS determinations in the East Caroline, Coral Sea, and South Fiji basins of the western South Pacific Ocean. Geochim Cosmochim Acta 1996;60:4631–44.
[325] Zhang SZ, Shan XQ, Yan X, Zhang HZ. The determination of rare earth elements in soil by inductively coupled plasma mass spectrometry. At Spectrosc 1997;18:140–4.
[326] Hu Y, Vanhaecke F, Moens L, Dams R, del Castilho P, Japenga J. Determination of the aqua regia soluble content of rare earth elements in fertilizer, animal fodder phosphate and manure samples using inductively coupled plasma mass spectrometry. Anal Chim Acta 1998;373:95–105.
[327] Kawamoto K, Sumino T, Takada J, Tanaka Y, Akaboshi M. Determination of rare earth and other elements in algae by ICP-MS and neutron activation analysis. J Radioanal Nucl Chem 1998;236:119–22.
[328] Kumagai H, Yamanaka M, Sakai T, Yokoyama T, Suzuki TM, Suzuki T. Determination of trace metals in sea-water by inductively coupled plasma mass spectrometry interfaced with an ion chromatographic separation system: effectiveness of nitrilotriacetate chelating resin as the column stationary phase for preconcentration and elimination of matrix effects. J Anal At Spectrom 1998;13:579–82.
[329] Tsuji H, Fujiwara S, Chayama K, Teranishi K, Isomura K. Determination of rare-earth elements in Arima spring water by ICP-MS with ion-exchange separation. Bunseki Kagaku 1998;47:699–705.
[330] Vicente O, Padro A, Martinez L, Olsina R, Marchevsky E. Determination of some rare earth elements in seawater by inductively coupled plasma mass spectrometry using flow injection preconcentration. Spectrochim Acta Part B 1998;53:1281–7.
[331] Zhang TH, Shan XQ, Liu RX, Tang HX, Zhang SZ. Preconcentration of rare earth elements in seawater with poly(acrylaminophosphonic dithiocarbamate) chelating fiber prior to determination by inductively coupled plasma mass spectrometry. Anal Chem 1998;70:3964–8.
[332] Cao XD, Yin M, Wang XR, Zhao GW. Determination of trace rare earth elements in soils by inductively coupled plasma-mass spectrometry after microwave digestion for sample preparation. Chinese J Anal Chem 1999;27:679–83.
[333] Wen B, Shan XQ, Xu SG. Preconcentration of ultratrace rare earth elements in seawater with 8-hydroxyquinoline immobilized polyacrylonitrile hollow fiber membrane for determination by inductively coupled plasma mass spectrometry. Analyst 1999;124:621–6.
[334] Yan XP, Kerrich R, Hendry MJ. Flow injection on-line group preconcentration and separation of (ultra)trace rare earth elements in environmental and geological samples by precipitation using a knotted reactor as a filterless collector for inductively coupled plasma mass spectrometric determination. J Anal At Spectrom 1999;14: 215–21.

[335] Cao XD, Chen Y, Gu ZM, Wang XR. Determination of trace rare earth elements in plant and soil samples by inductively coupled plasma-mass spectrometry. Int J Environ Anal Chem 2000;76:295–309.
[336] Gueguen C, Dominik J, Perret D. Use of chelating resins and inductively coupled plasma mass spectrometry for simultaneous determination of trace and major elements in small volumes of saline water samples. Fresenius J Anal Chem 2001;370:909–12.
[337] Ivanova J, Djingova R, Korhammer S, Markert B. On the microwave digestion of soils and sediments for determination of lanthanides and some toxic and essential elements by inductively coupled plasma source mass spectrometry. Talanta 2001;54:567–74.
[338] Ivanova J, Korhammer S, Djingova R, Heidenreich H, Markert B. Determination of lanthanoids and some heavy and toxic elements in plant certified reference materials by inductively coupled plasma mass spectrometry. Spectrochim Acta Part B 2001;56:3–12.
[339] Liu CM, Guo FJ, Guo YX. Preconcentration of trace rare earth elements in seawater with 8-hydroxyquinoline-5-sulfonic acid-cellulose microcolumn prior to determination by inductively coupled plasma-mass spectrometry. Chinese J Anal Chem 2001;29:984–1984.
[340] Willie SN, Sturgeon RE. Determination of transition and rare earth elements in seawater by flow injection inductively coupled plasma time-of-flight mass spectrometry. Spectrochim Acta Part B 2001;56:1707–16.
[341] Yabutani T, Chiba K, Haraguchi H. Multielement determination of trace elements in seawater by inductively coupled plasma mass spectrometry after tandem preconcentration with cooperation of chelating resin adsorption and lanthanum coprecipitation. Bull Chem Soc Japan 2001;74:31–8.
[342] Ai J, Hu SH, Qin S, Yu QW. Determination of trace rare earth elements, Sc, and Y in ground water by inductively coupled plasma mass spectrometry with preconcerntration method. Chinese J Anal Chem 2002;30:1226–30.
[343] Hirata S, Kajiya T, Aihara M, Honda K, Shikino O. Determination of rare earth elements in seawater by on-line column preconcentration inductively coupled plasma mass spectrometry. Talanta 2002;58:1185–94.
[344] Lee KH, Oshima M, Motomizu S. Inductively coupled plasma mass spectrometric determination of heavy metals in sea-water samples after pre-treatment with a chelating resin disk by an on-line flow injection method. Analyst 2002;127:769–74.
[345] Takaku Y, Kudo Y, Kimura J, Hayashi T, Ota I, Hasegawa H, Ueda S. Determination of rare-earth elements in environmental water using chelating resin disk separation/ICP-MS. Bunseki Kagaku 2002;51:539–44.
[346] Xu ZF, Liu CQ, Ma YJ, Zhang HX, Jin XD. Determination of trace rare earth elements in natural water by inductively coupled plasma mass spectrometry with solvent extraction. Chinese J Anal Chem 2002;30:1243–6.
[347] Casartelli EA, Miekeley N. Determination of thorium and light rare-earth elements in soil water and its high molecular mass organic fractions by inductively coupled plasma mass spectrometry and on-line-coupled size-exclusion chromatography. Anal Bioanal Chem 2003;377:58–64.
[348] Hirata S, Kajiya T, Takano N, Aihara M, Honda K, Shikino O, Nakayama E. Determination of trace metals in seawater by on-line column preconcentration inductively coupled plasma mass spectrometry using metal alkoxide glass immobilized 8-quinolinol. Anal Chim Acta 2003;499:157–65.
[349] Mito S, Ohata M, Furuta N. Determination of rare earth elements in river water by fully automated on-line column inductively coupled plasma mass spectrometry using iminodiacetate chelate resin as a column. Bunseki Kagaku 2003;52:575–82.
[350] Smirnova EV, Fedorova IN, Sandimirova GP, Petrov LL, Balbekina NG, Lozhkin VI. Determination of rare earth elements in black shales by inductively coupled plasma mass spectrometry. Spectrochim Acta Part B 2003;58:329–40.

[351] Smirnova EV, Fedorova IN, Sandimirova GP, Petrov LL, Lozhkin VI. Peculiarities in the behavior of rare-earth elements in their determination in black shales by inductively coupled plasma mass spectrometry. J Anal Chem 2003;58:533–41.
[352] Hennebruder K, Wennrich R, Mattusch J, Stark HJ, Engewald W. Determination of gadolinium in river water by SPE preconcentration and ICP-MS. Talanta 2004;63:309–16.
[353] Kajiya T, Aihara M, Hirata S. Determination of rare earth elements in seawater by inductively coupled plasma mass spectrometry with on-line column pre-concentration using 8-quinolinole-immobilized fluorinated metal alkoxide glass. Spectrochim Acta Part B 2004;59:543–50.
[354] Lee KH, Muraoka Y, Oshima M, Motomizu S. Determination of heavy metals and rare earth elements in environmental samples by ICP-MS after solid phase preconcentration with chelating resin fibers and anion exchanger filters. Anal Sci 2004;20:183–7.
[355] Gonzalez CH, Cabezas AJQ, Diaz MF. Preconcentration and determination of rare-earth elements in iron-rich water samples by extraction chromatography and plasma source mass spectrometry (ICP-MS). Talanta 2005;68:47–53.
[356] Liang P, Fa WJ. Determination of La, Eu and Yb in water samples by inductively coupled plasma atomic emission spectrometry after solid phase extraction of their 1-phenyl-3-methyl-4-benzoylpyrazol-5-one complexes on silica gel column. Microchim Acta 2005;150:15–9.
[357] Murakami M, Hamanaka E, Takada T. Novel sorbent extraction technique using a chelating agent impregnated porous PTFE filter tube: Preconcentration of rare earth elements (REEs) with bis(2-ethyl-hexyl)hydrogen phosphate (HDEHP) loaded porous PTFE filter tube prior to determination by ICP-MS. Anal Lett 2005;38:2449–61.
[358] Gabrielli P, Barbante C, Turetta C, Marteel A, Boutron C, Cozzi G, Cairns W, Ferrari C, Cescon P. Direct determination of rare earth elements at the subpicogram per gram level in antarctic ice by ICP-SFMS using a desolvation system. Anal Chem 2006;78:1883–9.
[359] Sucharova J, Suchara I. Determination of 36 elements in plant reference materials with different Si contents by inductively coupled plasma mass spectrometry: Comparison of microwave digestions assisted by three types of digestion mixtures. Anal Chim Acta 2006;576:163–76.
[360] Wang ZH, Yan XP, Wang ZP, Zhang ZP, Liu LW. Flow injection on-line solid phase extraction coupled with inductively coupled plasma mass spectrometry for determination of (ultra)trace rare earth elements in environmental materials using maleic acid grafted polytetrafluoroethylene fibers as sorbent. J Am Soc Mass Spectrom 2006;17:1258–64.
[361] Zhu YB, Itoh A, Fujimori E, Umemura T, Haraguchi H. Determination of rare earth elements in seawater by ICP-MS after preconcentration with a cheresin-packed minicolumn. J Alloy Compd 2006;408:985–8. lating.
[362] Fu Q, Yang LM, Wang QQ. On-line preconcentration with a novel alkyl phosphinic acid extraction resin coupled with inductively coupled plasma mass spectrometry for determination of trace rare earth elements in seawater. Talanta 2007;72:1248–54.
[363] Sun Y, Sun M. Determination of 42 trace elements in seawater by inductively coupled plasma mass spectrometry after APDC chelate co-precipitation combined with iron. Anal Lett 2007;40:2391–404.
[364] Yu KF, Kamber BS, Lawrence MG, Greig A, Zhao JX. High-precision analysis on annual variations of heavy metals, lead isotopes and rare earth elements in mangrove tree rings by inductively coupled plasma mass spectrometry. Nucl Instrum Method Phys Res Sect B 2007;255:399–408.
[365] Dick D, Wegner A, Gabrielli P, Ruth U, Barbante C, Kriews M. Rare earth elements determined in Antarctic ice by inductively coupled plasma - Time of flight, quadrupole and sector field-mass spectrometry: An inter-comparison study. Anal Chim Acta 2008;621:140–7.

[366] Sakamoto N, Kano N, Imaizumi H. Determination of rare earth elements, thorium and uranium in seaweed samples on the coast in Niigata Prefecture by inductively coupled plasma mass spectrometry. Appl Geochem 2008;23:2955–60.

[367] Chen SZ, Liu C, Lu DB, Zhu L. Determination of Trace Rare Earth Elements by ICP-MS after On-line Column Preconcentration and Separation with Single-Wall Carbon Nanotubes. At Spectrosc 2009;30:20–5.

[368] Spalla S, Baffi C, Barbante C, Turretta C, Cozzi G, Beone GM, Bettinelli M. Determination of rare earth elements in tomato plants by inductively coupled plasma mass spectrometry techniques. Rapid Commun Mass Spectrom 2009;23:3285–92.

[369] Takata H, Tagami K, Aono T, Uchida S. Determination of Trace Levels of Yttrium and Rare Earth Elements in Estuarine and Coastal Waters by Inductively Coupled Plasma Mass Spectrometry Following Preconcentration with Nobias-Chelate Resin. At Spectrosc 2009;30:10–9.

[370] Raju CSK, Cossmer A, Scharf H, Panne U, Luck D. Speciation of gadolinium based MRI contrast agents in environmental water samples using hydrophilic interaction chromatography hyphenated with inductively coupled plasma mass spectrometry. J Anal At Spectrom 2010;25:55–61.

[371] Zhou Y, Zheng PX. Study on determination of rare earth elements in multi-mineral phase soil by inductively coupled plasma mass. Geochim. Cosmochim. Acta 2010;74:A1230–A11230.

[372] Celo V, Dabek-Zlotorzynska E, Zhao JJ, Okonskaia I, Bowman D. An improved method for determination of lanthanoids in environmental samples by inductively coupled plasma mass spectrometry with high matrix introduction system. Anal Chim Acta 2011;706:89–96.

[373] Karadas C, Kara D, Fisher A. Determination of rare earth elements in seawater by inductively coupled plasma mass spectrometry with off-line column preconcentration using 2,6-diacetylpyridine functionalized Amber lite XAD-4. Anal Chim Acta 2011;689:184–9.

[374] Kumar SA, Pandey SP, Shenoy NS, Kumar SD. Matrix separation and preconcentration of rare earth elements from seawater by poly hydroxamic acid cartridge followed by determination using ICP-MS. Desalination 2011;281:49–54.

[375] Bulska E, Danko B, Dybczynski RS, Krata A, Kulisa K, Samczynski Z, Wojciechowski M. Inductively coupled plasma mass spectrometry in comparison with neutron activation and ion chromatography with UV/VIS detection for the determination of lanthanides in plant materials. Talanta 2012;97:303–11.

[376] Chandrasekaran K, Karunasagar D, Arunachalam J. Dispersive liquid-liquid microextraction for simultaneous preconcentration of 14 lanthanides at parts per trillion levels from groundwater and determination using a micro-flow nebulizer in inductively coupled plasma-quadrupole mass spectrometry. J Anal At Spectrom 2012;27:1024–31.

[377] Cho J, Chung KW, Choi MS, Kim HJ. Analysis of rare earth elements in seawater by inductively coupled plasma mass spectrometry after pre-concentration using TSK (TM)-HD-MW-CNTs (highly dispersive multi-walled carbon nanotubes). Talanta 2012;99:369–74.

[378] Birka M, Wehe CA, Telgmann L, Sperling M, Karst U. Sensitive quantification of gadolinium-based magnetic resonance imaging contrast agents in surface waters using hydrophilic interaction liquid chromatography and inductively coupled plasma sector field mass spectrometry. J Chromatogr A 2013;1308:125–31.

[379] Chen SZ, Cheng XL, He YY, Zhu SP, Lu DB. Determination of the rare earth elements La, Eu, and Yb using solidified floating organic drop microextraction and electrothermal vaporization ICP-MS. Microchim. Acta 2013;180:1479–86.

[380] Chen SZ, He YY, Lu DB, Guo XL. Determination of Trace Nd, Tb, and Ho in Environmental Samples by ICP-MS After Column Preconcentration and Separation With Titanium Dioxide Nanotubes. At Spectrosc 2013;34:73–7.

[381] Lindner U, Lingott J, Richter S, Jakubowski N, Panne U. Speciation of gadolinium in surface water samples and plants by hydrophilic interaction chromatography hyphenated with inductively coupled plasma mass spectrometry. Anal Bioanal Chem 2013;405:1865–73.

[382] Shaltout AA, Khoder MI, El-Abssawy AA, Hassan SK, Borges DLG. Determination of rare earth elements in dust deposited on tree leaves from Greater Cairo using inductively coupled plasma mass spectrometry. Environ Pollut 2013;178:197–201.

[383] Zereen F, Yilmaz V, Arslan Z. Solid phase extraction of rare earth elements in seawater and estuarine water with 4-(2-thiazolylazo) resorcinol immobilized Chromosorb 106 for determination by inductively coupled plasma mass spectrometry. Microchem J 2013;110:178–84.

[384] Su SW, Chen BB, He M, Hu B, Xiao ZW. Determination of trace/ultratrace rare earth elements in environmental samples by ICP-MS after magnetic solid phase extraction with Fe_3O_4@SiO_2@polyaniline-graphene oxide composite. Talanta 2014;119:458–66.

[385] Augagneur S, Medina B, Szpunar J, Lobinski R. Determination of rare earth elements in wine by inductively coupled plasma mass spectrometry using a microconcentric nebulizer. J Anal At Spectrom 1996;11:713–21.

[386] Huynh MPT, Carrot F, Ngoc SCP, Vu MD, Revel G. Determination of rare earth elements in rice by INAA and ICP-MS. J Radioanal Nucl Chem 1997;217:95–9.

[387] Zhou H, Liu JH. The determination of rare earth elements in plant foods by ICP-MS. At Spectrosc 1997;18:192–4.

[388] Cao XD, Zhao GW, Yin M, Li JX. Determination of ultratrace rare earth elements in tea by inductively coupled plasma mass spectrometry with microwave digestion and AG50W-x8 cation exchange chromatography. Analyst 1998;123:1115–9.

[389] Gomez MDM, Brandt R, Jakubowski N, Andersson JT. Changes of the metal composition in German white wines through the winemaking process. A study of 63 elements by inductively coupled plasma-mass spectrometry. J Agr Chem 2004;52:2953–61.

[390] Chen SZ, Xiao MF, Lu DB, Zhan XL. Carbon nanofibers as solid-phase extraction adsorbent for the preconcentration of trace rare earth elements and their determination by inductively coupled plasma mass spectrometry. Anal Lett 2007;40:2105–15.

[391] Rossano EC, Szilagyi Z, Malorni A, Pocsfalvi G. Influence of winemaking practices on the concentration of rare earth elements in white wines studied by inductively coupled plasma mass spectrometry. J Agric Food Chem 2007;55:311–7.

[392] Wang PL, Su XO. Study on determination of rare earth La and Ce in feed by using ICP-MS. Spectrosc Spect Anal 2007;27:2341–4.

[393] Cai DJ, Rui YK. Determination of rare earth elements in *Camellia oleifera* seeds from rare earth elements mining areas in Southern Jiangxi, China by ICP-MS. J Consum Prot Food Safe 2011;6:349–51.

[394] Cai DJ, Rui YK. Determination of Rare Earth Elements in Lotus Seed from Rare Earth Mining Area by ICP-MS Simultaneously. Asian J Chem 2011;23:2237–8.

[395] Zhang HX, Rui YK. Determination of Trace Elements, Heavy Metals and Rare Earth Elements in Corn Seeds from Beijing by ICP-MS Simultaneously. E J Chem 2011;8:782–6.

[396] Lagad RA, Alamelu D, Chaudhary AK, Aggarwal SK. Determination of Heavy Metals and Lanthanides in Indian Tea by Inductively Coupled Plasma Mass Spectrometry (ICP-MS). At Spectrosc 2012;33:109–16.

[397] Li Y, Yang JL, Jiang Y. Trace Rare Earth Element Detection in Food and Agricultural Products Based on Flow Injection Walnut Shell Packed Microcolumn Preconcentration Coupled with Inductively Coupled Plasma Mass Spectrometry. J Agric Food Chem 2012;60:3033–41.

[398] Cai DJ, Rui Y. Determination of Trace Elements and Rare Earth Elements in Gannan Navel Orange Fruit by ICP-MS. Asian J Chem 2013;25:579–80.

[399] Chen SZ, Zhu SP, Lu DB. Titanium dioxide nanotubes as solid-phase extraction adsorbent for on-line preconcentration and determination of trace rare earth elements by inductively coupled plasma mass spectrometry. Microchem J 2013;110:89–93.

[400] Ivanova-Petropulos V, Wiltsche H, Stafilov T, Stefova M, Motter H, Lankmayr E. Multielement analysis of Macedonian wines by inductively coupled plasma–mass spectrometry (ICO-MS) and inductively coupled plasma–optical emission spectrometry (ICP–OES) for their classification. Maced J Chem En 2013;32:265–81.
[401] Zhang HX, Rui YK, Li YX. Determination of Beneficial Elements, Heavy Metals and Rare Earth Elements in Northeastern Rice Brand of China by ICP-MS. Asian J Chem 2013;25:2315–6.
[402] Zhang Y, Ning ZH, Sui LZ, Zhong H. Analysis of Samarium and Europium in Wheat by ICP-MS after On-Line Separation and Preconcentration J`Chem Soc Pak 2013;35:1298–303.
[403] Hu SL, Xue J, Lin Y, Yu JP, Zhou JG. Determination of Rare Earth Elements in Navel Oranges from Different Geographical Regions of China by Inductively Coupled Plasma-Mass Spectrometry Anal Lett 2014;47:1400–8.
[404] Santos VS, Nardini V, Cunha LC, Barbosa F, Teixeira GHD. Identification of species of the Euterpe genus by rare earth elements using inductively coupled plasma mass spectrometry and linear discriminant analysis. Food Chem 2014;153:334–9.
[405] Allain P, Berre S, Premelcabic A, Mauras Y, Delaporte T. Concentrations of rare earth elements in plasma and urine of healthy subjects determined by inductively coupled plasma mass spectrometry. Clin Chem 1990;36:2011–2.
[406] Ramanujam VMS, Egger NG, Alcock NW, Anderson KE. Measurement of lanthanum and lutetium in metal-containing texaphyrins by inductively coupled plasma-mass spectrometry. At Spectrosc 1995;16:211–6.
[407] Fujimori E, Tomosue Y, Haraguchi H. Determination of rare earth elements in blood serum reference sample by chelating resin preconcentration and inductively coupled plasma mass spectrometry. Tohoku J Exp Med 1996;178:63–74.
[408] Inagaki K, Haraguchi H. Rare earth elements in human blood serum as determined by inductively coupled plasma mass spectrometry. Chem Lett 1997;8:775–6.
[409] Frame EMS, Uzgiris EE. Gadolinium determination in tissue samples by inductively coupled plasma mass spectrometry and inductively coupled plasma atomic emission spectrometry in evaluation of the action of magnetic resonance imaging contrast agents. Analyst 1998;123:675–9.
[410] Ming Y, Bing L. Determination of rare earth elements in human hair and wheat flour reference materials by inductively coupled plasma mass spectrometry with dry ashing and microwave digestion. Spectrochim Acta Part B 1998;53:1447–54.
[411] Vicente O, Pelfort E, Martinez L. Determination of rare earth elements in urinary calculi by ICP-MS. At Spectrosc 1998;19:168–71.
[412] Cao SQ, Chen HT, Zeng XJ. Determination of rare earth elements in biological samples by inductively coupled plasma mass spectrometry. Chinese J Anal Chem 1999;27:621–5.
[413] Fujimori E, Hayashi T, Inagaki K, Haraguchi H. Determination of lanthanum and rare earth elements in bovine whole blood reference material by ICP-MS after coprecipitation preconcentration with heme-iron as coprecipitant. Fresenius J Anal Chem 1999;363:277–82.
[414] Kawamoto K, Takada J, Tanaka Y, Akaboshi M. Determination of rare earth and other elements in standard biological samples by ICP-MS and activation analysis. J Radioanal Nucl Chem 1999;242:527–30.
[415] Chen HT, Cao SQ, Zeng XJ, Meng L, Zao DQ, Ni JZ. Determination of ultra-trace rare-earth elements in human plasma by inductively coupled plasma mass spectrometry. Spectrosc Spect Anal 2000;20:339–42.
[416] Inagaki K, Haraguchi H. Determination of rare earth elements in human blood serum by inductively coupled plasma mass spectrometry after chelating resin preconcentration. Analyst 2000;125:191–6.
[417] Wei ZG, Hong FS, Yin M, Li HX, Hu F, Zhao GW, Wong JWC. Off-line separation and determination of rare earth elements associated with chloroplast pigments of

hyperaccumulator *Dicranopteris dichotoma* by normal-phase liquid chromatography and ICP-MS. Anal Bioanal Chem 2004;380:677–82.
[418] Wang SJ, Cao L, Chang P, Hou TP, Hou Y. Determination of rare earth and microelements in Chinese herbal medicine by inductively coupled plasma mass spectrometry. Spectrosc Spect Anal 2006;26:1330–3.
[419] Chen SZ, Xiao MF, Lu DB, Zhan XL. Use of a microcolumn packed with modified carbon nanofibers coupled with inductively coupled plasma mass spectrometry for simultaneous on-line preconcentration and determination of trace rare earth elements in biological samples. Rapid Commun Mass Spectrom 2007;21:2524–8.
[420] Saussereau E, Lacroix C, Cattaneo A, Mahieu L, Goulle JP. Hair and fingernail gadolinium ICP-MS contents in an overdose case associated with nephrogenic systemic fibrosis. Forensic Sci Int 2008;176:54–7.
[421] Costas M, Lavilla I, Gil S, Pena. F.; de la Calle, I.; Cabaleiro, N.; Bendicho, C. Evaluation of ultrasound-assisted extraction as sample pre-treatment for quantitative determination of rare earth elements in marine biological tissues by inductively coupled plasma-mass spectrometry. Anal Chim Acta 2010;679:49–55.
[422] Handley SA, Raja KB, Sharpe C, Flanagan RJ. Measurement of serum lanthanum in patients treated with lanthanum carbonate by inductively coupled plasma-mass spectrometry. Ann Clin Biochem 2011;48:178–82.
[423] Shi Q. Analysis of Rare Earth Elements in *Sophora japonica* L by Inductively Coupled Plasma Mass Spectrometry. J Chem Soc Pak 2012;34:1096–100.
[424] Zhang L, Chen BB, He M, Hu B. Polymer monolithic capillary microextraction combined on-line with inductively coupled plasma MS for the determination of trace rare earth elements in biological samples. J Sep Sci 2013;36:2158–67.
[425] Skoog DA, Holler FJ, Crouch SR. Principles of Instrumental Analysis. 6th ed. Belmont, CA: Thomson Brooks/Cole; 2007. pp. 169–173.
[426] Karpinska J. Basic Principles and Analytical Application of Derivative Spectrophotometry, Macro to Nano Spectroscopy, Dr. Jamal Uddin (Ed.), ISBN: 978-953-51-0664-7, In-Tech, DOI: 10.5772/37673, 2012.
[427] Wylie AW. Spectrophotometric determination of praseodymium, neodymium, and samarium. J Soc Chem Ind London 1950;69:143–7.
[428] Medalia AI, Byrne BJ. Spectrophotometric Determination of Cerium (IV). Anal Chem 1951;23:453–6.
[429] Telep G, Boltz DF. Ultraviolet Spectrophotometric Determination of Cerium. Anal Chem 1953;25(6):971–3.
[427] Wylie AW. Spectrophotometric determination of praseodymium, neodymium, and samarium. J Soc Chem Ind London 1950;69:143–7.
[431] Gordon L, Feibush AM. Spectrophotometric Determination of Cerium after Oxidation to Cerium(IV) with Lead Dioxide. Anal Chem 1955;27:1050–1.
[432] Banks CV, Spooner JL, Olaughlin JW. Differential Spectrophotometric Determination of Neodymium in Neodymium-Yttrium Mixtures. Anal Chem 1956;28:1894–7.
[433] Conca N, Merritt C. Spectrophotometric Determination of Cerium in Carbonate Solution. Anal Chem 1956;28:1264–8.
[434] Greenhaus HL, Feibush AM, Gordon L. Ultraviolet Spectrophotometric Determination of Cerium(III). Anal Chem 1957;29:1531–4.
[435] Pro MJ, Nelson RA. Determination of added distinctive cations in whiskey. 9. Ultraviolet Spectrophotometric Determination of Cerium. J Assoc Off Agric Chem 1957;40:932–6.
[436] Bergstresser KS. Spectrophotometric determination of lanthanum in plutonium. Anal Chem 1958;30:1630–2.
[437] Dobkina BM, Maliutina TM. The spectrophotometric determination of cerium in lanthanum, neodymium and praseodymium specimens. Ind Lab 1958;24:442–4.
[438] Shibata S, Takeuchi F, Matsumae T. Spectrophotometric determination of lanthanum with neo-thorone. Anal Chim Acta 1959;21:177–81.

[439] Herrington J, Steed KC. Spectrophotometric determination of the rare earths yttrium and cerium by bromopyrogallol red. Anal Chim Acta 1960;22:180–4.
[440] Khopkar SM, De AK. Simultaneous Extraction and Spectrophotometric Determination of Cerium with 2-Thenoylfluoroacetone. Anal Chem 1960;32:478–80.
[441] Blatz LA. Precise difference method for the spectrophotometric determination of cerium(IV) at micromolar concentrations. Anal Chem 1961;33. 249-&.
[442] Culkin F, Riley JP. Spectrophotometric determination of microgram amounts of cerium. Anal Chim Acta 1961;24. 167-&.
[443] Malyutina TM, Dobkina BM, Chernikov YA. The determination of neodymium by differential spectrophotometric analysis. Ind Lab 1961;27:665–8.
[444] Misumi S, Nagano N. Spectrophotometric determination of microgram amounts of cerium with 2-methyl-8-quinolinol. Anal Chem 1962;34. 1723-&.
[445] Tonosaki K, Otomo M. Spectrophotometric determination of Ce(III) and some rare earths with xylenol orange. Bull. Chem Soc Japan 1962;35:1683–6.
[446] Ganopolskii VI, Krivonozhnikova LG, Shvarev VS. The use of complexone I in the Spectrophotometric Determination of Cerium. Ind Lab 1963;29:144–5.
[447] Onishi H, Banks CV. Spectrophotometric Determination of Cerium with Thenoyltrifluoroacetone. Anal Chem 1963;35. 1887-&.
[448] Onishi H, Toita Y. Spectrophotometric determination of cerium with thenoyltrifluoroacetone after cupferron-chloroform extraction. Anal Chem 1964;36. 1867-&.
[449] Poddar SN, Sengupta NR, Adhya JN. Spectrophotometric determination of cerium (4) with salicylhydroxamic acid. Indian J Chem 1965;3. 135-&.
[450] Roberts JE, Ryterban M. Ion exchange separation and spectrophotometric determination of cerium in cast iron. Anal Chem 1965;37. 1585-&.
[451] Sarma PL, Dieter LH. Spectrophotometric determination of cerium with sulphanilic acid. Talanta 1966;13. 347-&.
[452] Svoboda V, Chromy V. Reactions of metallochromic indicators on micelles-III Spectrophotometric determination of minute amounts of lanthanum with xylenol orange. Talanta 1966;13. 237-&.
[453] Lanning EW. Differential spectrophotometric determination of europium in yttrium europium vanadate using an expanded-scale absorption slidewire. Anal Chem 1969;41. 182-&.
[454] Vernon F. Spectrophotometric determination of cerium with methylene blue. Anal Chim Acta 1969;48. 425-&.
[455] Holland WJ, Veel AE, Gerrard J. 3-Thianaphthenoyltrifluoroacetone: A new reagent for the spectrophotometric determination of cerium(IV). Mikrochim. Acta 1970;58. 297-&.
[456] Malik WU, Sharma TC. A new reagent (*o-para*-toluene sulfonamide) aniline for spectrophotometric determination of cerium (IV). J Indian Chem Soc 1970;47. 167-&.
[457] Mori I, Enoki T. Spectrophotometric determination of cerium (III) with methyl xylenol blue and cetylpyridinium chloride. Yakugaku Zasshi 1970;90. 494-&.
[458] Shah VL, Sangal SP. Composition and stability constants of cerium-chromotrope 2R chelate and the spectrophotometric determination of cerium. Microchem J 1970;15. 548-&.
[459] Taketats T, Toriumi N. Spectrophotometric study and analytical application of rare earth tiron complexes-I Determination of neodymium, holmium and erbium. Talanta 1970;17:465–70.
[460] Cattrall RW, Slater SJE. Spectrophotometric determination of europium(iii) using chrome azurols. Microchem J 1971;16:602–10.
[461] Mazumdar AK, Chatterji A. Spectrophotometric determination of lanthanum, thorium and iron(III) with chromotropic acid azo dyes of the pyridine series. Talanta 1971;18:968–70.
[462] Poluekto NS, Ognichen SF, Ovchar LA, Lauer RS. Spectrophotometric determination of neodymium in a mixture with yttrium. Ind Lab 1971;37. 1139-&.

[463] Shah VL, Sangal SP. Spectrophotometric determination of praseodymium, neodymium, samaruim and europium using chromotrope 2 R as a chromogenic reagent. Chim Anal 1971;53. 47-&.
[464] Taketats T, Yamauchi T. Spectrophotometric study and analytical application of rare-earth Tiron complexes-II Determination of europium. Talanta 1971;18. 647-&.
[465] Ueda J. Spectrophotometric determination of gadolinium with 4-(2-Pyridylazo) resorcinol. Nippon Kagaku Zasshi 1971;92. 849-&.
[466] Das AK. Spectrophotometric determination of cerium and cobalt with sulfosalicyclic acid. Indian J Chem 1972;10. 306-&.
[467] Dhupar SC, Banerji SK, Srivasta K. Spectrophotometric determination of ytterbium(III) with chromotrope 2R. J Indian Chem Soc 1972;49. 425-&.
[468] Sharma CL, Tandon SN, Sabharwa A, Rai D. Spectrophotometric determination of Al(III), Ce(IV) and U(VI) using solochrome azurine B.S. Indian J Chem 1972;10. 744-&.
[469] Issa IM, Issa RM, Ahmed YZ. Complexometric and spectrophotometric determination of thorium(IV), cerium(III), and uranium(VI), using quinizarin sulphonic acid after separation with ion-exchange resins. Microchem J 1973;18:569–76.
[470] Poluekto N, Ovchar LA, Lauer RS. Spectrophotometric determination of erbium in a mixture with lanthanum by means of methylthymol blue and cetyltrimethylammonium bromide. Zh Anal Khim 1973;28:1958–61.
[471] Takahari T, Yamamoto Y, Nakamura M. Spectrophotometric determination of small amounts of cerium in stainless steel. Japan Anal 1973;22:383–9.
[472] Malik WU, Bembi R, Singh R. Spectrophotometric determination of cerium(IV) with crystal violet. Fresenius Z. Anal Chem 1974;272:364–1364.
[473] Poluektov NS, Ovchar LA, Lauer RS, Potapova SF. Extraction-spectrophotometric determination of dysprosium in its mixture with lanthanum. J Anal Chem Ussr 1974;29:1480–2.
[474] Romanov PN. Divided spectrophotometric determination of lanthanum and cerium with carboxynitrazo in presence of other rare earth elements. Zavodskaya Lab 1974;40:630–2.
[475] Elbeih IIM, Shawali AS, Aboutabl MA. Spectrophotometric Determination of of Cerium (IV) with 4-(Dimethylamino)azobenzene-4′-arsonic acid. Fresenius Z Anal Chem 1975;274:31–131.
[476] Kirillov AI, Turkina LA, Vlasov NA. Uv spectrophotometric determination of praseodymium by means of its diethylenetriaminepentaacetate complex. J Anal Chem Ussr 1975;30:1900–2.
[477] Poluektov NS, Ovchar LA, Potapova SF, Lauer RS. Extraction-spectrophotometric determination of terbium in its mixture with lanthanum. J Anal Chem Ussr 1975;30:1975–8.
[478] Agrawal YK, Kapoor HL. Solvent extraction and spectrophotometry determination of microgram quantities of cerium with *N*-*m*-tolyl-*m*-nitrobenzohydroxamic acid. Annal Chim 1976;66:117–26.
[479] Koroleva GN, Poluektov NS, Kirillov AI. Spectrophotometric determination of europium in the presence of praseodymium or erbium. Ind Lab 1976;42:178–80.
[480] Nepomnyashchaya NA, Menkov AA, Lenskii AS. Spectrophotometric determination of lanthanum by means of chlorphosphonazo in presence of scandium, yttrium, and yttrium subgroup rare-earths. J Anal Chem Ussr 1976;31:925–31.
[481] Agrawal YK. *N*-(*p*-Tolyl)-2-furohydroxamic acid is an analytical reagent for extraction and spectrophotometric determination of cerium(IV). Chem Anal (Warsaw) 1977;22:215–20.
[482] Agrawal YK. Spectrophotometric Determination of Cerium (IV) with N-p-tolylbenzohydroxamic acid. Fresenius Z Anal Chem 1977;285:258–1258.
[483] Gowda HS, Thimmaiah KN. Spectrophotometric determination of cerium(IV) with triflupromazine hydrochloride. J Indian Chem Soc 1977;54:869–71.

[484] Jercan E, Musat C, Mainerici M. Spectrophotometric determination of cerium using methyl green. Rev Chim 1977;28:470–4.
[485] Sankagowda H, Thimmaiah KN, Narayanaachar B. Spectrophotometric determination of palladium(II), cerium(IV), arsenic(III) and nitrite with fluphenazine hydrochloride. Rev Roum Chim 1977;22:745–51.
[486] Sankegowda H, Thimmaiah KN. Rapid spectrophotometric determination of cerium(IV), arsenic(III) and nitrite with promethazine hydrochloride. Indian J Chem Sect A 1977;15:763–4.
[487] Gowda HS, Thimmaiah KN. Spectrophotometric determination of cerium(IV), arsenic(III), and nitrite with promazine hydrochloride. Microchem J 1978;23:291–6.
[488] Mishchenko VT, Ovchar LA, Poluektov NS, Aleksandrova NN. Spectrophotometric determination of rare earths of the cerium subgroup in precipitates of oxalates. J Anal Chem Ussr 1978;33:52–7.
[489] Agrawal YK. Extraction and spectrophotometry determination of cerium in tracer levels. Bull Acad Polon Sci Ser Sci Chim 1979;27:681–7.
[490] Kharzeeva SE, Maltseva VS. Spectrophotometric determination of lanthanum and cerium subgroup elements by means of sulfonazo-ii and antipyrine. J Anal Chem Ussr 1979;34:790–1.
[491] Maltseva VS, Kharzeeva SE. Spectrophotometric method of neodymium determination in the presence of yttrium. Ind Lab 1979;45:242–3.
[492] Reddy VK, Reddy DV. Spectrophotometric determination of cerium (IV) with solo chrome black 6B. J Indian Inst Sci Sect B 1979;61:191–3.
[493] Arora HC, Venkateswarlu C, Rao GN. Extraction and spectrophotometric determination of cerium (IV) with 4-benzoyl-2,4-dihydro-5-methyl-2-phenyl-3*H*-pyrazol-3-one. Indian J Chem Sect A 1980;19:500–1.
[494] Bhattacharya S, Lyle SJ, Maghzian R. A spectrophotometric determination of europium in lanthanide and other mixtures by use of Methylene Blue. Talanta 1980;27:59–60.
[495] Gowda HS, Achar BN. Rapid spectrophotometric determination of microgram amounts of cerium(IV) and arsenic(III) with tetrasulfonated metal phthalocyanine complexes. Indian J Chem Sect A 1980;19:932–4.
[496] Jain RK, Agrawal YK. Para-chloro-substituted cinnamohydroxamic acids as analytical reagents for cerium- spectrophotometric determination with *N-p*-chlorocinnamohydroxamic acid. Croat Chem Acta 1980;53:471–5.
[497] Ovchar LA, Poluektov NS. Spectrophotometric determination of neodymium in its mixture with lanthanum by means of eosin and 2,2′-Dipyridyl. J Anal Chem Ussr 1980;35:1128–30.
[498] Davidescu L, Slivneanu I, Popa L. Determination of praseodymium in neodymium oxide by visible spectrophotometry. Rev Chim 1981;32:1198–201.
[499] Holzbecher Z, Kabrt L, Hoffmanova M. Spectrophotometric determination of traces of cerium in brasses containing manganese and chromium. Chem Listy 1981;75:543–5.
[500] Hsu CG, Hu CS, Jia XP, Pan JM, Chlorophosphonazo-m-NO2. a new reagent for the spectrophotometric determination of cerium sub-group rare earth elements in the presence of yttrium sub-group elements. Anal Chim Acta 1981;124:177–83.
[501] Bashir WA, Hagop GW, Falamerz S. Indirect spectrophotometric determination of cerium (IV). Microchem J 1982;27:21–5.
[502] Ishii H, Satoh K. Determination of Micro Amounts of Samarium and Europium by Analogue Derivative Spectrophotometry. Fresenius Z Anal Chem 1982;312:114–20.
[503] Lyle SJ, Zatar NA. Automatic spectrophotometric method for the determination of europium in mixtures with yttrium and other lanthanides. Fresenius Z Anal Chem 1982;313:313–5.
[504] Lyle SJ, Zatar NA. A selective spectrophotometric determination of europium in mixtures with other lanthanides, yttrium and scandium. Anal Chim Acta 1982;135:327–32.

[505] Azarova VN, Kirillov AI, Dubov RI. Separate spectrophotometric determination of gadolinium and dysprosium in binary mixtures by phthalexones using nomogram. J Anal Chem Ussr 1983;38:507–11.
[506] Cabreramartin A, Izquierdohornillos R, Quejidocabezas AJ, Peralfernandez JL. Spectrophotometric Determination of Cerium with Methylthymol Blue in the Presence of Oxalate and Cyanide as Masking Agents. Analyst 1983;108:534–7.
[507] Gowda HS, Gowda AT. Prochlorperazine bis(methanesulfonate) as a new spectrophotometric reagent for the determination of cerium(IV), arsenic(III) and nitrite. Natl Acad Sci Lett India 1983;6:407–11.
[508] Konno H, Takeyama S. Spectrophotometric Determination of Small Amounts of Cerium in Iron and Steels with Lanthanum Fluoride Coprecipitation-o-Tolidine. Bunseki Kagaku 1983;32:T41–3.
[509] Lyle SJ, Zatar NA. Automatic spectrophotometric method involving liquid-liquid extration for the determination of europium in the presence of other lanthanides, Yttrium and Scandium. Fresenius Z Anal Chem 1983;314:463–5.
[510] Perkov IG, Zaien KBS. Simultaneous spectrophotometric determination of lanthanum, cerium, praseodymium and neodymium in monazite sand. J Anal Chem Ussr 1983;38:1668–75.
[511] Abuzuhri AZ, Abedulhadi. K. sensitive spectrophotometric determination of La(III) with 1-(2-pyridylazo)-2-naphthol. Microchem J 1984;30:231–4.
[512] Abuzuhri AZ, Salim R. 2-(5-Bromo-2-pyridylazo)-5-diethylaminophenol as a reagent for the spectrophotometric determination of La(III), Y(III), and Ce(III). Microchem J 1984;29:126–31.
[513] Agrawal YK, John KT. New substituted hydroxamic acids: for spectrophotometric determination of cerium (IV). Bunseki Kagaku 1984;33:E137–42.
[514] Chandravanshi BS, Gupta VK, Extraction. Spectrophotometric Determination of Cerium(IV) with *N,P*-Chlorophenylcinnamohydroxamic Acid. Croat Chem Acta 1984;57:243–9.
[515] Garcia REG, Jimeno SA. Direct Spectrophotometric Determination of Erbium in Silicate-glasses. Anal Quim Serie B 1984;80:516–20.
[516] Hafez MB, Wassif MA. Spectrophotometric determination of cerium (IV) with tetraethylenepentaamineheptaacetic acid (TPHA). Microchem J 1984;29:304–6.
[517] Li XY, Chen ZH, He H, Zhang HS. Direct Spectrophotometric Determination of Cerium Sub-group Rare Earth Elements with m-Trifluoromethylchlorophosphonazo in the Presence of Yttrium Sub-group Elements in Nodular Cast Iron and Steel samples. Inorg Chim Acta 1984;94:137–9.
[518] Novak J, Vyhlidka P, Kohout V, Hajek B. Spectrophotometric determination of trivalent cerium with feriin in the presence of tetravalent cerium. Collect Czech Chem Commun 1984;49:2458–62.
[519] Rahim SA, Amin D, Bashir WA. Titrimetric and spectrophotometric methods for the determination of cerium(IV) in aqueous solution. Microchem J 1984;30:53–7.
[520] Barrero FA, Ceba MR, Ramirez AA. Spectrophotometric determination of La(III) by extraction with purpurin in methyl isobutyl ketone. Anal Quim Serie B 1985;81: 217–21.
[521] Chakrabarti A. K. spectrophotometric determination of cerium (IV) in aqueous medium with 4–sulpho–2–aminobenzenethiol and its studies. J Indian Chem Soc 1985;62:541–3.
[522] Gowda HS, Gowda AT, Gowda NMM. Rapid spectrophotometric determination of cerium(IV), arsenic(III), and nitrite with perphenazine. Microchem J 1985;31:385–91.
[523] Idriss KA, Harfoush AA, Elshahawy AS, Sedaira H. Spectrophotometric study of the complexation equilibria of lanthanum(III) with 1,4-bis(4′-methylanilino)anthraquinone and the determination of lanthanum(III). Analyst 1985;110:709–12.

[524] Azarova VN, Kirillov AI. Spectrophotometric determination of lanthanum, lutetium and yttrium with triphenylmethane group reagents and cetylpyridinium bromide. J Anal Chem Ussr 1986;41:1408–12.

[525] Gaokar UG, Eshwar MC. Rapid spectrophotometric determination of cerium(III) with 4-(2-thiazolylazo)resorcinol and 1-(2-thiazolylazo)-2-naphthol. Analyst 1986;111:1393–5.

[526] Ishii H, Satoh K. Determination of micro amounts of praseodymium by analogue derivative spectrophotometry. Bunseki Kagaku 1986;35:704–8.

[527] Lepine L, Gilbert R, Belanger G. Ultraviolet spectrophotometric determination of gadolinium in concentrated solutions of nitrate salts. Anal Chem 1986;58:1152–6.

[528] Liu CM, Chi XZ, Liang SQ, Liang SC. The additivity of the absorbance of europium and zinc complexes with xylenol orange and its application in spectrophotometric determination of europium and zinc. Acta Chim Sinica 1986;44:1011–4.

[529] Ramirez AA, Linares CJ, Barrero FA, Ceba MR. Study of the mixed-metal lanthanum-magnesium-purpurin complex: spectrophotometric determination of yttrium and lanthanides. Talanta 1986;33:1021–5.

[530] Rao TP, Sukumar R. Simultaneous determination of lanthanum and europium by higher order derivative molecular absorption spectrophotometry. Anal. Lett 1986;19:1731–46.

[531] Salim R, Abuzuhri A, Shalabi J Spectrophotometric determination of lanthanum(III) with 2,2′-Dipyridyl-2-pyridylhydrazone. Microchem J 1986;33:226–31.

[532] Aleksandrova NN, Mishchenko VT, Poluektov NS, Shilova LP. Determination of praseodymium, neodymium, dysprosium and thulium by the method of derivative spectrophotometry in solutions of chloride mixtures of rare earths. Ukr Khim Zh 1987;53:297–300.

[533] Aleksandrova NN, Mishchenko VT, Poluektov NS, Mukomel VL. Determination of gadolinium by derivative spectrophotometry. Ind Lab 1988;54:461–3.

[534] Bhagavathy V, Rao TP, Damodaran AD. Third order derivative molecular absorption spectrophotometry determination of traces of samarium with methyl thymol blue - CTAB or CPB. Anal Lett 1988;21:901–20.

[535] Mendez JH, Cordero BM, Pavon JLP. Simultaneous spectrophotometric determination of erbium and praseodymium with 1-(2-pyridylazo)-2-naphthol in the presence of octylphenol poly(ethyleneglycol) ether (TX-100). Talanta 1988;35:293–6.

[536] Merisov YI, Nuzhdenova LG. Direct spectrophotometric determination of neodymium in oxide compounds with ruthenium. Ind Lab 1988;54:1237–9.

[537] Mishchenko VT, Ulianova TM, Kononenko LI. An increase in selectivity of holmium and erbium determination by the method of derivative spectrophotometry. Ukr Khim Zh 1988;54:849–51.

[538] Pal BK, Mitra BK, Dasthakur SP. A rapid, selective and sensitive spectrophotometric method for determination of cerium(IV). J Indian Chem Soc 1988;65:862–3.

[539] Sukumar R, Rao TP, Damodaran AD. Determination of trace amounts of praseodymium by third-derivative molecular absorption spectrophotometry. Analyst 1988;113:1061–4.

[540] Chen DH, Cai RX, Zeng Y. Simultaneous spectrophotometric determination of lanthanum and calcium with an automated flow-injection analyzer. Anal Chim Acta 1989;222:205–9.

[541] Mishchenko VT, Aleksandrova NN. Determination of europium in oxide and oxysulfides luminophores and their waste products by direct and derivative spectrophotometry (exchange of experience). Ind Lab 1989;55:1252–3.

[542] Pavon JLP, Cordero BM, Mendez JH, Miralles JC. Micellar systems in flow injection for the spectrophotometric determination of neodymium in glasses. Analyst 1989;114:849–51.

[543] Vanderwalt TN, Coetzee PP, Fourie PJ. Spectrophotometric determination of europium in carrier-free Gd solutions. South Afr Chem 1989;42:13–6.

[544] Zhang PX, Ren Y. Spectrophotometric determination of lanthanum and sum of other light rare earths by flexible tolerance simplex method. Acta Chim Sinica 1989;47:688–93.

[545] Fernandez L, Olsina R. Sensitive spectrophotometric determination of lanthanum with 2-(3,5-dichloro-2-pyridylazo)-5-dimethylaminophenol. Anal Sci 1990;6:411–4.
[546] Milovanovic GA, Vucic N. Kinetic-spectrophotometric method for the determination of neodymium and mixtures of neodymium with lanthanum and cerium. Analyst 1990;115:791–3.
[547] Vyas PC, Mittal VP. Spectrophotometric determination of praseodymium (III) as its 1-(2-pyridylazo)-2-naphthol complexes with a mixture of benzophenone and naphthalene. Indian J Chem Sect A 1990;29:610–1.
[548] Elbanowski M, Makowska B, Ludwiczak E, Hoppe K. On the possibility of spectrophotometric determination of Ho(III) in Y_2O_3 and Gd_2O_3 with tiron. Chem Anal (Warsaw) 1991;36:587–9.
[549] Fragoso LO, Perezcaballero G. Spectrophotometric Determination of Lanthanum with Sulfonazo-III as Chromogenic Agent in Aqueous Ethanolic Solution. Abs Pap Am Chem Soc 1991;202. 74-ANYL.
[550] Morales A, Proal J, Perezcaballero G. Spectrophotometric Determination of Cerium(IV) with Sulfanilic Acid in the Presence of Cetyltrimethyl Bromide (CTAB). Abs Pap Am Chem Soc 1991;202:73–80.
[551] Fernandez L, Olsina R. Extraction and sensitive spectrophotometric determination of ytterbium with 2-(3,5-dichloro-2-pyridylazo)-5-dimethylaminophenol. Talanta 1992;39:1605–9.
[552] Hafez MB, Hafez N. Spectrophotometric Determination of Ce(IV), Fe(II1) and U(V1) using Desferrioxamine-B (desferal). J Chem Technol. Biotechnol 1992;54:267–9.
[553] Mori I, Fujita Y, Toyoda M, Kimura S. Improved spectrophotometric determination of dysprodium(III) with 4-(2-pyridylazo)resorcinol in the presence of Triton X 305 - hexadecylpyridinium chloride. Anal Lett 1992;25:1345–54.
[554] Mori I, Fujita Y, Toyoda M, Tanaka K. Spectrophotometric Determination of Gadolinium(III) with *O*-Hydroxyhydroquinonephthalein in Mixed Micelles of *N*-Hexadecylpyridinium Chloride and Brij 35. Anal Lett 1992;25(12):2377–88.
[555] Wang NX, Liang WA, Zhou SF, Qi P. Absorption spectra of 4f electron transitions of neodymium and erbium complexes with ferron and diethylamine. Determination of neodymium and erbium in rare earth mixtures by third-derivative spectrophotometry. Anal Chim Acta 1992;262:253–9.
[556] Bhagavathy V, Rao TP, Damodaran AD. Flotation-spectrophotometric determination of praseodymium with 5,7-dichloroquinolin-8-ol and Rhodamine 6G. Anal Chim Acta 1993;280:169–72.
[557] Bhagavathy V, Reddy MLP, Rao TP, Damodaran AD. Extractive spectrophotometric determination of neodymium with 5,7-dichlorooxine and rhodamine 6G. Indian J Chem Sect A 1993;32:463–4.
[558] Bobrowskagrzesik E, Grossman AM, Ciba J. Spectrophotometric determination of La, Ce, Pr and Nd in the presence of copper by use of arsenazo III. Fresenius J Anal Chem 1993;345:614–5.
[559] Martinez LD, Perino E, Marchevsky EJ, Olsina RA. Spectrophotometric determination of gadolinium(III) with 2-(5-bromo-2-pyridylazo)-5-diethylaminophenol (5-Br-PADAP). Talanta 1993;40:385–90.
[560] Wang NX, Wu QC, Shi JB, Qi P. Simultaneous determination of neodymium and erbium in rare earth mixture with 5,7-dibromo-8-hydroxyquinoline and TX-100 by third-derivative spectrophotometry. Mikrochim. Acta 1993;110:119–26.
[561] Baranowska I, Barszczweska K. Spectrophotometric method for the determination of europium with 2,2′-biquinoxalyl. Fresenius J Anal Chem 1994;350:568–9.
[562] Chen JF, Khalili FI, Mohammed AK, Choppin GR. Extraction and spectrophotometric determination of Nd(III), Th(IV) and U(VI) in synthetic brine using Chlorophosphonazo III. Anal Chim Acta 1994;284:593–7.

[563] Hsu CG, Hong L, Pan JM. Spectrophotometric determination of cerium subgroup rare earths in nickel-base alloys in the presence of yttrium with p-acetylchlorophosphonazo and mixed surfactants. Talanta 1994;41:1357–61.
[564] Rao CK, Reddy VK, Reddy TS. Rapid and sensitive spectrophotometric determination of cerium(IV) with 2,4-dihydroxy benzophenone benzoic hydrazone. Talanta 1994;41:237–41.
[565] Wang NX, Lan L. Derivative spectrophotometric determination of neodymium in rare earth mixtures using bromopyrogallol red and TX-100. Fresenius J Anal Chem 1994;350:365–7.
[566] Wang NX, Qi P, Jiang KZ. Fourth-derivative spectrophotometric determination of neodymium in mixed rare earths with methyl thymol blue and cetylpyridinium chloride. Mikrochim. Acta 1994;116:191–6.
[567] Yang W, Gao JZ, Kang JW, Wang ZM, Bai GB, Li CX. Quantitative determination of praesodymium(III)–neodymium(III)–holmium(III)–erbium(III) four-component systems by matrix–fourth derivative spectrophotometry. Analyst 1994;119:2463–6.
[568] Zhai QZ. Simultaneous determination of lanthanum and cerium in mixed rare earths with *p*-acetylarsenazo by spectrophotometry. Talanta 1994;41:703–6.
[569] Wang NX, Yang JH, Siu H, Qi P. Third-derivative spectrophotometric determination of neodymium and erbium in mixed rare earths with 2-(diphenylacetyl)indan-1,3-dione and dodecyl benzenesulfonic acid sodium salt. Analyst 1995;120:2413–6.
[570] Anbu M, Rao TP, Iyer CSP, Damodaran AD. Simultaneous determination of dysprosium, holmium and erbium in high purity rare earth oxides by second order derivative spectrophotometry. Chem Anal (Warsaw) 1996;41:781–5.
[571] Martinez LD, Gasquez JA, Olsina RA, Marchevsky EJ. Spectrophotometric determination of neodymium in high sensitivity conditions in rocks with 2-(5-bromo-2-pyridylazo)-5-diethylaminophenol(5-Br-PADAP). Chem Anal (Warsaw) 1996;41:275–82.
[572] Wang NX, Si ZK, Jiang W. Selective determination of the holmium in rare earth mixtures by second derivative spectrophotometry with 2-isobutylformyl-1,3-dione-indan and octylphenol poly-(ethyleneglycol)ether. Anal Lett 1996;29:2191–200.
[573] Wang NX, Si ZK, Jiang W, Qi ZC. Selective determination of holmium in rare earth mixtures by second-derivative spectrophotometry with benzoylindan-1,3-dione and cetylpyridinium chloride. Analyst 1996;121:1317–20.
[574] Wang NX, Si ZK, Yang JH, Du AQ, Li ZD. Simultaneous determination of neodymium, erbium and holmium in rare earth mixtures with 2-phenyltrifluoroacetone and octylphenol poly(ethyleneglycol) ether by third-derivative spectrophotometry. Talanta 1996;43:589–93.
[575] Amin AS, Moustafa MM, Issa RM. A rapid, selective and sensitive spectrophotometric method for the determination of Ce(III) using some bisazophenyl-beta-diketone derivatives. Talanta 1997;44:311–7.
[576] Silva MF, Fernandez L, Olsina RA, Stacchiola D. Cloud point extraction, preconcentration and spectrophotometric determination of erbium(III)-2-(3,5-dichloro-2-pyridylazo)-5-dimethylaminophenol. Anal Chim Acta 1997;342:229–38.
[577] Wang NX, Jiang W, Si ZK, Yang JH, Liang WA. Derivative spectrophotometric determination of Holmium in rare earth mixtures with 2-(diphenylacetyl)indan-1,3-dione and octylphenyl poly(ethyleneglycol) ether. Mikrochim. Acta 1997;126:251–5.
[578] Wang NX, Si SK, Yang JH, Jiang W, Liang WA, Li ZD, Du GY, Zhang GQ. Selective determination of neodymium and erbium in mixtures with other lanthanides by second-derivative spectrophotometry of complexes with benzoyl-indan-1,3-dione and cetylpyridinium chloride. Mikrochim. Acta 1997;127:71–5.
[579] Silva MF, Fernandez LP, Olsina RA. Monitoring the elimination of gadolinium-based pharmaceuticals. Cloud point preconcentration and spectrophotometric determination of Gd(III)-2-(3,5-dichloro-2-pyridylazo)-5-dimethylaminophenol in urine. Analyst 1998;123:1803–7.

[580] Vazquez FA, Martinez LD, Gazquez JA, Olsina RA, Marchevsky EJ. Sensitive spectrophotometric determination of Ytterbium(III) in geological samples and in bimetallic catalysts with 2-(5-bromo-2-pyridylazo)5-diethylaminophenol. J Trace Microprob. Tech 1998;16:353–62.
[581] D'Angelo J, Fernandez J, Martinez L, Marchevsky E. Sensitive spectrophotometric determination of erbium(III) in geological materials and glasses with 2-(5-bromo-2-pyridylazo)-5-diethylaminophenol. J Anal Chem 1999;54:586–90.
[582] Ensafi AA, Zomorrodian M. Catalytic spectrophotometric determination of cerium by ion exchange separation coupled to a flow injection system. J Anal Chem 1999;54: 442–5.
[583] Fernandez JE, Martinez L, Perino E, Olsina R, Marchevsky E. Sensitive spectrophotometric determination of holmium (III) in geological and lasers samples with 2-(5-bromo-2-pyridylazo)-5-diethylaminophenol. An Asoc Quim Argent 1999; 87:15–24.
[584] Qi P, Du GY, Ren XZ, Wang L, Jiang W, Wang NX. Simultaneous determination of neodymium and erbium with 8-hydroxyquinoline and TX-100 by derivative spectrophotometry. J Indian Chem Soc 1999;76:59–60.
[585] Wang NX, Wang L, Ren XZ, Jiang W, Si ZK, Zhou HB, Zhang M, Wang YZ. Derivative spectrophotometric determination of Nd, Ho and Er in rare earth mixtures with 1-ethyl-6,8-difluoro-7-(3-methyl-1-piperazinyl)-4-oxo-1,4-dihydro-3-quin oline carboxylic acid. Mikrochim Acta 1999;132:17–23.
[586] Soylak M, Turkoglu O. Spectrophotometric determination of samarium(III) with chrome azurol S in the presence of cetylpyridinium chloride. Talanta 2000;53:125–9.
[587] Wang NX, Ren XZ, Si ZK, Jiang W, Liu CB, Liu XR. Derivative spectrophotometric determination of praseodymium in rare earth mixtures with lomefloxacin. Talanta 2000;51:595–8.
[588] Vijayalekshmy B, Biju VM, Rao TP. Selective spectrophotometric determination of cerium(IV). Indian J Chem Sect A 2001;40:1365–6.
[589] Li ZJ, Yang YL, Liu JK, Pan JM, Tang J. Novel sensitive and selective method for spectrophotometric determination of cerium(III) in biological samples using 2,6-dibromo-4-chlorocarboxyarsenazo. Anal Lett 2002;35:1959–75.
[590] Masti SP, Seetharamappa J, Melwanki MB, Motohashi N. Spectrophotometric determination of cerium(IV) using a phenothiazine derivative. Anal Sci 2002;18:167–9.
[591] Revanasiddappa HD, Kumar TNK. Spectrophotometric determination of cerium with leuco xylene cyanol FF. Anal Sci 2002;18:1275–8.
[592] Shamsipur M, Yousefi M, Ghasemi Z, Hagiagha-Babaei L, Ganjali MR. Separation and pre-concentration of trace amounts of cerium(III) on octadecyl silica membrane discs modified with 1,3,5-trithiacyclohexane and its spectrophotometric determination by arsenazo(III). Sep Sci Technol 2002;37:3525–34.
[593] Turkoglu O, Soylak M, Colak H. Spectrophotometric determination of erbium(III) with murexide in synthetic samples. Asian J Chem 2002;14:363–6.
[594] Liawruangrath S, Sakulkhaemaruethai S. Flow injection spectrophotometric determination of europium using chlortetracycline. Talanta 2003;59:9–18.
[595] Revanasiddappa HD, Kumar TNK. A new spectrophotometric method for the determination of cerium using leuco disulphine blue. J Anal Chem 2003;58:922–6.
[596] Yu F, Wei J, Wang NX, Zhao WY. Derivative spectrophotometric determination of neodymium with neodymium-enoxacin-cetyltrimethylammonium bromide system. Chinese J Anal Chem 2003;31:92–4.
[597] Chamsaz M, Arbab-Zavar MH, Hosseini MS. Flotation spectrophotometric determination of La(III) as a ternary ion associate with alizarin and rhodamine B. Chem Anal (Warsaw) 2004;49:241–9.
[598] Zhai QZ, Zhang XX. Color-fading spectrophotometric determination of cerium with DBC-arsenazo. J Rare Earths 2004;22:440–2.

[599] Chamsaz M, Rounaghi G, Hossein M, Arbab-Zavar MH, Sovizi MR. Determination of lanthanum in human blood serum using the zero and first derivative spectrophotometry in micellar media. Asian J Chem 2005;17:1127–34.
[600] Zhai QZ, Hu WH, Zhang XX, Pang W. Color-fading spectrophotometric determination of cerium(IV) with DBC-chlorophosphonazo. Rare Metal Mat Eng 2005;34:1846–8.
[601] Garg A, Susila V, Kakkar LR. Spectrophotometric determination of cerium(IV) using potassium ferrocyanide as an analytical reagent. J Indian Chem Soc 2006;83:626–8.
[602] Shao J, Zhai QZ, Hu WH, Li JM, Zhang W. Solid phase spectrophotometric determination of trace amount of lanthanum with DBC-chlorophosphonazo. Rare Metal Mat Eng 2006;35:329–32.
[603] Revanasiddappa HD, Dayananda BP. A new reagent system for the sensitive and simple spectrophotometric determination of cerium. Bulg Chem Commun 2007;39:9–14.
[604] Chowdary PG, Reddy K, Reddy VK, Reddy PR. 2-Hydroxy-1-naphthaldehyde-p-hydroxybenzoic hydrazone as a spectrophotometric reagent for the determination of lanthanum (III). Indian J Chem Sect A 2008;47:1381–3.
[605] Zhu XS, Zhang YQ, Gong AQ, Zhang XW. Flow injection kinetic spectrophotometric determination of cerium based on the decolorization reaction between arsenazo III and cerium (IV). Spectro Lett 2008;41:237–41.
[606] Janghel EK, Pervez Y. Facile spectrophotometric determination of cerium using malachite green -iodide system in geological and environmental samples. J Sci Ind Res 2009;68:940–4.
[607] Liu YY, Wang P. Kinetic spectrophotometric method for the determination of cerium(IV) with naphthol green B. Rare Metals 2009;28:5–8.
[608] Li JM, Wei WT. Catalytic kinetic spectrophotometric determination of cerium. J Rare Earths 2010;28:387–90.
[609] Zhai QZ, Hu WH, Zhang Y. Kinetic Spectrophotometric Determination of Trace amount of Europium Based on its Inhibitory Effect of the Oxidation of Switzerland Pigment by Potassium Bromate. Instrum Sci Technol 2010;38:201–9.
[610] Dedkova VP, Shvoeva OP, Savvin SB. Sorption-Spectrophotometric Determination of Lanthanum, Gadolinium, and Ytterbium Using Chlorophosphonazo III on a PANF-Chel Adsorbent. J Anal Chem 2011;66:937–40.
[611] Kumar KG, Muthuselvi R, N N. ′-Bis(2-Aminobenzoyl)ethylenediamine as a Reagent for the Spectrophotometric Determination of Neodymium(III). Asian J Chem 2011;23:3620–2.
[612] Mathew A, Shyamala P, Satyanarayana A, Rao IM. Spectrophotometric determination of neodymium(III), samarium(III) in micellar medium - An alternative to solvent extraction procedures. J Indian Chem Soc 2011;88:995–1000.
[613] Mathew A, Kumar AVK, Shyamala P, Satyanarayana A, Rao IM. Spectrophotometric determination of Neodymium(III), Samarium(III), Gadolinium(III), Terbium (III), Dysprosium (III), Holmium (III) in micellar media. Indian J Chem Technol 2012;19: 331–6.
[614] Shahida S, Ali A. Determination of europium(III) in environmental samples using a flow injection analysis system with on-line preconcentration and spectrophotometric detection. J Radioanal Nucl Chem 2014;300:423–9.
[615] Zare-Dorabei R, Norouzi P, Ganjali MR. Design of a Novel Gadolinium Optical Sensor Based on Immobilization of (*Z*)-*N*′-((Pyridine-2-yl) Methylene) Thiophene-2-carbohydrazide on a Triacetylcellulose Membrane and Its Application to the Urine Samples. Anal Lett 2009;42:190–203.
[616] Zare-Dorabei R, Norouzi P, Ganjali MR. Design of a novel optical sensor for determination of trace gadolinium. J Hazard Mater 2009;171:601–5.
[617] Ganjali MR, Zare-Dorabei R, Norouzi P. Design and construction of a novel optical sensor for determination of trace amounts of dysprosium ion. Sens Actuators B 2009;143:233–8.

[618] Fouladgar M, Ensafi AA. A Novel Optical Chemical Sensor to Determine Samarium Ions in Aqueous Solutions. Sensor Lett 2009;7:1135–40.
[619] Ensafi AA, Fooladgar E. Selective lanthanum ions optical sensor based on covalent immobilization of 4-hydroxysalophen on a hydrolyzed triacetylcellulose membrane. J Anal Chem 2011;66:865–70.
[620] Echabaane M, Rouis A, Bonnamour. I, Ben Ouada H. Characterization of an azo-calix 4 arene-based optical sensor for Europium (III) ions. Mater. Sci. Eng. C 2012;32:1218–21.
[621] Valeur B, Berberan-Santos MN. A Brief History of Fluorescence and Phosphorescence before the Emergence of Quantum Theory. J Chem Educ 2011;88:731–8.
[622] Bunzli JCG. Rare earth luminescent centers in organic and biochemical compounds. In: Liu GK, Jacquier B, editors. Spectroscopic Properties of Rare Earths in Optical Materials, vol. 83. Berlin: Springer Verlag; 2005. Ch. 11.
[623] Ge P, Selvin PR. Thiol-reactive luminescent lanthanide chelates, part 2. Bioconjug. Chem 2003;14:870–6.
[624] Roy BC, Santos M, Mallik S, Campiglia AD. Synthesis of metal-chelating lipids to sensitive lanthanide ions. J Org Chem 2003;68:3999–4007.
[625] Lackowicz JR. Principles of Fluorescence Spectroscopy. 3rd edition Springer; 2006.
[626] Lucas J, Lucas P, Mercier, TL, Rollat A, Davenport W. Rare Earths Science, Technology, Production and Use. Elsevier book, ISBN: 978-0-444-62735-3, 2015, Ch.15.
[627] Sharma A, Schulman SG. Introduction to Fluorescence Spectroscopy. Wiley Interscience; 1999.
[628] Liang Z, Liu ZL, Gao YH. A selective colorimetric chemosensor based on calixarene framework for lanthanide ions-Dy^{3+} and Er^{3+}. Tetrahedron. Lett 2007;48:3587–90.
[629] Hosseini M, Ganjali MR, Veismohammadi B, Riahi S, Norouzi P, Salavati-Niasari M, Abkenar SD. Highly Selective Ratiometric Fluorescent Sensor for La(III) Ion Based on a New Schiff's Base. Anal Lett 2009;42:1029–40.
[630] Ganjali MR, Hosseini M, Hariri M, Faridbod F, Norouzi P. Novel erbium (III)-selective fluorimetric bulk optode. Sens. Actuators B 2009;142:90–6.
[631] Ganjali MR, Veismohammadi B, Hosseini M, Norouzi P. A new Tb^{3+}-selective fluorescent sensor based on 2-(5-(dimethylamino)naphthalen-1-ylsulfonyl)-*N*-henylhydrazinecarbothioam ide. Spectrochim. Acta Part A 2009;74:575–8.
[632] Ganjali MR, Veismohammadi B, Faridbod F, Norouzi P, Abkenar SD. Novel selective optode membrane for terbium ion based on fluorescence quenching of the 2-(5-(dimethylamino) naphthalen-1-ylsulfonyl)-*N*-henylhydrazinecarbothioamid. Sens Actuators B 2010;147:23–30.
[633] Hosseini M, Ganjali MR, Veismohammadi B, Faridbod F, Abkenar SD, Norouzi P. Determination of terbium in phosphate rock by Tb^{3+}-selective fluorimetric optode based on dansyl derivative as a neutral fluorogenic ionophore. Anal Chim Acta 2010;664:172–7.
[634] Das P, Ghosh A, Das A. Unusual specificity of a receptor for Nd^{3+} among other lanthanide ions for selective colorimetric recognition. Inorg Chem 2010;49:6909–16.
[635] Ganjali MR, Hosseini M, Hariri M, Norouzi P, Khandar AA, Bakhtiari A. Highly selective ratiometric fluorescence determination of Eu^{3+} ion based on (4*E*)-4-(2-phenyldiazenyl)-2((*E*)-(2-aminoethylimino)methyl)phenol. Mater Sci Engin C 2010;30:929–33.
[636] Zapata F, Caballero A, Espinosa A, Tarraga A, Molina P. A Selective Chromogenic and Fluorescent Molecular Probe for Yb-III Based on a Bichromophoric Azadiene. European J Inorganic Chem 2010;5:697–703.
[637] Hosseini M, Ganjali MR, Veismohammadi B, Norouzi P, Alizadeh K, Abkenar SD. A novel ratiometric fluorescent Yb^{3+} sensor based on a *N*′-(1-oxoacenaphthylen-2(1*H*)-ylidene)furan-2-carbohydrazide as a suitable fluorophore. Mater Sci Engin C 2010;30:348–51.
[638] Ganjali MR, Gupta VK, Hosseini M, Hariri M, Faridbod F, Norouzi P. Lanthanide recognition: A dysprosium(III) selective fluorimetric bulk optode. Sens Actuators B 2012;171:644–51.

[639] Ganjali MR, Hosseini M, Ghafarloo A, Khoobi M, Faridbod F, Shafiee A, Norouzi P. Selective recognition of Pr^{3+} based on fluorescence enhancement sensor. Mater Sci Engin C 2013;33:4140–3.

[640] Ganjali MR, Hosseini M, Khobi M, Farahani S, Shaban M, Faridbod F, Shafiee A, Norouzi P. A novel europium-sensitive fluorescent nano-chemosensor based on new functionalized magnetic core-shell Fe_3O_4–SiO_2 nanoparticles. Talanta 2013;115:271–6.

[641] Hosseini M, Ganjali MR, Rafiei-Sarmazdeh Z, Faridbod F, Goldooz H, Badiei A, Nourozi P, Ziarani GM. A novel Lu^{3+} fluorescent nano-chemosensor using new functionalized mesoporous structures. Anal Chim Acta 2013;771:95–101.

[642] Hosseini M, Ganjali MR, Aboufazeli F, Faridbod F, Goldooz H, Badiei A, Norouzi P. A selective fluorescent bulk sensor for lutetium based on hexagonal mesoporous structures. Sens Actuators B 2013;184:93–9.

[643] Shamsipur M, Mohammadi M, Taherpour A, Lippolis V, Montis R. Development of a novel PVC-membrane fluorescent sensor based on *N,N'*-bis(dansylamidoethyl)-*N,N'*-bis(2-pyridylmethyl)propylene-diamine as a new fluoroionophore for highly sensitive and selective monitoring of trace amounts of La3+ ions in aqueous solutions. Sens Actuators B 2014;192:378–85.

[644] Ganjali MR, Hosseini M, Karimi A, Haji-Hashemi H, Salavati-Niasari M, Norouzi P. Holmium(III)-selective fluorimetric optode based on *N,N*-bis (salicylidene)-naphthylene-1,8-diamine as a neutral fluorogenic ionophore. Spectrochim Acta A 2014;121:224–9.

[645] Hosseini M, Ganjali MR, Vaezi Z, Faridbod F, Arabsorkhi B, Sheikhha MH. Selective recognition of dysprosium(III) ions by enhanced chemiluminescence CdSe quantum dots. Spectrochim Acta A 2014;121:116–20.

[646] Faridbod F, Sedaghat M, Hosseini M, Ganjali MR, Khoobi M, Shafiee A, Norouzi P. Turn-on fluorescent chemosensor for determination of lutetium ion. Spectrochim Acta A 2015;137:1231–4.

CHAPTER

7

Other Techniques*

OUTLINE

In this chapter, some advanced instrumental techniques used for determination of lanthanide family will be briefly discussed. Methods such as mass spectroscopy (MS), neutron activation analysis (NAA), X-ray fluorescence (XRF), and chromatography have been considered.

MASS SPECTROMETRY

A mass spectrometer is an analytical instrument used to analyze the composition of samples through their ionization and determination of the abundance of gas-phase ions based on their mass to charge (m/z) ratios. The working mechanism of an MS is, as mentioned above, based on the ionization of the molecules leading to the formation of charged molecules

*Morteza Rezapour - IP Department, Research Institute of Petroleum Industry (RIPI), Tehran, Iran

Lanthanides Series Determination by Various Analytical Methods. http://dx.doi.org/10.1016/B978-0-12-804704-0.00007-4

or charged fragment of a molecule and the subsequent measurement of the different m/z values.

The ionized molecules and molecule fragments are differentiated based on their m/z values through different mechanisms. The commonest way to do so is to accelerate them over a path before subjecting them to a magnetic or electric field, which deflects the ions, based on their m/z values before they enter the detection system [1]. The detection system should naturally be able to determine the incident ions. Once the eventual result is produced in the form of a spectrum of the relative abundance of ions, the information can be used not only for quantitative analysis but also as a way to determine the molecular structure of unknown species through correlating the fragmented parts to some structures based on the possible fragmentation patterns.

An overall scheme of a mass spectrometer is illustrated in Fig. 7.1, where the major components, i.e. the ion source, the mass analyzer, and the detector can be seen. The ionizer is the segment of the instrument that ionizes a portion of the sample, and evidently different ionization techniques exist with different capabilities and efficiencies that are used depending on the nature and phase of the sample. The ionized sample is next fed into the mass analyzer. Different types of mass analyzers include the time of flight (TOF), quadrupole, multi-collector, magnetic sector, electric sector, and double focusing sector. All of these instruments differentiate and sort the ions based on their m/z values before feeding them in the detector. The different detectors used in MS systems are photographic plates, Faraday cups, electron multipliers, or CCD cameras and image plates. The detection system provides data on the relative abundances of the ionic species, but in the case of multichannel plate detectors, spatial information is also provided.

Maybe the key element of an MS instrument is its ionization technique, which can gravely influence the nature of the results. Different techniques are suited for different types of samples. Electron ionization (EI) and chemical ionization, for instance, are suitable for gaseous or volatile samples. Liquid or solid biological samples, on the other hand, are commonly

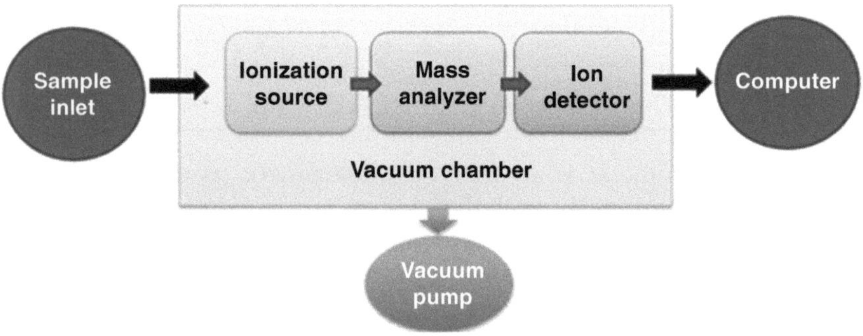

FIGURE 7.1 A simple schematic illustration of a MS instrument.

dealt with using electrospray ionization and matrix-assisted laser desorption/ionization systems.

There is no need to mention that each ionizer has certain disadvantages. EI, for instance, leads to a high degree of fragmentation, which can be of help in many cases and help gain invaluable structural information, yet the ionizer is not suitable for use in hyphenated techniques like HPLC-MS, because of the damages to its filaments by the atmospheric pressure introduced by the prior separation technique.

The ionizers are also categorized as being hard or soft based on the amount of residual energy left in the sample. The hard ionizers like EI leave high residual energies in the sample leading to extensive fragmentation, whereas soft ionization systems like fast atomic bombardment, chemical ionization, atmospheric-pressure chemical ionization, electrospray ionization, matrix-assisted laser desorption/ionization lead to far less fragmentation.

Another ionization system used for MS analysis is the inductively coupled plasma (ICP), which is primarily used for the analysis of cations because of its very high energy content, which can lead to the destruction of the molecular structure and atomization of molecular species. As mentioned in Chapter 6, the ICP is an electrically neutral phase composed of a substantial fraction of ions and electrons, which can help free the outer electrons from atoms. ICP systems usually use argon to generate the plasma, because the first ionization energy of this element is higher than that of almost all elements except for He, O, F, and Ne, and yet lower than the second ionization energy of most elements except for the most electropositive metals, providing a rather controlled energy range for the production of monovalent ions.

Further, some even less common ionization techniques are photoionization, glow discharge, field desorption, desorption/ionization on silicon, direct analysis in real time, atmospheric pressure chemical ionization, secondary ion mass spectrometry (SIMS), spark ionization, and thermal ionization mass spectrometry (TIMS) [1].

Determination of Lanthanides by MS Techniques

Different mass spectrometric techniques have been applied to the determination of lanthanides in various samples. The sample preparation techniques for ICP analyses described in Chapter 6 are applicable to MS analysis too. As mentioned in Chapter 6, a large number of applications of MS techniques in the determination of REEs has been in the form of ICP-MS [2-4]. Some examples of these techniques are given below.

In one work, the endogenous concentrations of the Gd and Dy in body tissues and fluids was reported using electrothermal vaporization–inductively coupled plasma–mass spectrometry (ETV-ICP-MS) [3]. The application of ETV-ICP-MS to such analysis can be difficult because the analytes have been known to form thermally stable oxides and refractory carbides, resulting in decreased sensitivity and severe memory effects. But the researchers solved the problems by using a fluorinating reagent (e.g., poly(tetrafluoroethylene) or trifluoromethane [Freon-23]) to form volatile lanthanide fluorides.

Inductively coupled plasma–mass spectrometry (ICP-MS) has also been employed for the determination of REE content of soil samples [4]. The samples were digested through a microwave digestion system using a mixture of HF, HCl, HNO_3, and EDTA before the ICP-MS analysis, and the standard reference materials (GSS-2, GSS-4) were also used for the purpose of validating the results.

There have also been reports on the application of isotope dilution mass spectrometry (IDMS), sector field mass spectrometry (SFMS), SIMS, and laser ablation inductively coupled plasma mass spectrometry (LA-ICP-MS) for the analysis of lanthanides. In more detail LA-ICP-MS and SIMS are capable of analyzing solid samples and hence provide unique analytical advantages like the elimination of the sample-preparation procedures, which in turn helps avoid the inherently time consuming dissolution/digestion and the consequent sample contamination and analyte loss.

Isotope Dilution Mass Spectrometry

IDMS is one of the mass spectrometry techniques that can be considered as a variation of internal standard addition. Basically, the method includes the addition of certain quantities of an isotopically enriched substance (IES) (which can be a stable or radioactive isotope) to the sample. Mixing of the isotopic standard (i.e., isotopically enriched form of analyte) with the sample dilutes the isotopic enrichment of the standard effectively and this forms the basis for the isotope dilution method. IDMS uses signal ratios for obtaining the required results. The method is applied when high-accuracy is required, and is hence used in the preparation of certified reference materials. Another advantage of IDMS is that it is applicable when low analyte recovery is a problem. IDMS has also been used in chemical separation of Ln fractions. By this way, absolute Ln abundances at the parts per million level with uncertainties of a few percent can be analyzed. Table 7.1 shows some of the reports on the determination of lanthanides through IDMS.

The hyphenation of ID with ICP-MS is another advance in the application of MS techniques for analysis of various samples, especially in the certification of environmental and biological CRMs, which require analytical values from numerous analytical methods with different analytical principles. Some distinct advantages of ID-ICP-MS over the conventional ICP-MS technique are the elimination of signal drifts or matrix effects in the former method. Further, strengths of the method are avoiding partial

TABLE 7.1 Some Reports on Determination of Ln by IDMS

Year	Analyte	Sample	Important Aspects	Ref.
1994	Nd	High purity lanthanum oxide	IDMS	[5]
1997	Nd, Sm	Silicate rocks and minerals	ID-ICP-MS	[6]
1999	Nd, Ce, Sm	Biological material (human hair and wheat flour samples)	ID-ICP-MS	[7]
2009	Lu	Geological sample	MC-ICP-MS	[8]
2011	Sm, Eu	Irradiated Eu-153 sample	Ion exchange chromatography -quadrupole ICP-MS with double spike isotope dilution technique	[9]
2011	Sm, Nd	Geological sample	MC-ICP-MS	[10]

loss of the analyte during sample preparation [5-7]. Alternatively, a number of works have proved that ICP-MS techniques can be a replacement method for IDMS [11-17].

In a report in 1994, because of the La/Nd ratio of $>10^6$, a chemical separation was used prior to the instrumental analysis. The challenge was the difficulty in the separation of two very similar elements. The researcher reported the use of the ICP-MS semiquantitative survey mode for the required separations and further measured Nd through IDMS, which does not require 100% recovery or absolute isolation of Nd. This resulted in the considerable flexibility in the design of preconcentration procedures. The method proved to be able to determine nanogram per gram quantities of Nd in high-purity La compounds [5].

One experiment on the determination of Nd and Sm in silicate rocks and minerals through ID-ICP-MS [6] compared the potentials obtained through this method with those of TIMS and showed that ID-ICP-MS can handle higher sample throughput. There are some reasons to this, the first being the direct measurement of the solutions obtained from acid digestion (in this case an HF-HNO_3 mixture), eliminating the need for a chemical separation. The second and third reasons were the sampling of rock solutions under ambient pressure and short time of analysis of about 10 min/sample, respectively. The method was also found to enjoy better tolerance to all the sources of signal drift and did not need matrix-matched standards.

Another instance of the application of ID-ICP-MS to fast chemical separation and highly reproducible determination of REEs is isotope dilution analysis having a multiple collector inductively coupled plasma mass spectrometer (MC-ICP-MS) [17,18]. A multiple collector (MC) is an ion analyzer used in ICP-MS to improve the precision of isotope-ratio measurements. MC-ICP-MS was found to be superior to compare to quadrupole ICP-MS and thermal ionization mass spectrometric isotope dilution techniques and found to excel them in terms of analytical reproducibility or speed of analysis. Application of a quadrupole analyzer only allows single-collector analysis, which limits the precision of ICP-MS measurements because of the nature of ICP, and isotope-ratio analysis through thermal ionization (TIMS) is too difficult in some cases such as the Hf-W and Lu-Hf.

In a procedure used by Baker et al. [18], the samples were digested using an HF-HNO_3 method and spiked the samples with two mixed spikes. The bulk REEs were eliminated using a cation exchange column, which separated the middle-heavy and light REEs into two groups, and also removing Ba from the light REES, rendering oxide interferences unimportant. The time required for the determination of the heavy (Er-Lu), middle (Eu-Gd), and light (La-Eu) REEs was 1–2 min.

Another report on using an MC-ICP-MS is the application of a sectional power-law (SPL) correction, to accurately and precisely measure $^{176}Lu/^{175}Lu$ ratios in geological samples [8]. The researchers adopted three independent power laws based on the $^{176}Lu/^{176}Yb$ ratios of the samples measured after chromatographic separation. By using isotope dilution and SPL correction, the results of measurements on the lutetium content were found to agree with the United States Geological Survey rock standards (BHVO-1, BHVO-2, BCR-2, AGV-1, and G-2).

Sector Field Mass Spectrometry (SFMS)

As mentioned earlier, ICP-MS is a well-established, powerful, and very sensitive multielement method that can be used for the determination of a wide range of samples, and the combination of this method with preconcentration and separation techniques has helped it overcome the problems caused by possible matrix effects. Additionally, the likely contamination of the sample during sample preparation, as well as the high values of naturally abundant elements, can lead to errors. These can be minimized through careful handling of the samples under ultraclean conditions and use of high-purity chemicals.

Further to these general problems, the possible mass-spectral interferences can hinder ultratrace determinations through ICP-MS, especially since the low differences in the masses and hence difficult separation of highly interfering monovalent isobaric atomic ions leads to serious determination problems. To handle such interferences, high mass resolutions like Fourier transform or ion trap mass spectrometers and sector field mass spectrometers (DFSF-MS) can be applied.

A sector field MS is based on the application of a static electric or magnetic sector or a combination of them as the mass analyzer (a popular combination of these sectors is BEB [magnetic-electric-magnetic]).

The majority of sector instruments are double-focusing because they focus the ion beams both in direction and velocity. The major components of a double-focusing magnetic sector mass spectrometer are an ion source that forms and accelerates the ions at energies of up to 10 kV, a magnetic sector with a magnetic field perpendicular to the motion path of the ion (to deflect ions), an electric sector exerting an electric field perpendicular to the motion path of the ion (to deflect ions), and a detector. The slits before the detector, the width of which is inversely proportional to the resolution, are placed to define positions and also energies of the ions. Further, electrostatic lenses are used to orient the ion beam and maximize the strength of the beam reaching the detector. A summary of some of the reports on the determination of lanthanides through SFMS is given in Table 7.2.

Secondary Ion Mass Spectrometry

SIMS can be regarded as the most powerful techniques of the mass spectrometry, for ultratrace surface and depth profile analyses. The technique uses a high-energy (0.25-30 keV) beam of ions (usually O^{2+}, O^-, Cs^+, Ar^+, Ga^+) or neutral species for bombarding the surface of the sample surface. The energy of the incident ion beam is transferred to atoms of the sample, displacing them and causing its further collision with atoms in the sample, leading to a chain of collisions, which may cause some of the surface atoms of the solid sample. This is also referred to as the sputtering process, which leads to the formation of secondary neutral species (SNs) as well as a portion of secondary ions (SIs) with kinetic energies that depend on the energy of the incident primary ions (in practice, this is usually in the order of 10 eV). The SIs are next focused through the application of electrical potentials and led to a mass analyzer, where they are separated based on their m/z values and finally detected using either of a photographic plate, a Faraday cup, an electron multiplier, or a CCD camera and image plate. In case, the analysis of SNs is desired (i.e., SN mass spectrometry), a so-called repeller voltage is applied to suppress the ionic species, and next the SNs are ionized using a post-sputtering means like an Ar plasma, laser, or electron impact ionization, before being driven into the mass spectrometer [31].

Like any other technique, SIMS suffers some disadvantages. One of the most important drawback of SIMS is that sputtered species not only consist of monoatomic ions but, depending on the nature of the samples, also of molecular species that can cause serious interferences, rendering the analysis of some elements impossible. The method also requires the application of proper standards (that are very close to the sample in nature), as well as empirical corrections to obtain quantitative information. This is particularity the case with isotopic analysis. Further, the sensitivity of the

TABLE 7.2 Some Reports on Determination of Ln by SFMS Technique

Year	Analyte	Sample	Important Aspects	Ref.
1993	Nd, Sm, Lu	In neodymium samarium mixtures, hafnium and hafnium lutetium mixtures	ICP source coupled to a double-focusing magnetic sector mass analyzer equipped with seven Faraday detectors	[19]
1997	REE	Tantalum matrix	Double-focusing ICP-SFMS	[20]
1997	Hf, Lu	Rock samples	Magnetic sector multiple collector ICP-MS	[21]
2002	REE	High-purity lanthanum oxide	HPLC-ICP-SFMS	[22]
2004	REE	High-purity gadolinium oxide	ICP-SFMS	[24]
2006	REE	Antarctic ice	ICP-SFMS	[25]
2006	REE	High-purity samarium oxide	HPLC-ICP-SFMS	[26]
2007	REE	Antarctic ice	ICP-SFMS, ICP-QMS, and ICP-TOFMS	[27]
2009	REE	Surface and subsurface waters	ICP-SFMS	[28]
2013	Gd	Surface waters	LC-ICP-SFMS	[29]
2013	Nd	Nuclear fuel	ICP-SFMS	[30]

method toward an element is a function of the matrix composition and the nature of the primary beam. Finally, it should be noted that the nature of the sample should be compatible with the high vacuums that are characteristic of the method.

As mentioned, an important issue encountered in SIMS analysis is the formation of isobaric interference caused by the unwanted molecular ions. To avoid this, a high-resolution ion microprobe is used. This is usually a double-focusing mass analyzer, the application of which makes it possible to focus the secondary ions using a set of Faraday cups.

SIMS, also referred to as an ion microprobe, is hence capable of analysis of isotopic and elemental abundances in very small solid samples of a few micrometers, and is hence used for the analysis of lanthanides in microscopic samples, or determination of lanthanides that are chemically separated from bulk samples.

As an example of such applications, the work of Sahijpal et al. [32] in 2003 can be mentioned. The researchers described the experimental and analytical procedures for measuring lanthanides as well as a plurality of other refractory trace elements using a Cameca ims-4f ion microprobe. They described the method as more versatile than conventional methods like NAA or IDMS because of its high spatial resolution that allows for the determination of domains of 10–20 μm within individual mineral phases. The ion microprobe measurements were performed at low mass-resolving powers, adopting the energy filtering technique for removal and suppression of unresolved complex molecular interferences, and synthetic standards were used for determining various instrument-specific parameters needed in the data deconvolution procedure adopted for obtaining REE abundances.

The results of the analysis of standards proved that the ion microprobe may be used for determining REE abundances in the parts-per-million range with uncertainties of ~10%–15%. Table 7.3 illustrates further reports on determination of lanthanides through SIMS [33-50].

Laser Ablation Inductively Coupled Plasma Mass Spectrometry (LA-ICP-MS)

Laser ablation is the process of removing material from a solid (or occasionally liquid) surface by irradiating it with a laser beam. At low laser flux, the subjected material is heated by the absorbed laser energy and evaporates or sublimates. At high laser flux, the material is typically converted to plasma. In technical jargon, laser ablation is used to refer to removing material with a pulsed laser, but it is possible to ablate material with a continuous-wave laser beam if the laser intensity is high enough.

The depth over which the laser energy is absorbed and thus the amount of material removed by a single laser pulse, depend on the optical properties of the material and the laser wavelength, as well as the pulse length. The total mass ablated from the target per laser pulse is usually referred to as the ablation rate.

Laser pulses can vary over a very wide range (from milliseconds to femtoseconds) and fluxes and can be precisely controlled. This makes laser ablation very valuable for both research and industrial applications.

In LA-ICP-MS, the sample is directly analyzed by ablating with a pulsed laser beam. The created aerosols are transported into the core of inductively coupled argon plasma, which generates temperatures of approximately 8000°C. The plasma in ICP-MS is used to generate ions that are then introduced to the mass analyzer. These ions are subsequently separated and collected based on their m/z values. The constituents of a sample can then be identified and measured.

Any type of solid sample can be ablated for analysis in this technique, and there are no sample-size requirements and no sample preparation

TABLE 7.3 Some Reports on Determination of Ln by SIMS

Year	Analyte	Sample	Important Aspects	Ref.
1999	REE	Rock samples	SIMS	[33]
2000	La	Metallurgical, environmental, and geological samples	SIMS	[34]
2000	REE	Mantle-derived minerals	SIMS	[35]
2001	REE	Zircon	Ion microprobe SIMS	[36]
2002	REE	Alkaline rocks	SIMS	[37]
2002	REE	Minerals	SIMS	[38]
2002	REE	Silicate glass, apatite and zircon	Ion microprobe SIMS	[39]
2002	La, Ce, Nd, Sm, Dy, Er and Yb	Zircon	SIMS	[40]
2003	REE	Apatite and carbonatite	SIMS and LA-ICP-MS	[41]
2003	REE	Meteorites from hot and cold deserts	SIMS	[42]
2004	REE	Zircon	Ion microprobe SIMS	[43]
2004	REE	Silicate REE-minerals	SIMS	[44]
2005	REE	Garnet	SIMS	[45]
2008	REE	Gadolinite	SIMS	[46]
2009	REE	Bussyite-(Ce) mineral	Electron-microprobe and SIMS	[47]
2013	REE	Monazite and xenotime U-Th-Pb geochronology	Ion microprobe SIMS	[48]
2014	REE	Meteoritic zircon	SIMS	[49]

procedures required. Chemical analysis using laser ablation requires a smaller amount of sample (i.e., in micrograms) than that required for solution nebulization (which requires milligram ranges of samples). Depending on the analytical measurement system, very small amounts of sample quantities may be sufficient for this technique. In addition, a focused laser beam permits spatial characterization of heterogeneity in solid samples, with typically micrometer resolution both in terms of lateral and depth conditions.

Ever since the advent of this technique by Gray in 1985 [50], the method has gradually improved over the years and the commercial LA-ICP-MS instruments have become available, changing the method to a well-established technique that is used in different laboratories. Today the method is widely used in various scientific applications, especially in petrology or geochronology, as well as in the environmental sciences. However, to achieve reliable results, the availability of suitable and well-characterized reference materials is of outmost importance. Table 7.4 shows some reports on determination of Ln by LA-ICP-MS [51-73].

TABLE 7.4 Some Reports on Determination of Ln by LA-ICP-MS

Year	Analyte	Sample	Important Aspects	Ref.
1993	REE	Single mineral grains	LA-MP-ICP-MS	[51]
1993	REE	Rock samples	LA-ICP-MS	[52]
1995	La	Ferromanganese oxide deposits	LA-ICP-MS	[53]
1996	REE	Apatite	LA-ICP-MS	[54]
1998	REE	Rock samples	A double-focusing, magnetic sector ICP-MS with ultraviolet (UV) laser ablation (LA)	[55]
2000	Sm, Nd	Sm-Nd garnet geochronology	LA-ICP-MS	[56]
2001	REE	Gem-quality zircon crystal	LA-ICP-MS	[57]
2003	REE	Silicate rock samples	LA-ICP-MS	[58]
2005	REE	Geological glass reference materials	UV (266 nm) laser ablation ICP-MS	[59]
2006	Nd isotope	Geological materials (geochronology)	LA-MC-ICP-MS	[60]
2007	REE	Carbonate	LA-ICP-MS	[61]
2010	Gd	Tumor tissue	LA-ICP-MS	[62]
2010	U-Pb, Sr, Nd and Hf isotope	Eudialytes (a group of complex Na-Ca zirconosilicate minerals that generally occur in peralkaline agpaitic syenites)	LA-(MC)-ICP-MS	[63]

(*Continued*)

TABLE 7.4 Some Reports on Determination of Ln by LA-ICP-MS (*cont.*)

Year	Analyte	Sample	Important Aspects	Ref.
2011	REE	Water sample (lake and sea water)	LA-ICP-MS	[64]
2011	REE	Zircon samples	Multiple spot LA-ICP-MS	[65]
2013	Nd isotope ratios	Glasses and minerals	LA-MC-ICP-MS	[66]
2013	Lu, Hf	Zircon samples	LA-MC-ICP-MS	[67]
2013	Gd	Tissue	LA-ICP-MS	[68]
2014	Sm-Nd isotopes	Monazite	LA-MC-ICP-MS	[69]
2014	REE	Geological samples	LA-ICP-MS	[70]
2014	REE	Impurities and basic components of lanthanum gallium silicate	LA-ICP-MS	[71]
2014	Eu	Geological glasses	LA-ICP-MS	[72]
2015	Nd isotope	Geological materials	LA-ICP-MS	[73]

In an example of the application of the technique, the effect of the matrix on the analysis of REEs through LA-ICP-MS was tested using carbonate and NIST glass standards [61]. According to the researchers, a UV 213-nm Nd:YAG laser coupled to an ICP-MS was used. The laser ablation was performed out in He and Ar atmospheres to evaluate the effect of the ablation gas on the analytical results, while in both experiments a small amount of N_2 was added to the carrier gas to enhance the signal intensities. The researchers used synthetic $CaCO_3$ standards, doped with REEs, as well as NIST glasses (NIST SRM 6 10 and 612) for calibration. Pure carbonate, composed of pure calcite, was analyzed as the carbonate sample.

Further, the extent of matrix effect on the results was evaluated through comparing the results with those of synthetic $CaCO_3$ and NIST glass standards. The results revealed that in the case of laser-ablation in a He atmosphere, the differences between the results of analyzing the real and standard samples were less than 10% for all REEs, except for those of La (with a 25% difference), whereas in an Ar atmosphere, the difference reached 25%–40%.

In another example LA-ICP-MS was used in the postmortem imaging of gadolinium (Gd) spatial distribution in a mouse tumor model post administration of PEGylated Gd liposomal nanoparticles. The PEGylated liposomal nanoparticles were formulated using a paramagnetic lipid incorporating Gd, in addition to a fluorescent lipid, and were injected IV to Balb/C nude mice having IGROV-1 tumors. Two hours after the injection the tumors and selected tissues were scanned by magnetic resonance imaging (MRI). The results proved the presence of Gd in tumor tissues and the kidneys, and after comparison with histology results, the concentration was found to be higher in regions of higher vascularity. This novel use of LA-ICP-MS for the spatial detection of Gd in tumor tissue proved the technique to be valuable in providing spatio-specific information of MRI contrast agents (e.g., Gd).

In another report in 2011 [64], a simple simultaneous preconcentration and matrix reduction technique was used for the analysis of REEs in water samples through LA-ICP-MS. The effects of several parameters like the pH, the amount of magnesium, the shaking time, the efficiency of Ba removal, and the sample matrix were optimized, and repetitive laser ablation was used to remove the dried-droplet samples from the filter medium. The enrichment factors ranged from 8 to 88 and the detection limit, for an enrichment factor of 32, ranged from 0.03 to 0.20 pg/mL. The reported relative standard deviations for the determination of REEs at a concentration of 1 ng/mL for 40-mL samples were 2.0% to 4.8%.

Finally, a novel calibration method for multi-elemental analysis and U-Pb dating of zircon samples through LA-ICP-MS coupled with galvanometric optics was reported in 2011 [65]. The galvanometric optics helped shorten the laser ablation of two or more sample materials to about 10 ms. The sample aerosols prepared through the LA step were mixed and homogenized within the sample cell, before being transported into the ICP ion source. The researchers used multispot laser ablation to enable direct spiking of analytes or internal standard elements into the solid samples, consequently making standard addition calibration method possible for solid samples. The researchers determined the REE abundances of two zircon samples through the technique using a multispot laser ablation of the glass standard material (NIST SRM612), and the result showed that the nonspectroscopic interferences on 14 REEs could be significantly reduced by the proposed technique.

NEUTRON ACTIVATION ANALYSIS

Neutron activation analysis (NAA) is one of the sensitive instrumental analytical techniques. Based on its mechanism of action, NAA allows for the determination of more than 60 elements. Basically, only the nuclei of the mentioned elements can undergo nuclear reactions that converts them from stable nuclei to radioactive ones, through the neutron capture reaction. Next of the resulting radioactivity of the nuclei or their daughters can be measured. The short-living compound nuclei that form emit so-called prompt γ-rays, the measurement of which can be used for the determination of the elements. In this case, the technique is referred to as prompt-γ neutron activation analysis, but if the deexcitation route is dominant, the nuclei decay in some seconds or less or much longer, emitting electrons or positrons, or through electron capture, which can be accompanied by the emission of delayed γ-rays (Fig. 7.2) [74-77].

Thus, technically, based on the time of measurement, NAA can be divided into two categories:

1. Prompt gamma-ray neutron activation analysis (PGNAA), where measurements take place during irradiation, or
2. Delayed gamma-ray neutron activation analysis (DGNAA), where the measurements follow radioactive decay. It should be noted that the DGNAA mode is a more common technique; thus, when one mentions NAA, it is generally assumed that measurement of the delayed gamma rays is intended.

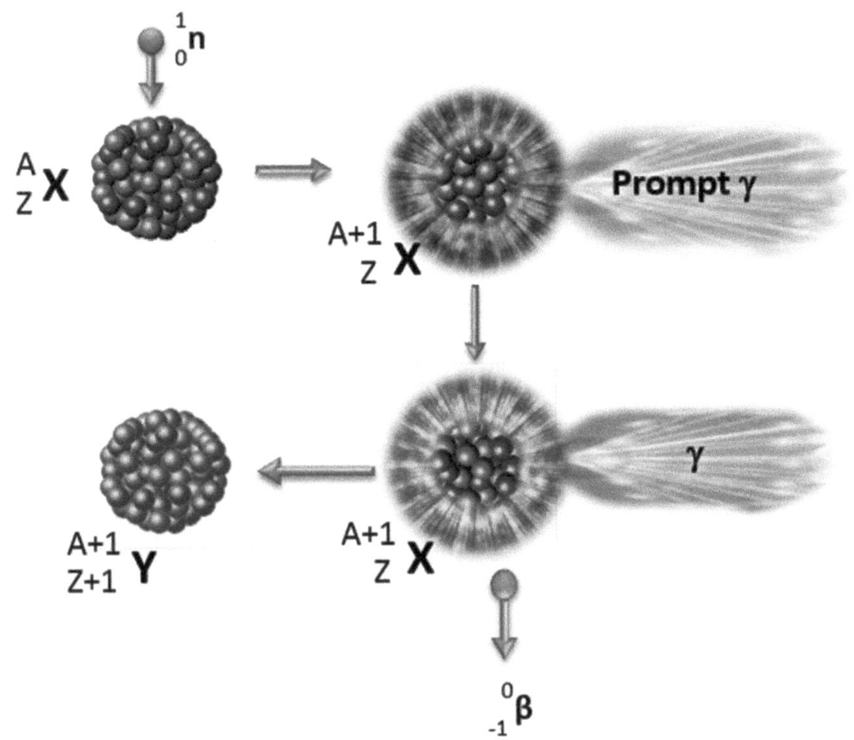

FIGURE 7.2 Scheme of the process of neutron capture by a target nucleus and the emission of gamma rays.

The NAA is a sensitive technique that can be used for the qualitative and quantitative evaluation of several elements. A major difference between NAA and the spectroscopic techniques is that it is based in nuclear rather than electronic transitions, through the bombardment of the sample nuclei with neutrons to produce radioisotopes of the analyte.

A neutron activation analyzer is generally composed of the sections mentioned below:

1. Neutron source
   The sources of neutron used in NAA systems include nuclear reactors, actinoid-like californium which emits neutrons, mixtures of alpha sources like radium or americium with beryllium which produce neutrons through a (α,^{12}C+n) reaction, and D-T fusion reaction in gas discharge tubes.
2. Reactor
   The reactors used for the NAA are chosen from the common reactors used for neutron irradiation of samples and production of radioisotopes in different applications. In general, the sample is placed in a so-called irradiation container, which is placed in the reactor. In the case using epithermal neutrons for irradiation, the neutrons are filtered by cadmium.
3. Fusor
   The commonest fusor used in NAA is the well-known Farnsworth–Hirsch device, which can be used to generate neutrons and also enjoy the advantage of being compact, small, and easy to operate, but also suffer the drawback that it cannot produce neutron fluxes that are comparable to those produced by reactor.
4. Isotope source
   The application of a reactor is often refrained from due to the high costs, and neutron sources based on the combination of an alpha emitter and beryllium are used instead, but of course at the cost of being weaker than reactors.
5. Gas discharge tube
   Gas discharge tubes serve to create pulses of neutrons in cases where the target isotope produced decays very fast, as in the case of oil wells.
6. Detector
   NAA systems use several types of γ detectors. The most conventional examples are the gas ionization, scintillation, and semiconductor detectors, among which the last two are the most common. Other detecting systems used in NAA include the planar detector that is used for prompt γ neutron activation analysis and the well detector which is applied to DGNAA. The planar detectors have a flat and large collecting surface and can even serve as a lid to close the sample, while the well type only surrounds the sample. The second

group of detectors are the semiconductor ones, which are based on a germanium positive-intrinsic-negative diode. The detectors show signals that are proportional with the energy of the incident radiation. A measure taken to reduce dark current and detector noise is cooling the semiconducting element down to about 77 K using liquid N_2. The detector can be subdivided to the two groups of lithium-drifted germanium or pure germanium. Some less sensitive devices include silicon as the semiconducting element. There is no need to mention that because of the higher atomic number of Ge, devices based on the element have higher efficiencies for absorbing high-energy γ-rays and producing the signal. Pure Ge detectors have further been found to be more temperature independent. Another type of sensitive and stable detectors with good resolution, that is, the scintillation detectors, radiation-sensitive crystals like sodium iodide doped with thulium is used. On receiving a γ photon, the crystal emits light. Another less preferred family of detectors used in NAA are particle detectors that are used for detecting the emission of α and β particles that often accompany the γ emission. The reason behind the low preferability of the detectors is that only the surface of the sample emits α and β particles, apart from the fact that they are absorbed by atmospheric gases. This mandates the application of high vacuums, which is not required in the case of devices that can detect γ rays.

NAA measurements are conducted through some steps. The first step, that is, sample preparation, usually involves the common routines in other analytical techniques like heating, dry-freezing, crushing, fractionating, evaporation, preconcentration, sieving, homogenizing, etc. The storage and preservation of the samples is also of importance in this regard. The next stage involves the activation of the sample using the neutron beams. The measurements are based on the measurement of the energy of the γ rays that enjoy the advantage of a very high penetration power. The distinct energies of the γ rays makes them superior to other forms of radiation like β rays that have continuous energy distributions, giving γ rays the potential for higher selectivity. In cases where the γ rays lead to ionization of the sample signals can be manifested as electrical currents that can be recorded. In the case of γ rays, semiconductor detectors are used to detect the signals that are fed to computer-based multichannel analyzers. Finally, the acquired γ spectra is evaluated and analyzed for obtaining the analytical data on the concentration of the trace elements.

Modes of Neutron Activation Analysis

The method can be used in a variety of modes, including the following [75-77].

Instrumental Neutron Activation Analysis (INAA)

In this mode, which is the simplest mode of NAA, no chemical treatment of the sample is performed, and on the irradiation of the sample, the γ-ray measurement is performed on the radionuclides. The mode is used for a variety of matrices and is very common in NAA. If the activity of the neutron activation products of the matrix does not hinder the analysis of the analyte, INAA can be applied. However, in case the target is present in trace or ultratrace levels, then the activity of the radionuclides of the matrix should be eliminated, using other processes like chemical ones in the so-called chemical modes.

Chemical modes for eliminating the interference by radionuclides are radiochemical and molecular neutron activation analysis (RNAA MNAA) modes. RNAA involves the decomposition of the sample followed by elimination of the interfering species through a proper separation method like extraction, ion exchange, precipitation, and distillation, before the NAA analysis. To keep the analyte at trace or ultratrace concentrations, inactive species can be added to the sample. The yield of the separation process can also be determined through radio tracing, reactivation of the separated portion with the added carrier, etc., which can naturally add to the costs and time requirements of the method [76,78,79]. In MNAA mode, on the other hand, the elements causing interferences in INAA or those to be determined can also be chemically separated prior to irradiation. In this approach, however, one of the main advantages of NAA for trace and ultratrace analysis – the essentially blank-free nature – may be lost. Also, there is little control over the separation yield. On the other hand, there is also no time limitation for performing chemical separation, and the radiation burden of personnel is even lower than in INAA. Another advantage is the possibility to perform speciation analysis.

Physical approaches of elimination of interfering radionuclides are Cyclic and Prompt γ neutron activation analysis (CNAA) and (PGNAA) modes. Cyclic and Prompt γ neutron activation analysis can be of help in the case of elements which form short-lived radionuclides, and helps optimize the detection limits. Based on this approach, the irradiation and measurements on the sample are recurrently performed short instances. Next the spectra acquired for each runs are summed, leading the modification of the counting statistics of peak areas. The repetition of the runs can be done to the extent that total activity of long-living radionuclides is not amplified, which can be mitigated through the application of fresh samples. In PGNAA the prompt-γ radiation of the target species is measured simultaneously with the irradiation and requires special add-ons for the reactor, including a neutron beam guide as well as an adjacent γ-ray detector. This mode can be used to acquire information about elements like H, B, C, N, P, S, Cd, Pb, and lanthanides like Sm and Gd, which cannot be determined through classical NAA [80,81]. The modes discussed above are summarized in Fig. 7.3.

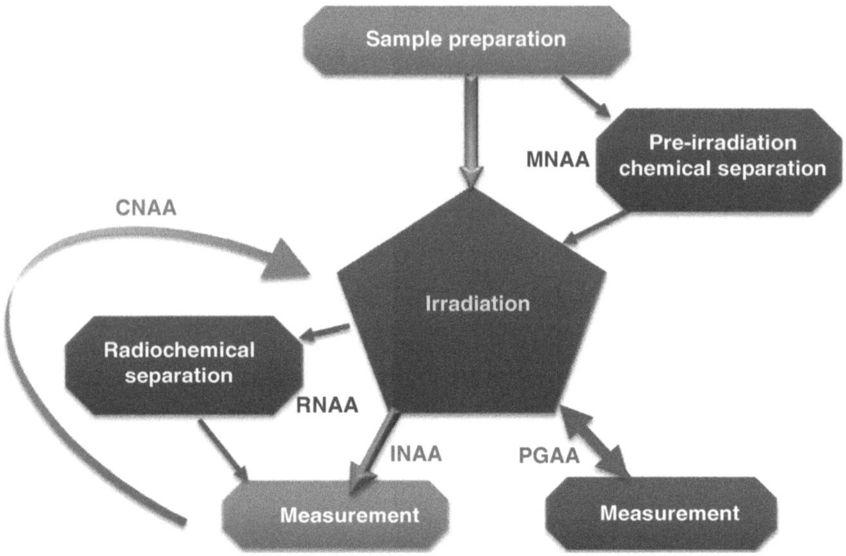

FIGURE 7.3 Various modes of NAA.

Selective Activation Modes

As the name indicates, the method aims at avoiding the activation of undesired species [76,82]. The different approaches used in this mode include, first, a thermal neutron activation analysis (TNAA), where thermal neutrons are used for the activation of the sample. The method requires the use of a moderator block (like a graphite block) at a neutron source. On the passage of neutrons, it is decelerated and used for activation.

In epithermal neutron activation analysis (ENAA), however, the irradiation beam is composed of 0.55 eV, which lead to the preferred activation of the nuclei with high resonances for the radiation neutron capture with neutrons that have the mentioned energy. Here the thermal neutrons are shielded off using cadmium or boron filters.

Another class of selective activation modes is the fast neutron activation analysis (FNAA), which is based on nuclear reaction like (n, p), (n, 2n), and (n, α) with fast neutrons that are obtained by filtering the thermal neutrons or through other sources such as 14-MeV neutron generators or cyclotrons.

Selective Counting Methods

Another mode of the NAA is the selective counting method in which unimportant signals from the activated matrix and the spectrometric interferences of γ rays are filtered off through selective counting methods like Compton suppression counting (CSC), coincidence counting (CC), and measurement of delayed neutrons for the determination of fissile elements [83,84].

Applications of NAA in Determination of Lanthanides

NAA has found various applications in areas like archeology, forensics, medicine, or quality control and assurance (QC and QA) for the analysis of a wide range of samples, including amber, bone, ceramics, gems, metals, paints, raw materials, soils, stone and clays, rain and water, aerosols and airborne particles, fuels and combustion residue, waste and sewage sludge, asbestos, bore hole specimens, oil and hydrocarbons, dusts, sediments, minerals, alloys, catalysts, ceramics and refractory materials, dyes and fertilizers or more intricate biomedical specimens like healthy or cancerous tissues, tracers, foods, bile, blood, dialysis fluids, and herbs, or venoms. NAA is also used for evaluations required for element content certificates and homogeneity tests [85,86].

The major advantages of the technique are its being nondestructive and applicable to a wide range of samples; low detection limits of about 10^{-6} mg/kg; not requiring analytical blanks and low matrix interferences, high specificity; simplicity in terms of results and the nature of errors; the adjustability of the energy and density of neutrons in a range. This final advantage allows for obtaining simple optimization through measures like the selective activation of species [76,78,87,88].

As in the case of other methods, NAA suffers certain disadvantages, including the need for a nuclear reactor and working with radioactive materials, and the subsequent need for the safe disposal of radioactive wastes; rather limited range of industrial applications; the method's inability of determination of elements that do not form radionuclides (e.g., Pb); and the need for drying or dry-freezing samples to avoid damages and even explosion due to radiolysis of water, being time-consuming based on the lifetime of the radionuclides as well as sophisticated instrumentation [76,78,87].

Table 7.5 contains the information on the approximate detection limits for determination of lanthanides assuming interference free spectra [89].

The sensitivity of NAA is dependent on the

1. irradiation parameters, such as neutron flux, irradiation, and decay times;
2. measurement conditions, such as measurement time and detector efficiency; and
3. nuclear parameters of the elements being measured, such as isotope abundance, neutron cross section, half-life, and gamma-ray abundance.

Usually, the accuracy of an individual NAA measurement ranges from 1% to 10% of the reported value.

Table 7.6 provides a summary of the reports on the determination of lanthanides in various samples by NAA method [90-172]. Based on the data provided, the method has been used for the determination of lanthanides in a variety of geological, industrial, environmental, and

TABLE 7.5 Detection Limits of NNA for Lanthanides and Other Elements Using γ-ray INAA (At a Reactor Neutron Flux of 1×10^{13n} cm^{-2} s^{-1})

Sensitivity (picograms)	Elements
1	Dy, Eu
1–10	In, Lu, Mn
10–100	Au, Ho, Ir, Re, Sm, W
100–1000	Ag, Ar, As, Br, Cl, Co, Cs, Cu, Er, Ga, Hf, I, La, Sb, Sc, Se, Ta, Tb, Th, Tm, U, V, Yb
1000–10^4	Al, Ba, Cd, Ce, Cr, Hg, Kr, Gd, Ge, Mo, Na, Nd, Ni, Os, Pd, Rb, Rh, Ru, Sr, Te, Zn, Zr
10^4–10^5	Bi, Ca, K, Mg, P, Pt, Si, Sn, Ti, Tl, Xe, Y
10^5–10^6	F, Fe, Nb, Ne
10^7	Pb, S

biological specimens, ever since the application of its less sensitive forms back in the 1960s [90-100].

Some of the most recent reports on the application of NAA for the analysis of lanthanides are reviewed in more detail. In 2004, REE contents of water samples from Periyar River, as well as seawater from Mumbai, India, was evaluated through NAA [155], after separation and preconcentration of the analytes in the pre-irradiation stage through ion exchange. The researchers used a standard addition procedure, and radiotracer studies were performed to validate the analytical methodology.

TABLE 7.6 Some Reports on Determination of Ln by NAA

Year	Analyte	Sample	Important aspect	Ref.
1960	Eu	Stone meteorites	Simultaneous NAA	[90]
1961	REE	Chondrite, achondrite, iron meteorites	NAA	[91]
1964	Ce, Pr	Ores, minerals	NAA	[92]
1966	Tb, Er, Yb	-	NAA	[93]
1967	Dy	Bone of chicken salmon fingerlings	NAA	[94]
1968	Dy, Er, Sm, Gd	High pure yttrium oxide	NAA	[95]
1968	La,Sm,Dy,Eu	Lavas of chaine des pure	Use of chromatography and computer in NAA	[96]

TABLE 7.6 Some Reports on Determination of Ln by NAA (*cont.*)

Year	Analyte	Sample	Important aspect	Ref.
1968	Dy	Rock sample	Using Ge(Li) detector after chemical separation	[97]
1969	Sm, Gd	Rare earth ores	Neutron capture gamma-ray activation analysis	[98]
1969	Gd	CdF_2	NAA	[99]
1969	Sm	Rocks	Epithermal NAA	[100]
1970	REE	Europium oxide	NAA	[101]
1970	Ho, La	Tungsten Bronzes	Thermal NAA	[102]
1971	La	Metallurgical materials	NAA	[103]
1971	Eu, Lu	-	NAA	[104]
1971	Dy	Graphites	NAA	[105]
1972	Eu	Photographic film	NAA	[106]
1972	Dy	-	NAA	[107]
1973	La	Steel	NAA	[108]
1973	La, Ce, Eu	Monazite sand	NAA	[109]
1973	Ce	-	NAA	[110]
1974	La	Refractory materials	Gamma-spectrometric in NAA	[111]
1974	La	-	NAA using high-resolution gamma spectrometry	[112]
1974	Ce	Seawater	NAA	[113]
1975	Dy	Hay and feces	NAA	[114]
1977	La, Sm	Nimroc standards	NAA	[115]
1979	Gd	Manganese bath solution	NAA	[116]
1980	Ce, Sm	Biological materials	NAA	[117]
1980	Dy, Ce	Dysprosium, cerium, and chromium marker determination	NAA	[118]
1981	La, Eu	Carbonaceous shale and monazite	Epithermal NAA	[119]

(*Continued*)

TABLE 7.6 Some Reports on Determination of Ln by NAA (*cont.*)

Year	Analyte	Sample	Important aspect	Ref.
1982	Nd	Nd:YAG laser crystal	NAA	[120]
1983	Yb	Simultaneous determination of selenium and ytterbium	NAA	[121]
1984	Gd, Ce, Eu	Fluorite deposits	Electron paramagnetic resonance and NAA	[122]
1984	La	Steel	NAA	[123]
1985	Yb	Organs of mice	NAA	[124]
1986	Ce, La, Sm,	Duodenal digesta	NAA	[125]
1986	Gd	Geological sample s	Prompt gamma NAA (PGNAA)	[126]
1986	La, Ce, Nd	Lateritic materials	Energy dispersive X-ray fluorescence and NAA	[127]
1986	Eu, La	Asbestos	NAA	[128]
1987	Eu, Gd, Pr, Sm	Standard rock GS-N	NAA	[129]
1987	Dy, Eu	Standard samples of rocks	NAA	[130]
1987	Sm, Gd	Spectral interference of Gd in Sm	NAA	[131]
1987	Sm, Gd	23 geological reference materials	Neutron-capture prompt gamma-ray AA	[132]
1987	Sm, Gd	Rock samples	Prompt gamma-ray NAA	[133]
1988	REE	Biological materials	NAA	[134]
1989	Pr, Nd, Er,	Hot spring and crater lake waters	NAA incorporating the standard addition	[135]
1989	Eu	Fattened chicken	NAA	[136]
1992	REE	High-purity lanthanum oxide	NAA	[137]
1993	REE	Environmental samples near a coal power station	K_0-standardized NAA with counting in a LEPD	[138]

TABLE 7.6 Some Reports on Determination of Ln by NAA (*cont.*)

Year	Analyte	Sample	Important aspect	Ref.
1993	REE	Composition of Gabbro GOG-1 reference sample	NAA and comparison with SIMS and ICP-AES data	[139]
1993	La, Ce, Sm, Eu, Tb, Yb, Lu	15 French geochemical reference samples	NAA	[140]
1993	La	Phosphate rock and its products	NAA	[141]
1994	Ce	High-purity lanthanum oxide	NAA	[142]
1995	Lu	Alloyed wires and foils	NAA	[143]
1995	Ce, Yb	Atmosphere of some major urban and industrial areas of Punjab, Pakistan	NAA	[144]
1997	REE	Natural plants	NAA	[145]
1999	La	Zirconium dioxide	HPLC, X-ray fluorescence, and NAA	[146]
2000	Gd	Human blood plasma	PAA and NAA	[147]
2001	REE	Binding protein in human liver	INAA	[148]
2001	REE	Sediments of a polluted marine environment	INAA	[149]
2002	La, Ce, Nd, Sm, Eu, Tb, Yb, and Lu	Wheat powder and human hair reference materials	Radiochemical neutron activation analysis (RNAA)	[150]
2002	La, Ce, Nd, Sm	Metallic gallium	NAA	[151]
2003	REE	Carbonatites	INAA	[152]
2003	Gd	Kidney and liver tissues of rats	NAA	[153]
2003	REE	Lake sediments	NAA	[154]
2004	REE	River water and seawater	NAA after preconcentration via ion exchange	[155]

(*Continued*)

TABLE 7.6 Some Reports on Determination of Ln by NAA (*cont.*)

Year	Analyte	Sample	Important aspect	Ref.
2004	REE	Carbonatites	INAA	[156]
2005	REE	Altered ultramafic rocks	INAA	[157]
2005	REE	Pallasite meteorites	NAA with extraction chromatography	[158]
2006	REE	Sediments	NAA	[159]
2006	Sm, Gd	Geological Survey of Japan reference materials	PGNAA	[160]
2007	REE	Agricultural crops	Instrumental and radiochemical NAA	[161]
2008	Dy	Ghanaian cereals and vegetables	NAA, short-lived nuclides and Compton suppression spectrometry	[162]
2009	Sm^{153}	Chinese medicinal herbs	Epithermal NAA (ENAA)	[163]
2009	Gd	Biological samples	PGNAA	[164]
2010	La, Ce, Sm, Eu, Tb, Dy, Lu and Yb	Surface marine sediments	INAA	[165]
2011	Gd	Kidney liver and leg muscle	PGNAA	[166]
2012	Ce, La, Sm	Industrial corridor	NAA	[167]
2013	REE	Kpong carbonatite	INAA	[168]
2013	REE	Clay	INAA	[169]
2014	Gd	Tissues	PGNAA	[170]
2014	REE	Surface mangrove sediments	INAA	[171]
2014	REE	Biological and environmental reference samples	INAA	[172]

In another study, the composition of pallasite meteorites, known as stony-iron meteorites, was evaluated by NAA in 2005 [158], to evaluate the details of their formation process based on trace elemental abundances in pallasite. The NAA analysis was performed after an extraction chromatography (EXC) stage and the results revealed the method to have the potential of quantitative determination of parts-per-million levels of La, Ce, Nd, Sm, Eu, Gd, Tb, and Lu in only 100 mg of the sample. In another study in 2006 [159], NAA was used to determine the REE, Fe, Ta, and Th content of sediment cores from Catingueiro Barrinha and Rio das Pedras marginal lagoons of the Moji-Guacu River basin. The method consisted of using Fe, Sc, Ta, and Th as conservative elements, to perform the double normalization technique to assess the distribution pattern of Ce, Eu, La, Sm, and Th for each core.

In 2007 [161], Th and U were measured in soil samples from crops grown in the area polluted by emissions from a phosphate fertilizer plant, using neutron activation analysis (INAA) and radiochemical neutron activation analysis (RNAA) after alkaline-oxidative fusion of the irradiated samples followed by precipitation of REE oxalates. The results indicated high levels of lanthanides Th and U, in some samples wheat chaff and parsley.

In another research in 2009 [163], the optimum conditions for the simultaneous epithermal neutron activation analysis (ENAA) of As, Sb, and Sm in Chinese medicinal herbs, standardized powder, lichen, and tomato leaves were evaluated and the lower detection limits for As-76, Sb-122, and Sm-153 in the medicinal herb specimens were found to depend on the weight of the irradiated sample, irradiation, and the decay durations.

In 2011 [166], an in vivo prompt gamma neutron activation analysis (PGNAA) was applied to the analysis of Gd in kidney, liver, and the leg muscle tissues, reaching initial detection limits of 7.2 ± 0.3 ppm for the kidney samples, 3.0 ± 0.1 ppm for the liver samples, and 2.33 ± 0.08 ppm for the lower leg muscle samples. The researchers reported that the detection limits were significantly improved after system optimizations and reported technique as a promising means for in vivo studies of gadolinium retention.

Further in 2013 [168], INAA was applied to the determination of REE concentrations in carbonatite from Kpong, southeastern Ghana, and the results revealed total rare earth elements (TREEs) in the range of 540 to 705 mg/kg, and a total of 11 elements, namely, La, Ce, Nd, Sm, Eu, Tb, Dy, Ho, Tm, Yb, and Lu in the samples, revealing high enrichments of light rare earth element (LREE) deposits in the samples, marking them as a potential source of REEs.

In a recent study in 2014 [170], prompt gamma neutron activation analysis (PGNAA) was used as an in vivo, noninvasive method for the determination of Gd in tissues, to evaluate gadolinium (Gd)–based contrast agents that are routinely applied in magnetic resonance imaging (MRI) procedures. A set of 3 NAA leg scans were performed. A very low effective radiation dose of 0.6 μSv was used to make multiple scans possible. The results indicated equal calibration phantom and in vivo detection limits of 0.58 ppm. Gd was detectable after exposure to the contrast agent Gadovist (R) at greater than 99% of confidence, in 9 participants within 1 h of contrast administration. The concentration of Gd in the samples ranged from 2.0 to 17.3 ppm (6.9–56 uncertainties different from zero), whereas no detectable Gd signals were observed in participants in the third neutron scan (conducted 0.7–5.9 days postcontrast).

As a final example, we refer to a very recent study in 2014 [171] where the concentrations of REEs present in surface mangrove sediments from 10 locations throughout the west coast of Malaysia were evaluated using INAA. In this procedure, the samples were dried and crushed to a powder. Then, approximately 150 mg of the sample was irradiated for a "short irradiation time" and another 200 mg was subjected to a "long irradiation time" as calibration and quality control measures. Next, blank samples and a standard reference material (SL-1) were irradiated at a thermal neutron flux of 4×10^{12} cm^{-2} s^{-1} at the MINT TRIGA Mark II research reactor, operating at 750 kW using a pneumatic transport facility. The REEs present in the surface sediment samples were reported to be Dy, Sm, Eu, Yb, Lu, Tb, La, and Ce, which were present in the concentration range of 0.35–117.4 mg/kg. The researchers also evaluated the geochemical behavior of REEs in surface sediments and normalized pattern (chondrite and shale) and computed the degree of sediment contaminations using an enrichment factor varying in the range of 0.75–6.75.

X-RAY FLUORESCENCE SPECTROMETRY

XRF is a spectrometric technique based on the emission characteristic of X-ray photons from inner electronic shells of atoms, which can be used for the measurement or identification. The first application of XRF for quantitative analysis dates back to late 1920s, and the works of Glocker and Schreiber [173]; however, the technique was not commercially available until the 1950s because of the lack of proper detection systems. Further improvements to XRF were made in 1970 [174]. Because of the nature of excitation and emissions in the XRF, the technique can be used for a wide range of samples in varied forms ranging from solid to liquid, in a rather wide range of concentrations.

To have an overview of the mechanism of the XRF phenomenon, one needs to have some information about the interactions of the inner electronic orbitals of elements with photons. If a high-energy electron beam is radiated on a sample, emission of the photons with a wide range of energies is observed. It is called the bremsstrahlung radiation (or braking radiation), which is as a result of the deceleration of high-energy electrons in the sample. In a parallel phenomenon, photoelectrons are dropped out of the inner orbitals of the atoms. The energy of these photoelectrons is less than that of the incident electron beam (E) (because some energy is spent on overcoming the binding energy of the inner orbital electron [φ]). In the latter case, that is, the ejection of an electron from the inner orbital, an electron vacancy (hole) is formed that is filled by the electrons in higher energy levels; as a result of such transition, a fluorescence phenomenon occurs that is termed X-ray emission. The energy of the x-ray photon equals the difference between the energy of the higher electron and its energy in the lower state. This energy is naturally characteristic of the nature of the atom and its electronic structure and hence can be used for its identification [174,175]. Because all of the inner orbital electrons are prone to being ejected upon being struck by the electron beam, and also different higher electrons may replace them, a range of transitions can be allowed to occur based on quantum mechanics, and consequently, different X-ray energies may be observed.

The X-ray lines are roughly classified into the K, L, or M series based on the orbital where the hole is formed. Because K represents the lowest orbitals, the spectral lines in this series are characterized as having the highest energies. The next lines, in terms of energy, include L and M. Another set of labeling symbols are α, β, and γ, which indicate the energy level of the dropping electron, as well as a number that shows the quantum state of the dropping electron in its initial state (e.g., $K\alpha_2$). Further detail on the notations and energies can be found in different references [174]. One important practical consideration is that as a result of technical limitations, only the $K\alpha$ lines are used in experimental work [176]. The overall process is known as X-ray fluorescence and is the basis for the quantitative and qualitative studies through the XRF spectrometry technique.

One advantage of the technique is that it is free from chemical shifts, because the signals are not affected by chemical bonding or sample media, as a result of the fact that inner orbitals are involved in the XRF phenomenon. This is especially the case with heavier elements.

A further point is that under the mentioned conditions, other relaxation mechanisms, like the Auger phenomenon, absorption, transmission, and scattering of the incident beam may occur, which reduce the fluorescence yield and cause limitations by creating strong background radiation [174,175].

XRF instruments can be categorized into two major categories of energy dispersive (ED) and wavelength dispersive (WD). In the former class of XRF instruments (Fig. 7.4) [177], the entire spectrum of the sample reaches a detector, which can distinguish among the energies of the incident photons, whereas in the WD system (Fig. 7.5), the instrument functions based on Bragg's diffraction law, and one or more analyzing crystals are used to separate the beams reaching the detector. It is possible to rotate the analyzing crystal with the detector assembly to be able to scan through the wavelengths. In case all or a wide spectral region is to be resolved, a plurality of crystals should be present. The ED-XRF systems are more complicated in terms of electronics and computer software but can be used more simply.

The X-ray source in both ED and WD systems is a vacuum tube containing a cathode (usually a tungsten wire) and an anode. The cathode is

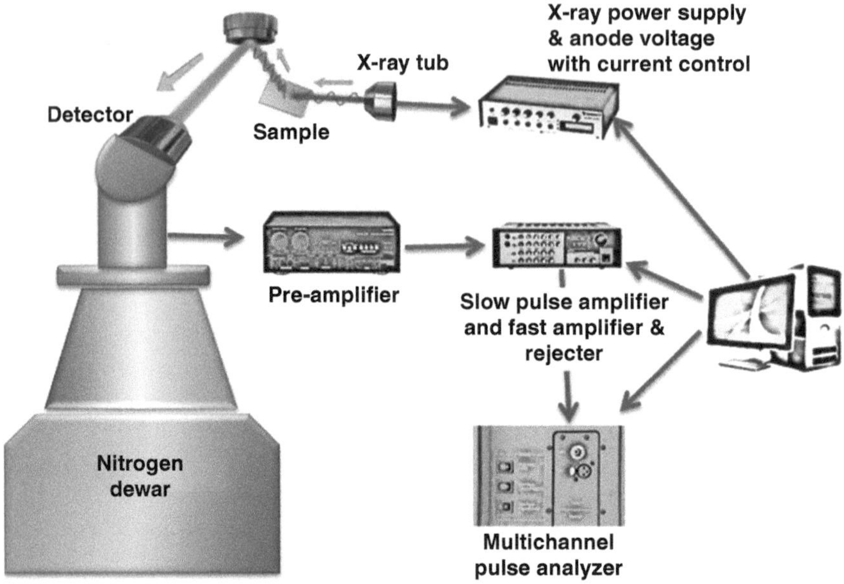

FIGURE 7.4 Schematic illustration of a typical ED-XRF instrument.

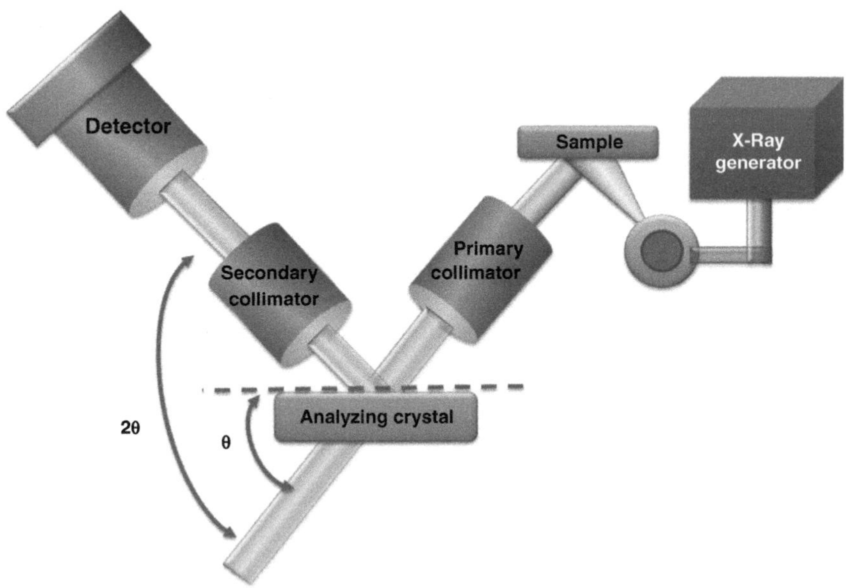

FIGURE 7.5 Schematic illustration of a typical WD-XRF instrument.

heated and a potential difference is applied between the two electrodes, as a result of which the cathode releases thermal electrons that move toward and strike and accelerate toward the anode, leading to the emission of the bremsstrahlung radiation in addition to the characteristic X-ray lines of the anode. A beryllium window is built on one side of the tube to let the photons leave. The choice of the anode and the power of the source are important. The anode materials should be chosen considering the wavelength required for exciting the sample. Some systems are equipped with secondary anodes or can automatically change anodes for multielemental analysis. The power of the source, on the other hand, can vary from less than a watt to tens of kilowatts, depending on the type of the instrument and practical considerations.

The detection systems of WD-XRF and ED-XRF are totally different. ED-XRF systems depend on semiconductor-type detectors, based on germanium, Si(Li), or lithium-drifted silicon. The detectors receive the entire emitted spectrum from the sample and decode it into a histogram of number of counts versus photon energy. WD spectrometers, however, use an analyzing crystal to disperse the emitted photons based on their wavelengths and place the detector in the correct physical location to receive X-rays of a given energy. Such systems use detectors like gas-flow proportional systems that are suitable for low-energy radiations (>0.15 nm), or thallium-doped sodium iodide (NaI(Tl)) scintillation detectors for radiations with high energies. The detectors can also be used in combination, and also in predetermined fixed locations to extend the capabilities of the instruments and simplify the operations.

Nowadays, development of portable XRF (pXRF) instruments has greatly expanded the range of samples suitable for analysis. There is no longer need to fit a sample into a small chamber. In the case of pXRF, it is possible to analyze the samples with the instrument in a stand or the instrument can be moved to the sample, as in the case of analyzing an exposed rock outcrop or a large painting.

Sample Pretreatment

Because of the nature of the XRF phenomenon, which deals with lower orbitals that are not affected by the chemical environment of the elements; most of samples do not need any pretreatment. This has been regarded as a virtue in applications like archeology and forensics where the preservation of the sample can be crucial. However, because clean, thick, and flat samples can lead to better results, pretreatment of the samples can become a necessary measure in many applications. These pretreatment and preparation stages are mainly physical in nature. The inhomogeneity of samples, for example, the formation of oxide layers on the surface of metallic samples, can lead to the absorption of certain photons by the oxide layer, which can be used as a technique for gaining phase information, but can lead to errors too [177].

Another issue to be considered is the depth of analysis, which becomes rather important in case of nonhomogenous or thin samples, depending on the element of interest. The depth of analysis depends on the nature of the sample and the energy of the incident x-ray photon and is increased upon increases in the latter.

The simplest form of pretreatment can be smoothing the sample surface, which can act as an independent phase to the device, if very rough. Other common pretreatments are pelletizing of powders or ground samples; this process requires the addition of known amounts of binding compounds made up of organic materials like cellulose, starch, and polyvinyl alcohol to the homogenized sample, before pressing it into smooth pellets. In the case of ground particles, homogeneity is a very effective element and hence its effect has been the subject of some studies [175]. Another approach to homogenization of samples is melting and solidification in the presence of reducing agents like sodium or lithium [176], but the approach can lead to the formation of unanticipated chemical species even in inert environments.

Applications of XRF Spectrometry in Lanthanide Determination

Because of rapidness and simplicity in analytical procedure, XRF analysis has been widely used to determine whole composition of major and trace elements in various samples. However, in the case of lanthanide, acceptable analytical precision of measurements was not achieved for many samples, because of a lack in analytical sensitivity. In order to achieve precise elemental analysis of trace elements, including REE, more sensitive techniques should be used. However, sometimes, the preconcentration methodologies applied prior to the determination of trace elements in different samples increase the potential of the instrumental analysis [178]. Table 7.7 shows some reports on determination of Ln by an XRF spectrometer.

A simple XRF spectrometry for the determination of Cd, Hf, Hg, and Gd in uranium solution was developed in 1993 by Qadeer [179]. Wavelength-dispersive X-ray fluorescence spectrometer (WD-XRF) was used in the studies for measuring the concentration of the above-mentioned elements in a uranium solution. The sample solutions were introduced to the XRF spectrometer directly in simple polyethylene bottles. This is unlike the earlier sample pretreatment techniques such as pelletization, bead formation, deposition, and precipitation on a substrate. The matrix effects have been rectified using standards of similar composition. Statistical evaluation has established it as a fast and reliable method applicable to trace analysis.

The development of semiconductor technology sets high requirements for the control of the synthesis process and for the quantitative determination of the composition of the final product that may substantially differ

TABLE 7.7 Some Reports on Determination of Ln by XRF Spectrometer

Year	Analyte	Sample	Important Aspects	Ref.
1993	Cd, Hf, Hg, and Gd	Uranium solution	WD-XRF	[179]
1999	Cr, Co, Ni, Cu, Zn, Ga, Se, Sb, Yb	Microsamples of multielement mono- and semiconductor polycrystals	XRF with thin layer method	[180]
1999	Sm, Gd, Eu	Reference materials	Preconcentration on a new chelating resin loaded with thorin and XRF analysis	[181]
1999	REE	Fossilized bone	Synchrotron radiation induced X-ray microfluorescence	[182]
2000	Cr, Mn, Ni, Cu, Zn, Ga, As, Se, Sb, Yb	Microsamples of multielement mono- and semiconductor polycrystals	XRF with thin layer method	[183]
2000	Sm, Gd, Eu	Geochemical reference samples	Preconcentration and analysis by XRF with chemofiltration	[184]
2001	REE	NIST standard reference glasses	Preconcentration and analysis by XRF with chemofiltration	[185]
2003	Eu	Synthetic host material	XRF	[186]
2005	REE	Apatites	WD-XRF with digestion method and dropping the sample on a membrane filter	[187]
2008	La	(Pb,La)(Zr,Ti)O-3 type ferroelectric ceramics	WD-XRF by making pellets with boric acid	[188]
2014	La	Sediments of deep-sea rare-earth elements and yttrium (REY)-rich mud	Micro-focused XRF using high-energy synchrotron radiation	[189]
2014	REE	Geological samples	Handheld (portable) X-ray fluorescence (pXRF)	[190]
2014	REE	Phosphate rocks	pXRF	[191]
2015	Gd	In vivo detection of gadolinium (Gd) in bone	XRF	[192]

from the assumed composition. A simple and quick method of durable sample preparation by the thin layer method through direct digesting of the analyzed material on the substrate has been presented for synthetic semiconductor material analysis [180]. Interelement influences are one of the main sources of errors in XRF analysis. There are lots of methods, consisting both in mathematical correction and in special preparation of samples, that allow minimizing the matrix effects. The thin layer method belongs to this second category. At a sufficiently small thickness (mass per unit area), decreasing of both absorption effects and secondary amplification allows to obtain a linear relationship between the intensity of fluorescent radiation and the concentration of the element being determined. Unfortunately, problems connected with preparing a sample in this technology might substantially reduce its practical application.

A few methods of thin samples preparation are used. Powder (dry) methods are used, in which the analyzed powder is deposited on a substrate in various ways, also in the form of suspensions in water or appropriate organic solvents as well as wet methods, where a sample is initially digested and the obtained solution is dropped onto the substrate. But in the first case, a granularity effect as well as frequently a mineralogical effect, resulting from a diverse chemical composition of the analyzed grains, must be taken into account. In the second case, the sample should be initially digested, which substantially extends the analytical process and often requires larger weighted amounts of the studied material, and as a result of recrystallization a granularity effect may occur. Here the possibility of using the thin layer method in XRF analysis has been studied using digestion of samples directly on the substrate (carrier of the sample). Four- and three-component mono- and polycrystals have been analyzed. Standards have been used in calibration containing Cr, Co, Ni, Cu, Zn, Ga, Se, Sb, and Yb. To improve the correlation between the concentration and the fluorescent radiation, models of mathematical corrections have additionally been used: multiple linear regression, Lucas-Tooth-Pyne model, and de Jongh model.

A very stable chelating resin was prepared by adsorption of *o*-[3,6-disulfo-2-hidroxy-1-naphthylazo]-benzenearsonic acid (thorin) on a macroporous resin Amberlite XAD-7 [181]. The optimal conditions for preparing it were obtained through the study of the adsorption properties of the resin and the thermodynamic quantities of the adsorption processes. Likewise, the behavior of the REE-loaded resin was studied (pH of retention, sorption kinetics, etc.). The conditions to prepare a thin film with this system were also evaluated. The loaded resin was successfully used for the separation and preconcentration of Sm(III), Eu(III), and Gd(III) before their determination by XRF spectrometry. The preconcentration factor obtained was 500 and the concentrations at low detection limits were 13.8, 17, and 15.7 μg/L for Sm, Eu, and Gd, respectively.

In 2000 [184], a preconcentration method for subsequent determination of REE by XRF spectrometry was developed. The method is based on using *o*-[3,6-disulfo-2-hydroxy-1-naphthylazo]-benzenearsonic acid (Thorin) as a complexing agent, which is retained on a polyamide membrane by a chemofiltration process. The pH dependence of the chemofiltration of these metal ions on the membrane and other variables, such as flow rate, contact time, kinetic of complex formation, etc., were determined. The membrane containing the chemofiltrate formed a thin film, which eliminated the interelemental effects when measured by XRF. The detection limits were 23, 23, and 49 ng/mL for Sm(III), Eu(III), and Gd(III), respectively. High enrichment factors were obtained. The method was successfully applied to the preconcentration of Sm(III), Eu(IU), and Gd(III) from different samples.

In a study in 2003 [186], the preconcentration method was use for Eu based on the formation of a ternary complex with 1,8-dihydroxy-2-[4-sulfophenylazo]-naphthalen-3,6-disulfonic acid and a surfactant agent. The complex is collected on a nitrocellulose filter, obtaining a thin film. The Eu complex retained on the membrane is measured by XRF spectrometry. The principal advantage of this preconcentration method is that the concentrated ternary complex is retained on the membrane surface, forming a thin film, so it is possible to obtain large enrichment factors and low detection limits in XRF determinations. The development of the preconcentration methodology implied the study and optimization of the ternary complex formation in terms of pH, surfactant agent, reagent concentration, etc. The preconcentration factor was 2×10^4, and the lower detection limit for the preconcentration of 25 mL of aqueous solution was 22 ng/mL The method was successfully applied to the determination of Eu in synthetic host material. The detection of trace rare earths in diverse materials and the importance of their determination have led to the search for different preconcentration methods for their quantification through instrumental analysis.

In 2014, handheld (portable) X-ray fluorescence (pXRF) instruments were designed for use in the exploration for base metals, precious metals, and specialty metals (e.g., REEs, Ta, and Nb) and allow rapid decision making directly in the field [190]. This study indicated that the pXRF was capable of producing readings for Si, K, Al, Fe, Ca, Ti, Pb, Zn, Sr, Ag, Cd, Th, Sb, P, S, Mo, Mn, Mg, As, Nb, Rb, La, Ce, Pr, Nd, and Y within 10 %RSD of the reported certified value in at least one of the three standards measured, depending on the concentrations within that standard. The instrument was able to provide relatively accurate and precise data for Nd, Pr, Ce, La, and Y (%diff = 17; %RSD = 3.7) when the concentrations were greater than 1000 ppm (e.g., rare earth ore CGL 124). It generated lower-quality data (in terms of %diff and %RSD) when concentrations of these elements were in the hundred parts per million range or lower. This indicates that the instrument is suitable for use in the exploration and development of carbonatite-related REE deposits. A pXRF without a radioactive source cannot be used to analyze for most of the heavy lanthanides; however, Y can be used as a pathfinder for these elements. Limited but effective use of the pXRF in exploration for ion adsorption clay deposits with grades as low as 500 ppm total REE appears possible.

In a recent study [192], the feasibility of using a Cd-109 gamma-ray–induced K X-ray fluorescence (K-XRF) system for the in vivo detection of gadolinium (Gd) in bone has been investigated. The K-XRF bone measurement system employs an array of four detectors and is normally used for the noninvasive study of bone lead levels. The system was used to measure bone-simulating phantoms doped with varying levels of gadolinium and fixed amounts of sodium (Na), chlorine (Cl), and calcium (Ca). The detection limits for bare bone phantoms, using a source of activity 0.17 GBq, were determined to be 3.9 and 6.5 ppm (μg Gd per gram phantom) for the $K\alpha_1$ and $K\ \alpha_2$ Gd X-ray peaks, respectively. This leads to an overall detection limit of 3.3 ppm (μg Gd per gram phantom). Layers of plastic were used to simulate overlying soft tissue, and this permitted prediction of a detection limit using the current strength of our radioisotope source, of 6.1–8.6 ppm (μg Gd per gram phantom) for fingers with 2–4 mm of overlying tissue. With a new source of activity 5 GBq, we predict that this system could achieve a detection limit of 4–5.6 μg Gd/g Ca. This is within the range of levels (2–30 μg Gd/g Ca) previously found in the bone of patients receiving Gd-based contrast imaging agents. The technique is promising and warrants further investigation.

CHROMATOGRAPHIC METHODS

It is well known that all lanthanides (except for promethium), as discussed in first chapter, are present in the earth's crust, and many of these elements are rather abundant. The abundance of the elements, however, is rather varied, with the most and least abundant elements being cerium and thulium, respectively. However, the nature and extent of its concentration in mineral deposits influence its commercial availability. Given the similarity among the chemical properties of these elements, most of their applications, which range from medicine and pharmacy, agriculture, photography, space, oil, lubrication, polymers, alloys and magnets, catalysts, to phosphors, etc., are similar. These applications, in addition to the recent increase in the use of lanthanides in various industries, have led to the increased emission of these elements and led to the accumulation of the elements in the environment. In the light of the above, and regarding the importance of the analysis of these elements in different samples, one significant problem could be the interference of different coexisting species. The interfering effects of species like Fe, Al, alkali, alkaline earth, and transition metals necessitate different pretreatment measures prior to analysis through conventional techniques like NAA, MS, inductively coupled plasma mass spectrometry and atomic emission spectroscopy (ICP-MS and ICP-OES), XRF, etc. Such measures usually tend to eliminate the disturbing matrix to some extent.

One of the candidates for sample pretreatment is removing the matrix through chromatographic techniques, which are well known as very powerful and robust separation techniques. One drawback could be the complexity of the chromatographic process and the need for long spans of time, but the strength of the techniques overrides such disadvantage, and hence the analytical and even preparative application of chromatographic techniques for the separation of lanthanides is very common. One advantage of chromatographic techniques is their flexibility, in terms of the possibility of using various applicable mobile and stationary phases, as well as the instrumental and operating conditions, which can help increase the separation quality and time.

A survey through the literature indicated that during the past 80 years, various chromatographic techniques have been used for the separation and preconcentration of lanthanides prior to determination. The reports started with techniques like ion exchange and planar chromatographic methods using various stationary phases and eluent systems, but the majority of such applications have been conducted through techniques like high-pressure ion chromatography (HPIC), high-performance liquid chromatography (HPLC), and the rather recent countercurrent chromatography (CCC), which have been described in excellent reviews [193-195]. The basic and detail principles, definitions, and separation mechanisms of chromatographic methods have been presented in various texts and hence shall not be covered here [196-198].

Ion exchange chromatography (IEC), could be mentioned as the first chromatographic technique used for treating lanthanide samples. This revolutionary step reduced the pretreatment and separation times from some years (required by the primary separation techniques like fractional crystallization-precipitation) to some days or even hours. Subsequently, other forms of chromatography like paper and thin layer chromatography (PC and TLC), liquid chromatography (LC), as well as HPLC were used to the same end, and later more specialized derivatives of the technique like micellar chromatography (MC), electrokinetic chromatography (EC), gel-permeation chromatography, and CCC were applied.

Classification of Chromatographic Methods

The principles of chromatography are very well described in different references [196-198]; however, the different classes of chromatographic techniques used in the determination of lanthanides shall be described to some degree in the following lines. In general, chromatographic techniques used in the determination of lanthanides can be classified into the two groups of planar and column methods.

Planar Methods

Planar methods include the techniques in which the stationary phase is in the form of a plane that can be made of paper or a layer of solid particles spread on the planar support like a glass plate. The different mobilities of the species on the plane, based on the nature and strength of their interactions with the stationary phase, leads to the separation of species that can be identified based on their specific retention factors (R_f) to some extent, which gives them a semiquantitative and yet flexible and cost-effective nature. The planar methods are further classified into the groups of paper and thin layer methods based on the nature of the stationary phase.

Paper chromatography (PC) is a rather primitive technique that is based on placing one or several dots of the sample on the paper stationary phase that is made of cellulose (a polar material). The stationary phase is next placed in a container on top of a layer of solvent. The container is then sealed to let the solvent rises through the stationery phase. The traveling solvent (which is the eluent) interacts and moves the species present in the sample after reaching the dot(s) and starts moving them with it. The polar nature of the stationary phase leads to the faster movement of the non- or less polar species, while the polar ones travel slower, and hence different species are separated based on their polarity.

In the thin-layer chromatographic (TLC) methods, on the other hand, the stationary phase is a homogenous thin layer composed of substances like silica gel, alumina, or cellulose powder. These adsorbing materials are evenly spread on a flat, inert support. One can assume that instead of a layer of cellulose (i.e., paper) that is used in PC, layers of a wider range of materials can be used in TLC, and hence this gives the method higher flexibility in terms of the speed and resolution of separation in comparison to PC. The method was also developed into the high-performance thin layer chromatography (HPTLC), which of course did not become very widespread as a result of the advent of column methods.

Column Methods

As the name implies, column chromatographic techniques are based on the application of tubular structures filled with different stationary phases, which are referred to as columns. The columns fall into two groups of packed and open tubular. In the packed columns, the stationary phase is used either in the form of particles, which fill the columns, or as the coating of such particles. The open tubular columns on the other hand are hollow columns, the inner walls of which are directly or indirectly coated with the stationary phase to increase the chances of interaction between the stationary and mobile phases.

Once again, it is noteworthy that because the text tends to focus on the popular techniques used for the separation and determination of lanthanides, the focus is on the techniques used to this end, which as already mentioned can be classified as ion exchange chromatography (IEC) (cation exchange chromatography [CEC] and lesser anion exchange chromatography [AEC]), HPIC, HPLC, EXC, and CCC. Two rather new chromatographic techniques that have been used in the separation and determination of lanthanide ions include ultra–high-performance liquid chromatography (UHPLC) and capillary column high-pressure ion chromatography.

Ion Exchange Chromatography

In the 1930s, coal-based and phenolic resins were introduced as ion exchange materials for industrial applications. This was followed by the advent of polystyrene-based resins with sulfonate or amine groups that gave the structure cation and anion exchange properties, respectively. These ion-exchanging materials are supplied in the form of dense beads with no distinct pores (e.g., gel resins) or with porous structures. Apart from the nature of the resins, the internal core of the beads plays a significant role in its activity and behavior. In the case of the former class, the ions should diffuse through the particle to reach the ion-exchange groups, which can be regarded as a shortcoming; however, they are more mechanically stable, have faster preparation and application reactions, and enjoy higher capacities. The latter form, also known as macroporous resins, has comparatively higher surface areas, which facilitate the ion exchange phenomenon and can be used rather independently of the size of the ions and nature of the solvent.

Structurally speaking, ion exchange polymers contain covalently bound ionic sites that are coupled with counter-ions with opposite charges, for the sake of electroneutrality. Because of the so-called Donnan exclusion, in the presence of other counter-ions in a solution of an inert solvent, the counter-ions can be exchanged with other ions from the surrounding medium, based on the relative affinity of the ionic species to the covalently bound ionic sites. The nature of the counter ions and the interaction between this ion and the covalently bound ionic sites have been used as the bases for the classification of these compounds to the classes of cationic and anionic, as well as strong and weak, exchangers. The exchanging resins containing quaternary ammonium groups have been found to act as strong anion exchange resins, whereas those containing sulfonic acid ion exchange sites act as strong cation exchangers. Resins containing ammonium chloride or hydroxide or carboxylic sites are classified as weak anionic and cationic exchangers, respectively. Another class of ion exchangers is the so-called mixed-bed resin, which is actually a blend of anion and cation resins, or an ampholytic resin, which is a resin containing a mixture of anionic and cationic sites in its structure. More specifically tailored resins with specific complexation properties have also been prepared for selective exchange of different ions.

In the case of lanthanides, the similarity of the physico-chemical properties of this group of elements with themselves as well as with many transition metals necessitates additional measures like the incorporation of chelating agents (e.g., ethylenediaminetetraacetic acid [EDTA], pyridine-2,6-dicarboxylic acid [PDCA], 1,2-diaminocyclohexanetetraacetic acid [DCTA], diethylenetriaminepentaacetic acid [DTPA], ethylene glycol tetraacetic acid [EGTA], α-hydroxyisobutyric acid [HIBA], or even simpler species like oxalic acid, tartaric acid, or citric acid) in the eluent or the use of chelating ion exchangers, to achieve acceptable ion exchange efficiencies. Addition of chelating agents to the eluent changes the properties of the species, which can help with their discrimination by the stationary phase. This measure can be performed either before the sample enters the separation path (off-line mode) or during its contact with the stationary phase (online mode).

As an example, in the case of environmental samples, where lanthanides are prone to co-elution effects due to the presence of high concentrations of transition metals, the difference of the charges of the complexes formed between pyridine-2,6-dicarboxylic acid (PDCA) with transition metals and lanthanides has been found to increase the separation efficiency. Lanthanide complexes can be retained in the column, while the transition metal–PDCA complexes are eluted. Next, the lanthanide species are eluted with a phase like oxalic or diglycolic acid.

Changes in other operating conditions like the pH is also used to increase the separation efficiency through altering the nature of the species present. For instance, if ethylenediaminetetraacetic (EDTA) is used as the chelating agent, changes in the pH can change the charge of the formed species. This is because the net charge of the EDTA complexes formed at different pH values is different (i.e., at alkaline pH values, the complexes are negatively charged, whereas acidic media give rise to the formation of neutral species).

High-Performance Liquid Chromatography and High-Pressure Ion Chromatography

Regarding the need for sample pretreatment, in the case of analysis of lanthanides, HPIC and HPLC are also methods that can be used for the efficient separation of desired species. There is no need to mention that the efficiency of HPLC or HPIC separation is highly dependent on the choice of the proper instrumental and operational conditions, including the column packing, dimensions and temperature, the nature and flow rate of the eluent, sample injection volume, etc.

With different samples, the surface area of the column packing is very important. For instance, samples with complex matrixes require packings with very high surface areas, which can be achieved by using fine packing material that can be chosen from spherical or amorphous materials. This unfortunately leads to close packing of the material, which hinders the passage of the eluent from the column. To avoid this problem, high pressures are applied to the solvent to drive it through the column, making it possible to perform an HPLC of the sample. The nature of the mobile and stationary phases, in terms of polarity, also influences the separation properties, and hence this liquid chromatographic technique is also divided into the wide categories of normal-phase HPLC, where the polarity of the eluent is higher than that of the stationary phase, and reverse phase HPLC, where the opposite is true.

The elements of any HPLC or HPIC system include the following:

- The eluent supply system
- The sample introduction system
- The column
- The detector
- The data acquisition and processing system

In general, the working mechanism of an HPLC or HPIC system (Figs. 7.6 and 7.7), includes the introduction of the eluent into the system at a given flow rate, with a high-pressure mechanical pump. In case one eluent is used, which is the most conventional mode used in HPLC or HPIC, the separation mode is called isocratic. This rather simple and inexpensive mode, however, suffers certain drawbacks, including poor resolution and low sensitivity for early and later eluting peaks, respectively, and long separation periods and contamination of the column because it is not fully eluted.

To overcome such problems, especially in the case of more complicated samples requiring better separation, two pumps are used to vary the composition of the eluent by mixing two different eluents during the separation. This mode, which leads to much better elution and separation, is referred to as the binary gradient eluent mode. In this mode, the composition of the mobile phase is gradually altered during the chromatographic separation, usually increasing the elution power, whereas the flow rate is constant, leading to improved discrimination of the species. More complicated systems are capable of applying more complex mixing modes using more than two mobile phases for one or a plurality of separation systems. There is no need to mention that more complicated systems are more costly and complicated.

After the system reaches an equilibrium state, the sample is introduced into the column either manually, using a microsyringe, or automatically, using an electropneumatic valve. This is the beginning of the separation process as a result of which the species are discriminated through their interactions with the stationary phase and sequentially detected by the

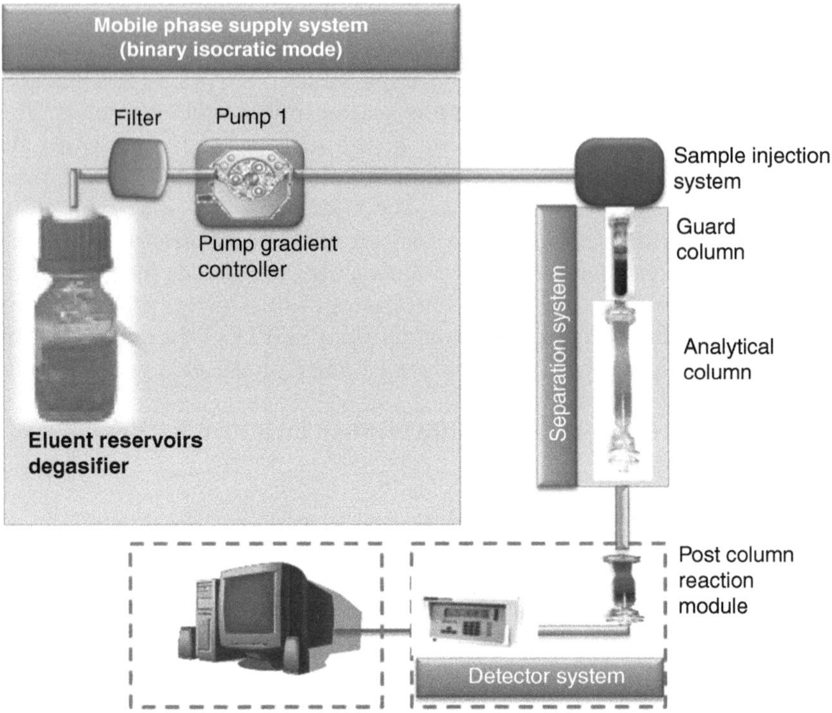

FIGURE 7.6 Schematic diagram of HPLC or HPIC setup with an isocratic mobile phase.

detection system. In the cheapest HPLC and HPIC systems, UV detection is used, but other spectrometric and electrochemical methods are used in these methods.

1. The eluent supply system
   Although the pumps can be regarded as the heart of the eluent supply system, because the eluents must be free from particles, contaminants and dissolved gases filtering and degassing systems should be mentioned as the first components of this system. The degassing is performed through external vacuum or systems or on-line systems in the pumps, and the filters are usually membranes with 0.20–0.45-μm pores.

   The major component of the eluent supply system, as mentioned earlier, is one or some high-pressure mechanical pumps used to supply the eluent, or a mixture thereof. This should be done at a precise and pulse-free flow. To decrease the instrumental costs, some systems use low-pressure mixing module (LPMM) for performing the required alterations in the composition of the eluent during the separation, eliminating the need for a plurality of pumps that

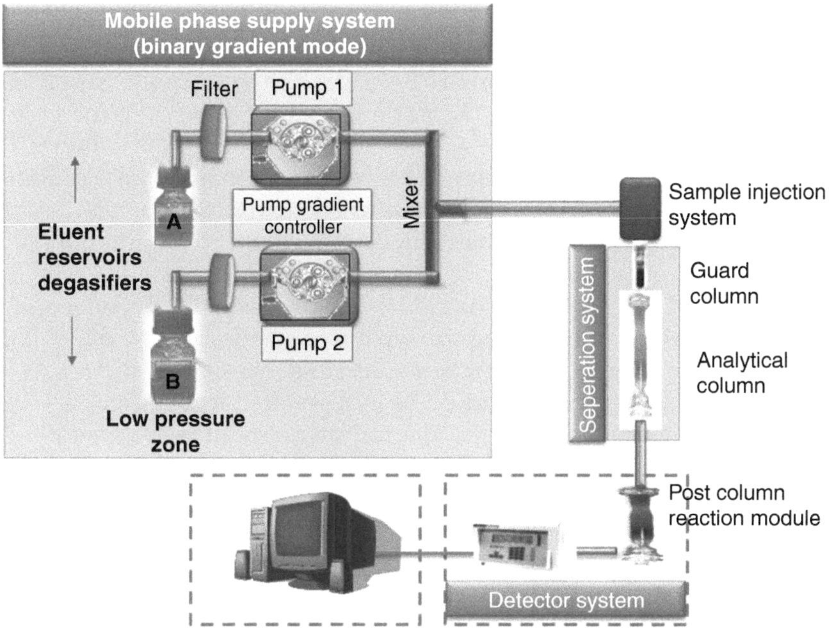

FIGURE 7.7 Schematic diagram of HPLC or HPIC setup with a binary gradient mobile phase.

is a requirement of gradient elution systems. The function of an LPMM is based on time-programmed micro-proportioning valves, the open or close period of which can be planned to reach desired eluent compositions. The pumps used in HPLC or HPIC are conventionally stainless steel, poly-ether-ether-ketone (PEEK), or polytetrafluoroethylene (Teflon) or ceramic standard-bore pumps capable of generating constant flow rates of 0.1–10 mL/min of the mobile phase, which requires the exertion of pressures up to 6000 psi.

2. The sample introduction system
   Introduction of samples into the analysis system is performed through a manual or automatic solenoid valve. The manual introduction of the samples is based on the injection of 5–100 μL of the sample through a septum using a microsyringe. At this stage, the valve should be in Load position. Next, the valve is turned to the Inject position to release the sample into the mobile phase.
3. The column
   The column can be named as the most important part of any chromatography system, since it is where the separation of the components takes place. HPLC and HPIC are stainless steel, PEEK, or Teflon tubes of 100–250 mm in length and 3–9 mm in internal diameter that are packed with stationary phases. The packing materials (the

stationary phases) used in HPLC and HPIC usually consist of silica or polymeric backbones grafted with functionalized groups. These material are usually in the form of 5–13-μm spheres. Smaller particles increase the surface area and separation capabilities but also increase the pressures required for the solvent to travel through the column. To avoid damages to the expensive columns, one or more guard columns precede the analytical columns. These are actually smaller versions of the main column, which protect it in case of eluent contamination.

4. The detector
An HPLC or HPIC detector should be capable of qualitatively and quantitatively monitoring the separated species over wide concentration ranges, with low noise and high sensitivity and precision, and its response should be independent from other external or internal parameters. The common detection systems used with HPLC and HPIC are the UV–vis absorbance, fluorescence, conductivity detectors, or more developed detection systems like MS, ICP-MS, and ICP-OES, among which UV–vis spectrometry, ICP-MS, or ICP-OES have been used for the analysis of lanthanides more frequently. HPLC and HPIC systems may also include post-column reaction systems intended for increasing the sensitivity of detection through the addition of chemical reagents after the separation, since the detection systems are set to be more sensitive toward a physical or chemical characteristic of the outlet of the post-column reaction module. This characteristic can be the increased absorption of light by the resulting mixture for UV–vis detectors.
5. The data acquisition and processing system
The final element of an HPLC or HPIC is the interface between the detector and the indicator, which can be a computer screen or a printer.

UHPLC and Capillary-HPIC

Waters Corporation has rather recently introduced UHPLC, in which the operating pressure is actually doubled (up to 15,000 psi), to decrease the analysis time, while achieving identical or improved separation resolution. A more advanced integrated capillary high-pressure ion chromatograph (HPIC) (Thermo Scientific Dionex ICS-4000) was also introduced in 2012 [199]. The instrument is equipped with integrated conductivity and electrochemical detection systems as well as a so-called charge detector introduced by the producing company. The device has been claimed to enjoy to be able to operate at up to 5000 psi leading to fast, high-resolution analyses, without using chemical reagents.

Countercurrent Chromatography

Another chromatographic method, which has been widely used for the separation and determination of lanthanide ions, is the CCC [200].

The method is based on a continuous and repeated mixing of two liquid phases (both the mobile and stationary phases are liquids), to increase their contact and facilitate the transport of the analyte from one to the other, letting the system rest, and their separation. The separation (which is actually an extraction process) depends on the partition coefficient of the analyte, which is the ratio of the equilibrium amount of the species in each of the two phases. The overall mixing and settling of the mobile and stationary phases happens in the column, and hence one of the major problems in CCC is the loss (or bleeding) of the stationary phase from the column. Therefore, a good CCC instrument should be able to withhold the stationary phase. Another important factor influencing the retention of the stationary phase is the settling time, which depends on the nature of the solvent and sample matrix. CCC has a range of subclasses, which are described in the following paragraphs.

I. Droplet countercurrent chromatography (DCCC) In DCCC, which is the oldest and simplest form of CCC, the eluent travels through the steeped or vertical column (and hence the stationary phase) because of gravity. In case the eluent–sample system has a higher density than the stationary phase, it moves down the column as a result of gravitational effects, and hence this configuration is referred to as the descending mode. Naturally, the opposite is also possible and in case the stationary phase has a higher density. If so, the eluent ascends the column while gradually passing through the stationary phase, and the method is referred to as ascending DCCC.

II. Hydrodynamic countercurrent chromatography (HCCC) One can simply guess that DCCC is a very slow method and it cannot achieve good mixing of the phases. Another less evident disadvantage of DCCC was the fact that it was more a preparative technique in nature requiring rather high amounts of the sample.

This negatively influences both the time and quality of the analyses and hence led to the advent of the high-performance and high-speed CCC, and high-performance CCC categorized as hydrodynamic techniques.

To increase the speed and quality of CCC, two major modes are used. The first mode is based on using columns whereas the second uses cartridges or discs.

In the former case, the column is a coil-shaped tube made of the inert polytetrafluoroethylene (PTFE) or Tezel etc., with a bobbin at the center of the coil. The movement and variable gravitational field created this way leads to increased speed of the separations, and that is why the method is also known as high-speed countercurrent chromatography (HSCCC). A partitioning step takes place during each rotation. HSCCC techniques need much less sample for the separation procedure and the tests can be carried out with a few milligrams of the samples. HSCCC has been further developed into the so-called high-performance countercurrent chromatography (HPCCC). This has been majorly due to the use of instruments that can create gravitational accelerations of 240*g*, which is three times that of HSCCC devices (around 80*g*). HPCCC systems also use columns with larger bores as compared to HSCCC instruments. Both these alterations have helped increase the sample capacity of the CCC devices, in addition to better partitioning and hence separation of the sample as a result of improved eluent flow rates.

Another variation of the instrument is the toroidal countercurrent chromatography, in which the bobbin only spins around its axis while the rotational movement is not performed. In some other simple CCC systems, which do not use any spinning or shaking measures, capillary tubes are used instead of the large-bore ones, leading to the more efficient mixing of the two phases.

Countercurrent systems using cartridges or discs work based on creating a constant gravitational field through single-axis rotations. The eluent is fed into the disks through the rotating seals. After some time, a hydrostatic equilibrium similar to a static coil dominates.

III. Hydrostatic countercurrent chromatography (HCCC)

Centrifugal Partition Chromatography (CPC) CPC uses a design in which the rotor is a special disk where several interconnected small cells are engraved. The connection of each cell with the previous one is through a head duct, while it also leads to a next cell through its bottom. This can be imagined as a long column with sequential expansions (cells) where the chromatographic separation takes place and is repeated. The stationary phase is kept in the cells because of the torque and centrifugal force created by the rotations whereas the eluent is pumped through the system. After a period of time, a hydrostatic regime is established in the system, which is why the method is classified under hydrostatic techniques. Like DCCC, CPC is also performed in the descending or ascending modes.

CCC Modes

CCC can be performed in normal or reverse-phase modes. The other modes of its application are elution extrusion, pH zone refining, and dual-mode elution. Each mode is chosen based on the properties of the separation required. For example, in the case of ionizable species, the pH zone refining mode is used. Because charged species prefer polar and especially aqueous media by varying the pH, different analytes are extracted into the aqueous media based on their acidity constants.

The elution extrusion mode, on the other hand, is used in cases when the sample is completely separated in the column. According to this mode, when once the separation is complete and all the analyte is extracted into the stationary phase, the pumping of eluent is stopped and instead the stationary phase in pumped into the column to extrude the column contents. This way, less solvent is also consumed. For instance, when using ethyl acetate as the stationary phase and water as the eluent, after the separation is completed, the flow of water is stopped and ethyl acetate is pumped into the system. This switching of solvents can be performed by installing a valve at the beginning of the column or using a system with more than one pump each for a solvent system.

Another mode that is used to achieve better separation is the application of two eluents and is therefore called the dual-mode elution. Based on this mode, initially an aqueous eluent is used, while the eluent is changed to an organic solvent after a certain time. The switching can be repeated to reach the desired separation. Another variation of this mode is the gradient mode, in which the composition of the eluent is gradually changed by changing the composition of one of its component. As the definition implies, a mixture of miscible solvents like methanol and water are used in this mode.

Displacement Chromatography

The basic principle of displacement chromatography is as follows: A molecule with a high affinity for the chromatography matrix (the displacer) will compete effectively for binding sites and thus displace all molecules with lesser affinities. There are distinct differences between displacement and elution chromatography. In the elution mode, substances typically emerge from a column in narrow, Gaussian peaks. Wide separation of peaks, preferably to baseline, is desired in order to achieve maximum purification. The speed at which any component of a mixture travels down the column in elution mode depends on many factors. But for two substances to travel at different speeds, and thereby be resolved, there must be substantial differences in some interaction between the biomolecules and the chromatography matrix. Operating parameters are adjusted to maximize the effect of this difference. In many cases, baseline separation of the peaks can be achieved only with gradient elution and low column loadings. Thus, two drawbacks to elution mode chromatography, especially at the preparative scale, are operational complexity, due to gradient solvent pumping, and low throughput, due to low column loadings. Displacement chromatography has advantages over elution chromatography in that components are resolved into consecutive zones of pure substances rather than "peaks." Because the process takes advantage of the nonlinearity of the isotherms, a larger column feed can be separated on a given column, with the purified components recovered at significantly higher concentrations.

Extraction Chromatography

A rather recent, yet very versatile, technique is the EXC, which has found use in the separation and preconcentration of metallic ions. The technique, as its name implies, is a hybrid of solvent extraction and chromatography.

As one can guess, the separation of species takes place based on liquid-liquid extraction, but based on the nature of the species, as well as the nature and composition of the solvents, different phenomena including ion pair formation, ion exchange, coordination, or a combination thereof can account for the separation. The technique uses a liquid extracting phase adsorbed on a porous support as the stationary phase. Technically, a wide range of materials can be used as the extracting stationary phase. The support is a hydrophobic material like silica gel, PTFE, etc. usually as 50–150-μm particles. The eluent, on the other hand, is a solvent containing a dissolved base, acid, or salt.

EXC has been used for the separation of lanthanides, using phosphate-based stationary phases (like dihexyl [(diethyl-carbamoyl)methyl] phosphonate, bis(l-methyl-heptyl) hydrogenphosphate, and 1-methylheptyl-isoalkyl phosphonate) as the stationary phase. Although other stationary phases (e.g., tributyl phosphate) have been used to this end, the acidic phosphate–based phases have been found to yield better separation factors. For anionic species, the stationary phases are usually chosen from tertiary and quaternary amine with long chains.

Another important factor in EXC, like any other form of extraction, is the choice of the eluent. The solvents used as the eluent in the EXC of lanthanides are solutions of hydrochloric, hydrobromic, nitric, or sulfuric acid. Hydrochloric acid solutions further containing other acidic compounds like ammonium salts or nitric acid have also been used to this end. To enhance the selectivity of the eluent, complexants like oxalic or hydrofluoric acid are also used in the composition of the eluent where necessary. The most frequently used solvents have been hydrochloric acid solutions with or without ammonium chloride. Different compositions of HCl/NH_4Cl solutions further containing Na_2SO_4 have been used for separating different mixtures of Sm, Eu, Gd, Ho, Y, and Er in different matrices. Other eluents used have been ascorbic and sulfosalicylic acids in an acetate or dichloroacetate buffer.

Ion-Pair Chromatography (IPC)

IPC can be regarded as a variation of EXC where extraction of the ionic solutes between the phases is facilitated through the formation of ion pairs. The application of this method in chromatography dates back to the 1970s. In the early years after its development, IPC was limited to normal-phase modes using silica or cellulose particles coated with a polar stationary phase. The stationary phase further incorporated an ion-pair forming reagent. Later works, however, led to the development of reverse-phase IPC using nonpolar stationary phases like bonded microparticulate alkyl silicas as the stationary phase.

The reverse-phase IPC has been tested as a tool for the separation of lanthanides, in a limited number of articles. One example is the application of a reverse-phase C18 column and tetrabutylammonium chloride as the ion pairing agent, in a water/methanol eluent to separate the anionic EDTA complexes of lanthanides [194,201,202]. The method did not show good resolution for adjacent lanthanides, and the concentrations of the ion pairing agent and methanol were found to greatly influence retention and peak shapes.

A very effective family of ion-paring agents used during the reverse-phase IPC separations of lanthanides includes hydroxyisobutyric acid (HIBA) complexes with organic sulfonates or sulfates (e.g., octane-1-sulfonate), using a C18 phase [203]. The combination has been reported to lead to complete separation of all lanthanides in a short time. The good resolution has been attributed to larger hydration radii of the complexes of the smaller lanthanides leading to their weaker ion pairs as compared to larger lanthanides. This leads to higher partition ratios of earlier lanthanides increasing their retention. Changes in the concentration of octane-1-sulfonate were found to influence the effective capacity and selectivity of the stationary phase.

Application of Chromatographic Methods in the Separation, Preconcentration and Determination of Lanthanide Ions

Conventionally, spectroscopic techniques have been used in the determination of lanthanide species; however, if the concentration of the species is very low, spectroscopic techniques even as powerful as ICP-OES face some limitations. This is also true for samples with complex matrixes, which lead to spectroscopic interference from elements in the matrix, especially when their concentrations are rather high. Consequently, pretreatment and preconcentration of samples is required. This is where chromatographic methods are needed. Once the chromatographic procedures separate and preconcentrate the different species present in the sample, a detection system is used for the quantification. The nature of the detection system is also very important in terms of the nature of the samples as well as the instrumentation used for feeding the sample into the detection system.

Because of the nature of lanthanide species in various samples, IEC has been the most widely used technique for their separation. IEC columns are packed with cationic as well as anionic resin exchanger, and in some cases, complexes of lanthanide cations with negatively charged ligands like EDTA are formed and separated through the anionic columns. The most common detectors hyphenated with chromatographic systems include ICP-OES and ICP-MS. More recently, HPLC and HPIC are dominantly used in the area of separation of lanthanides, whereas some newer techniques like CCC and EXC have been used to this end. In the following pages, an overview of the chromatographic methods used for the determination of lanthanides in geological, industrial, environmental, biological, and food samples shall be presented. In all cases, the principles of the sample preparation procedures are the same as those previously described for ICP-OES methods in Chapter 6.

Cation and Anion Exchange Chromatography

In a report in 1998, the concentration of ultratrace amounts of lanthanides in Chinese tea samples were determined after microwave digestion of the sample. The digested sample was next analyzed by using a AG50W-x8 cation-exchange chromatograph equipped with an ICP-MS detection system. This way, the anticipated spectroscopic interferences from the polyatomic BaO^+ and $BaOH^+$ ions were eliminated and hence the LOD values of between 0.0039 and 0.0003 ng/cm^3 were reached [204]. In a later work, an LC-ICP-MS system was used for the determination of lanthanide content of geochemical standard reference samples (i.e., GSJ JB-1 and Pond Sediment; NIES No. 2), using a strong ion-exchange resin stationary phase together with ammonium lactate as the eluent [205]. A multi-elemental detection was performed using the ICP-MS system in peak hopping mode (PHM) to measure all desired chromatograms at the corresponding mass numbers.

Four years later in 2003, Premadas [206] used a simple cation-exchange chromatograph (CEC) equipped with ICP-OES detection for the determination of lanthanides in refractory minerals such as zircon, ilmenite, rutile, columbite-tantalite, garnet, and sillimanite. The method was reportedly capable of removing milligram amounts of zirconium, titanium, niobium, and tantalum from microgram amounts of lanthanides, which is normally not viable using oxalate or fluoride precipitation separation. The sample preparation included fusing zircon, itmenite, and rutile samples with a flux (powdered mixture of 3:1 ratio of potassium bifluoride and sodium fluoride), and next treating the cooled melt with sulfuric acid for removing fluoride. The sample was then taken in a 0.25 mol/L oxalic acid solution and the acidity of the solution was maintained around 1 N, before starting the chromatographic analysis. Columbite-tantalite samples were fused with potassium bi-sulfate, and the solution was prepared in oxalic acid for the IEC determination, whereas garnet and sillimanite minerals were acid digested using hydrofluoric, hydrochloric, and perchloric acids.

In a more recent work, negatively charged lanthanide complexes were isocratically separated through anion exchange chromatography [207]. To form the desired complexes, solutions of ethylenediaminetetraacetic acid were used as the mobile phase. The detection system was an ICP-MS system. The results showed that because the stoichiometry and the charge of the EDTA complexes varies with eluent pH, from pH values of 3.5 to 7.5, the retention increases from La-EDTA to Sm-EDTA complexes and then decreases to Lu-EDTA complexes. At pH values greater than 8.5, the retention of the light lanthanides considerably increased. This change in elution behavior, by changing the pH, was hence suggested for particular applications like the elimination of most isobaric and polyatomic interferences in ICP-MS determination of lanthanides between 5.5 and 7.5.

Very recently in 2014 [208], cation and anion exchange techniques involving a high-resolution inductively coupled plasma mass spectrometer (HR-ICP-MS) were used for the separation of lanthanides and Fe in magnetites from banded iron formations (BIFs). The results showed that 10 mL of concentrated HCl was sufficient for the complete separation of lanthanides from Fe using anion exchange chromatography, whereas in the case of the cation exchange technique, 30 mL of 6 mol/L HCl was needed.

HPLC and HPIC

HPLC and HPIC have been widely used as cheap, relatively rapid, precise, and accurate separation and quantification analytical techniques for the determination of REEs. A brief review of the literature on the applications of HPLC and HPIC shows that determination of REEs in synthetic solutions, as well as real samples like nuclear fuels and geochemical materials, has been conducted using these techniques [193]. Combination of the microwave digestion for sample preparation, HPIC, and on-line sample solution concentration, for instance, has been used for the determination of lanthanides in coal samples. The acid dissolution of powdered coal under microwave irradiation was used instead of dry ashing, which is intended for the elimination of organic carbon. This way, the method became less time-consuming and also the risks of sample loss and contamination were minimized. The HPIC analysis was performed through chelation and gradient elution, and the results showed the ability of the method for the separation and determination of 12–14 lanthanides in coal samples with detection limits of between 10 and 50 ng/g, and the requirement for small sample sizes of about 0.2 g. The results gained for testing coal standard reference materials (i.e., NBS-1632a, SARM-18, SARM-19, and SARM-20) were found to be very satisfactory [209].

Another example is the application of a narrow-bore HPLC for the simultaneous separation of free gadolinium diethylenetriaminepenta-acetic acid (GdDTPA) and its monomethylamide (GdDTPA-MMA) and the bis-methylamide (GdDTPA-BMA) compounds in human serum and urine. After the separation phase, the species were detected at 658 nm after post-column derivatization with Arsenate III. The preparation of the serum samples included ultrafiltration while the urine samples were centrifuged and diluted. The method requires injecting a volume of 10 μL on a 2.1-mm-internal diameter reversed-phase column, yielding LODs of 0.3 and 1.1 μM for GdDTPA-BMA in serum and urine, respectively, with precisions ranging from 1% to 4% [210].

In 2002 [211], HR ICP-MS was used for the separation of lanthanides helping avoid spectral interferences during the detection phase. Sc, Y, and

14 lanthanides were selectively analyzed by the HPLC/ICP-SFMS system at concentration gradient regimes and detection limits of 0.5–10 pg/mL with RSDs of 2.5% and 4.5%. The recovery values ranged from 90% to 100% for the various elements.

Another instance of application of HR ICP-MS is the trace analysis of REEs in highly pure lanthanum oxide (La_2O_3) produced in the Instituto de Pesquisas Energeticas e Nucleates (IPEN) as compared to certified references.

In 2003, Ishikawa et al. [212] reported the analysis of REEs in rock samples using an IEC by a conventional ion-exchange column, followed by HPIC. They used alpha-hydroxyisobutyric acid and Arsenazo III as eluents and a postcolumn reagent for the HPIC. This way they could analyze all 14 REEs, eliminating the interference from Y and other transition metals during the ICP-MS analyses. The method was reported to be capable of handling rock samples as small as <50 mg.

Rather recently, another report discussed the application of bare core-shell silica (1.7 μm) modified with iminodiacetic acid functional groups via standard silane chemistry as the stationary phase [213]. The *N*-hydroxyethyliminodiacetic acid (HEIDA) functionalized core-shell stationary phase was used in the high-performance chelation ion chromatography of alkaline earth and transition and heavy metal cations, and the effects of the concentration and pH of nitric acid (eluent), addition of dipicolinic acid (the complexing agent), and column temperature on the performance of the columns were evaluated. The results showed that the separation efficiency of the proposed column during the separation of transition and heavy metal cations were identical to or better than the results acquired using fully porous silica-based chelation stationary phases, and a similarly modified monolithic silica column, ranging from similar to 15–56-μm HETP. It was reported that increasing the ionic strength of the mobile phase through adding KNO_3 (0.75 mol/L) and heating the columns to 70°C facilitates the isocratic separation of a mixture of 14 lanthanides and yttrium in under 12 min, with HETP averaging 18 μm (7 μm for Ce(III)).

Countercurrent Chromatography

The first example of the application of CCC for the preseparation of REEs in rock samples dates back to 1993 [214]. A later work in 1995 by Pukhovskaya et al. [215] reported the application of CCC-ICP-OES for the preconcentration and determination of REEs in geological samples. The results showed the method to be capable of rapid and reliable separation and preconcentration of light REE and Y using a 0.5 mol/L di-2-ethylhexylphosphoric acid (D_2EHPA) system in different concentrations of *n*-decane-hydrochloric acid using a planetary centrifuge as the CCC instruments. It was further reported that Tm, Nb, and Lu were partially retained in the stationary phase. Three other two-phase liquid systems using trioctylphosphine oxide (TOPO), D_2EHPA and TOPO mixtures and diphenyl (dibutylcarbamoylmethylphosphine)oxide (Ph(2)-Bu(2)) as the extractants were tested to evaluate their ability for the separation of all REEs. The results showed partial loss of La and Ce using a 0.1 mol/L solution of TOPO in isobutylmethylketone (IBMK) (stationary phase)–1 mol/L NH_4NO_3–6 mol/L HCl aqueous solutions (mobile phase), whereas the complete isolation of the REEs was reportedly viable in two systems (i.e., 0.3 mol/L D_2EHPA + 0.02 ml/L TOPO in the solvents mixture (3:1) of *n*-decane + IBMK, respectively (stationary phase)–1 mol/L NH_4NO_3–6 mol/L HCl aqueous solution (mobile phase); and 1.0 mol/L Ph(2)-Bu(2) solution in chloroform (stationary phase) 3 mol/L HNO_3 aqueous solution (mobile phase)). The researchers finally recommended the D_2EHPA + TOPO mixture as the more economic and accessible system that can be used to this end.

Later, the same goal was pursued using a CCC-ICP-MS system using a solution of diphenyl[dibutylcarbamoylmethyl]phosphine oxide in chloroform (0.5 mol/L) as the reagent for the extraction and preconcentration of trace rare-earth elements from aqueous 5% $CaCl_2$ solution, 3 mol/L in HNO_3, and 0.1 mol/L in $HClO_4$. During the process, the analytes were back-extracted into a small volume of water, and the aqueous eluate was subjected to ICP-MS measurements. The method was reported to be applicable to the determination of ultratrace REEs [216].

Another example of the application of a CCC-ICP-MS system has been the determination of REEs and gadolinium diethylenetriaminepentaacetate (GdDTPA) in river water through CCC [217]. The method used the CCC prior to the ICP-MS detection because the concentrations of Gd-DTPA and REEs in river water are often below in the LODs of quadrupole ICP-MS on applying pneumatic nebulization. Ethylhexylphosphates was used as the stationary phase, and the results were compared with those of solid-phase extraction (SPE) using the same samples' acidified aqueous samples (pH 2) and ethylhexylphosphates as the stationary phase.

In 2013, Krattli et al. [218] tried to show the possibility of using continuous CCC to decrease the purification cost. In particular, they studied a model mixture of three REEs (i.e., praseodymium, cerium, and lanthanum) and tried to purify the elements up to >95% in a single-step, high-yield process. They achieved a yield of 75.4% and a productivity increase of 5- to 15-fold with respect to batch chromatography, but the productivity value remained approximately 1 g/L stationary phase and day, which is too low for large-scale applications. This was attributed to the low loading caused by the limited solubility of the REEs in the adopted buffer used.

Extraction Chromatography

Other group F techniques used for the separation of lanthanides are the agent extraction and liquid/solid separation processes which allow for the separation/refinement of high-quality REEs due to their excellent selection and separation performance.

As an example, in 1995, the adsorption behavior of octadecylsilylated silica gel (ODS) particles loaded with bis(2-ethylhexyl) hydrogenphosphate (B_2EHP) was studied for a series of REEs (except for Pm^{3+}) in dilute HCl solutions. Each lanthanide element was separated through flash column chromatography using ODS particles loaded with B_2EHP as the solid phase. Pairs of neighboring ions in the atomic number such as La^{3+}-Ce^{3+}, Ce^{3+}-Pr^{3+}, Sm^{3+}-Eu^{3+}, Eu^{3+}-Gd^{3+}, and Tb^{3+}-Dy^{3+} were completely separated from each other by developing with 0.20–1.2 mol/L HCl at the eluting rate of 100 cm^3/h and 45°C. At the smaller eluting rates, Pr^{3+} was separated from Nd^{3+}. The ions heavier than Ho^{3+} were separated from their neighboring species in the value of distribution ratio by developing with 1.8–3.5 mol/L of HCl at the eluting rate of 30 cm^3/h and 45°C. Further, separation of the Y^{3+}-Er^{3+} pair was found to be incomplete, with a resolution of 0.52, coinciding with its smallest separation factor, 1.10, calculated from the present data of distribution ratios, among all the pairs [219].

An analytical method was introduced for the rapid analysis of REEs in 2012 [220]. According to this procedure, the Sc and Y content of meteoritic and geological samples are analyzed through multi-collection ICP-MS detection after a single EXC step. The measurement of the REEs was conducted using a dynamic multi-collector cup configuration through a desolvating nebulizer and standard-sample bracketing technique.

Finally, the work of Kim in 2014 [221] included the application of Amberlite XAD-7HP as the solvent-impregnated resin and D_2EHPA as an agent to separate REEs from standard sample solutions containing heavy REEs (e.g., La and Ba) and light REEs (e.g., Eu, Tb, and Y) each at about 100 ppm in DI water through extraction chromatography ICP-AES. The effects of changing HCl concentration between 0.1 and 3 mol/L, as well as the flow rate between 0.9 and 2.6 mL/min were assessed, and the optimal values were found to be between 1.6 and 2.6 m/min, respectively.

References

[1] Sparkman OD. Mass spectrometry desk reference. Pittsburgh: Global View Pub; 2000. ISBN 0-9660813-2-3.

[2] He M, Hu B, Jiang ZC. Electrothermal vaporization inductively coupled plasma mass spectrometry for the determination of trace amount of lanthanides and yttrium in soil with polytetrafluroethylene emulsion as a chemical modifier. Anal Chim Acta 2005;530:105–12.

[3] Buseth E, Wibetoe G, Martinsen I. Determination of endogenous concentrations of the lanthanides in body fluids and tissues using electrothermal vaporization inductively coupled plasma mass spectrometry. J Anal Atom Spectrom 1998;13:1039–49.

[4] Cao XD, Yin M, Wang XR, Zhao GW. Determination of trace rare earth elements in soils by inductively coupled plasma-mass spectrometry after microwave digestion for sample preparation. Chinese J Anal Chem 1999;27:679–83.

[5] Beary ES, Paulsen PJ. Determination of ultratrace neodymium in high-purity lanthanum compounds by high-accuracy isotope-dilution inductively-coupled plasma-mass spectrometric analysis with chemical preconcentration. Anal Chem 1994;66:431–9.
[6] D'orazio M, Tonarini S. Simultaneous determination of neodymium and samarium in silicate rocks and minerals by isotope dilution inductively coupled plasma mass spectrometry: results on twenty geochemical reference samples. Anal Chim Acta 1997;351:325–35.
[7] Li B, Sun YL, Yin M. Determination of cerium, neodymium and samarium in biological materials at low levels by isotope dilution inductively coupled plasma mass spectrometry. J Anal Atom Spectrom 1999;14:1843–8.
[8] Yuan HL, Gao S, Zong CL, Dai MN. Sectional power-law correction for the accurate determination of lutetium by isotope dilution multiple collector-inductively coupled plasma-mass spectrometry. Spectrochim Acta Part B 2009;64:1228–34.
[9] Bourgeois M, Isnard H, Gourgiotis A, Stadelmann G, Gautier C, Mialle S, Nonell A, Chartier F. Sm isotope composition and Sm/Eu ratio determination in an irradiated Eu-153 Sample by ion exchange chromatography-quadrupole inductively coupled plasma mass spectrometry combined with double spike isotope dilution technique. J Anal Atom Spectrom 2011;26:1660–6.
[10] Yang YH, Chu ZY, Wu FY, Xie LW, Yang JH. Precise and accurate determination of sm, nd concentrations and nd isotopic compositions in geological samples by Mc-Icp-Ms. J Anal Atom Spectrom 2011;26:1237–44.
[11] Braverman DS. Determination of rare-earth elements by liquid-chromatographic separation using inductively coupled plasma mass-spectrometric detection. J Anal Atom Spectrom 1992;7:43–6.
[12] Brenner IB, Taylor HE. A critical-review of inductively coupled plasma-mass spectrometry for geoanalysis, geochemistry, and hydrology 1. Analytical performance. Critic Rev Anal Chem 1992;23:355–67.
[13] Kane JS, Beary ES, Murphy KE, Paulsen PJ. Impact of inductively-coupled plasma-mass spectrometry on certification programs for geochemical reference materials. Analyst 1995;120:1505–12.
[14] Panday VK, Becker JS, Dietze HJ. Trace impurities in zircaloys by inductively-coupled plasma-mass spectrometry after removal of the matrix by liquid-liquid-extraction. Atom Spectrosc 1995;16:97–101.
[15] Pin C, Telouk P, Imbert JL. Direct determination of the samarium – neodymium ratio in geological-materials by inductively-coupled plasma quadrupole mass-spectrometry with cryogenic desolvation – comparison with isotope-dilution thermal ionization mass-spectrometry. J Anal Atom Spectrom 1995;10:93–8.
[16] Montero P, Bea F. Accurate determination of Rb-87/Sr-86 and Sm-147/Nd-144 ratios by inductively-coupled-plasma mass spectrometry in isotope geoscience: an alternative to isotope dilution analysis. Anal Chim Acta 1998;358:227–33.
[17] Yang YH, Wu FY, Chu ZY, Xie LW, Yang JH. High-precision simultaneous determination of Sm-147/Nd-144 And Nd-143/Nd-144 ratios in Sm-Nd mixtures using multi-collector inductively coupled plasma mass spectrometry and its comparison to isotope dilution analysis. Spectrochim Acta Part B 2013;79–80. 82-87.
[18] Baker J, Waight T, Ulfbeck D. Rapid and highly reproducible analysis of rare earth elements by multiple collector inductively coupled plasma mass spectrometry. Geochim Cosmochim Acta 2002;66:3635–46.
[19] Walder AJ, Platzner I, Freedman PA. Isotope ratio measurement of lead, neodymium and neodymium–samarium mixtures, hafnium and hafnium–lutetium mixtures with a double focusing multiple collector inductively coupled plasma mass spectrometer. J Anal Atom Spectrom 1993;8:19–23.
[20] Becker JS, Dietze HJ. Double-focusing sector field inductively coupled plasma mass spectrometry for highly sensitive multi-element and isotopic analysis - Invited lecture. J Anal Atom Spectrom 1997;12:881–9.

[21] BlichertToft J, Chauvel C, Albarede F. Separation of Hf and Lu for high-precision isotope analysis of rock samples by magnetic sector multiple collector ICP-MS. Contrib Mineral Petr 1997;127:248–60.
[22] Pedreira WR, Sarkis JES, Rodrigues C, Tomiyoshi IA, Queiroz CAD, Abrao A. Determination of trace amounts of′ rare earth elements in high pure lanthanum oxide by sector field inductively coupled plasma mass spectrometry (HR ICP-MS) and high-performance liquid chromatography (HPLC) techniques. J Alloy Compd 2002;344:17–20.
[23] Pedreira WR, Sarkis JES, Queiroz CAD, Rodrigues C, Tomiyoshi IA, Abrao A. Determination of trace amounts of rare-earth elements in highly pure neodymium oxide by sector field inductively coupled plasma mass spectrometry (ICP-SFMS) and high-performance liquid chromatography (HPLC) techniques. J Solid State Chem 2003;171:3–6.
[24] Pedreira WR, Queiroz CAD, Abrao A, Pimentel MM. Quantification of trace amounts of rare earth elements in high purity gadolinium oxide by sector field inductively coupled plasma mass spectrometry (ICP-MS). J Alloy Compd 2004;374:129–32.
[25] Gabrielli P, Barbante C, Turetta C, Marteel A, Boutron C, Cozzi G, Cairns W, Ferrari C, Cescon P. Direct determination of rare earth elements at the subpicogram per gram level in antarctic ice by ICP-SFMS using a desolvation system. Anal Chem 2006;78:1883–9.
[26] Pedreira WR, Queiroz CA, Abrao A, Rocha SM, de Vasconcellos ME, Boaventura GR, Pimentel MM. Trace amounts of rare earth elements in high purity samarium oxide by sector field inductively coupled plasma mass spectrometry after separation by HPLC. J Alloy Compd 2006;418:247–50.
[27] Dick D, Wegner A, Gabrielli P, Ruth U, Barbante C, Kriews M. Rare earth elements determined in Antarctic ice by inductively coupled plasma – time of flight, quadrupole and sector field-mass spectrometry: an inter-comparison study. Anal Chim Acta 2008;621:140–7.
[28] Chung CH, Brenner I, You CF. Comparison of microconcentric and membrane-desolvation sample introduction systems for determination of low rare earth element concentrations in surface and subsurface waters using sector field inductively coupled plasma mass spectrometry. Spectrochim Acta Part B 2009;64:849–56.
[29] Birka M, Wehe CA, Telgmann L, Sperling M, Karst U. Sensitive quantification of gadolinium-based magnetic resonance imaging contrast agents in surface waters using hydrophilic interaction liquid chromatography and inductively coupled plasma sector field mass spectrometry. J Chromatogr A 2013;1308:125–31.
[30] Krachler M, Alvarez-Sarandes R, Van Winckel S. Cross-validation of analytical procedures for the reliable determination of Nd concentrations in nuclear fuel using ICP-OES and sector field ICP-MS. J Anal Atom Spectrom 2013;28:114–20.
[31] Pisonero J, Fernandez B, Gunther D. Critical revision of GD-MS, LA-ICP-MS and SIMS as inorganic mass spectrometric techniques for direct solid analysis. J Anal Atom Spectrom 2009;24:1145–60.
[32] Sahijpal S, Marhas KK, Goswami JN. Determination of rare earth and refractory trace element abundances in early solar system objects by ion microprobe. Proc Indian Acad Sci (Earth Planet Sci) 2003;112:485–98.
[33] Tribuzio R, Tiepolo M, Vannucci R, Bottazzi P. Trace element distribution within olivine-bearing gabbros from the Northern Apennine ophiolites (Italy): evidence for post-cumulus crystallization in MOR-type gabbroic rocks. Contrib Mineral Petr 1999;134:123–33.
[34] Rao TP, Biju VM. Trace determination of lanthanides in metallurgical, environmental, and geological samples. Crit Rev Anal Chem 2000;30:179–220.
[35] Roden MF, Shimizu N. Trace element abundances in mantle-derived minerals which bear on compositional complexities in the lithosphere of the Colorado Plateau. Chem Geol 2000;165:283–305.

[36] Hanchar JM, Finch RJ, Hoskin PWO, Watson EB, Cherniak DJ, Mariano AN. Rare earth elements in synthetic zircon: Part 1. Synthesis, and rare earth element and phosphorus doping. Am Mineral 2001;86:667–80.
[37] Busa T, Clochiatti R, Cristofolini R. The role of apatite fractionation and REE distribution in alkaline rocks from Mt. Etna, Sicily. Miner Petrol 2002;74:95–114.
[38] James OB, Floss C, McGee JJ. Rare earth element variations resulting from inversion of pigeonite and subsolidus reequilibration in lunar ferroan anorthosites. Geochim Cosmochim Acta 2002;66:1269–84.
[39] Sano Y, Terada K, Fukuoka T. High mass resolution ion microprobe analysis of rare earth elements in silicate glass, apatite and zircon: lack of matrix dependency. Chem Geol 2002;184:217–30.
[40] Thomas JB, Bodnar RJ, Shimizu N, Sinha AK. Determination of zircon/melt trace element partition coefficients from SIMS analysis of melt inclusions in zircon. Geochim Cosmochim Acta 2002;66:2887–901.
[41] Klemme S, Dalpe C. Trace-element partitioning between apatite and carbonatite melt. Am Mineral 2003;88:639–46.
[42] Crozaz G, Floss C, Wadhwa M. Chemical alteration and REE mobilization in meteorites from hot and cold deserts. Geochim Cosmochim Acta 2003;67:4727–41.
[43] Whitehouse MJ. Multi-collector SIMS determination of trace lanthanides in zircon. Geostand Geoanaly Res 2004;28:195–201.
[44] Ottolini L, Camara F, Devouard B. Title: New SIMS procedures for the characterization of a complex silicate matrix, Na-3(REE,Th,Ca,U)Si6O15 center dot 2.5H(2)O (sazhinite), and comparison with EMPA and SREF results. Microchim Acta 2004;145: 139–46.
[45] Tirone M, Ganguly J, Dohmen R, Langenhorst F, Hervig R, Becker HW. Rare earth diffusion kinetics in garnet: Experimental studies and applications. Geochim Cosmochim Acta 2005;69:2385–98.
[46] Camara F, Oberti R, Ottolini L, Della Ventura G, Bellatreccia F. The crystal chemistry of Li in gadolinite. Am Mineral 2008;93:996–1004.
[47] Grice JD, Rowe R, Poirier G, Pratt A, Francis J. Bussyite-(Ce), a new beryllium silicate mineral specie s from Mont Saint-Hilaire, Quebec. Can Mineral 2009;47:193–204.
[48] Li QL, Li XH, Lan ZW, Guo CL, Yang YN, Liu Y, Tang GQ. Monazite and xenotime U-Th-Pb geochronology by ion microprobe: dating highly fractionated granites at Xihuashan tungsten mine, SE China. Contrib Mineral Petr 2013;166:65–80.
[49] Roszjar J, Whitehouse MJ, Bischoff A. Meteoritic zircon – Occurrence and chemical characteristics. Chem Erde–Geochem 2014;74:453–69.
[50] Gray AL. Solid sample introduction by laser ablation for inductively coupled plasma-mass spectrometry. Analyst 1985;110:551–6.
[51] Chenery S, Cook JM. Determination of rare-earth elements in single mineral grains by laser ablation microprobe inductively coupled plasma-mass spectrometry – preliminary-study. J Anal Atom Spectrom 1993;8:299–303.
[52] Jarvis KE, Williams JG. Laser-ablation inductively-coupled plasma-mass spectrometry (LA-ICP-MS) – a rapid technique for the direct, quantitative-determination of major, trace and rare-earth elements in geological samples. Chem Geol 1993;106:251–62.
[53] Decarlo EH, Pruszkowski E. Laser ablation ICP-MS determination of alkaline and rare earth elements in marine ferro-manganese deposits. Atom Spectros 1995;16:65–72.
[54] Chenery S, Williams T, Elliott TA, Forey PL, Werdelin L. Determination of rare earth elements in biological and mineral apatite by EPMA and LAMP-ICP-MS. Mikrochim Acta 1996;13:259–69.
[55] Odegard M, Dundas SH, Flem B, Grimstvedt A. Application of a double-focusing magnetic sector inductively coupled plasma mass spectrometer with laser ablation for the bulk analysis of rare earth elements in rocks fused with Li2B4O7. Fresen J Anal Chem 1998;362:477–82.

[56] Prince CI, Kosler J, Vance D, Gunther D. Comparison of laser ablation ICP-MS and isotope dilution REE analyses - implications for Sm-Nd garnet geochronology. Chem Geol 2000;168:255–74.
[57] Jain JC, Neal CR, Hanchar JM. Problems associated with the determination of rare earth elements of a "Gem" quality zircon by inductively coupled plasma-mass spectrometry. *Geostand. Newslett.* J Geostand Geoanal 2001;25:229–37.
[58] Orihashi Y, Hirata T. Rapid quantitative analysis of Y and REE abundances in XRF glass bead for selected GSJ reference rock standards using Nd-YAG 266 nm UV laser ablation ICP-MS. Geochem J 2003;37:401–12.
[59] Strnad L, Mihaljevic M, Sebek O. Laser ablation and solution ICP-MS determination of rare earth elements in USGS BIR-1G, BHVO-2G and BCR-2G glass reference materials. Geostand Geoanal Res 2005;29:303–14.
[60] Foster GL, Vance D. In situ Nd isotopic analysis of geological materials by laser ablation MC-ICP-MS. J Anal Atom Spectrom 2006;21:288–96.
[61] Tanaka K, Takahashi Y, Shimizu H. Determination of rare earth element in carbonate using laser-ablation inductively-coupled plasma mass spectrometry: An examination of the influence of the matrix on laser-ablation inductively-coupled plasma mass spectrometry analysis. Anal Chim Acta 2007;583:303–9.
[62] Kamaly N, Pugh JA, Kalber TL, Bunch J, Miller AD, McLeod CW, Bell JD. Imaging of gadolinium spatial distribution in tumor tissue by laser ablation inductively coupled plasma mass spectrometry. Mol Imag Biol 2010;12:361–6.
[63] Wu FY, Yang YH, Marks MAW, Liu ZC, Zhou Q, Ge WC, Yang JS, Zhao ZF, Mitchell RH, Markl G. In situ U-Pb, Sr, Nd and Hf isotopic analysis of eudialyte by LA-(MC)-ICP-MS. Chem Geol 2010;273:8–34.
[64] Hsieh HF, Chen YH, Wang CF. A magnesium hydroxide preconcentration/matrix reduction method for the analysis of rare earth elements in water samples using laser ablation inductively coupled plasma mass spectrometry. Talanta 2011;85:983–90.
[65] Yokoyama TD, Suzuki T, Kon Y, Hirata T. Determinations of rare earth element abundance and u-pb age of zircons using multispot laser ablation-inductively coupled plasma mass spectrometry. Anal Chem 2011;83:8892–9.
[66] Kimura JI, Chang Q, Kawabata H. Standardless determination of Nd isotope ratios in glasses and minerals using laser-ablation multiple-collector inductively coupled plasma mass spectrometry with a low-oxide molecular yield interface setup. J Anal Atom Spectrom 2013;28:1522–9.
[67] Payne JL, Pearson NJ, Grant KJ, Halverson GP. Reassessment of relative oxide formation rates and molecular interferences on in situ lutetium-hafnium analysis with laser ablation MC-ICP-MS. J Anal Atom Spectrom *2013*;28:1068–79.
[68] Sussulini A, Wiener E, Marnitz T, Wu B, Muller B, Hamm B, Becker JS. Quantitative imaging of the tissue contrast agent Gd(DTPA) (2-) in articular cartilage by laser ablation inductively coupled plasma mass spectrometry. Contrast Media Mol I 2013;8:204–9.
[69] Goudie DJ, Fisher CM, Hanchar JM, Crowley JL, Ayers JC. Simultaneous in situ determination of U-Pb and Sm-Nd isotopes in monazite by laser ablation ICP-MS. Geochem Geophys Geosystem 2014;15:2575–600.
[70] Kaygusuz A, Arslan M, Siebel W, Sipahi F, Ilbeyli N, Temizel I. LA-ICP MS zircon dating, whole-rock and Sr-Nd-Pb-O isotope geochemistry of the Camibogazi pluton, Eastern Pontides, NE Turkey: implications for lithospheric mantle and lower crustal sources in arc-related I-type magmatism. Lithos 2014;192:271–90.
[71] Khvostikov VA, Karandashev VK, Burmii ZP, Buzanov OA. Control of the composition of lanthanum gallium silicate by laser ablation inductively coupled plasma mass spectrometry. J Anal Chem 2014;69:495–501.
[72] Tang M, McDonough WF, Arevalo R. High-precision measurement of Eu/Eu* in geological glasses via LA-ICP-MS analysis. J Anal Atom Spectrom 2014;29:1835–43.

[73] Xu L, Hu ZC, Zhang W, Yang L, Liu YS, Gao S, Luo T, Hu SH. In situ Nd isotope analyses in geological materials with signal enhancement and non-linear mass dependent fractionation reduction using laser ablation MC-ICP-MS. J Anal Atom Spectrom 2015;30:232–44.
[74] Pollard AM, Batt CM, Stern B, Young SMM. Neutron activation analysis. Analytical Chemistry in Archaeology. Cambridge: Cambridge University Press; 2007. pp. 123–136, Chapter 6.
[75] Blaauw, M. The holistic analysis of gamma-ray spectra in instrumental neutron activation analysis. The Interfaculty Reactor Institute Delft, Netherlands, 1993.
[76] [online] available at: http://www.naa-online.net/.
[77] Bode, P. Instrumental and organizational aspects of a neutron activation analysis laboratory. The Interfaculty Reactor Institute Delft, Netherlands, 1996.
[78] Use of research reactors for neutron activation analysis. IAEA-TECDOC-1215, IAEA Vienna, 2001.
[79] Kučera J, Zeisler R. Do we need radiochemical separation in activation analysis? J Radioanal Nucl Chem 2004;262:255–60.
[80] Molnár GL. Handbook of prompt gamma activation analysis. Dordrecht, The Netherlands: Kluwer Academic Publishers; 2004.
[81] Lindstrom RM. Using k0 in PGAA: perspectives. J Radioanal Nucl Chem 2003;257:557–60.
[82] De Soete D, Gijbels R, Hoste J. Neutron activation analysis. London, New York, Sydney, Toronto: Wiley-Interscience; 1972.
[83] Freitas MC, Pacheco AMG, Bacchi MA, Dionísio I, Landsberger S, Braisted J, Fernandes EAN. Compton suppression instrumental neutron activation analysis performance in determining trace- and minor-element contents in foodstuff. J Radioanal Nucl Chem 2008;276:149–56.
[84] Glasgow DC. Delayed neutron activation analysis for safeguards. J Radioanal Nucl Chem 2008;276:207–11.
[85] Hamidatou L, Dakar AS, Boukari S. k0-NAA quality assessment in an Algerian laboratory by analysis of SMELS and four IAEA reference materials using Es-Salam Research reactor. J Nucl Instru Methods Phys Res A 2012;682:75–8.
[86] Hamidatou, L.; Slamene, H.; Akhal, T.; Zouranen, B. Concepts, instrumentation and techniques of neutron activation analysis, imaging and radioanalytical techniques in interdisciplinary research - fundamentals and cutting edge applications. Faycal Kharfi, editor., ISBN: 978-953-51-1033-0, InTech, DOI: 10.5772/53686. Ch. 6, 2013.
[87] Kučera J, Anda Z. The present role of neutron and photon activation analysis in determination of trace elements. 3rd Symp. on Nuclear Chemistry, Halifax, NS, Canada, 11–14 June, 2001.
[88] Kučera J, Byrne AR. Role of the self-validation principle of NAA in the quality assurance of bioenvironmental studies and in the certification of reference materials. IAEA-SM-344/8, IAEA Vienna, 1997.
[89] [online] available at: http://archaeometry.missouri.edu/naa_overview.html.
[90] Bate GL, Potratz HA, Huizenga JR. Scandium, chromium and europium in stone meteorites by neutron activation analysis. Geochim Cosmochim Acta 1960;18:101–7.
[91] Mosen AW, Vasilevskis J, Schmitt RA. A procedure for the determination of the rare-earth elements lanthanum through lutetium in chondritic, achondritic, and iron meteorites by neutron-activation analysis. Anal Chim Acta 1961;25:10–24.
[92] Cuypers M, Menon MP. Determination of cerium + praseodymium in ores+minerals using 14.5 MeV neutron activation analysis. Trans Am Nucl Soc 1964;7:330.
[93] Graber FM, Lukens HR, Heydorn K. Determination of terbium, erbium, ytterbium and yttrium by neutron activation analysis. Trans Am Nucl Soc 1966;9:87–187.
[94] Babb AL, Miller WP, Wilson WE, Woodruff GL, Novotny AJ. Neutron activation analysis of stable dysprosium biologically deposited in the bone of chinook salmon fingerlings. Trans Am Nucl Soc 1967;10:449–60.

[95] Alimarin IP, Jakovlev YV, Miklisha Az, Dogadkin NN, Stepanet OV. Determination of dysprosium, europium, samarium and gadolinium in yttrium oxide of high purity by means of neutron activation analysis. J Radioanal Chem 1968;1:139–45.
[96] Coulomb R, Coquema C, Goldszte M, Schiltz JC. Use of chromatography and computers in neutron activation analysis-application to quantitative analysis of La Sm Dy Eu in lavas of chaine des puys. Bull Soc Franc Mineral Cristallogr 1968;91:75.
[97] Tomura K, Higuchi H, Onuma N, Hamaguch H. Rapid determination of dysprosium in rock samples by neutron activation analysis with a ge(li) detector after chemical separation. Anal Chim Acta 1968;42:389–95.
[98] Lombard SM, Isenhour TL. Determination of samarium and gadolinium in rare earth ores by neutron-capture gamma-ray activation analysis. Anal Chem 1969;41:1113–6.
[99] Marsh RH, Allie W. Precision determination of gadolinium in cadmium fluoride by neutron activation analysis. Anal Chim Acta 1969;45:179–82.
[100] Steinnes E. Determination of samarium in rocks by epithermal neutron-activation analysis. Freseniu Zeitschrif Anal Chem 1969;247:23–7.
[101] Alimarin IP, Miklisha Az, Yakovlev YV. Neutron activation analysis of rare earth impurities in europium oxide. J Radioanal Chem 1970;4:75–80.
[102] Menapace LM, Voigt AF. Thermal neutron activation analysis for potassium, holmium and lanthanum in tungsten bronzes. J Radioanal Chem 1970;6:219–26.
[103] Jaskolsk H, Rowinska L, Radwan M. Search for internal isotopic tracers in metallurgical materials. 2. Determination of W, As, Au, La, In, Sct. Ir and Ta in metallurgical materials by the neutron activation analysis. J Radioanal Chem 1971;8:235.
[104] Vasilev SS, Velyus LM, Starchik LP, Kodiri S, Abbosov O. Neutron activation analysis of enriched isotopes of europium, indium, iridium, and lutetium. Sov Atom Energy-USSR 1971;31:1022–4.
[105] Zmijewsk W, Sorantin H. Non-destructive determination of dysprosium in graphites by neutron activation analysis. J Radioanal Chem 1971;8:83–8.
[106] Adams F, Bouten P. Neutron activation analysis for europium in photographic film. Anal Chim Acta 1972;61:307–10.
[107] Young MC, Theurer B, Wacks ME, Hale WH, Ogden PR, Haskin FE. Neutron activation analysis of dysprosium - potential inert marker. J Anim Sci 1972;34:898–905.
[108] Foldzins A, Dybczyns R. Neutron activation analysis of steel for sub ppm amounts of lanthanum and gold without determining the chemical yield. J Radioanal Chem 1973;17:229–36.
[109] Pung TC, Wu SC, Tsai HT. Internal-reference method for determination of lanthanum, cerium and europium in monazite sand by neutron activation analysis. J Chinese Chem Soc 1973;20:221–6.
[110] Rouchaud JC, Revel G. Systematic analysis of Ce by neutron activation. J Radioanal Chem 1973;17:167–72.
[111] Foldzins A, Dybczyns R. Radiochemical scheme for the quantitative isolation and gamma-spectrometric determination of La, Ga. Hf and Sc in neutron activation analysis of refractory materials. J Radioanal Chem 1974;21:507–17.
[112] Lavi N. Determination of lanthanum in the presence of uranium by neutron activation analysis using high resolution gamma spectrometry. Radiochim Acta 1974;21:73–4.
[113] Sastry VN, Krishnam Tm, Doshi GR, Sarma TP. Manganese dioxide as sorber in the determination of Ce, Ru, Zr, and Cs fission nuclides from seawater by neutron activation analysis. Indian J Chem 1974;12:200–2.
[114] Young MC, Haskin FE, Wacks ME, Theurer B, Ogden PR. Neutron activation analysis of dysprosium (a potential inert marker) in hay and feces. J Anim Sci 1975;41:178–84.
[115] Brits RJN, Greeff GJ. Determination of Ga, La, and Sm in the NIMROC standards by NAA. Mikrochim Acta 1977;1:105–12.
[116] Axton EJ, Cross C, Goodier IW, Risebrowsmith GH. Assay of trace quantities of gadolinium in a manganese bath solution using neutron activation analysis. Int J Appl Radiat Is 1979;30:379–1379.

[117] Cercasov V, Heller R. Application of 14 MeV neutron activation analysis to the determination of Ce and Sm markers in biological materials. J Radioanal Chem 1980;60: 453–9.

[118] Kennelly JJ, Apps MJ, Turner BV, Aherne FX. Dysprosium, cerium and chromium marker determination by instrumental neutron activation analysis. Can J Anim Sci 1980;60:749–61.

[119] Chen SG, Tsai HT, Wu SC. An internal-standard method for the determination of uranium, thorium, lanthanum and europium in carbonaceous shale and monazite by epithermal neutron activation analysis. Radiochem Radioanal Lett 1981;49: 149–60.

[120] Rosseel TM, Hulett LD, Northcutt KJ. Analysis of a Nd:YAG laser crystal by X-ray fluorescence, scanning electron microscopy and neutron activation analysis. J Radioanal Chem 1982;75:259–64.

[121] Czauderna M, Samochocka K, Kalicki A. Practical aspects of simultaneous determination of selenium and ytterbium by neutron activation analysis. J Radioanal Chem 1983;76:181–92.

[122] Meary A, Galland D, Chatagnon B, Diebolt J. A study of Gd, Ce and Eu impurities in fluorite by electron paramagnetic resonance and neutron activation analysis. Phys Chem Miner 1984;10:173–9.

[123] Sugimoto K, Taniguchi M, Narita K. Trace analysis of Al, V, Mn, CU, As, Se, Sb, La, Ta and W in steel by neutron activation. T. Iron Steel Jap 1984;24:B303–B1303.

[124] Samochocka J, Czauderna M, Konecki J, Wolna M. The incorporation of selenium and ytterbium into some organs of mice determined by neutron activation analysis. J Radioanal Nucl Chem 1985;91:329–35.

[125] Arambel MJ, Bartley EE, Higginbotham JF, Simons GG, Nuzback LJ. Neutron activation analysis of cerium, chromium, lanthanum and samarium in duodenal digesta. Nut Rep Int 1986;33:861–8.

[126] Bergeron M, Heaman LM. Determining the abundance of gadolinium in geological samples by prompt-gamma neutron activation analysis. Chem Geol 1986;54:333–7.

[127] Labrecque JJ, Beusen JM, Vangrieken RE. Determination of barium, lanthanum, cerium and neodymium in lateritic materials by various energy-dispersive X-ray fluorescence techniques and neutron activation analysis. X-Ray Spectrom 1986;15:13–8.

[128] Teherani DK. Application of neutron activation analysis for determination of mercury, iron, europium, lanthanum and potassium in asbestos. J Radioanal Nucl Ch-Lett 1986;104:231–40.

[129] Figueiredo AMG, May S, Pinte G. "Determination of the concentration values ol" dysposlum, europium gadolinium, praséodyme, samarium and lanthane in the standard rock GS-N by neutron activation analysis. Analysis 1987;15:179–82.

[130] Kabirova GM, Smetannikov VV, Solodukhin VP. Instrumental neutron-activation analysis of dysprosium and europium in standard of rocks. Geokhimiya 1987;5: 739–41.

[131] Landsberger S, Simsons A. Quantification of uranium, thorium and gadolinium spectral interferences in instrumental neutron activation analysis of samarium. Chem Geol 1987;62:223–6.

[132] Vittoz P, Oliver RA, Vivier G, Kerr SA, Hoyler F. Determination of B, Sm and Gd in 23 Geological reference materials by neutron-capture prompt gamma-ray activation analysis (PGAA). Geostand Newslett 1987;11:43–7.

[133] Zaghloul R, Gantner E, Mostafa M, Ache HJ. Prompt gamma-ray neutron activation analysis facility testing by Sm, Gd and Mn determination in rock samples. J Radioanal Nucl Ch-Ar 1987;109:309–19.

[134] Tsukada M, Yoshikawa H, Yanaga M, Endo K, Nakahara H. DY-156 as an activable yield tracer for determination of rare earth elements in biological materials by neutron activation analysis. J Radioanal Nucl Ch-Ar 1988;125:351–65.

[135] Honda T, Oi T, Ossaka T, Nozaki T, Kakihana H. Determination of praseodymium, neodymium and erbium in hot spring and crater lake waters by neutron activation analysis incorporating the standard addition technique. J Radioanal Nucl Ch-Ar 1989;134:13–25.

[136] Teherani DK, Badawi N. Instrumental neutron activation analysis for determination of europium distribution in fattened chicken. J Radioanal Nucl Chem-Lett 1989;136: 1–7.

[137] Vansuc N, Desai HB, Parthasarathy R, Gangadharan S. Rare earth impurities in high purity lanthanum oxide determined by neutron activation analysis. J Radioanal Nucl Chem-Lett 1992;164:321–5.

[138] Freitas MC, Decorte F, Gouveia MA, Prudencio MI, Cabral JMP. Determination of the REE in environmental samples near a coal power station based on k 0 -standardized NAA with counting in a LEPD. J Radioanal Nucl Chem-Lett 1993;169:219–28.

[139] Meloni S, Oddone M, Bottazzi P, Ottolini L, Vannucci R. Neutron activation analysis investigations in updating REE composition of gabbro GOG-1 reference sample: a comparison with SIMS and ICP-AES data. J Radioanal Nucl Ch-Ar 1993;168:115–23.

[140] Reddy GR, Pant DR. La, Ce, Sm, Eu, Tb, Yb and Lu content of 15 french geochemical reference samples determined by neutron activation analysis. Geostand Newslett 1993;17:113–6.

[141] Solacherios M, Ordonezregil E, Iturbe JL. Determination of lanthanum in phosphate rock and in its products by neutron activation analysis. J Radioanal Nucl Chem-Lett 1993;176:471–5.

[142] Suc NV. Neutron activation analysis for determination of cerium trace impurity in high purity lanthanum oxide. J Radioanal Nucl Chem-Lett 1994;187(3):237–41.

[143] Decorte F, Masumoto K, Dewispelaere A, Bellemans F. A dedicated NAA method to determine the neutron spectrum monitor elements Co, Au and Lu with high accuracy in alloyed wires and foils. J Radioanal Nucl Ch-Ar 1995;192:91–6.

[144] Qadir MA, Iqbal MZ, Zaidi JH. Neutron activation analysis of Ce, Yb, Se and Cr in the atmosphere of some major urban and industrial areas of Punjab, Pakistan. J Radioanal Nucl Chem-Lett 1995;201:347–54.

[145] Wang YQ, Sun JX, Chen HM, Guo FQ. Determination of the contents and distribution characteristics of REE in natural plants by NAA. J Radioanal Nucl Chem 1997;219:99–103.

[146] Franca YV, Leitao F, Shihomatsu HM, Scapin WS, de Moraes NMP, Salvador VL, Figueiredo AMG, Muccillo ENS, Muccillo R. Determination of yttrium and lanthanum in zirconium dioxide by HPLC, X-ray fluorescence and neutron activation analyses. Chromatographia 1999;49:91–4.

[147] Cantone MC, Cornolti E, de Bartolo D, Giussani A, Tsipenyuk YM, Firsov VI, Gorbunov AV. Proton, photon and neutron activation analysis for the determination of stable isotopes of gadolinium in human blood plasma. J Radioanal Nucl Chem 2000;243:707–14.

[148] Chen CY, Zhang PQ, Chai ZF. Distribution of some rare earth elements and their binding species with proteins in human liver studied by instrumental neutron activation analysis combined with biochemical techniques. Anal Chim Acta 2001;439:19–27.

[149] Yusof AM, Akyil S, Wood AKH. Rare earth elements determination and distribution patterns in sediments of a polluted marine environment by instrumental neutron activation analysis. J Radioanal Nucl Chem 2001;249:333–41.

[150] Cao L, Tian W, Ni B, et al. Radiochemical neutron-activation analysis of uncertified ultra-trace rare earth elements in two biological certified reference materials. Anal Bioanal Chem 2002;372:397–400.

[151] Figueiredo AMG, Avristcher W, Masini EA, Diniz SC, Abrao A. Determination of lanthanides (La, Ce, Nd, Sm) and other elements in metallic gallium by instrumental neutron activation analysis. J Alloy Compd 2002;344:36–9.

[152] Ohde S. Determination of rare earth elements in carbonatites from the Kangankunde Mine, Malawi by instrumental neutron activation analysis. J Radioanal Nucl Chem 2003;257:433–5.

[153] Bartolini ME, Pekar J, Chettle DR, McNeill F, Scott A, Sykes J, Prato FS, Moran GR. An investigation of the toxicity of gadolinium based MRI contrast agents using neutron activation analysis. Magn Reson Imag 2003;21:541–4.

[154] Oliveira SMB, Larizzatti FE, Favaro DIT, et al. Rare earth element patterns in lake sediments as studied by neutron activation analysis. J Radioanal Nucl Chem 2003;258:531–5.

[155] Kayasth S, Swain K. Determination of rare-earth elements in Periyar River water (Kerala, India) and seawater (Mumbai, India) using neutron activation analysis. J Radioanal Nucl Chem 2004;262:191–4.

[156] Ohde S. Instrumental neutron activation analysis of carbonatites from Homa Mountain, Kenya. J Radioanal Nucl Chem 2004;260:213–8.

[157] Dampare SB, Asiedu DK, Osae S, et al. Determination of rare earth elements by neutron activation analysis in altered ultramafic rocks from the Akwatia District of the Birim diamondiferous field, Ghana. J Radioanal Nucl Chem 2005;265:101–6.

[158] Minowa H. A study of the chemical evolution of pallasite meteorites in a small planetary body based on neutron activation analysis data. Bunseki Kagaku 2005;54:183–4.

[159] Ferreira JR, Fernandes EAN, Franca EJ, Tavares GA, da Silva NC, Taddei MHT. Rare earth elements in sediment profiles from marginal lagoons of the Moji-Guacu River basin, Brazil. J Radioanal Nucl Chem 2006;270:15–20.

[160] Sano T, Hasenaka T, Sawahata H, Fukuoka T, Determination of Ti K. Sm and Gd values in Geological Survey of Japan reference materials by prompt gamma neutron activation analysis. Geostand Geoanal Res 2006;30:31–7.

[161] Kucera J, Mizera J, Randa Z, et al. Pollution of agricultural crops with lanthanides, thorium and uranium studied by instrumental and radiochemical neutron activation analysis. J Radioanal Nucl Chem 2007;271:581–7.

[162] Nyarko BJB, Akaho EHK, Fletcher JJ, Chatt A. Neutron activation analysis for Dy, Hf, Rb, Sc and Se in some Ghanaian cereals and vegetables using short-lived nuclides and Compton suppression spectrometry. Appl Radiat Isotopes 2008;66:1067–72.

[163] Chen CY. Simultaneous determination of As-76, Sb-122 and Sm-153 in Chinese medicinal herbs by epithermal neutron activation analysis. Phytochem Anal 2009;20:50–7.

[164] Khokhlov VF, Zaitsev KN, Beliayev VN, Kulakov VN, Lipengolts AA, Portnov AA. Prompt gamma neutron activation analysis of B-10 and Gd in biological samples at the MEPhI reactor. Appl Radiati Isotopes 2009;67:S251–3.

[165] Rezaee Kh, Saion EB, Wood A. Khalik, et al. Rare earth elements determination and distribution patterns in surface marine sediments of the South China Sea by INAA, Malaysia. J Radioanal Nucl Chem 2010;283:823–9.

[166] Grafe JL, McNeill FE, Byun SH, Chettle DR, Noseworthy MD. The feasibility of in vivo detection of gadolinium by prompt gamma neutron activation analysis following gadolinium-based contrast-enhanced MRI. Appl Radiat Isotopes 2011;69:105–11.

[167] Isaac-Olive K, Solis C, Martinez-Carrillo MA, Andrade E, Lopez C, Longoria LC, Lucho-Constantino CA, Beltran-Hernandez RI. Tillandsia usneoides L, a biomonitor in the determination of Ce, La and Sm by neutron activation analysis in an industrial corridor in Central Mexico. Appl Radiat Isotopes 2012;70:589–94.

[168] Hayford MS, Akiti TT, Serfor-Armah Y, Dampare SB. Investigation of Kpong carbonatite as a potential source for rare earth elements (REEs) using instrumental neutron activation analysis (INAA). Radiochim Acta 2013;101:739–44.

[169] Waheed S, Siddique N, Faiz Y. Rare earth. High field-strength elements in the multani mitti clay: a study using INAA. Geostand Geoanal Res 2013;37:197–205.

[170] Graefe JL, McNeill FE, Noseworthy MD, Chettle DR. Gadolinium detection via in vivo prompt gamma neutron activation analysis following gadolinium-based

contrast agent injection: a pilot study in 10 human participants. Physiol Meas 2014;35: 1861–72.
[171] Kumar K, Saion E, Halimah MK, Yap CK, Hamzah MS. Rare earth element (REE) in surface mangrove sediment by instrumental neutron activation analysis. J Radioan Nucl Chem 2014;301:667–76.
[172] Grogan KP, O'Kelly DJ. Analytical applications of delayed and instrumental neutron activation analysis. J Radioan Nucl Chem 2014;299:543–9.
[173] Glocker R, Schreiber H. Quantitative rontgen spectrum analysis by means of cold excitation of the spectrum. Ann Physik 1982;85:1089.
[174] Jenkins R. X-Ray Fluorescence spectrometry. John Wiley & Sons, Inc. QD96.X2J47; 1988. ISBN 0-471-83675-3.
[175] Anzelmo JA, Lindsay JR. X-ray fluorescence spectrometric analysis of geologic materials Part 1. Principles and instrumentation. J Chem Educ 1987;64:A181–5.
[176] Skoog DA, Holler FJ, Nieman TA. Principles of instrumental analysis. 5th ed. Thomson Learning, Inc; 1998. ISBN 0-03-002078-6.
[177] Jenkins R, Gould RW, Gedcke D. Quantitative X-ray spectrometry. 2nd ed. Marcel Dekker, Inc. QD96.X2 J46; 1995. ISBN 0-8247-9554-7.
[178] Buhrke VE, Jenkins R, Smith DK. A practical guide for the preparation of specimens for X-ray fluorescence and X-ray diffraction analysis. John Wiley & Sons, Inc. QD96. X2P73; 1998. ISBN 0-471-19458-1.
[179] Qadeer R, Hanif J, Khan M. XRF estimation of Cd, Hf, Hg, and Gd in uranium solution. J Radioan Nucl Ch-Ar 1993;172:137–43.
[180] Jurczyk J, Sitko R, Zawisza B, Buhl F, Malicka E. XRF analysis of microsamples of semiconductor type multielement materials by the thin layer method. Determination of Cr, Co, Ni, Cu, Zn, Ga, Se, Sb, Yb. Mikrochim Acta 1999;132:41–7.
[181] De Vito IE, Masi AN, Olsina RA. Determination of trace rare earth elements by X-ray fluorescence spectrometry after preconcentration on a new chelating resin loaded with thorin. Talanta 1999;49:929–35.
[182] Janssens K, Vincze L, Vekemans B, Williams CT, Radtke M, Haller M, Knöchel A. The non-destructive determination of REE in fossilized bone using synchrotron radiation induced K-line X-ray microfluorescence analysis. Fres J Anal Chem 1999;363: 413–20.
[183] Jurczyk J, Sitko R, Zawisza B. XRF analysis of powder microsamples of multielement mono- and polycrystals. Determination of Cr, Mn, Ni, Cu, Zn, Ga, As, Se, Sb, Yb. Chem Anal – Warsaw 2000;45:415–28.
[184] De Vito IE, Olsina RA, Masi AN. Enrichment method for trace amounts of rare earth elements using chemofiltration and XRF determination. Fres J Anal Chem 2000;368:392–6.
[185] De Vito IE, Olsina RA, Masi AN. Preconcentration and elimination of matrix effects in XRF determinations of rare earth elements by preparing a thin film through chemofiltration. J Anal Atom Spectrom 2001;16:275–8.
[186] Bonfiglioli TA, Olsina RA, Fernandez LP, Masi AN. Europium preconcentration by filtration of ternary complexes prior their determination by XRF spectrometry. J Anal Atom Spectrom 2003;18:1263–6.
[187] Sitko R, Zawisza B, Czaja M. Fundamental parameters method for determination of rare earth elements in apatites by wavelength-dispersive X-ray fluorescence spectrometry. J Anal Atom Spectromet 2005;20:741–5.
[188] Sitko R, Zawisza B, Plonka M, Malicka E. Determination of chemical composition of (Pb,La)(Zr,Ti)O-3 type ferroelectric ceramics by WD-XRF and ED-EPMA. Chem Anal – Warsaw 2008;53:123–31.
[189] Kashiwabara T, Toda R, Fujinaga K, Honma T, Takahashi Y, Kato Y. Determination of host phase of lanthanum in deep-sea REY-rich mud by XAFS and mu-XRF using high-energy synchrotron radiation. Chem Lett 2014;43:199–200.

[190] Simandl GJ, Stone RS, Paradis S, Fajber R, Reid HM, Grattan K. An assessment of a handheld X-ray fluorescence instrument for use in exploration and development with an emphasis on REEs and related specialty metals. Miner Deps 2014;49:999–1012.
[191] Simandl GJ, Fajber R, Paradis S. Portable X-ray fluorescence in the assessment of rare earth element-enriched sedimentary phosphate deposits. Geochem Exp Environ Anal 2014;14:161–9.
[192] Mostafaei F, McNeill FE, Chettle DR, Noseworthy MD. A feasibility study to determine the potential of in vivo detection of gadolinium by X-ray fluorescence (XRF) following gadolinium-based contrast-enhanced MRI. Physiol Meas 2015;36: N1–N13.
[193] Verma SP, Santoyo E. High-performance liquid and ion chromatography: separation and quantification analytical techniques for rare earth elements. Geostand Geoanal Res 2007;31:161–84.
[194] Robards K, Clarke S. Advances in the analytical chromatography of the lanthanides, a review. Analyst 1988;113:1757–79.
[195] Kumar M. Recent trends in chromatography procedures for separation and determination of rare earth elements, a review. Analyst 1994;119:2013–24.
[196] Nesterenko PN, Jones P, Paull B. High performance chelation ion chromatography. Royal Society of Chemistry 2010; UK.
[197] Snyder LR, Kirkland JJ. Introduction to modern liquid chromatography. 3rd ed. Wiley; 2009. ISBN-13: 978-0470167540,.
[198] Fanali S, Haddad PR, Poole C, Schoenmakers P, Lloyd DK. Liquid chromatography. Elsevier Ltd; 2013. ISBN: 978-0-12-415806-1,.
[199] Pittcon. 2012 Conferences, HPLC Systems and Components; 2012. Orlando, USA.
[200] Hu R, Pan Y. Recent trends in counter-current chromatography. Trends Anal Chem 2012;40:16–27.
[201] Goetze HJ, Bialkowski D. Separation of some rare earths EDTA complexes by ion-pair chromatography. Fres Z Anal Chem 1985;320:370.
[202] Goetze HJ, Bialkowski D. Flüssig-chromatographische Trennung von Lanthanoid-Komplexen mittels Ionenpaar-Chromatographie (HPLC). Fres Z Anal Chem 1986;323:350–4.
[203] Knight CH, Cassidy RM, Recoskie BM, Green LW. Dynamic ion exchange chromatography for determination of number of fissions in thorium-uranium dioxide fuels. Anal Chem 1984;56:474–8.
[204] Cao XD, Zhao GW, et al. Determination of ultratrace rare earth elements in tea by inductively coupled plasma mass spectrometry with microwave digestion and AG50W-x8 cation exchange chromatography. Analyst 1998;123:1115–9.
[205] Itoh A, Hamanaka T, et al. Multielement determination of rare earth elements in geochemical samples by liquid chromatography/inductively coupled plasma mass spectrometry. Anal Sci 1999;15:17–22.
[206] Premadas A. Cation exchange chromatographic group separation and ICP-AES determination of rare earth elements and yttrium in refractory minerals zircon, ilmenite, rutile, columbite-tantalite, garnet, and sillimanite. Atomic Spectrosc 2003;24:149–58.
[207] Fernandez RG, Alonso JIG. Separation of rare earth elements by anion-exchange chromatography using ethylenediaminetetraacetic acid as mobile phase. J Chromatogr A 2008;1180:59–65.
[208] Li WJ, Jin XD, Gao BY, Wang CL, Zhang LC. Analysis of ultra-low level rare earth elements in magnetite samples from banded iron formations using HR-ICP-MS after chemical separation. Anal Methods 2014;6:6125–32.
[209] Watkins RT, Ridley MK, et al. Determination of rare-earth elements in coal using microwave digestion and gradient ion chromatography. Chem Geol 1995;121:273–83.
[210] Hvattum E, Normann PT, et al. Detection and quantitation of gadolinium chelates in human serum and urine by high-performance liquid-chromatography and

postcolumn derivatization of gadolinium with arsenazo-III. J Pharm Biomed Anal 1995;13:927–32.
[211] Pedreira WR, Sarkis JES, Queiroz CAD, Rodrigues C, Tomiyoshi IA, Abrao A. Determination of trace amounts of rare-earth elements in highly pure neodymium oxide by sector field inductively coupled plasma mass spectrometry (ICP-SFMS) and high-performance liquid chromatography (HPLC) techniques. J Solid State Chem 2003;171:3–6.
[212] Ishikawa T, Sugimoto K, Nagaishi K. Determination of rare-earth elements in rock samples by an improved high-performance ion chromatography. Geochem J 2003;37:671–80.
[213] McGillicuddy N, Nesterenko EP, et al. A new *N*-hydroxyethyliminodiacetic acid modified core-shell silica phase for chelation ion chromatography of alkaline earth, transition and rare earth elements. J Chromatogr A 2013;1321:56–64.
[214] Pukhovskaya VM. Inductively coupled plasma atomic emission spectroscopic determination of rare earth elements in geological samples after preconcentration by countercurrent chromatography-I. Spectrochim Acta B 1993;48:1365–70.
[215] Pukhovskaya VM, Grebneva ON, et al. Inductively-coupled plasma-atomic emission spectroscopic determination of rare-earth elements in geological samples after preconcentration by countercurrent chromatography. Spectrochim Acta B 1995;50:5–12.
[216] Ignatova SN, Maryutina TA, et al. Group separation of trace rare-earth elements by countercurrent chromatography for their determination in high-purity calcium chloride. Fres J Anal Chem 2001;370:1109–13.
[217] Hennebruder K, Engewald W, Stark HJ, Wennrich R. Enrichment of rare-earth elements (REE) and Gd-DTPA in surface water samples by means of countercurrent chromatography (CCC). Anal Chim Acta 2005;542:216–21.
[218] Krattli M, Muller-Spath T, Ulmer N, Strohlein G, Morbidelli M. Separation of lanthanides by continuous chromatography. Indus Eng Chem Res 2013;52:8880–6.
[219] Yoshida I, Nagai H, et al. Column separation of rare-earth-metal ions by flash chromatography using octadecylsilylated silica-gel particles loaded with bis(2-ethylhexyl) hydrogenphosphate. Nippon Kagaku Kaishi 1995;7:518–22.
[220] Pourmand A, Dauphas N, Ireland TJ. A novel extraction chromatography and MC-ICP-MS technique for rapid analysis of REE, Sc and Y: revising CI-chondrite and Post-Archean Australian Shale (PAAS) abundances. Chem Geol 2012;291:38–54.
[221] Kim JG. Separation of heavy rare earth elements with extraction chromatography. Curr Nanosci 2014;10:11–5.

Subject Index

H

I

R

Kolibri 360 GmbH
Ferdinand-Jühlke-Straße 7
99095 Erfurt, Deutschland
produktsicherheit@kolibri360.de